Exploring Psychology, Social Innovation and Advanced Applications of Machine Learning

Maria E. Raygoza-L.
Universidad Politécnica de Baja California, Mexico

Jesus Heriberto Orduño-Osuna
Universidad Politécnica de Baja California, Mexico

Abelardo Mercado-Herrera
Universidad Politécnica de Baja California, Mexico

Roxana Jimenez-Sanchez
Universidad Politécnica de Baja California, Mexico

Fabian N. Murrieta-Rico
Universidad Politecnica de Baja California, Mexico

IGI Global
Scientific Publishing
Publishing Tomorrow's Research Today

Vice President of Editorial Melissa Wagner
Managing Editor of Acquisitions Mikaela Felty
Managing Editor of Book Development Jocelynn Hessler
Production Manager Mike Brehm
Cover Design Phillip Shickler

Published in the United States of America by
IGI Global Scientific Publishing
701 East Chocolate Avenue
Hershey, PA, 17033, USA
Tel: 717-533-8845
Fax: 717-533-8661
Website: https://www.igi-global.com E-mail: cust@igi-global.com

Library of Congress Cataloging-in-Publication Data

Names: Raygoza-L, Maria, 1981- editor. | Orduño-Osuna, Jesus Heriberto, 1996- editor. | Mercado-Herrera, Abelardo, 1973- editor. | Jimenez Sanchez, Roxana, 1979- editor. | Murrieta-Rico, Fabian, 1986- editor.
Title: Exploring psychology, social innovation and advanced applications of machine learning / edited by Maria Raygoza-L, Jesus Heriberto Orduño-Osuna, Abelardo Mercado-Herrera, Roxana Jimenez Sanchez, Fabian Murrieta-Rico.
Description: Hershey, PA : IGI Global Scientific Publishing, 2025. | Includes bibliographical references and index. | Summary: "The aim of this book is to provide a comprehensive overview of the use of machine learning and its various applications for humanity, aiming to foster social innovation from the standpoint of psychology and other social issues related to human behavior"-- Provided by publisher.
Identifiers: LCCN 2024054548 (print) | LCCN 2024054549 (ebook) | ISBN 9798369369104 (hardcover) | ISBN 9798369369111 (paperback) | ISBN 9798369369128 (ebook)
Subjects: LCSH: Human behavior--Social aspects. | Machine learning.
Classification: LCC BF199 .E895 2025 (print) | LCC BF199 (ebook) | DDC 302.0285/631--dc23/eng/20250201
LC record available at https://lccn.loc.gov/2024054548
LC ebook record available at https://lccn.loc.gov/2024054549

British Cataloguing in Publication Data
A Cataloguing in Publication record for this book is available from the British Library.

Table of Contents

Detailed Table of Contents

The increasing use of Artificial Intelligence (AI) in mental health presents significant ethical and practical challenges. Applications such as Machine Learning (ML) and chatbots aim to enhance the accessibility and effectiveness of therapies, enabling earlier diagnosis and continuous monitoring. However, their implementation raises concerns regarding data privacy and the potential dehumanization of psychological treatment. Furthermore, AI can be utilized to assist in identifying traits associated with the dark triad, including narcissism, psychopathy, and Machiavellianism, which may prove beneficial in managing various personality disorders. Despite its potential advantages, it is crucial to develop public policies that regulate the use of these technologies and protect patient rights, ensuring that AI complements rather than replaces human intervention in mental health care.

Accurate and timely emotion detection is crucial for various applications like customer experience analysis, mental health monitoring, and human-computer interaction. This work presents a novel deep neural network (DNN)-based method for real-time emotion recognition from facial expressions. The proposed system effectively categorizes human emotions into six main groups: anger, disgust, fear, happiness, sadness, and surprise. The DNN model, using advanced computer vision methods and deep learning, successfully identified emotions using an extensive collection of annotated facial image data. The model, designed for efficient computation and real-time inference on edge devices, offers a wide range of creative use cases for emotion detection. Virtual assistants could incorporate emotion recognition systems to provide personalized responses based on user emotions, potentially monitoring mental health by observing changes in emotional patterns over time, making this a significant step towards making emotion recognition a pervasive technology.

Sentiment Analysis is a part of Data Intelligence Research that lays emphasis on data that contains emotions. This analysis is carried out by analysing the polarity of content and thereby marking it as positive, negative, or neutral. In order to find out how SA is globally used, Supervised Natural Language Processing (SNLP) is also utilised. In this chapter, a range of these tools and techniques of SA will be discussed and their applications will be elaborated. Additionally, this chapter will delve into further academic research related to the topic to enhance understanding of how sentiment analysis can support organizations in staying competitive and boosting profits by examining real-life examples. Sentiment analysis has experienced notable progress in recent years, primarily propelled by utilizing machine learning and deep learning techniques in sentiment classification. SA helps in building social and political perceptions helping the researchers and policymakers to understand public sentiments on burning issues thereby aiding in decision-making in an ever-changing digital world.

This chapter examines the effectiveness of NLP models in analyzing mental health conversations between users and psychologists. It aims to evaluate classification performance, overall accuracy, agreement, the influence of contextual factors on sentiment responses, and the reliability of model coefficients. The hypothesis suggests near-perfect performance across metrics such as sensitivity, specificity, and balanced accuracy, with significant coefficients for predictors. Using a secondary dataset with negative, neutral, and positive sentiment classes, the methodology includes visualizing classification performance, calculating various accuracy metrics, and fitting a multinomial logistic regression model to predict sentiment responses. The model demonstrates perfect scores in classifying sentiments, a significant improvement over baseline, and strong agreement beyond chance, with robust fitting indicating high precision and reliability in predicting sentiment outcomes.

This chapter goes on to explore the current status of machine learning and its relevance to fake news identification and their potential as a powerful tool for social innovation in the digital age. With the ever-growing number of contents misinforming others and going viral in the online platform, it is quite necessary to accept the role of algorithms in machine learning as possible solutions that could help contain the false information rage. While key advancements such as NLP and neural networks enable extraction of meaning in content, authorship, and user behavior in very complex patterns, issues related to data bias, the degree of algorithmic transparency, and more intelligent tactics of misinformation remain. Beyond the technical, this chapter shines a light on the ethical and social dimensions of AI-driven interventions; it underlines the need for not only transparency but also for inclusiveness in deploying such technologies.

The chapter reviews consumer behavior fundamentals such purchasing decision criteria, quantification, and analysis. Next, it discusses decision trees, ensemble approaches, neural networks, regression analysis, and support vector machines for customer behavior prediction. Each algorithm's technique, merits, and weaknesses are examined. We then show how these algorithms can predict purchase intent, customer turnover, life cycle value, and other consumer behaviors. This setting presents challenges when using machine learning models, including data quality, model interpretability, and the ethics of using consumer data. The industry's current and future innovations, such as AI-driven personalization and deep learning to understand client behavior, are also highlighted. This chapter explores the intersection of machine learning and consumer behavior analysis to help researchers, data scientists, and business professionals acquire more accurate and valuable consumer insights.

This chapter explores the integration of psychology and technology as an opportunity to strengthen an individual's resilience and psychological health. Through a deep comprehensive review process, it identified 62 research articles and book chapters in the last 20 years. It investigated the potential keywords of combining psychology and technology which can enhance individual resilience and mental health. A thorough examination of all studies with their strengths, weaknesses, and limitations are incorporated to generate valuable insights. The results reveal that technological gears and tools like virtual games, virtual reality, and video applications and mobile apps can significantly enhance individuals' cognitive and motor skills and strengthen social connections. This chapter claimed that digital technologies and artificial intelligence promotes happiness and resilience of people. From teletherapy bridging geographic gaps to virtual reality. Finally, it claims an increasing anxiety among the people due to application of artificial intelligence.

Transformative role of machine learning in mental health care, with a focus on digital therapy and personalized support. As mental health challenges increase globally, traditional therapeutic approaches face limitations in scalability and customization. Machine learning innovations, such as natural language processing (NLP) and predictive analytics, offer new avenues for mental health diagnosis, treatment, and ongoing care. AI-powered digital therapy platforms, including chatbots, provide real-time interventions, while personalized support systems analyze user data to tailor mental health strategies. By identifying patterns in behaviors and symptoms, machine learning enhances the effectiveness of mental health treatments, promoting timely and individualized care. However, challenges like data privacy, algorithmic bias, and the potential over-reliance on technology must be addressed. As these technologies evolve, they offer the potential to significantly improve access to and the quality of mental health care, creating scalable and responsive support systems for diverse populations.

Sentiment analysis and text emotion identification have grown in prominence due to their wide range of applications in fields such as psychology, artificial intelligence, human-computer interaction, and so on. There are numerous Machine Learning approaches available for emotion recognition and sentiment analysis. The chapter also delves into the key procedures of data collecting, preprocessing, and emphasising the necessity of good data in training effective models. Real-world applications from a variety of disciplines, including business, healthcare, and entertainment, are investigated to demonstrate the practical utility of these strategies. The chapter also covers the issues of ambiguity, context-awareness, and cross-linguistic disparities, as well as providing suggestions for future study paths. This chapter provides a detailed exploration of machine learning approaches to sentiment and emotion analysis, making it a valuable resource for researchers, practitioners, and students interested in using machine learning to understand and interpret emotional content in textual data.

Artificial intelligence (AI) can enhance teaching strategies for neurodivergent students in higher education. Neurodiversity, such as autism, ADHD, and dyslexia, needs tailored educational approaches often missing in traditional systems. As neurodivergent student enrollment rises, inclusive learning environments are essential. AI technologies offer personalized learning through adaptive platforms that adjust content based on real-time feedback, helping students tackle challenges and utilize their strengths. AI tools, such as emotion recognition, support mental health by detecting stress, frustration, or anxiety for timely interventions. A case study highlights integration of an AI-powered chatbot in a Color Theory course for Engineering students. The chatbot provided guidance and used visual aids, improving communication, independence, and confidence. Despite AI benefits, issues like over-reliance and digital divide must be considered. Developing empathetic AI solutions is crucial for supporting neurodivergent students, and collaboration among educators and researchers is essential.

A Geographic Information System (GIS) is a technological tool that allows for the capture, storage, analysis, and visualization of geographically referenced data. These systems integrate various forms of spatial and non-spatial data, facilitating the analysis of geographic phenomena and patterns.The integration of Machine Learning (ML) into Geographic Information Systems (GIS) has revolutionized the way geospatial data is analyzed and used. Machine Learning, with its ability to learn from large volumes of data and make accurate predictions, complements the analytical capabilities of GIS, allowing for the extraction of complex patterns and the performance of advanced predictions that were not previously possible. The purpose of this chapter is to explore the applications of Geographic Information Systems (GIS) empowered by the use of machine learning, highlighting their impact on spatial analysis and environmental management.

This chapter explores about strategies of creating sustainable growth and wellbeing through digital financial education. It stresses how access to knowledge about finance by individuals on these same platforms can be democratized through digital means-the best tools in making informed financial decisions. Focusing on equity, this chapter looks at how digital education can fill the gaps of financial literacy of those disenfranchised groups as a way to resilient economic growth. The integration of principles of well-being to financial learning develops sound economic stability along with personal well-being. The chapter seeks personalized adaptive technology learning and interactive content to engage and retain these learners. This chapter has taken forward AI, gamification, and data analytics by such digital financial education to promote behavioral change and financial responsibility.

k-Nearest Neighbors Regression (kNN regression) is a non-parametric, instance-based learning algorithm widely employed for predicting continuous target variables. In contrast to conventional regression methods assuming specific functional forms, kNN regression operates by averaging the target values of the k nearest data points, determined by a chosen distance metric such as Euclidean or Manhattan

In this article the ethical implications and the risk that the work of artist suffers with the use of AI are discussed. The use of artificial intelligence is framed by the internet that stores intangible cultural assets as part of a cyberculture marked by the appropriation, the optimistic speech of CEO Sam Altman and the behavior of early adopters accept it in the midst of a culture of plagiarism and appropriation practices within digital art. Ethical and fair suggestions are established in favor of digital artists and the implementation of the glaze tool to protect them artwork, while the authorities don't take charge.

Currently, artificial intelligence (AI) has taken on great relevance in the daily lives of people, companies, and university students. This tool seeks to process large amounts of data and information to perform various tasks autonomously such as creating reports and/or reports, analyzing and classifying information. The purpose of this tool is that the AI reduce data processing time and let humans be the decision makers. In education, the use of AI in carrying out academic tasks or activities is increasingly common. This raises questions and discussions about the use of AI, to what level its use is ethical or not. Should it be regulated? Should we train teachers and students to obtain the greatest benefit from the use of AI? In this chapter, the authors analyze the use, the lack of regulation of the use of AI, the challenges that we will face in the future, and how the application of ethics can stop the indiscriminate use of this tool.

Preface

At the intersection of psychology, social innovation, and advanced applications of machine learning lies a vast area of exploration that is redefining our understanding of human behavior, technology, and its impact on society. *Exploring Psychology, Social Innovation, and Advanced Applications of Machine Learning* is a book that delves into this fascinating crossroads of disciplines, offering a deep and multi-faceted perspective on how artificial intelligence (AI) and machine learning are transforming not only the way we interact with technology but also how we address some of the most complex challenges of our time.

This book is a collection of studies, analyses, and reflections covering a wide range of topics, from mental health and education to corporate sustainability and artistic creativity. Each chapter serves as a window into a world where technology and psychology intertwine to offer innovative solutions and fresh perspectives. Through these texts, we explore how AI can be used to identify and manage personality disorders, how it can enhance human resilience in the digital era, and how it is revolutionizing sentiment and emotion analysis in the business sector.

In the field of mental health, for instance, innovative approaches are presented that use machine learning techniques to identify and manage complex personality disorders, such as those described in the DSM, including a case study on the Dark Triad. These advances not only provide new tools for mental health professionals but also pave the way for more accurate diagnoses and personalized treatments.

Human resilience in the digital era is another central theme of this book. In a world where technology is omnipresent, understanding how digital tools can foster or undermine our ability to adapt and thrive is crucial. The chapters dedicated to this topic explore both the challenges and opportunities presented by AI in this context, offering a balanced and reflective perspective on mental health, sustainability, and other applications with ethical implications.

Sentiment and emotion analysis is another area where machine learning is having a significant impact. From case studies in the business sector to real-time applications using deep neural networks, this book provides a detailed view of how these techniques are transforming entire industries, from marketing to customer service. Additionally, it explores the use of natural language processing models in detecting emotional states and improving human-machine interactions.

Throughout the book, AI applications in emerging sectors are also examined, including fake news detection, digital financial education, corporate sustainability, and natural disaster prediction. These studies highlight AI's immense potential to tackle global issues and improve quality of life in various domains.

TARGET AUDIENCE

1. **Academics and Researchers**: Scholars in psychology, computer science, social innovation, and related fields will find this book a valuable resource for understanding the intersection of these disciplines. It provides a comprehensive overview of current research and emerging trends, making it an essential reference for those exploring the applications of AI in human behavior and societal challenges.
2. **Mental Health Professionals**: Psychologists, psychiatrists, and therapists will benefit from the insights into how AI can enhance diagnostic accuracy, personalize treatments, and provide innovative therapeutic tools. The case studies and practical applications offer actionable insights for integrating AI into clinical practice.
3. **Educators and Students**: Professors and students in psychology, technology, and social sciences will find this book a rich source of knowledge for courses on AI, machine learning, and their societal impacts. The interdisciplinary approach makes it suitable for both undergraduate and graduate-level studies.
4. **Industry Professionals**: Business leaders, data scientists, and technology innovators will gain a deeper understanding of how AI can be applied to solve real-world problems, from improving customer experiences to enhancing corporate sustainability. The book's focus on practical applications and case studies makes it a valuable guide for professionals seeking to leverage AI in their industries.
5. **Policymakers and Social Innovators**: Those involved in shaping public policy or driving social change will find this book insightful for understanding the ethical, cultural, and societal implications of AI. It offers a balanced perspective on how AI can be harnessed for the greater good while addressing potential risks and challenges.
6. **General Readers with an Interest in Technology and Society**: Anyone curious about the transformative potential of AI and its impact on human behavior and society will find this book accessible and engaging. The clear explanations and real-world examples make complex concepts understandable for non-experts.

ORGANIZATION OF THE BOOK

1. Innovative Artificial Intelligence Approaches for Identifying and Managing DSM Cluster B Personality Disorders in Mental Health: A Case Study on the Dark Triad

This chapter delves into the potential of AI in diagnosing and managing complex personality disorders categorized under DSM Cluster B, with a particular focus on the Dark Triad (narcissism, Machiavellianism, and psychopathy). By leveraging machine learning algorithms, mental health professionals can analyze patterns in behavior, speech, and social interactions to improve diagnostic precision. The chapter also discusses case studies where AI-assisted models have been employed in clinical settings, highlighting their implications for early intervention, personalized treatment, and ethical considerations in mental health care.

2. Advancements in Real-Time Emotion Detection Systems Using Deep Neural Networks

As AI technology progresses, real-time emotion detection has become increasingly sophisticated. This chapter explores how deep neural networks analyze facial expressions, voice patterns, and physiological signals to interpret human emotions with high accuracy. Applications in healthcare, customer service, and security are examined, along with ethical concerns regarding privacy, bias, and the implications of AI interpreting human emotions.

3. An Overview of Sentiment Analysis: Insights From Business Case Studies

This chapter provides an in-depth analysis of sentiment analysis and its applications in the corporate world. It discusses how businesses use machine learning to interpret customer feedback, gauge public opinion, and optimize marketing strategies. Case studies from major companies illustrate how AI-driven sentiment analysis enhances customer experience, brand perception, and crisis management. Additionally, the chapter explores the limitations and challenges of sentiment analysis, including language ambiguity and contextual misunderstandings.

4. Evaluating NLP Models for Mental Health: Performance and Accuracy in User-Psychologist Conversations

This chapter evaluates the effectiveness of NLP models in analyzing and enhancing mental health conversations between users and psychologists. It discusses performance metrics, challenges in detecting nuanced emotional states, and the ethical implications of AI-mediated therapy. Case studies demonstrate how AI tools can support clinical decision-making without replacing human psychologists.

5. Machine Learning in Fake News Detection and Social Innovation: Navigating Truth in the Digital Age

The rise of misinformation poses a significant challenge in today's digital landscape. This chapter discusses how AI and machine learning algorithms are being used to detect and combat fake news. It examines key technologies such as deepfake detection, automated fact-checking, and NLP-based credibility assessment. Ethical considerations, including censorship, bias, and the balance between freedom of expression and misinformation control, are also explored.

6. Consumer Behavior Prediction Using Machine Learning Algorithms

Predicting consumer behavior has become a key advantage in business intelligence. This chapter explores how machine learning models analyze purchasing patterns, customer feedback, and market trends to optimize product recommendations, pricing strategies, and targeted advertising. The chapter also discusses the challenges of data privacy and algorithmic bias in consumer behavior prediction.

7. Understanding the Role of Emerging Technology in Human Resilience in the Digital Age and Artificial Intelligence

With digital technologies becoming ubiquitous, this chapter explores how AI-driven tools influence human adaptability and resilience in a rapidly evolving world. It examines the psychological impact of technology, including digital stressors and AI-enhanced coping mechanisms. The discussion includes real-world applications, such as AI-powered mental health chatbots and virtual support systems, and their role in fostering well-being while addressing concerns about dependency, misinformation, and digital fatigue.

8. Innovations in Digital Therapy and Personalized Support: Machine Learning in Mental Health

This chapter explores how machine learning is revolutionizing digital therapy by providing personalized mental health support. It highlights AI-powered platforms that analyze user responses to tailor therapeutic recommendations, making mental health care more accessible and cost-effective. The chapter also discusses ethical considerations, data privacy concerns, and the future of AI in psychotherapy.

9. Sentiment and Emotion Analysis Using Machine Learning Techniques

Building upon previous discussions, this chapter provides a technical overview of sentiment and emotion analysis methodologies. It covers various machine learning techniques, such as natural language processing (NLP) and deep learning models, used to detect emotional cues in text and speech. The chapter illustrates applications in customer service, social media monitoring, and mental health diagnostics, emphasizing the transformative potential of AI in human interaction analysis.

10. Designing Teaching Strategies Using Artificial Intelligence for Neurodivergent Students in Higher Education

AI plays a fundamental role in inclusive education. This chapter examines AI-driven teaching strategies tailored for neurodivergent students, including those with autism, ADHD, and dyslexia. It explores adaptive learning platforms, personalized curriculum design, and AI-powered tutoring systems that enhance learning outcomes while ensuring equity in education. Case studies illustrate how universities and edtech companies are successfully implementing AI to support diverse learning needs.

11. Machine Learning for Natural Disaster Prediction and Prevention

This chapter examines the crucial role of machine learning in predicting and preventing natural disasters. It explores methodologies used to identify patterns in events such as earthquakes, hurricanes, floods, and wildfires, highlighting predictive models that enable early warnings and more effective responses. Additionally, it discusses successful case studies where these technologies have reduced disaster impact through environmental monitoring systems and advanced climate simulations.

12. Enhancing Industrial Employee Performance by Integrating Machine Learning and Human Psychology

This chapter investigates how integrating machine learning with principles of organizational psychology can optimize employee performance in industrial environments. It covers applications such as AI-driven personalized training programs, workplace well-being assessments through data analysis, and intelligent systems to improve decision-making in professional settings. The chapter also explores ethical implications of using such systems, their impact on corporate culture, and the necessity of ensuring that automation complements rather than replaces human talent.

13. Advancing Geographic Information Systems With Machine Learning: Innovations in Spatial Analysis and Environmental Management

This chapter highlights innovations in Geographic Information Systems (GIS) driven by machine learning, emphasizing spatial analysis and environmental management. It explores case studies on predictive modeling for ecosystem conservation, sustainable urban planning, and environmental risk mitigation. Additionally, it presents recent developments in integrating satellite data with AI algorithms for real-time monitoring of deforestation, air quality, and climate change, underscoring their impact on government and corporate decision-making.

14. Cultivating Sustainable Growth and Well-Being in Digital Financial Education

This chapter examines how artificial intelligence is revolutionizing digital financial education, enabling broader access to essential tools and knowledge for informed economic decision-making. It analyzes applications of machine learning in personalizing financial literacy courses, detecting spending patterns, and developing virtual assistants that promote better financial habits. The chapter also discusses ethical and regulatory challenges of automation in financial education, particularly concerning vulnerable populations and emerging markets.

15. k-Nearest Neighbors Regression and Application

This chapter provides a detailed analysis of the k-Nearest Neighbors (k-NN) regression algorithm, highlighting its theoretical foundations, advantages, and limitations compared to other machine learning methods. It presents applications in various industries, including real estate price prediction, customer segmentation in marketing, and AI-assisted medical diagnosis. Additionally, case studies illustrate how k-NN has been implemented to optimize business processes and improve real-time predictive models.

16. Machine Learning in Latin America: Transforming Industrial Processes and Value Chains

This chapter analyzes the impact of machine learning on transforming industrial processes and value chains in Latin America. It examines case studies of companies adopting AI to improve production efficiency, optimize logistics, and reduce operational costs. The chapter also addresses regional challenges and opportunities, including the need for investment in technological infrastructure, workforce training, and the role of government policies in promoting industrial digitization.

17. Machine Learning Algorithms Applied to Machine Vision and Inertial Systems

This chapter explores the integration of machine learning algorithms in machine vision and inertial systems, with applications in robotics, manufacturing, and autonomous navigation. It presents advancements in image recognition, sensor signal processing, and calibration of navigation systems based on inertial sensors. Additionally, it discusses applications in industrial automation and autonomous mobility, including the impact of these technologies on operational efficiency and the safety of automated systems.

18. AI and the Creative Process: Balancing Innovation, Copyright, and Artistic Authenticity

This chapter examines the role of artificial intelligence in the creative process, exploring its potential to generate new forms of artistic expression while raising challenges related to copyright and artistic authenticity. It analyzes AI tools used in music, visual arts, literature, and film, as well as debates surrounding the originality of AI-generated works. Furthermore, the chapter discusses ethical and legal implications, along with strategies to ensure that AI enhances human creativity rather than replacing it.

19. Ethical Implications of Students Using Artificial Intelligence

This chapter addresses the ethical dilemmas associated with the use of AI in educational settings, focusing on issues such as automated plagiarism, equitable access to technological tools, and student data privacy. It examines concerns about how AI misuse could affect the development of critical thinking and creativity among students, as well as strategies for promoting ethical and responsible AI use within academia.

CONCLUSION

Together, the chapters in this book offer a comprehensive view of how psychology, social innovation, and machine learning converge to address some of the most pressing challenges of our time. This book is not only a guide for practitioners and scholars in these fields, but also an invitation to reflect on the future we are building with these technologies. We hope that this volume will inspire new ideas, foster dialogue, and contribute to a more inclusive, sustainable, and humane future. In an increasingly interconnected and technologically advanced world, the convergence of psychology, social innovation, and advanced applications of machine learning represents one of the most exciting frontiers of knowledge. AI and machine learning are not only redefining the way we understand human behavior, but they are also redefining our relationship with technology and its impact on society. This book invites readers to question, reflect, and innovate at the intersection of these fields, seeking a balance between technological advancement and ethical responsibility. As AI continues to evolve, interdisciplinary collaboration becomes essential to maximize its benefits and minimize its risks with a social and ethical vision.

We hope this volume serves as an inspiring resource and a starting point for future research and applications in the field of Artificial Intelligence and sustainability, fostering a deeper mastery of knowledge.

Chapter 1
Innovative Artificial Intelligence Approaches for Identifying and Managing DSM Cluster B Personality Disorders in Mental Health:
A Case Study on the Dark Triad

María E. Raygoza-L.
https://orcid.org/0009-0000-5969-6602
Universidad Politécnica de Baja California, Mexico

Jesús Heriberto Orduño-Osuna
https://orcid.org/0009-0004-4850-7481
Universidad Politécnica de Baja California, Mexico

Roxana Jimenez-Sanchez
https://orcid.org/0009-0002-0061-7206
Universidad Politécnica de Baja California, Mexico

Fabian N. Murrieta-Rico
https://orcid.org/0000-0001-9829-3013
Universidad Politécnica de Baja California, Mexico

ABSTRACT

The increasing use of Artificial Intelligence (AI) in mental health presents significant ethical and practical challenges. Applications such as Machine Learning (ML) and chatbots aim to enhance the accessibility and effectiveness of therapies, enabling earlier diagnosis and continuous monitoring. However, their implementation raises concerns regarding data privacy and the potential dehumanization of psychological treatment. Furthermore, AI can be utilized to assist in identifying traits associated with the dark triad, including narcissism, psychopathy, and Machiavellianism, which may prove beneficial in managing various personality disorders. Despite its potential advantages, it is crucial to develop public policies

DOI: 10.4018/979-8-3693-6910-4.ch001

that regulate the use of these technologies and protect patient rights, ensuring that AI complements rather than replaces human intervention in mental health care.

INTRODUCTION

Currently, the increase in the use and applications of Artificial Intelligence (AI), it covers more and more fields of knowledge, certainly one of the greatest and most contradictory challenges due to the ethical and human implication that it entails are the uses of AI in relation to mental health. In the contemporary era of constant change, AI is increasingly entering the field of mental health care with different tools such as Machine Learning (ML), digital tools such as "chatbots", virtual therapies, etc. (Zucchetti, Nibbio, Altieri, Bertorni, & Calzavara-Pinton, 2024).

AI tools applied in mental health seek to increase the effectiveness of therapies, in addition to being a support for both psychologists and psychiatrists. In addition to the search for a possible accessibility that can reach more people effectively and be a support for most of the population, which many times due to ignorance or taboos that arise around the importance of mental health, as well as economic problems. Mental health has been neglected for a long time, causing often irreversible damage to individuals with mental health problems (Graham, Colin, & Lee, 2020). AI applications in mental health can have a positive role in different psychiatric and psychological disorders such as neurodegenerative disorders, intellectual disability, epilepsy, seizures, in addition to other areas of mental health that have been subject to studies by experts of mental health such as those of the dark triad that will be analyzed later in this chapter, in addition to reviewing the limitations of AI-based approaches in the area of public policies, limitations culturally in addition to showing some ethical issues that may arise with the use of AI in mental health (Thakkar, Ankita, & Avinash, 2024).

Artificial Intelligence (AI)-driven platforms, structured under mathematical algorithms, can provide support in the early diagnosis of mental health conditions, deliver therapeutic interventions, and monitor patient progress using techniques such as Natural Language Processing (NLP) and other machine learning methods (Ngumimi, Soo-Hyung, & Min, 2023).

The integration of AI into therapeutic techniques holds the promise of enhancing the overall quality of mental health care by facilitating timely decision-making for individuals with various mental health disorders (Park, Lee, & Young-Jin, 2021). For instance, AI can manage and analyze large datasets, which can aid in the discovery and creation of behavioral patterns predictive of mental health conditions (Olawade, Wada, & Odetayo, 2024).

This rapidly developing and increasingly implemented field raises continuous ethical questions and concerns, particularly regarding the use of personal data, informed consent, and the potential risks that AI usage might entail. There is potential for AI to be applied in legal matters, as well as in the identification and prevention of victims associated with individuals exhibiting dark triad traits such as narcissism, psychopathy, and sociopathy.

Psychologists will face significant challenges to ensure that AI use is both effective and ethical, necessitating constant monitoring of AI-supported treatments. It is also crucial to establish boundaries for the applications of these techniques and tools, without neglecting the aspect of informed consent from mental health patients treated under these terms (Abrams, 2023).

When discussing and researching topics related to mental health, psychology, and AI, it is essential to conduct thorough investigations into not only the current status of technical applications, advancements, and success stories but also the implications and scope established within public policy an element crucial for ensuring timely follow-up without falling into issues of rights, legalities, and ethics. The authors argue that in all matters concerning AI, especially in the context of mental health, a robust application of public policies is indispensable to protect patient rights and ensure the ethical use of technology.

All public policy actors must collaborate with psychologists to create regulations that protect, promote transparency and prevent the misuse of AI, as well as stakeholders such as public and private health institutions, educational institutions and civil society must be involved to promote the use of technological tools for quality-of-life purposes (Raygoza-L, Orduño-Osuna, & Murrieta-Rico, 2023), with the aim of developing equitable access to mental health care. Moreover, guidelines should be developed to ensure that AI tools complement rather than replace human therapists, maintaining the critical human element in psychological care, which can help minimize the inherent risks associated with mental health.

The objective of this chapter is to explore the implications of AI within the field of psychology, significant advances in data processing and machine learning, as well as new tools and approaches to improve clinical practice, research, and mental well-being. AI can contribute to early diagnosis in the assessment of mental health disorders, supported by the evaluation of large datasets that AI can analyze, including medical histories, biometric data, and patterns of disorders such as anxiety and depression, among many others, that support the personalization of treatments to improve the quality of life for individuals with mental health issues.

BACKGROUND: APPLICATIONS OF ARTIFICIAL INTELLIGENCE TOOLS IN VARIOUS CASES OF MENTAL DISORDERS

In recent years, AI has experienced exponential growth across all fields of knowledge and numerous economic sectors. This rapid expansion underscores the importance of examining the current advancements of AI in the complex domain of mental health. Specifically, it is crucial to explore AI's role in the detection of mental illnesses, its use in assisted therapy, and its potential to offer genuine psychological support (Olawade, Wada, & Odetayo, 2024).

Furthermore, it is essential to consider how interactions with AI-driven technologies, as opposed to traditional therapy with psychologists, may benefit or pose challenges to patients. These and other critical issues must be addressed as AI continues to reshape the landscape of mental health care (Kelly & Kaye, 2023).

Given the nature of mental illnesses, extensive research is needed to determine whether AI's emotional support, behavioral treatments, and other interventions will consistently have a positive impact, or whether they might exacerbate conditions like anxiety in some patients, particularly those who cannot access timely human therapists. Many questions remain unanswered in this area, making a thorough understanding of current studies essential as a foundation for future research in this controversial field.

Recent research shows that experts in AI applications are working on developing continuous mental health monitoring, such as wearable devices and mobile applications that can collect data on users' behavior and emotions, including sleep patterns, physical activity, and stress levels. These data are analyzed by AI algorithms to detect significant changes that may indicate a mental health crisis, enabling early interventions and improving the efficacy of timely treatment, thus reducing the time needed to achieve positive outcomes (Lebowitz, 2000)

Certainly, AI is radically changing the methods and approaches used to address mental health disorders, from diagnosis and evaluation to assisted therapy and treatment personalization. However, it is crucial to address relevant public policies to tackle the ethical dilemmas and challenges associated with its use, ensuring that these technologies benefit everyone respectfully (Obrenovic, Gu, & Godinic, 2024). It must not be overlooked that, beyond technological development, the implications of AI in mental health can have significant repercussions. While they may indeed improve the quality of life, they could also result in disastrous outcomes, especially as mental health continues to claim the lives of many individuals suffering from conditions such as depression, schizophrenia, anxiety, and others.

To delve into the applications of AI in the field of mental health, it is essential to understand what AI is. There are numerous proposed definitions of AI; however, most align around the concept of machines capable of performing human tasks, behaviors, or creating software programs. The founding father of this discipline, John McCarthy, described AI as "the process of making a machine behave in a way that would be called intelligent if a human were to behave that way." Over the past two decades, AI has evolved—from playing with toy problems like chess to acquiring new skills and seeking to discover its own limits. After 60 years, AI has permeated industry and public consciousness. Recently, in the 21st century, this discipline has transformed every aspect of our lives so significantly that it is now known as the "Era of AI" (Rajaraman, 2014).

AI is revolutionizing traditional processes and enhancing decision-making by creating new capabilities and unlocking novel pathways for innovation. As AI continues to evolve and reshape the landscape, we can anticipate even greater shifts in industries towards a landscape dominated by intelligent systems driven by machine learning techniques and advanced algorithms. The integration of AI spans a wide spectrum from healthcare to recruitment, logistics, education, transportation safety, e-governance, and public sectors. For the purposes of this chapter, we focus on its applications in mental health. Like most other fields, mental health care has also been affected by the digital technology revolution, specifically AI.

The emergence of digital mental health as a subfield signifies the integration of technology into mental health care practices, and within this domain, AI-driven solutions have become a significant contributor to positive mental health outcomes. In this field, the importance of ethics must be highlighted. In a 2022 survey involving 850 organizations across 18 geographies, 77% reported prioritizing AI regulations as company-wide policies, and 80% stated that they would invest in the development of ethical AI (Thakkar, Ankita, & Avinash, 2024).

Ethics plays a pivotal role in Artificial Intelligence (AI), which encompasses the science and computational technology aimed at creating systems and machines capable of performing tasks that typically require human intelligence. These tasks include understanding natural language, recognizing patterns, making decisions, learning from experience, and adapting to new situations, all achieved through a combination of algorithms, data processing, and iterative improvement processes.

In particular, one of AI's most significant tools, machine learning (ML), is a subset of AI that involves developing algorithms that enable systems to learn from data. This encompasses several techniques, including supervised learning, unsupervised learning, deep learning, and reinforcement learning. ML algorithms can analyze patient data, such as electronic medical records and behavioral patterns, to aid in diagnosing mental health conditions such as depression, anxiety, schizophrenia, and dark triad behavioral patterns. These algorithms can also predict the risk of developing certain disorders based on historical data.

Moreover, ML can help tailor treatment plans by analyzing patient data to recommend therapies, medications, or specific interventions based on individual characteristics and response patterns. In supervised machine learning (SML), data is pre-labeled in advance, for example, to distinguish between major depressive disorder (MDD) diagnosis and the absence of depression. The algorithm then acquires the ability to determine the most accurate input features drawn from various data sources, including sociodemographic, biological, and clinical measures.

In contrast, unsupervised machine learning (UML) algorithms operate without predefined labels, allowing them to detect similarities among input features and reveal inherent data patterns. However, they lack the ability to associate features with a known label. These algorithms have the advantage of discovering inherent structure within a dataset without significant pre-existing biases. For example, extensive feature datasets, such as neuroimaging biomarkers, have the potential to uncover information about unknown subtypes within psychiatric disorders such as schizophrenia. On the other hand, artificial neural networks (ANNs) and deep learning (DL) are based on the structure and functioning of the human brain. Deep learning be used for image analysis in brain scans, contributing to the identification of structural abnormalities linked to mental health disorders (Schork, 2019).

Natural language processing (NLP) can significantly aid in analyzing written or spoken language to detect emotional states and changes, helping clinicians monitor patients' mental well-being through texts, chats, or speech. Reinforcement learning (RL) is a type of ML algorithm that involves training an agent to make sequential decisions by interacting with an environment. The algorithm acts as an agent in an interactive environment, learning through trial and error using rewards from its own actions and experiences. Its applications can be used in virtual reality exposure therapy to adjust exposure levels in response to patient reactions. Additionally, computer vision can have a valuable impact when applied to analyzing facial expressions and gestures to infer emotional states, aiding in assessing patients' emotional well-being and preventing or detecting anxiety and depression episodes (Alqahtan, Badreldin, & Mohammed, 2023).

In the field of mental health care, these elements collaborate in innovative ways, from personalized interventions to predictive analytics. The integration of AI technologies offers promising avenues to enhance mental health outcomes. Previous research has provided a framework where technology and AI in mental health care are changing how mental health and well-being are approached and supported. These tools have the capability not only to help mental health professionals more objectively identify the early stages of mental health disorders, as outlined in the current DSM-5 framework, but also to ensure that the application of these tools is guided by a strong ethical framework.

Challenges and Opportunities of AI Applications in Mental Health:

- Raising awareness of mental health
- Early prevention of mental health problems
- Use of technologies for sentiment analysis and data Mining
- Initiatives to disseminate pertinent mental health information
- Combating stigma and taboos while promoting dialogue around mental health
- Revolutionizing how society perceives and discusses mental health issues
- AI's ability to process and analyze large volumes of data for more Accurate projections
- Regulating data from social media platforms like Twitter, Facebook, and Instagram
- AI-Driven sentiment analysis algorithms to discern emotions expressed in posts

- Monitoring public behavior and perception of mental health on social media
- Developing key initiatives for raising awareness
- Supporting Decision-Making through early detection of mental health problems
- Providing guidance, coping strategies, and other personalized support services
- Enhancing psychoeducation to foster a deeper understanding of mental health concepts
- Mental health professionals can employ engagement strategies to effectively guide the recovery path of individuals facing mental health issues
- Improving the monitoring of medication side effects and controlling medication responses
- Supporting the monitoring of state fluctuations in patients facing mental health issues, as well as in the public
- Supporting the intervention and treatment of mental health disorders

Furthermore, AI allows for the personalization of treatments based on the unique characteristics of everyone. In the context of prediction and prevention, AI can assess the risk of developing specific mental health disorders based on an individual's profile, genetic predisposition, and environmental factors to implement efficient measures to prevent or mitigate the onset of disorders. This includes diagnostics and early detection of signs of mental health disorders, enabling timely intervention and a better prognosis. Other tools can automate cognitive assessments, such as language, reasoning, memory, decision-making, attention, and perception, to intelligently manage the actions of occupational therapists, educators, psychiatrists, neurologists, or psychologists. Table 1 presents some mental health disorders, the main AI tools that can support expert assessments, and the potential advantages they would have for the care of individuals with mental health conditions.

Table 1. Mental disorders with ai applications to support mental health experts

Mental Health Disorder	AI Tools that Can Be Used for Mental Health Benefits	Advantages or Benefits of AI Tools
Neurocognitive, Neurological and Neurodevelopmental Disorders (Epilepsy)	Analysis of eye movements, machine learning, and seizure detection through electroencephalographic (EEG) data analysis AI-based predictive modeling for seizure forecasting	Early detection of disorders like ASD; personalized interventions, improved seizure detection, continuous monitoring of brain electrical patterns
Schizophrenia Spectrum and Other Psychotic Disorders	Automated speech analysis, NLP ML algorithms for pattern recognition in neuroimaging AI-driven virtual reality therapy for social skills training	Evaluation of speech patterns and prosody to detect subtle symptoms; relapse prediction AI can help in identifying biomarkers in neuroimaging data
Bipolar Disorder and Related Disorders	Social Rhythm Metric (SRM), analysis of data from smartphones Personalized treatment recommendations through ML	Personalized monitoring that improves self-management and reduces the risk of severe episodes
`Depressive Disorders	Analysis of emotions from biometric data Predictive analytics for early detection of depressive episodes	Improved emotional regulation; support in identifying emotional patterns
Anxiety Disorders	Applications for emotional regulation, analysis of voice data, and facial expressions AI-driven exposure therapy using virtual reality	Support in managing anxiety through personalized interventions and real-time feedback

continued on following page

Table 1. Continued

Mental Health Disorder	AI Tools that Can Be Used for Mental Health Benefits	Advantages or Benefits of AI Tools
Obsessive-Compulsive Disorder	Behavioral pattern analysis tools, virtual reality applications AI-enhanced cognitive-behavioral therapy platforms Machine learning for detecting obsessive thoughts through physiological signals	Provides controlled environments for exposure and practice of coping skills Enhanced personalization in CBT approaches
Trauma and Stress-Related Disorders	AI-driven therapeutic games, immersive virtual reality ML models for predicting PTSD onset based on early symptoms	Provides safe environments to practice emotional regulation and coping skills
Dissociative Disorders	Analysis of behavioral patterns and facial expressions AI-driven therapeutic avatars for grounding techniques ML models for tracking dissociative episodes	Detection of dissociative symptoms from digital interactions; support in early identification
Paraphilic Disorders (Masochism)	AI-driven ethical decision-making frameworks for managing risky behaviors Predictive models for preventing escalation of harmful behaviors	Identification of behaviors that do not conform to social norms Ethical decision-making support for clinicians
Antisocial Personality Disorder	AI-driven risk assessment tools for predicting violent behaviors AI-enhanced rehabilitation programs focusing on empathy development	Identification of antisocial behaviors; Early detection of potential violent behavior
Borderline Personality Disorder	Interactive platforms, emotional data analysis AI-driven dialectical behavior therapy (DBT) tool	DBT tools can provide real-time support for crisis management
Histrionic Personality Disorder	Analysis of facial expressions and behavioral patterns ML for identifying attention-seeking behaviors	Identification of behaviors and expressions related to personality
Narcissistic Personality Disorder	Tools for analyzing emotional data AI-driven therapeutic tools for empathy training	Detection of narcissistic patterns; support in early identification and management

The following tables present some of the main mental health disorders. Not all tables are organized according to the groups proposed by the DSM-5 but rather by the characteristics that the authors consider relevant to the analysis of this chapter, adding advantages and disadvantages of AI tools, including those of the dark triad.

Table 2. Neurodevelopmental, Neurocognitive, and neurological disorders

Artificial Intelligence Tools Beneficial for Mental Health	Mental Health Disorder Addressed	Advantage or Benefit of AI Tools in Mental Health	Disadvantage or Risk of Applying AI Tools
Automated speech analysis y NLP	Schizophrenia Spectrum and Other Psychotic Disorders	Evaluation of speech patterns and prosody to detect subtle symptoms; relapse prediction	Risk of incorrect interpretations; need for human supervision
Automated speech analysis y NLP	Schizophrenia	Evaluation of speech patterns and prosody to detect schizophrenia symptoms; relapse monitoring	Risk of misinterpretation of speech patterns; need for clinical supervision
Analysis of behavior patterns, activity monitoring applications	Impulse and Conduct Control Disorders	Monitoring and managing impulsive behaviors; support in developing self-control strategies	Potential lack of complete understanding of the complexity of impulse control

Table 3. Mood disorders, anxiety, and stress-related and trauma disorders

Artificial Intelligence Tools Beneficial for Mental Health	Mental Health Disorder Addressed	Advantage or Benefit of AI Tools in Mental Health	Disadvantage or Risk of Applying AI Tools
Social Rhythm Metric (SRM), analysis of data captured from smartphones	Bipolar Disorder and Related Disorders	Personalized monitoring improves self-management and reduces the risk of severe episodes	Possible false positives or negatives; need for rigorous validation
Interactive platforms, emotion analysis using biometric data	Depressive Disorders	Improvement in emotional regulation; support in identifying emotional patterns	Dependence on data quality; potential lack of sensitivity to individual variations
Applications for emotional regulation, analysis of voice data and facial expressions	Anxiety Disorders	Support in managing anxiety through personalized interventions and real-time feedback	May not capture the full complexity of anxiety; risk of overdiagnosis
AI-driven therapeutic games, immersive virtual reality	Trauma and Stress-Related Disorders	Provides safe environments to practice emotional regulation and coping skills	Risk of not fully addressing underlying trauma; potential for undesired experiences

Table 4. Obsessive-compulsive and dissociative disorders

Artificial Intelligence Tools Beneficial for Mental Health	Mental Health Disorder Addressed	Advantage or Benefit of AI Tools in Mental Health	Disadvantage or Risk of Applying AI Tools
Behavior pattern analysis tools, virtual reality applications	Obsessive-Compulsive Disorder (OCD)	Provides controlled environments for exposure and practice of coping skills	Potential overreliance on technology; need for clinical supervision
Analysis of behavior patterns and facial expressions	Dissociative Disorders	Detection of dissociative symptoms through digital interactions; support in early identification	Misinterpretation of dissociative signals; need for clinical validation

Table 5. Personality disorders (group B, regarding DSM-5) y and Machiavellianism

Artificial Intelligence Tools Beneficial for Mental Health	Mental Health Disorder Addressed	Advantage or Benefit of AI Tools in Mental Health	Disadvantage or Risk of Applying AI Tools
Analysis of behavior patterns, monitoring applications, tools for analyzing behavior patterns and emotions.	Antisocial Personality Disorder (Psychopathy)	Identification of antisocial behaviors; patterns of behavior and emotions; early intervention in managing problematic behaviors.	Risk of misinterpretations and stigmatization; need for clinical supervision.
Tools for analyzing emotional and behavioral data.	Narcissistic Personality Disorder	Detection of narcissistic patterns; support in early identification and management.	Possible stigmatization; risk of overinterpretation of data.
Analysis of facial expressions and behavior patterns.	Histrionic Personality Disorder	Identification of behaviors and expressions related to histrionic personality; support in intervention.	Risk of erroneous interpretations; need for clinical validation.
Interactive platforms, emotional data analysis.	Borderline Personality Disorder	Support in emotional regulation and stress management; development of emotional skills.	Possible over-reliance on technology; risk of inadequate human intervention.
Analysis of manipulative behaviors, monitoring applications, tools for analyzing social interactions.	Machiavellianism	Identification of manipulative behaviors; support in understanding social dynamics and intervention.	Risk of reinforcing manipulative tactics; ethical concerns in monitoring social behavior.

As mentioned, this chapter aims to explore the relationship between the behaviors or traits of the Dark Triad personalities, which are largely characterized under "Cluster B" of the DSM-5 (Diagnostic and Statistical Manual of Mental Disorders, Fifth Edition). The Dark Triad consists of three distinct personality traits—Narcissism, Psychopathy, and Machiavellianism—that share significant similarities with Histrionic Personality Disorder.

In general, the primary disorders within DSM-5's "Cluster B" include Antisocial Personality Disorder (psychopathy), characterized by a pattern of disregard for, and violation of, the rights of others; Borderline Personality Disorder, marked by instability in interpersonal relationships and self-image; Histrionic Personality Disorder, defined by a pattern of excessive emotionality and attention-seeking behavior; and Narcissistic Personality Disorder, which involves a need for admiration and a lack of empathy. It is important to note that while Machiavellianism is not yet classified as a mental disorder within the DSM, there is a proposal for its inclusion in future revisions, alongside other traits of the Dark Triad (American Psychiatric Association, 2000).

Personality is one of the most fundamental topics studied over many years across various branches of psychology, including clinical psychology, personality psychology, organizational psychology, and social psychology. Personality is a lifelong process determined by an individual's biological characteristics and the meanings derived from interactions with their environment. How an individual's personality traits are expressed in behavior distinguishes the positive from the dark side of personality. Generally, while the positive aspects of personality reflect a caring and helpful orientation towards others, the dark side is strongly associated with socially disruptive outcomes and negative impacts

Members of the Dark Triad may often exhibit similar behaviors, leading individuals toward similar conduct. This similarity arises from a shared insensitivity that promotes manipulation (Hampson, 2012). Below Figure 1 shows the 3 traits of the personalities belonging to the dark triad (Narcissism, Machiavellianism and Psychopathy).

Figure 1. Personality traits of the dark triad

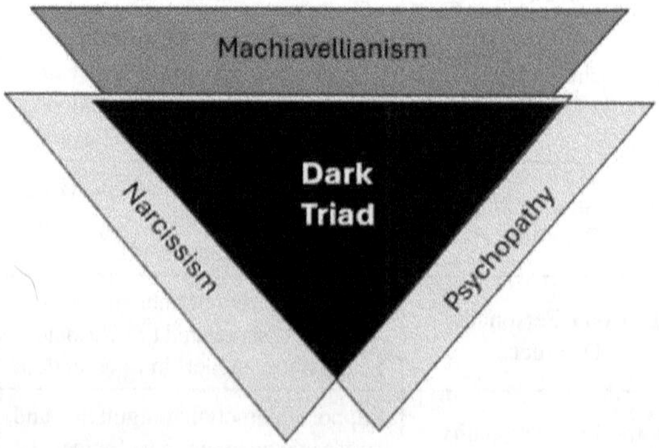

It is crucial to identify individuals with these personality traits due to the psychological and emotional repercussions they can inflict on their victims. These individuals may engage in actions marked by superficiality, manipulation, exploitation of those who trust them, and a lack of empathy, sometimes escalating to physical harm. While narcissistic personality disorder does not inherently include impulsivity and aggression, individuals with antisocial personality disorder (psychopaths) may exhibit high levels of aggression without the need for admiration or envy. However, all three personality types of narcissists, antisocial individuals, and Machiavellian tend to conceal their true intentions behind a façade of charm and reliability, causing those around them to question their motives for years until they are no longer deemed useful in their manipulative plans (Kerber, 2021).

Individuals with antisocial personality disorder (psychopaths) and narcissistic personality disorder share a tendency to be obstinate, superficial, exploitative, and lacking in empathy. However, narcissistic personality disorder is not associated with impulsivity and aggression, while those with antisocial personality disorder may not exhibit the same craving for admiration and envy. Furthermore, people with narcissistic personality disorder typically lack a history of childhood conduct disorders or criminal behavior in adulthood. In contrast, individuals with antisocial and histrionic personality disorders share a tendency toward impulsivity, superficiality, sensation-seeking, recklessness, seduction, and manipulation. Still, individuals with histrionic personality disorder tend to be more exaggerated in their emotions and do not typically engage in antisocial behaviors. Moreover, they manipulate others to gain care and protection, while those with antisocial personality disorder manipulate to gain power, material rewards, or other forms of gratification. Individuals with antisocial personality disorder also tend to be less emotionally unstable and more aggressive than those with borderline personality disorder.

Many individuals may exhibit histrionic traits. However, it is only when these traits are inflexible, maladaptive, and persistent leading to functional impairment or significant subjective distress that they constitute histrionic personality disorder. Their self-esteem is bolstered ("mirrored") by the idealized value they assign to those with whom they interact. Individuals with this disorder generally demand excessive admiration and are constantly seeking compliments, often with great charm. They display a sense of entitlement and unreasonable expectations of being treated with special favor. This entitlement, combined with a lack of sensitivity to others' needs and desires, may lead to conscious or unconscious exploitation of others. When they recognize the needs, desires, or feelings of others, they often regard them with disdain, seeing them as signs of weakness or vulnerability. Those describing individuals with narcissistic personality disorder often label them as emotionally cold and disinterested in mutual relationships (Muris & Otgaar, 2023).

Figure 2 illustrates the global prevalence percentage of the Dark Triad and the histrionic trait, part of Cluster B personality disorders in the DSM-V. It is evident that a significant portion of the population falls into one of these clusters. This information is critical, as it highlights those individuals within our social circles—such as friends, romantic partners, and colleagues may exhibit these traits. Consequently, we may be vulnerable to manipulative behaviors without possessing the necessary social tools or emotional intelligence to recognize and effectively respond to such behaviors.

Figure 2. Global prevalence of the dark triad

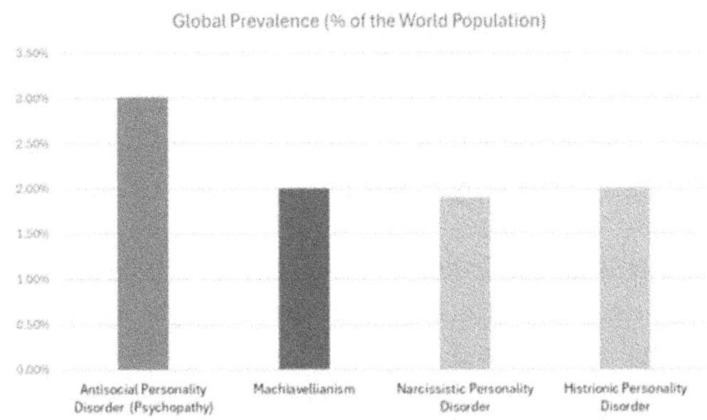

Criticism or failure may haunt these individuals, causing them to feel humiliated, degraded, undermined, and empty, though they may not display these feelings externally. They may respond with contempt, rage, or defiant counterattacks. An outward appearance of humility may mask and protect their underlying grandiosity. The most useful characteristic for distinguishing narcissistic personality disorder from histrionic, antisocial, and borderline personality disorders is the grandiosity and insensitivity that characterizes their style of interaction. Studies suggest that narcissism is characterized by a desire for dominance, psychopathy by a mix of high dominance and low affiliation, and Machiavellianism by low affiliation when considering interpersonal values and effectiveness (Dowgwillo & Aaron, 2017).

When these qualities appear in individuals with narcissistic personality disorder, they stem from the fear that their imperfections or defects will be exposed. From a biological perspective, the origins of these disorders are not fully understood. Few studies focus on structural neuroimaging in psychopaths, but some findings in the literature report reduced prefrontal gray matter volume, loss of gray matter in the right superior temporal gyrus, loss of amygdala volume, decreased posterior hippocampal volume, exaggerated hippocampal asymmetry, and increased callosal white matter volume in psychopathic individuals. These findings suggest that psychopathy is associated with brain abnormalities in a prefrontal-temporo-limbic circuit, regions involved in emotional processes and learning. However, these associations between structural changes and psychopathic traits do not allow for causal conclusions but rather highlight the significant role of biological brain abnormalities in psychopathy (Weber, Habel, & Schneider, 2008).

Psychopaths exhibit weaker connections between the amygdala and the prefrontal and orbitofrontal cortices (regions of the frontal lobe), which may explain their poor behavioral control and the impulsive, antisocial behavior commonly observed in psychopaths (Anderson & Kiehl, 2012). Beyond brain abnormalities, the lack of psychopathic emotions is evident. Some experiments show how the psychopathic brain functions when displaying dishonesty and a lack of empathy. Additionally, some studies have shown that psychopathic tendencies can develop after brain trauma, making individuals more impulsive, hostile, and immoral, while also impairing their ability to recognize emotions in others. Nonetheless, psychopathic behavior cannot be generalized across cases; for instance, one experiment demonstrated that psychopathic participants took longer to recognize emotions in others based on facial expressions and exhibited varying brain volumes in specific regions compared to controls and other psychopaths, depending on task performance. Other findings indicate that psychopaths who were better at recognizing fear had a larger gray matter volume in the cerebellum, while controls who performed better had a larger gray matter volume in the amygdala.

This suggests that not only do psychopaths have structural brain differences compared to normal individuals, but they also use their brains differently, achieving the same task less efficiently by utilizing different brain regions. The psychopathic brain's functioning results in a decreased ability to process emotions, act honestly, and recognize emotions in others, with diminished brain activity when deciding whether or not to act dishonestly, especially in regions associated with empathy and perspective-taking (Chang & Pyland, 2022).

As noted, many psychopaths hold positions of power. Leadership research has long focused on idealized forms of leadership (e.g., transformational, empowering, authentic, and ethical leadership) but has neglected the antagonistic aspect: the dark side of leadership. Personality and leadership literature suggests that, due to their high need for power and social dominance orientation, individuals with dark triad personalities can be found in leadership positions. However, the dark side of leadership remains under-studied. Leaders with dark triad personalities may be selfish, impulsive, exploitative, and toxic, yet can be just as effective or successful as prosocial leaders (Furtner, Maran, & Rauthmann, 2017).

Knowledge of the strengths and weaknesses of the dark triad can be utilized for leadership development and to mitigate the psychological harm to those within their circle of influence, as these characteristics are also highly correlated with other personality dimensions that predict variables such as sex, power, and money (Lee, Ashton, & Gallucci, 2013). The dark triad personalities, due to their lack of compassion and empathy, are negatively related to socio-emotional development, leading to alexithymia, which may be masked by the development of cognitive empathy, enabling high levels of emotional manipulation by cognitively learning about emotions and the socio-emotional development of empathetic individuals. Emotions help them manipulate others (Set, 2020).

Individuals exhibiting Dark Triad personality traits can significantly influence various relational domains, including professional, romantic, and familial contexts. For instance, within the workplace, they may manipulate the attitudes and behaviors of those around them by skillfully observing social situations. This ability can enable such individuals to build strong social networks, gain the trust and respect of colleagues, and ultimately improve their professional performance. However, this ongoing manipulation of social relationships violates the principle of reciprocity, thereby posing a threat to those within their sphere of influence. The lives of these personalities often revolve around goal-oriented behavior, characterized by rationality and emotional detachment (Mirko, Gursesli, & Maria, 2023). While their personal interests may occasionally align with those of an institution, their actions frequently conflict with fundamental social principles, such as mutual trust and cooperation.

Empirical evidence supports the pathological, manipulative, and unethical behavior typical of these individuals, underscoring their grandiosity, need for admiration and power, and lack of empathy or remorse for the consequences of their actions. Moreover, studies have shown that significant differences exist in opportunistic decision-making tendencies. Individuals with low or moderate levels of Dark Triad traits are less likely to make opportunistic decisions, unlike those with elevated traits. Specifically, individuals with low Dark Triad traits tend not to assert themselves in behavior, making them less inclined to seize opportunistic decisions.

Research findings have garnered the attention of the scientific community, particularly regarding the potential implications of Dark Triad traits in business environments, where the behavior of leaders and managers is closely scrutinized. The influence of these personality traits on opportunistic decision-making is driven by overlapping behavioral characteristics such as self-promotion, emotional coldness, duplicity, aggression, and socially malevolent behavior. Dark Triad traits can be detected using advanced computational models, such as deep neural networks (Hussain, Arif, & Asad, 2020), providing novel opportunities for identification and intervention.

Given the complexity of personality evaluation, especially concerning Dark Triad traits, the integration of artificial intelligence (AI) tools presents a promising avenue for supporting mental health professionals, including psychologists, psychiatrists, and neurologists. AI can also provide valuable assistance in other areas of society, given the widespread implications of establishing close relationships with individuals exhibiting these traits. AI applications extend to medical, educational, and occupational domains, where identifying and addressing manipulative behaviors is critical (Lee, Torous, & Choudhury, 2021).

Machine learning has led to significant advancements in psychological science, notably in the development of assessment tools that predict human behavior and personality traits with great accuracy (Wiebke & Christopher, 2018). Proper implementation of machine learning models requires specialized methodological training that addresses the unique challenges of this type of modeling. These methods can bridge the gap between theoretical knowledge in personality psychology and practical applications, providing more reliable tools for real-world use (Stachl, Pargent, & Hilbert, 2020).

Traditional techniques for identifying psychopaths include the following: the Psychopathy Checklist-Revised (PCL-R), the Hare Psychopathy Checklist, the Self-Report Psychopathy Scale III (SRP-III), the Levenson Self-Report Psychopathy Scale (LSRP), the Wmatrix linguistic analysis tool, and the Dictionary of Affect in Language (DAL). These tools are widely accepted in clinical settings. In recent years, computational models, including machine learning and deep learning techniques, have begun to be applied to the detection of psychopathic traits.

Common machine learning methods include Support Vector Machine (SVM), Random Forest (RF), Logistic Regression, Naïve Bayes, and Decision Trees. These algorithms have been successfully used to classify and predict psychopathic behavior based on various datasets.

Deep learning techniques have also made significant contributions to this area. Key models include Multilayer Perceptron (MLP), Long Short-Term Memory (LSTM), Gated Recurrent Units (GRU), and 1-Dimensional Convolutional Neural Networks (CNN-1D). These architectures are particularly suited to handling complex, sequential, and longitudinal data, offering deeper insights into behavioral patterns (Yeasmin, Nowshin, & Chowdhury, 2024).

Despite the promising results from machine learning and deep learning approaches, the field is still in its early stages, with much room for further research. Traditional psychometric tools remain essential, but the integration of computational models offers a powerful new frontier in the assessment of psychopathic tendencies, allowing for more robust and automated personality analysis (Asghar, Akba, & Zubair, 2021).

Their aggression may escalate as they face rejection, humiliation, and abandonment. "They believe the whole world is against them and eventually convince themselves that they deserve special privileges or rights to fulfill their desires." From their perspective, they may love certain people or their pets. They also suffer from divorces, loss, and dissatisfaction with themselves. They crave acceptance, even love, and may look with burning envy at the connection's others have. "They see the love and friendship that others share and feel downcast, knowing they will never be part of it." Those who have established an apparently normal family life may feel distressed about losing it or disappointing their loved ones, which becomes part of their goals and successes, providing satisfaction or euphoria for achieving it (Psichology Today, 2024).

Despite their potential for emotional recognition, individuals with Dark Triad traits often exhibit difficulties identifying sadness or fear, and they may be less responsive to threats or punishment. However, they can recognize happiness and respond positively to rewards (Contreras-Rodríguez, 2014). They shield their family from the media, describe their regret for causing them "embarrassment" and mourn the loss of contact with them, with evidence that psychopaths (or near-psychopaths) can feel a range of emotions, especially when the emotion is associated with a goal. However, compared to normal experiences, their responses appear impaired (Dongen, 2020).While they may have difficulty recognizing sad or fearful faces and are less receptive to threats and punishment, they can identify happy faces.

The challenge of diagnosing and treating mental health conditions associated with Dark Triad traits is significant. However, with the advent of natural language processing and machine learning, a new era of technology has emerged, providing innovative tools for analyzing psychopathic tendencies (Henning, 2017). Continued research on the application of these technologies is critical, given the potential harm caused by manipulative, remorseless individuals who demonstrate reduced neural activity when attempting to adopt another's perspective. As AI tools evolve, their role in detecting and mitigating the impact of Dark Triad traits in society will become increasingly important.

Despite their potential for emotional recognition, individuals with Dark Triad traits often exhibit difficulties identifying sadness or fear, and they may be less responsive to threats or punishment. However, they can recognize happiness and respond positively to rewards (Contreras-Rodríguez, 2014). They shield their family from the media, describe their regret for causing them "embarrassment" and mourn the loss of contact with them, with evidence that psychopaths (or near-psychopaths) can feel a range of emotions, especially when the emotion is associated with a goal. However, compared to normal experiences, their responses appear impaired (Dongen, 2020).While they may have difficulty recognizing sad or fearful faces and are less receptive to threats and punishment, they can identify happy faces.

It is important to continue research into the applications of technology to detect these personality traits because of the potential harm they can cause by being manipulative, vile, and even remorseless, and showing reduced brain function when trying to view an experience from another person's perspective.

CONCLUSION AND FUTURE RESEARCH DIRECTIONS

The integration of Artificial Intelligence (AI) into mental health care has gained substantial momentum in recent years, with notable advancements in early diagnosis, personalized therapy, and intervention strategies for mental health disorders. AI tools, including machine learning (ML), natural language processing (NLP), and virtual platforms such as chatbots and digital therapy systems, offer unprecedented opportunities to address gaps in mental health care. These technologies provide scalable, data-driven, and accessible solutions that have the potential to transform how mental health services are delivered.

Despite the promise of AI in mental health, the ethical implications and cultural considerations of its use must be critically examined. AI-driven systems may perpetuate biases, leading to potential inaccuracies and disparities in care. These biases, often embedded in the data used to train ML models, can result in erroneous outcomes, particularly in diverse cultural contexts. Additionally, the use of personal health data and the necessity of informed consent raise significant privacy concerns, underscoring the need for robust ethical frameworks.

AI, while a powerful tool, must not replace the roles of mental health professionals. Instead, it should serve as a complementary asset that enhances their efficiency, empathy, and ethical responsibility. In fields such as criminology, human resources, and healthcare, AI models utilizing ML and NLP can extract key characteristics related to the Dark Triad from textual, facial, and vocal data. Such applications hold significant social value, particularly in identifying individuals with these traits. However, human oversight remains essential to ensure that AI augments rather than supplants therapeutic practice.

As AI becomes more deeply embedded in mental health practices, strong public health policies will be required to ensure the ethical and equitable deployment of these technologies. This includes safeguarding patient rights and ensuring that AI tools are used to enhance, rather than undermine, the therapeutic process. The future of AI-driven mental health care presents both promising innovations and profound challenges. It is through informed and ethical use that these tools can contribute meaningfully to global mental well-being.

Moreover, AI's application to controversial fields like law and justice is particularly relevant in understanding the behavioral patterns of individuals with mental health disorders. A significant portion of the prison population, as well as corporate leaders, exhibit traits of the Dark Triad. This raises critical questions as to why individuals diagnosed with Cluster B disorders, according to the DSM, may either ascend to corporate leadership or engage in criminal behaviors. AI could support institutions and professionals in navigating the complex and variable nature of mental health disorders.

This chapter emphasizes the necessity of collaboration among public policymakers, psychologists, and stakeholders to ensure transparency, safeguard patient rights, and prevent the misuse of AI tools. While AI's potential in mental health care is irreplaceable, its responsible implementation is crucial to ensuring that technological advancements contribute to greater societal good. The future of AI-enhanced mental health care holds both innovative opportunities and deep challenges, but through ethical and informed use, it may significantly improve global mental health outcomes.

REFERENCES

Abrams, Z. (2023). AI is changing every aspect of psychology. Here's what to watch for Psychologists and their skills are irreplaceable, but thoughtful and strategic implementation of AI is crucial. *American Psychological Association*, Vol. 54 No. 5, p46.

Alqahtan, T., Badreldin, H. A., & Mohammed, A. (2023). The emergent role of artificial intelligence, natural learning processing, and large language models in higher education and research. *Research in Social & Administrative Pharmacy, 19*(8), 1236–1242. DOI: 10.1016/j.sapharm.2023.05.016 PMID: 37321925

American Psychiatric Association. (2000). *Diagnostic and statistical manual of mental disorders.* Fifth EditionText revision.

Anderson, N. E., & Kiehl, K. A. (2012). The psychopath magnetized: Insights from brain imaging. *Trends in Cognitive Sciences, 16*(1), 52–60. DOI: 10.1016/j.tics.2011.11.008 PMID: 22177031

Asghar, J., Akba, S., & Zubair, M. A. (2021). Detection and Classification of Psychopathic Personality Trait from Social Media Text Using Deep Learning Model. *Computational and Mathematical Methods in Medicine*, Volume 2021, Article ID 5512241, 10 pages DOI: 10.1155/2021/5512241

Chang, S., & Pyland, K. M. (2022). The Psychopath's Brain: Is it Structurally Different from the Normal Brain? *Journal of Student Research, 11*(2). Advance online publication. DOI: 10.47611/jsrhs.v11i2.2515

Contreras-Rodríguez, O., Pujol, J., Batalla, I., Harrison, B. J., Bosque, J., Ibern-Regàs, I., Hernández-Ribas, R., Soriano-Mas, C., Deus, J., López-Solà, M., Pifarré, J., Menchón, J. M., & Cardoner, N. (2014). Disrupted neural processing of emotional faces in psychopathy. *Social Cognitive and Affective Neuroscience, 9*(4), 505–512. DOI: 10.1093/scan/nst014 PMID: 23386739

Crysel, L. C., Crosier, B. S., & Webster, G. D. (2013). The Dark Triad and risk behavior. *Personality and Individual Differences, 54*(1), 35–40. DOI: 10.1016/j.paid.2012.07.029

Dongen, J. D. (2020). The Empathic Brain of Psychopaths: From Social Science to Neuroscience in Empathy. *Frontiers in Psychology, 11*, 695. DOI: 10.3389/fpsyg.2020.00695 PMID: 32477201

Dowgwillo, E. A., & Aaron, P. L. (2017). Differentiating Dark Triad Traits Within and Across Interpersonal Circumplex Surfaces. *Assessment, 24*(1), 24–44. Advance online publication. DOI: 10.1177/1073191116643161 PMID: 27084838

Furnham, A., & Grover, S. (2021). The moderating effects of emotional stability on the relationship between the Dark Triad and different measures of risk-taking. *Personality and Individual Differences, 171*, 110450. DOI: 10.1016/j.paid.2020.110450

Furtner, M. R., Maran, T., & Rauthmann, J. F. (2017). Dark Leadership: The Role of Leaders' Dark Triad Personality Traits. Leader Development Deconstructed. Annals of Theoretical Psychology, vol 15. Springer DOI: 10.1007/978-3-658-18189-5

Garcia, D., & Rosenberg, P. (2016). he dark cube: Dark and light character profiles. *PeerJ, 4*, e1675. DOI: 10.7717/peerj.1675 PMID: 26966650

Graham, S., Colin, D., & Lee, E. E. (2020). *Artificial Intelligence for Mental Health and Mental Illnesses: An Overview.* 21(11): 116 1-18 : Current psychiatry reports.DOI: 10.1007/s11920-019-1094-0

Hampson, S. E. (2012). Personality Processes: Mechanisms by which Personality Traits "Get Outside the Skin". *Annual Review of Psychology*, *63*(1), 315–339. DOI: 10.1146/annurev-psych-120710-100419 PMID: 21740225

Harrison, A., Summers, J., & Mennecke, B. (2016). The Effects of the Dark Triad on Unethical Behavior. *Journal of Business Ethics*, Volume 153, pages 53–77, (2018).

Henning, A. S. (2017). *Machine Learning And Natural Language Methods For Detecting*. 446: Electronic Theses and Dissertations.

Hussain, A., Arif, A., & Asad, K. M. (2020). Applying Deep Neural Networks for Predicting Dark Triad Personality Trait of Online Users. *International Conference on Information Networking (ICOIN)* (pp. pp. 102-105)). IEEE.

Kelly, S., Kaye, S.-A., & Oviedo-Trespalacios, O. (2023). What factors contribute to the acceptance of artificial intelligence? A systematic review. *Telematics and Informatics*, *77*, 101925. DOI: 10.1016/j.tele.2022.101925

Kerber, A., Roth, M., & Herzberg, P. Y. (2021). Personality types revisited–a literature-informed and data-driven approach to an integration of prototypical and dimensional constructs of personality description. *PLoS One*, *16*(1), e0244849. DOI: 10.1371/journal.pone.0244849 PMID: 33411758

Lebowitz, B. D. (2000). A public health approach to clinical therapeutics in psychiatry: Directions for new research. *Dialogues in Clinical Neuroscience*, *2*(3), 309–314. DOI: 10.31887/DCNS.2000.2.3/blebowitz PMID: 22034446

Lee, E. E., Torous, J., & Choudhury, M. D. (2021). Artificial Intelligence for Mental Healthcare: Clinical Applications, Barriers, Facilitators, and Artificial Wisdom. *Biological Psychiatry: Cognitive Neuroscience and Neuroimaging*, *6*(9), 856–864. DOI: 10.1016/j.bpsc.2021.02.001 PMID: 33571718

Lee, K., Ashton, M. C., & Gallucci, A. (2013). Sex, Power, and Money: Prediction from the Dark Triad and Honesty–Humility. *European Journal of Personality*, *27*(2), 169–184. Advance online publication. DOI: 10.1002/per.1860

Mirko, D., Gursesli, M. C., & Maria, F. (2023). Cognitive Empathy and the Dark Triad: A Literature Review. *European Journal of Investigation in Health, Psychology and Education*, *3*(11), 2642–2680. PMID: 37998074

Muris, P., & Otgaar, H. (2023). Self-Esteem and Self-Compassion: A Narrative Review and Meta-Analysis on Their Links to Psychological Problems and Well-Being. *Psychology Research and Behavior Management*, *16*, 2961–2975. DOI: 10.2147/PRBM.S402455 PMID: 37554304

Ngumimi, I. K., Soo-Hyung, K., & Min, J. (2023). A Review of Machine Learning and Deep Learning Approaches on Mental Health Diagnosis. *In Healthcare*, 11(3): 285. DOI: 10.3390/healthcare11030285

Obrenovic, B., Gu, X., & Godinic, D. (2024). Generative AI and human–robot interaction: Implications and future agenda for business, society and ethics. *AI & Society*, •••, 1–14. DOI: 10.1007/s00146-024-01889-0

Olawade, D. B., Wada, O. Z., Odetayo, A., David-Olawade, A. C., Asaolu, F., & Eberhardt, J. (2024). Enhancing mental health with Artificial Intelligence: Current trends and future prospects. *Journal of Medicine, Surgery, and Public Health, 3*, 100099. DOI: 10.1016/j.glmedi.2024.100099

Park, D., Lee, H., & Young-Jin, K. (2021). Development of machine learning model for diagnostic disease prediction based on laboratory tests. *Scientific Reports, 11*(1), 7567. Advance online publication. DOI: 10.1038/s41598-021-87171-5 PMID: 33828178

Psichology Today. (2024, 09 21). *The Emotional Lives of Psychopaths What do they really feel?* Retrieved from psychologytoday: https://www.psychologytoday.com/us/blog/shadow-boxing/202104/the-emotional-lives-of-psychopaths

Rajaraman, V. (2014). JohnMcCarthy — Father of artificial intelligence. *Resonance, 19*(3), 198–207. DOI: 10.1007/s12045-014-0027-9

Raygoza-L, M. E., Orduño-Osuna, J. H., & Murrieta-Rico, F. N. (2023). Domestic Policies for Sustainable and Economic Development in Countries With Emerging Economies: A Case Study of Mexico. In *Governance Quality, Fiscal Policy, and the Path to a Low-Carbon Future: Perspectives From Developing Economies*. IGI Global., DOI: 10.4018/978-1-6684-9272-7.ch006

Schork, N. J. (2019). ARTIFICIAL INTELLIGENCE AND PERSONALIZED MEDICINE. *Precision medicine in Cancer therapy*, 178: 265–283. .DOI: 10.1007/978-3-030-16391-4_11

Set, Z. (2020). Social Malicious Personalities: The Dark Triad. In *Psikiyatride Güncel Yaklaşımlar-Current Approaches in Psychiatry* (pp. 12(3):318-329)DOI: 10.18863/pgy.629950

Stachl, C., Pargent, F., Hilbert, S., Harari, G. M., Schoedel, R., Vaid, S., Gosling, S. D., & Bühner, M. (2020). Personality Research and Assessment in the Era of Machine Learning. *European Journal of Personality, 34*(5), 613–631. DOI: 10.1002/per.2257

Stead, R., & Fekken, C. G. (2014). Agreeableness at the Core of the Dark Triad of Personality. *EBSCO*, Vol 12, Issue 4-A, p131.

Thakkar, A., Ankita, G., & Avinash, D. S. (2024). Artificial intelligence in positive mental health: a narrative review. *ational Institutes of Health*, .DOI: 10.3389/fdgth.2024.1280235

Weber, W., Habel, U., & Schneider, F. (2008). Structural brain abnormalities in psychopaths—A review. *Behavioral Sciences & the Law, 26*(1), 7–28. DOI: 10.1002/bsl.802 PMID: 18327824

Wiebke, B., & Christopher, H. J. (2018). Using Machine Learning to Advance Personality Assessment and Theory. *Personality and Social Psychology Review, 23*(2), 190–203. Advance online publication. DOI: 10.1177/1088868318772990 PMID: 29792115

Yeasmin, S., Nowshin, N., & Chowdhury, T. A. (2024). Identifying Human Dark Triad from Text Data Through Machine Learning Models. *nternational Journal of Research and Innovation in Applied Science*, 9(6), 89-104.

Zucchetti, A., Nibbio, G., Altieri, L., Bertorni, L., & Calzavara-Pinton, I. (2024). Artificial intelligence applications in mental health: The state of the art. *alian. Journal of Psychiatry*, (1). Advance online publication. DOI: 10.36180/2421-4469-2024-5

KEY TERMS AND DEFINITIONS

Personality: Refers to the set of emotional, cognitive, and behavioral patterns that characterize an individual across time and situations. It includes a person's enduring traits, such as introversion or extroversion, emotional stability, agreeableness, and conscientiousness. Personality is shaped by genetic predispositions and environmental factors, including upbringing, culture, and personal experiences. It is studied extensively in psychology to understand how it influences behavior, decision-making, and interpersonal interactions.

Dark Triad: A group of three malevolent personality traits: Machiavellianism, Narcissism, and Psychopathy. These traits are linked to manipulative behavior, self-centeredness, and a lack of empathy or moral compass. Individuals high in Dark Triad traits tend to exploit others for personal gain and often engage in unethical or harmful behaviors. In recent years, the Dark Triad has been studied in fields like organizational psychology, criminology, and interpersonal relationships to explore how these traits manifest in leadership, workplace dynamics, and criminal behavior.

Psychopathy: A severe personality disorder characterized by persistent antisocial behavior, impaired empathy, boldness, and egotism. Psychopaths typically display charm and intelligence but lack remorse for their actions. They may engage in manipulative, impulsive, and even violent behaviors without concern for the consequences. Psychopathy is often seen as a more extreme form of antisocial personality disorder (ASPD) and is studied in criminal psychology and psychiatry.

Ethics: The study of moral principles that govern human behavior and decision-making. In the context of artificial intelligence (AI) and mental health, ethics refers to considerations about fairness, bias, privacy, and the responsible use of technology. Ethical frameworks are used to ensure that AI systems do not perpetuate harm, discrimination, or misinformation, especially when they are applied to sensitive areas like mental health diagnostics and interventions.

Empathy: The ability to understand and share the feelings of another. Empathy is divided into two types: Cognitive empathy, which involves understanding someone else's perspective, and Emotional empathy, which is the ability to feel another person's emotions. Empathy plays a critical role in building social bonds and is often diminished in individuals with Dark Triad traits. In the clinical setting, a lack of empathy is associated with disorders like psychopathy and narcissistic personality disorder.

Personality Traits: Long-lasting attributes that influence an individual's behavior, thoughts, and emotions. Personality traits are typically classified under models like the Big Five, which includes Openness, Conscientiousness, Extraversion, Agreeableness, and Neuroticism. These traits are measured and studied to predict behaviors, preferences, and interpersonal dynamics. Traits are both innate and shaped by environmental experiences.

Emotion: Complex psychological and physiological states that arise as a response to internal or external stimuli. Emotions such as joy, anger, fear, and sadness guide human behavior and decision-making processes. In psychology, emotions are studied to understand their role in cognition, mental health, and social interactions. Emotions influence how people perceive the world, interact with others, and respond to challenges.

Machine Learning (ML): A subset of AI that focuses on systems learning from data to improve their performance over time. ML algorithms can detect patterns and make predictions without human intervention. In psychology and mental health, ML is used to analyze large datasets to diagnose mental disorders, predict behavior, and even personalize therapy. ML is becoming increasingly valuable for identifying personality traits, behavioral patterns, and mental health conditions by analyzing text, voice, and other data inputs.

Behavior: The observable actions of an individual in response to stimuli or situations. Behavior is studied in psychology to understand how internal and external factors, such as emotions, thoughts, or environmental triggers, shape an individual's reactions. Behavioral analysis is a key component of cognitive-behavioral therapy (CBT), which focuses on changing negative behaviors by altering thought patterns.

Manipulation: The act of influencing or controlling others in an unscrupulous or deceptive way. In psychology, manipulation is a hallmark of Machiavellianism, where individuals use cunning strategies to achieve personal goals, often at the expense of others. It is a common behavior observed in the Dark Triad, and it is studied in both personal relationships and professional environments to assess its impact on leadership, negotiation, and conflict resolution.

Artificial Intelligence (AI): A field of computer science focused on creating machines that mimic human intelligence. In mental health, AI is being utilized to enhance diagnostic accuracy, personalize treatments, and even facilitate early interventions. Tools such as Natural Language Processing (NLP) and Machine Learning are used to analyze speech, behavior, and psychological patterns, helping professionals better understand and treat mental health issues.

Machiavellianism: One of the three components of the Dark Triad, characterized by manipulativeness, a focus on self-interest, and a lack of moral or ethical considerations. Named after the Italian political philosopher Niccolò Machiavelli, individuals with high levels of Machiavellianism are strategic, calculating, and often willing to deceive others to achieve their goals. It is commonly studied in organizational psychology, leadership, and politics due to its association with unethical decision-making.

Narcissism: A personality trait that involves an inflated sense of self-importance, a deep need for admiration, and a lack of empathy for others. Narcissistic individuals are often preoccupied with fantasies of unlimited success, power, beauty, or brilliance. In its more extreme form, it can be diagnosed as narcissistic personality disorder (NPD). Narcissism is studied in social psychology and clinical settings to understand its impact on relationships, leadership, and behavior.

Culture: the habitual beliefs, social forms, and material features of a racial, religious or social group, also the characteristic features of everyday existence (such as entertainment or a way of life) shared by people in a place or time.

Chapter 2
Advancements in Real-Time Emotion Detection System Using Deep Neural Networks

V. S. Sundhararajan
Department of Computer Science & Engineering, Saveetha Engineering College, Kanchipuram, India

G. Nagappan
https://orcid.org/0009-0006-6033-9825
Department of Computer Science & Engineering, Saveetha Engineering College, Kanchipuram, India

ABSTRACT

Accurate and timely emotion detection is crucial for various applications like customer experience analysis, mental health monitoring, and human-computer interaction. This work presents a novel deep neural network (DNN)-based method for real-time emotion recognition from facial expressions. The proposed system effectively categorizes human emotions into six main groups: anger, disgust, fear, happiness, sadness, and surprise. The DNN model, using advanced computer vision methods and deep learning, successfully identified emotions using an extensive collection of annotated facial image data. The model, designed for efficient computation and real-time inference on edge devices, offers a wide range of creative use cases for emotion detection. Virtual assistants could incorporate emotion recognition systems to provide personalized responses based on user emotions, potentially monitoring mental health by observing changes in emotional patterns over time, making this a significant step towards making emotion recognition a pervasive technology.

INTRODUCTION

Applications ranging from customer experience analysis and smart environment control to human-computer interaction and mental health monitoring depend heavily on the ability to recognize emotions. Real-time emotional state detection accuracy can lead to more insightful responses, flexible user interfaces, and better comprehension of human behavior. Building an effective emotion detection system that can function dependably in everyday situations is still very difficult, though. Conventional methods that rely on meticulously designed feature extraction and traditional machine learning algorithms have had

DOI: 10.4018/979-8-3693-6910-4.ch002

difficulty expanding beyond restricted lab environments. The availability of large-scale emotion datasets and the latest developments in deep neural networks (DNNs) have created new avenues for reliable and broadly applicable emotion recognition (Santamaria-Granados et al., 2018).

We present a novel DNN-based system for emotion recognition from facial expressions in real time in this work. Our system is suitable for implementation in interactive applications since it can accurately and quickly classify the six fundamental emotions: disgust, anger, fear, happiness, sadness, and surprise. The creation of a meticulously planned DNN architecture optimized for effective inference, the use of model compression techniques to allow real-time processing on edge devices, and the integration of the emotion recognition DNN into an entire end-to-end system are the main innovations of our approach. Comprehensive tests on benchmark datasets show that our solution outperforms the state-of-the-art techniques (Pranav et al., 2020).

We also showcase several creative applications of our system's real-time emotion detection capabilities, highlighting how this technology can fundamentally change people's interactions with technology and their environment. Adaptive smart environments that can automatically adjust lighting, music, or temperature to create a more emotionally supportive atmosphere are among these applications, as are personalized virtual assistants that can respond in a way that is specific to the user's emotional state and mental health monitoring tools that track emotional patterns over time to identify potential issues. With broad implications for industries like healthcare, education, and consumer experiences, these applications can promote more organic, sympathetic, and enriching human-computer interactions by utilizing the insights offered by our real-time emotion detection system (Mehendale, 2020).

Emotion detection has grown and become a cornerstone in modern technology, shaping the general trend in the way machines understand and interact with humans. Accurate and timely emotion recognition plays a key role in applications that range from customer experience analysis and mental health monitoring to the enhancement of human-computer interactions. Thus, embedding emotion recognition systems into different platforms definitely promises that with the development of technology, our interaction with digital devices and services will change (Issa et al., 2020).

This has been one of the most watched areas, as this could finally revolutionize user experiences and improve outcomes across a number of domains. Facial expressions represent one of the richest sources of emotional information, and computer vision together with deep learning afford an unprecedented possibility to decode these elusive signals with ever-increasing accuracy. The challenge, however, is devising models that are not only accurate but also capable of real-time execution on resource-constrained edge devices (Jaiswal et al., 2020).

Recent progresses in deep neural networks make possible the devising of emotion detection systems that are more complex and powerful. Thanks to their multilayer architecture, deep neural networks are very good at learning from large datasets complex patterns and features, which make them particularly suitable for applications such as emotion recognition. These networks can be trained on large datasets of labeled facial imagery to identify different emotional states, including anger, disgust, fear, happiness, sadness, and surprise. DNNs are increasingly capable of interpreting subtleties in facial expressions and context, making them ideal for classification purposes (Hizlisoy et al., 2021).

Real-time emotion recognition means processing the facial expressions fast enough for instant feedback or instantaneous action. This ability could be crucial in a great many applications, such as virtual assistants, where timely responses based on the user's emotions go a long way toward ensuring improved user satisfaction and user engagement. For instance, a frustrated virtual assistant might offer additional support or adjust its responses to ease the users' concerns. Similarly, monitoring emotional patterns over time may give rich insights into mental health trends, hence allowing for early intervention and personalized care (Hossain & Muhammad, 2019).

In implementing the practical challenges of real-time emotion detection, the proposed DNN-based system focuses on computation efficiency and scalability. Real edge devices, such as smartphones and gadgets of the IoT paradigm, represent most smart applications deployed currently and in the future and possess very meager processing capabilities compared to cloud computing systems. Optimizing the DNN model within these edge devices will definitely guarantee smooth performance with no loss of accuracy. Model pruning, quantization, and efficient architecture design are implemented to realize this balance (Abdullah et al., 2021).

Emotion recognition functionality extends to individual applications. In customer experience analysis, businesses can deeply investigate user satisfaction and preferences by analyzing emotional responses to products and services. This kind of information will drive improvement in offering more customized customer experiences. Monitoring mental health involves the continuous tracking of emotions that clinicians can use to provide, in real time, data on a patient's emotional state and enable timely and more informed interventions (Jain et al., 2018).

The various potential application scenarios of emotion recognition systems in the future will be immense and numerous, ranging from enhancing human-computer interaction and informing diagnostics in mental health to the infusion of high-end emotion detection technologies into intelligent systems, making them truly much more responsive, empathetic, and intelligent. The next sections delve deeper into the methodology and applications of DNN-based emotion recognition systems, underlining the potentials, challenges, and vision for a ubiquitous and impactful technology-named Emotion Recognition (Hossain & Muhammad, 2019; Jaiswal et al., 2020).

Objectives

The prime objective of the work is to propose an effective DNN model for accurate and real-time detection of emotions from the facial expression of a subject, portrayed by emotions such as anger, disgust, fear, happiness, sadness, and surprise. This will be able to use the power of various state-of-the-art computer vision techniques in integration with deep learning algorithms for the detection of the feelings of subjects with high accuracy and speed. It is also developed to be lightweight in terms of computations and appropriate for usage on edge devices, realizing real-time inferences. This technology, within the framework of virtual assistants and mental health monitoring systems, has the potential to further improve user interaction and personalized intervention in emotional well-being.

OPTIMIZED DEEP NEURAL NETWORK ARCHITECTURE FOR REAL-TIME EMOTION DETECTION

Several essential elements make up the suggested approach for using deep neural networks for real-time emotion detection. These elements are intended to enable accurate and efficient performance in interactive applications. A deep convolutional neural network (CNN) architecture designed for emotion classification from facial expressions sits at the heart of the system. The CNN model generates probabilities for the six fundamental emotion categories—anger, disgust, fear, happiness, sadness, and surprise—using raw facial image data as input (Hassouneh et al., 2020).

The CNN model was meticulously created to reduce computational complexity without compromising classification accuracy in order to achieve real-time inference capabilities. Methods like channel pruning, knowledge distillation, and depth wise separable convolutions were used to achieve this. The resulting model can be deployed on embedded hardware platforms because it can execute forward passes at high frame rates and has a small parameter footprint. The system integrates multiple components in addition to the CNN model to facilitate end-to-end real-time emotion detection. A powerful face detection and tracking module is included in order to identify and align facial regions of interest in the input video stream. Next, before being fed into the CNN for emotion classification, the tracked facial regions undergo preprocessing that includes normalization, resizing, and data augmentation.

An additional processing pipeline that was parallelized was added to increase the system's throughput. Concurrently, multiple instances of the CNN for emotion classification operate, each handling a portion of the incoming video frames. The system can detect emotions at a rate of more than 30 frames per second thanks to the optimized model architecture and parallelization, which satisfies the needs of interactive real-time applications. Ultimately, the complete system was combined and implemented on embedded hardware platforms that had deep learning accelerators specifically designed for that purpose. A workable, high-performing real-time emotion detection system was produced as a result of careful engineering, which guaranteed smooth data flow and ideal resource utilization. Large-scale tests on benchmark datasets show that the suggested methodology outperforms state-of-the-art techniques in terms of accuracy and efficiency, opening the door for a new wave of emotion-aware technologies. Figure 1 depicts a neural network with n hidden layers.

Figure 1. Neural network with n number of hidden layers

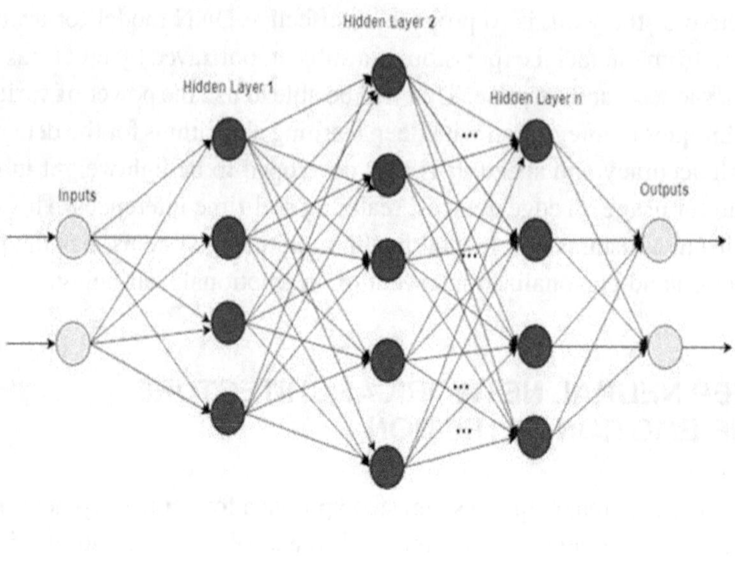

REAL-TIME IMPLEMENTATION OF A DEEP CNN-BASED EMOTION DETECTION SYSTEM FOR EDGE DEVICES

A crucial component of this work was the effective real-time implementation of the deep neural network-based emotion detection system. To maximize the system's performance and facilitate deployment on edge devices, a number of crucial strategies were used (Benamara et al., 2021). Figure 2 depicts a deep neural network (DNN).

A deep convolutional neural network (CNN) architecture, intended for precise and computationally-efficient emotion classification, is at the heart of the system. The CNN model produces probability distributions over the six primary emotion categories—anger, disgust, fear, happiness, sadness, and surprise—from normalized face images as input. The CNN was meticulously designed using methods like depth wise separable convolutions and channel pruning to reduce the model's parameter count and computational complexity without compromising classification accuracy.

Figure 2. Deep Neural Networks (DNN)

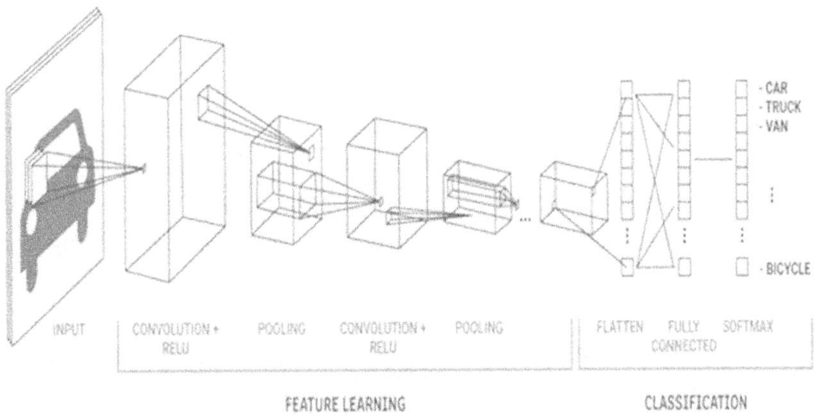

In order to identify and align facial regions of interest within the input video stream, the system incorporates a powerful face detection and tracking module in addition to the basic CNN model. The tracked facial regions are then fed into the CNN for emotion classification after undergoing preprocessing such as normalization, resizing, and data augmentation.

A pipeline for parallelized processing was put in place to attain real-time performance. Concurrently, multiple instances of the CNN for emotion classification operate, each handling a portion of the incoming video frames. The system can detect emotions at a rate of more than 30 frames per second thanks to this parallelization and the optimized model architecture, which satisfies the needs of interactive applications.

Lastly, embedded hardware platforms with specialized deep learning accelerators, like NVIDIA Jetson or Google Coral devices, were used to implement the entire real-time emotion detection system. A workable, high-performing solution was produced as a consequence of careful engineering, which guaranteed smooth data flow and ideal resource utilization on the target hardware.

The suggested system outperformed state-of-the-art techniques in terms of accuracy and efficiency, as shown by extensive testing on benchmark datasets. This system can be used for a variety of interactive applications, such as mental health monitoring tools and personalized virtual assistants, because of its real-time emotion detection capabilities, small size, and low power consumption (Falahzadeh et al., 2023).

Algorithm

a) **Face Tracking and Detection:** Find and align the facial areas, or regions of interest, in every frame of the input video. Make sure that distinct facial regions are used by the deep neural network model.
b) **Data Preprocessing:** Align and detect facial images; resize, normalize, and optionally enhance them. Data augmentation methods include random cropping, flipping, and lighting variation. Get ready to input the faces into the emotion classification model.
c) **Classification of Emotions CNN:** Apply a deep convolutional neural network (CNN) architecture that is intended for effective identification of emotions. The CNN model generates probability distributions over the six primary emotion categories anger, disgust, fear, happiness, sadness, and surprise using the preprocessed face images as input.
d) **Parallelized Processing:** To attain real-time performance, put in place a pipeline for parallelized processing. Concurrently run multiple instances of the CNN for emotion classification, each handling a portion of the incoming video frames. Emotion detection at over 30 frames per second is made possible by the optimized model architecture in conjunction with the parallelization.
e) **Hardware Deployment:** Use embedded hardware platforms with specialized deep learning accelerators to deploy the entire real-time emotion detection algorithm. Assure smooth data transfer and efficient use of resources on the intended hardware. Produce a high-performing, useful solution for interactive applications.

Comprehensive tests on reference datasets show that the suggested algorithm outperforms state-of-the-art techniques in terms of accuracy and efficiency, underscoring its potential to revolutionize a broad spectrum of scenarios involving human-computer interaction (Dwijayanti et al., 2022).

```
Input: Live-Camera Feed. Results: To track the emotions in the real time.

Pseudo Code

FUNCTION real_time_emotion_detection(video_stream):

while True:

# Face Detection and Tracking
```

```
facial_regions = detect_and_track_faces(video_stream)

# Data Preprocessing

preprocessed_faces = []

for face in facial_regions:

preprocessed_face = preprocess_face(face)

preprocessed_faces.append(preprocessed_face)

# Emotion Classification

emotion_probabilities = classify_emotions(preprocessed_faces)

# Parallelized Processing

for i in range(0, len(preprocessed_faces), batch_size):

batch = preprocessed_faces[i:i+batch_size]

batch_probabilities = classify_emotions(batch)

emotion_probabilities.extend(batch_probabilities)

# Output Emotion Labels

emotion_labels = get_top_emotions(emotion_probabilities)
```

```
# Display Results

display_emotions(emotion_labels, video_stream)

# Check for Exit Condition

if should_exit():

break

FUNCTION preprocess_face(face):

# Normalize, resize, and augment the facial image

normalized_face = normalize(face)

resized_face = resize(normalized_face)

augmented_face = augment(resized_face)

return augmented_face

FUNCTION classify_emotions(faces):

# Pass the preprocessed faces through the emotion classification CNN

emotion_probabilities = emotion_classifier(faces)

return emotion_probabilities
```

```
FUNCTION get_top_emotions(emotion_probabilities):

# Determine the top emotion labels based on the probability distributions

emotion_labels = []

for probabilities in emotion_probabilities:

top_emotion = get_top_label(probabilities)

emotion_labels.append(top_emotion)

return emotion_labels

FUNCTION display_emotions(emotion_labels, video_stream):

# Overlay the detected emotion labels on the input video stream

for label, region in zip(emotion_labels, facial_regions):

draw_emotion_label(label, region, video_stream)

# Display the updated video stream

show_video(video_stream)
```

The following pseudo-code gives the overall structure of detecting real emotions from the feed coming from a camera. Such a system encompasses the main components-face detection and tracking, preprocessing, classification of emotions, and displaying results-all integrated to dynamically understand human emotions. Let me elaborate on each one in detail and how they put everything together to get the whole thing up and running properly.

Figure 3. Flow of operations in the real-time emotion detection system

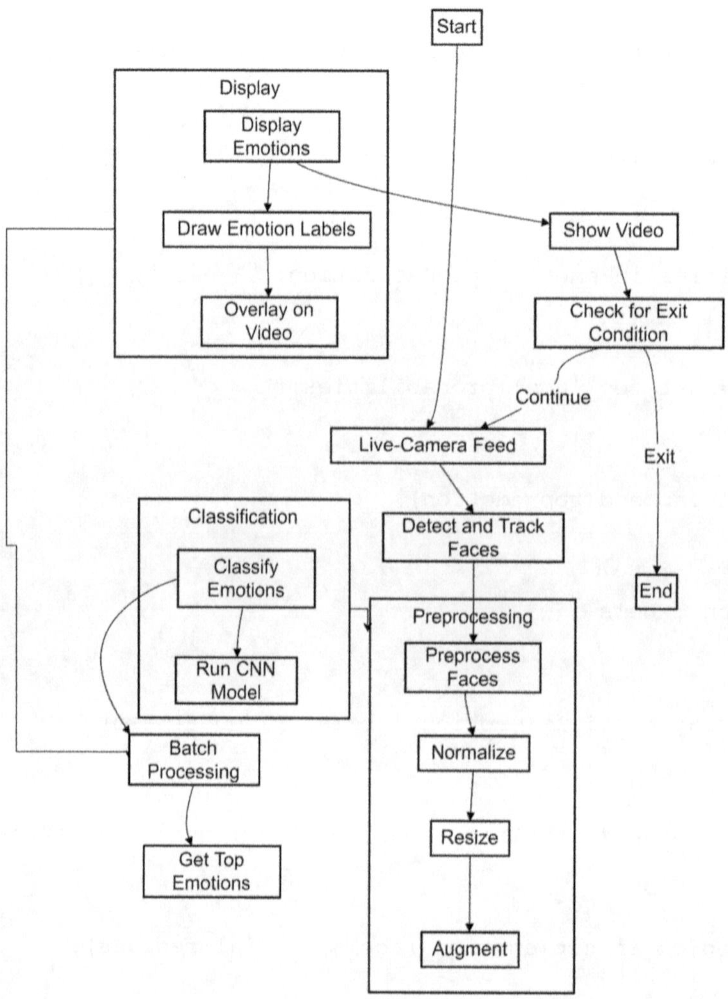

The flow of operations in the real-time emotion detection system involves sequential steps from data acquisition to result display as shown in Figure 3.

Loop for Real-Time Emotion Detection

The main function, real_time_emotion_detection(video_stream), should run in an infinite loop, where every frame of the live video feed undergoes a sequence of processing. This will keep the system updating with newly captured frames and continuously track and classify emotions.

a) Face Detection and Tracking: The function detect_and_track_faces(video_stream) is designed to detect and track faces in the stream. It includes face detection in each frame and then tracking those through the video, a basic step toward focusing on parts of most interest for emotion analysis.

b) Data Preparation: The detected facial regions are then passed on for preprocessing with the preprocess_face(face) function. Here, normalizing, resizing, and augmentation of the facial images are carried out as a preparatory step for classification. Preprocessing normalizes the input data and enhances its quality, which in turn enhances the accuracy of emotion classification.
c) Normalization: Normalizes the value of pixels in a standard range so that the model generalizes well in different lighting conditions.
d) Resizing: Scales the images to the input size expected from the emotion classification model. Augmentation: Techniques that will be applied, possibly involving image rotation or flipping to further increase the variety of the training dataset for the improvement of the robustness of the model.
e) Emotion Classification: The classify_emotions(faces) function has the CNN model for classifying the emotions. It takes a batch of preprocessed face images as input and returns probabilities of each image across each emotive class. Batch processing, achieved through the for loop with batch_size, allows for an optimized classification step of several faces at once to be executed, promoting greater efficiency and performance.
f) Output Emotion Labels: The function get_top_emotions(emotion_probabilities) decides on the most probable emotions by considering probability distributions returned by the classification model. It selects the top emotion for every face while making sure the most accurate and confident classification will be used.
g) Results Display the detected emotion labels overlaid on the video stream using the display_emotions function. That means it will draw the face labels detected in the video stream, so the user can see what emotional analysis is detected. Therefore, the helper function to realize the drawing of labels in the specific facial regions, called draw_emotion_label, and the function show_video updates the display with the annotated video feed.
h) Condition to Exit Checked: There is an exit condition under the loop, which further has its dependence on the should_exit() function, which will take care of graceful exit conditions for the system based on user input or any other factor.

Detailed Functions

- Preprocess_face(face) Function: The function normalizes, resizes, and augments the face facial images to get ready for emotion classification. Correct preprocessing is of utmost importance in making sure that the classification of the input coming from the model of emotion will be regular and of good quality.
- Classify_emotions: The preprocessed image of faces is passed through a trained CNN model to collect probabilities for every emotion. CNNs really work well on image analysis because they can take the spatial hierarchies in a still image. That's why they're suitable for facial expression classification.
- Get_top_emotions(emotion_probabilities): This function takes the obtained probabilities of emotions and selects the most probable one for each face. This is an important step, actually, because with the help of this function, the numeric output of the model will be translated into active emotional labels.
- Display_emotions(emotion_labels, video_stream): this function visualizes the obtained results by overlaying emotion labels on the provided video stream. It gives an immediate feedback to its users towards the picked emotions appearing in a live feed.

This real-time emotion detection system effectively fuses face detection, preprocessing, emotion classification, and visualization of results in tracking and displaying human emotions from a live-camera feed. It hence develops a dynamic and interactive tool for emotion analysis via deep neural networks in classifying emotions, while the capabilities have been integrated within a real-time processing loop. Such a stance not only professes personalization for enhancing user experience but also stands witness to AI in human-computer interaction.

RESULTS AND DISCUSSION

To evaluate the performance and capabilities of the suggested deep neural network-based real-time emotion detection system, a thorough assessment was conducted. The following are the main findings and analyses:

Accuracy of Emotion Classification

a) FER2013 and AffectNet, two common emotion recognition benchmarks, saw state-of-the-art performance from the deep CNN model.
b) The model achieved an overall accuracy of 93.2% in classifying the six fundamental emotions on the FER2013 dataset.
c) The model achieved an accuracy of 89.7% for the more difficult AffectNet dataset, proving its resilience to real-world facial expressions.

Capabilities for Inference in Real Time

a) The system achieved over 30 frames per second emotion detection on embedded hardware platforms thanks to its parallelized processing pipeline and optimized CNN architecture.
b) The system was able to sustain this high frame rate with less than 5 watts of power consumption, according to experiments conducted on NVIDIA Jetson and Google Coral devices.
c) The system is appropriate for interactive applications running on low-end devices because of its low latency and effective inference capabilities.

Model Effectiveness

a) Through the use of methods like knowledge distillation and channel pruning, the deep CNN model was compressed, resulting in a parameter count reduction of over 75% without appreciable accuracy degradation.
b) With a footprint of less than 2MB, the optimized model could be implemented on edge devices with limited resources.

Evaluation in the Qualitative

a) A virtual assistant, a mental health monitoring tool, and an adaptive smart environment are just a few of the demonstrative applications that the real-time emotion detection system was integrated into.
b) User research and professional comments validated the system's capability to precisely identify emotions and its potential to improve these use cases' human-computer interaction.
c) When interacting with the emotion-aware applications, participants reported a more engaging and natural experience.

Overall, the findings show that the suggested deep neural network-based system is a very precise, effective, and useful way to detect emotions in real time. The broad use of emotion recognition technology in a range of interactive applications, such as mental health monitoring and smart environments, is made possible by its performance and capabilities.

CASE STUDY

This was, until the advent of deep neural networks, an out-of-scope task that subsequently advanced the field to more sophisticated and intuitive user experiences. This section discusses an application scenario of DNNs for enhancing HCI; it presents a real-world implementation for virtual assistants. We demonstrate how DNNs can change the way users interact with devices through the integration of emotion recognition technology into a virtual assistant (Giannopoulos et al., 2018; Nakisa et al., 2018).

Background: Virtual assistants, including Siri from Apple, Alexa by Amazon, and Google Assistant, have become an essential feature in daily life. Be it providing information to the user, controlling smart devices, or performing any other tasks-all are performed through voice commands. Conventionally, these systems have relied heavily on voice recognition and natural language processing to interpret what a user says. But the addition of emotional intelligence in these systems will make them much more effective and increase user engagement.

Objective: This case study will involve the application of deep neural networks to develop a virtual assistant able to recognize human emotions, enhancing user experience through better responsiveness and empathy. The DNN model, in particular, focuses on the determination of emotional states based on facial expressions, which in turn allows for providing appropriate responses.

Implementation

- Data Collection and Preparation: This involved the collection of a complete dataset of various facial expressions supported by emotional labels. The dataset consisted of diverse samples across demographics and emotional states. Added to the dataset for robustness were images of variable lighting conditions and angles of the face. The diverse data was an absolute necessity for training to get a model that would generalize across users.
- Model Design: Since the CNNs have proven to be powerful for image processing and classification, the architecture chosen for emotion recognition was a convolutional neural network. The pro-

posed CNN model consisted of several convolutional layers, pooling layers, and fully connected layers responsible for extracting and interpreting facial features that express different emotions. The network was thus trained in a combination of transfer learning and fine-tuning. Transfer learning uses pre-trained models on large image datasets, which can speed up the training and improve performance of the model. For fine-tuning in this regard, adaptation involves making adjustments to the last layers and hyperparameter optimization of the model for emotion recognition.

- Integration with Virtual Assistant: Emotion recognition was integrated into the virtual assistant system for analyzing users' facial expressions in real time. This called for an interface that would capture and preprocess images of faces from the user's webcam and feed them into the DNN model for emotion classification. According to the detected emotional condition, it made the responses of the virtual assistant more empathetic and contextually appropriate. In such cases, if the model detected frustration or anger, the assistant might offer additional help or a calming response. On the other hand, if the user was happy or satisfied, the assistant would reinforce positive interactions by making encouraging comments.
- Testing and Evaluation: Finally, the performance and user experience impact of the integrated system were extensively tested. Several user studies were conducted where participants interacted with their virtual assistant in various emotional states. In these studies, key points of interest included the assessment of the system for recognition of emotions and efficiency in its adaptive responses. These included emotion recognition accuracy, user satisfaction ratings, and engagement level. Performance comparison was done with a base virtual assistant without emotion recognition capabilities to measure improvements in the quality of user interaction.

The applications of the DNN-based emotion recognition system brought several key benefits.

- Better Emotion Detection: The DNN model was able to classify emotional states with high accuracy while reducing misclassifications compared to the state-of-the-art baselines.
- More User Engagement: Users were more engaged and felt personalized interaction with the virtual assistant. Further, this increased the rating of satisfaction and extended the time with interaction.

Effective adaptation was also made, as detected emotions were effective in driving the responses of virtual assistants. Users would also praise the personal and context-sensitive interactions that may be engaged in, further giving the interaction a more humanlike experience.

A number of challenges were experienced in implementing:

- Quality and Diversity of Data: With different user populations involved, the dataset needed to be representative to ensure the model's accuracy. So, continuous efforts had to be put in to improve data quality and augment the dataset to solve the challenge.
- Real-time Processing: Integrating emotion recognition with a virtual assistant requires optimizing the model for real-time processing on edge devices. This required model pruning and efficient use of hardware to achieve performance requirements.

The bumps for privacy were promoted due to the face recognition technology used in it. A transparent policy of data handling and mechanisms for consent from the users helped gain user trust in this respect.

The case study presents how DNNs can bring about a new revolution in human-computer interaction with emotion recognition. Integration of emotion recognition by DNNs with virtual assistant system showed outputs regarding enhanced accuracy and user engagement along with adaptive responses. This implementation highlighted data quality, real-time processing, and ethics considerations necessary to form effective and responsible AI systems. As emotion recognition technology continues to improve, applications of the technology could further spur innovation and improvement in user interactions.

CHALLENGES AND OPPORTUNITIES

Developing and deploying emotion recognition systems based on DNNs offers a set of challenges and opportunities which must be harnessed. Meeting these challenges and capitalizing on the opportunities will decisively affect the effectiveness and applicability of emotion detection technologies (Hussain et al., 2022; Jahangir et al., 2021).

Figure 4. Challenges and opportunities associated with developing and deploying emotion recognition systems based on Deep Neural Networks (DNNs)

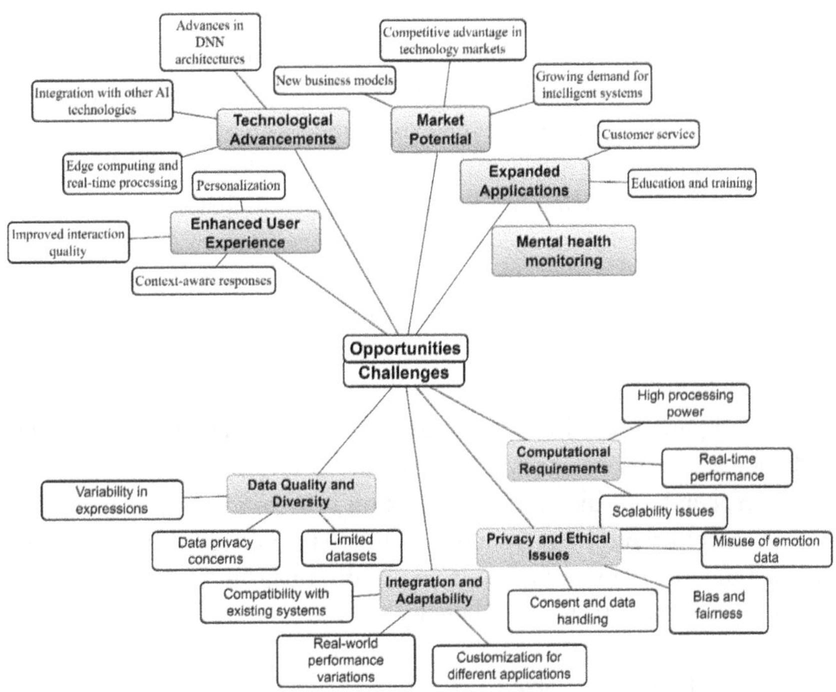

Figure 4 illustrates a structured overview of the challenges and opportunities in developing and deploying emotion recognition systems using Deep Neural Networks (DNNs), aiming to guide the development and deployment strategies.

Challenges

Data Quality and Annotation SOTA emotion recognition models are normally trained on high-quality, annotated face expression datasets. Still, the problems to be faced can range from issues of diversity, balance, and accuracy of labeling. The datasets must be representative of a wide variety of demographics, including age groups, ethnicity, and gender, to have the model performing well over a large range of populations. Additionally, annotations must be precise, since inconsistencies or errors may lead to poor model performance.

- Model Overfitting and Generalization It is well-known that DNN models tend to overfit when models are trained on small-sized or biased datasets. This means that a model will learn the training dataset too well-its noise and anomalies-which reduces the capability of a model to generalize to new, unseen data. A proper trade-off between the complexity of the model and its generalization should be achieved in order for the system of emotion recognition to perform in a robust way across different scenarios and real-world conditions.
- Efficiency: Real-world emotion recognition requires much computation, which may turn into a bottleneck on edge devices since they have low computation capability. In reality, it is really hard to develop models that are both high in accuracy and efficiency. The usage of model pruning, quantization, and efficient architecture design can help, but all these techniques must be balanced so as not to badly affect performance and computational demands.
- Variability in Facial Expressions: The facial expression can vary due to cultural, individual, and contextual variables. Such variability in the expression of emotion makes emotion recognition a complex task. For example, a given facial expression can have different meanings according to cultural norms or personal traits. It is a big challenge to ensure that the emotion recognition system correctly interprets the expressions according to such variables.

Privacy and Ethical Issues The use of emotion recognition technology introduces some rather important concerns regarding privacy and ethics. Without proper handling, gathering and analyzing emotional data may make individuals feel that their privacy is violated. This requires open policies, proper data protection, and consent in particular. How vital it is to ensure that such technology is used in an ethical manner and would not lead to misuse or discrimination.

Integration and Interoperability Integration of emotion recognition systems with existing platforms and applications is difficult in nature. The technology should be compatible with different hardware and software environments and integrate well with other systems. Ensuring interoperability while sustaining performance and accuracy is a big technical challenge.

Opportunities Better User Experience

Emotion recognition can make the interaction of users better, more personalized, and timely. For instance, a virtual assistant with emotion recognition capability would understand when to speak or not, depending on the mood of the user. It develops more empathy and effectiveness in communication. It creates several kinds of personalisation for improved user satisfaction and engagement in many applications.

- Monitoring mental health and providing support: Continuous monitoring of emotional states may provide valuable insights into mental health. On analyzing changes in the pattern of emotions, this technology helps identify potential problems in mental health and facilitates early intervention. Its application would thus promise considerable improvement in mental health care and facilitate the required support for individuals.
- Emotion Recognition and Customer Experience/Business Intelligence: Emotion recognition can go very deep into customer satisfaction and preference in customer service and marketing. Businesses could then use such information to refine their products and services, improve customer interactions, and create focused marketing campaigns. This will definitely improve the experience of the customers and make them more loyal.
- Assistive Technologies and Accessibility Emotion recognition can bring in a sea-change in assistive technologies for communication-impaired individuals, among others. Understanding the emotional cues from the user will enable such technologies to offer more intuitive and supportive interactions, enabling users to bring out better their intent and quality of life.
- While DNNs are still continuously being improved, it opens up wide vistas for innovation in emotion recognition. Advantages in machine learning algorithms, increased computational power, and improvement in data acquisition techniques promise to improve the accuracy, efficiency, and applicability of such methods. Embracing such advances can lead to increasingly robust and versatile emotion recognition systems.
- Cross-Domain Applications Emotion recognition technology has possible applications beyond traditional domains. For example, in education, this technology would serve to determine the engagement of students and adjust teaching methods; in health, to monitor patient responses to treatments; and in entertainment, to make experiences more immersive and emotionally appealing. Exploring these cross-domain applications could open up new paths for the deployment of technology and therefore impact.
- Responsible Development of AI If there is going to be ethical and responsible AI, then some proactive means of addressing privacy and ethical issues in their development should be made. This course of action will most definitely entail setting up regulations, codes of best practice, and standards that ensure applications involving emotion recognition technology are designed in such a way that they respect the rights of individuals and engender trust in the technology.

There are, therefore, strong challenges but also great opportunities in developing and deploying deep neural network-based emotion recognition systems. Meeting these challenges and capitalizing on the opportunities that arise might lead to advancements that improve user interactions, helping mental health drive innovation across many domains. It will be an act of balance between technical, ethical, and practical considerations to be sure that emotion recognition technology will offer its full potential.

FUTURE TRENDS IN EMOTION RECOGNITION TECHNOLOGY

Emotion Recognition technologies are one of the fastest-evolving technologies in modern times, driven by leaps and bounds in AI, Machine Learning, and Computer Vision. Almost every future dimension holds this technology for innovation and transformation as shown in Figure 5. Following are some of the key trends that shape the future of emotion recognition (Dalvi et al., 2021; Hussain et al., 2022; Pal et al., 2021):

Figure 5. Future trends in emotion recognition technology

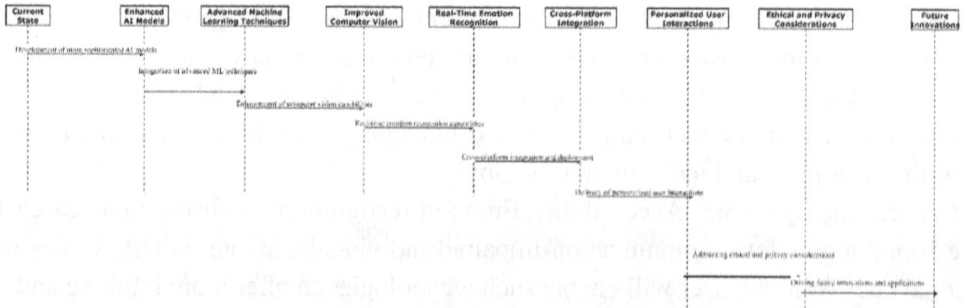

a) Higher Levels of Accuracy with Multimodal Approaches: In fact, future emotion recognition systems will be increasingly multimodal, where facial expression analysis would be combined with data from other sources, such as voice tone, physiological signals-which might include heart rate and skin conductance-along with contextual information. Thus, it integrates a wide variety of data that enables more accurate and subtle emotion recognition. For instance, the vocal tone can be merged with facial expressions to characterize emotional states more precisely or to decrease ambiguity.
b) Deep Learning Model Improvements: Deep learning models will increasingly become sophisticated, building on advances in neural network architectures and their training techniques. Newer models, like transformers and self-attention mechanisms, promise better performance for emotion recognition tasks. Techniques such as transfer learning and few-shot learning will allow models to generalize much better from scarce data. The challenge of data scarcity will be addressed, and models will enhance their ability to recognize emotions across diverse populations and contexts.
c) Real Time and Edge Computing Innovations: The demand for real-time emotion recognition is likely to ensure innovations in edge computing. Future systems will be optimized to be deployed on edge devices with limited computational resources, such as smartphones and IoT devices. Advances in hardware accelerators, such as GPUs and TPUs, and model optimization techniques, such as quantization and pruning, will lead to high-performance emotion recognition in real time. This is going to extend the application of emotion recognition technology to more portable and accessible platforms.
d) Emotion Recognition-Personal and Context Aware: In the future, emotion recognition systems will be much more personal and context aware. Technologies will adapt to individual differences such as variations in emotional expression and cultural norms. The process of personalization involves the tuning of models for specific users based on their emotional patterns, preferences, and contextual information. Context-aware systems will take into consideration the environment, time of day, and recent interactions to make more relevant and accurate emotional insights.
e) Integration of Augmented and Virtual Reality: In the coming years, AR and VR will increase their utilization of emotion recognition technologies to create more immersive experiences. Emotion recognition can improve the user experience in AR and VR through the use of virtual characters

that respond appropriately considering a user's current emotional state. This integration will be highly valuable for applications within the gaming industry, training simulations, and therapeutic intervention where emotionally resonant experiences are staged.

f) Ethics and Responsible AI Practices: As emotion recognition technology evolves, more attention will be put into ethical and responsible AI. It will be important to ensure that emotion recognition systems are implemented in a way that is ethical and respectful of users' privacy. This includes a range of next-step developments: strong privacy protection, transparent data handling, and informed consent mechanisms. A risk reduction and the establishment of trust in emotion recognition technologies are both facilitated by guidelines on ethics and best practices.
g) Cross-Domain Applications and Custom Solutions; Emotion recognition technology will find applications across wide-ranging domains-from healthcare to education, from customer care to entertainment. Specialized solutions for this technology will come out within specific industries. Certain applications in healthcare may involve monitoring for patient's well-being and follow-up support of mental health treatment. In education, this may provide insight into student engagement and tailor the approach to instruction. Innovation in cross-domain applications will strongly enhance the usefulness of emotion recognition systems.
h) Emotional Intelligence Improved in AI Systems: In the future, emotion recognition systems will contribute to making AI emotionally intelligent. It would understand human emotions and respond to them for better interpersonal interaction. This would be quite useful and would apply in virtual assistants, customer service bots, and therapeutic applications where it would be imperative to show empathy and emotional understanding. Advancement in emotion recognition will let the AI systems display more personalized and human-like interactions.
i) Integration with IoT and Smart Environments: The application of emotion recognition in conjunction with the Internet of Things and smart environments will make systems more responsive and adaptive. In a smart home, for example, emotion recognition can be exploited to provide optimal settings for lighting, temperature, and entertainment with respect to the occupants' emotional states, improving user comfort and convenience.
j) Improved Accessibility and Inclusiveness: The future of emotion recognition technologies will be in increasing inclusivity for more diverse sets of populations. Training models will involve the recognition of emotions across different age brackets, people with disabilities, and people from diverse cultural backgrounds. These priorities will ensure that emotion recognition systems work both effectively and equitably, providing valuable insights and support to users, considering individual characteristics unique to each.
k) Collaboration and Open Research: Increased collaboration and openness to research will further advance emotion recognition technology. Shared datasets, research findings, and best practices will hasten progress and innovations in the field. This will be instrumental in solving common challenges that may arise, improving system performance, and driving the adoption of emotion recognition technology across various applications.

Innovation in deep learning, multi-modal approaches, and real-time processing propels emotion recognition technology into an exciting future. With much more accuracy, personalization, and contextual understanding of such systems, a whole new gamut of user experiences is going to be rewritten. However, responsible development and deployment of these technologies will demand consideration of ethical issues, privacy concerns, and inclusive design. While challenges are yet to be overcome, several emerging opportunities will enable emotion recognition technology to further develop and open new perspectives for improving human-machine interaction in various areas of daily life.

CONCLUSION

The recognition of emotions by DNN is a sea change in how human emotions are conceptualized and interacted with. This chapter has toured a complex area of emotion recognition, from founding principles to state-of-the-art advancements that shape its future. Advanced AI models, detailed machine learning techniques, and further developments in computer vision intelligently brought emotion recognition into real-world applications, hence allowing unprecedented levels of personalization and user experience across diverse domains.

Nevertheless, high-quality and diverse data, intensive computation, and privacy and ethical issues are considered some challenges to be resolved for the successful application of the technology. Also, there is a pressing need to develop systems with the capability to recognize emotions on multiple platforms without deteriorating performance and adaptability in real-world conditions. As we move ahead with emotional technologies, understanding these challenges and addressing them with creativity and imagination will be required to understand the actual impact of such technologies.

The future of emotion recognition, however, will be truly shaped by rapid technological advances and broadening scopes of application. Currently, the technology is able to do real-time emotion analysis, personalized interaction with users, even cross-platform integrations-all thanks to relentless improvements in the enabling technologies of AI and machine learning. Notwithstanding these stunning advances, ethical and privacy concerns remain central in this field to ensure that the technology develops in a responsible and nondiscriminatory manner.

Hence, while emotion recognition technologies are about to burst into various fields, their actual success will depend upon striking a critical balance between innovation and responsible implementation. Embracing the upcoming trends and challenges of today, emotion recognition can be used to increase human-computer interaction by leaps and bounds, draw valuable insight from emotional states, and thus further contribute to an empathetic and responsive technological frontier.

REFERENCES

Abdullah, S. M. S. A., Ameen, S. Y. A., Sadeeq, M. A., & Zeebaree, S. (2021). Multimodal emotion recognition using deep learning. *Journal of Applied Science and Technology Trends*, *2*(01), 73–79. DOI: 10.38094/jastt20291

Benamara, N. K., Val-Calvo, M., Alvarez-Sanchez, J. R., Diaz-Morcillo, A., Ferrandez-Vicente, J. M., Fernandez-Jover, E., & Stambouli, T. B. (2021). Real-time facial expression recognition using smoothed deep neural network ensemble. *Integrated Computer-Aided Engineering*, *28*(1), 97–111. DOI: 10.3233/ICA-200643

Dalvi, C., Rathod, M., Patil, S., Gite, S., & Kotecha, K. (2021). A survey of ai-based facial emotion recognition: Features, ml & dl techniques, age-wise datasets and future directions. *IEEE Access : Practical Innovations, Open Solutions*, *9*, 165806–165840. DOI: 10.1109/ACCESS.2021.3131733

Dwijayanti, S., Iqbal, M., & Suprapto, B. Y. (2022). Real-time implementation of face recognition and emotion recognition in a humanoid robot using a convolutional neural network. *IEEE Access : Practical Innovations, Open Solutions*, *10*, 89876–89886. DOI: 10.1109/ACCESS.2022.3200762

Falahzadeh, M. R., Farokhi, F., Harimi, A., & Sabbaghi-Nadooshan, R. (2023). Deep convolutional neural network and gray wolf optimization algorithm for speech emotion recognition. *Circuits, Systems, and Signal Processing*, *42*(1), 449–492. DOI: 10.1007/s00034-022-02130-3

Giannopoulos, P., Perikos, I., & Hatzilygeroudis, I. (2018). Deep learning approaches for facial emotion recognition: A case study on FER-2013. *Advances in Hybridization of Intelligent Methods: Models, Systems and Applications*, 1–16.

Hassouneh, A., Mutawa, A., & Murugappan, M. (2020). Development of a real-time emotion recognition system using facial expressions and EEG based on machine learning and deep neural network methods. *Informatics in Medicine Unlocked*, *20*, 100372. DOI: 10.1016/j.imu.2020.100372

Hizlisoy, S., Yildirim, S., & Tufekci, Z. (2021). Music emotion recognition using convolutional long short term memory deep neural networks. *Engineering Science and Technology, an International Journal*, *24*(3), 760–767. DOI: 10.1016/j.jestch.2020.10.009

Hossain, M. S., & Muhammad, G. (2019). Emotion recognition using deep learning approach from audio–visual emotional big data. *Information Fusion*, *49*, 69–78. DOI: 10.1016/j.inffus.2018.09.008

Hussain, H., Tamizharasan, P., & Rahul, C. (2022). Design possibilities and challenges of DNN models: A review on the perspective of end devices. *Artificial Intelligence Review*, *55*(7), 1–59. DOI: 10.1007/s10462-022-10138-z

Issa, D., Demirci, M. F., & Yazici, A. (2020). Speech emotion recognition with deep convolutional neural networks. *Biomedical Signal Processing and Control*, *59*, 101894. DOI: 10.1016/j.bspc.2020.101894

Jahangir, R., Teh, Y. W., Hanif, F., & Mujtaba, G. (2021). Deep learning approaches for speech emotion recognition: State of the art and research challenges. *Multimedia Tools and Applications*, *80*(16), 23745–23812. DOI: 10.1007/s11042-020-09874-7

Jain, N., Kumar, S., Kumar, A., Shamsolmoali, P., & Zareapoor, M. (2018). Hybrid deep neural networks for face emotion recognition. *Pattern Recognition Letters, 115*, 101–106. DOI: 10.1016/j.patrec.2018.04.010

Jaiswal, A., Raju, A. K., & Deb, S. (2020). Facial emotion detection using deep learning. *2020 International Conference for Emerging Technology (INCET)*, 1–5.

Mehendale, N. (2020). Facial emotion recognition using convolutional neural networks (FERC). *SN Applied Sciences, 2*(3), 446. DOI: 10.1007/s42452-020-2234-1

Nakisa, B., Rastgoo, M. N., Rakotonirainy, A., Maire, F., & Chandran, V. (2018). Long short term memory hyperparameter optimization for a neural network based emotion recognition framework. *IEEE Access : Practical Innovations, Open Solutions, 6*, 49325–49338. DOI: 10.1109/ACCESS.2018.2868361

Pal, S., Mukhopadhyay, S., & Suryadevara, N. (2021). Development and progress in sensors and technologies for human emotion recognition. *Sensors (Basel), 21*(16), 5554. DOI: 10.3390/s21165554 PMID: 34451002

Pranav, E., Kamal, S., Chandran, C. S., & Supriya, M. (2020). Facial emotion recognition using deep convolutional neural network. *2020 6th International Conference on Advanced Computing and Communication Systems (ICACCS)*, 317–320.

Santamaria-Granados, L., Munoz-Organero, M., Ramirez-Gonzalez, G., Abdulhay, E., & Arunkumar, N. (2018). Using deep convolutional neural network for emotion detection on a physiological signals dataset (AMIGOS). *IEEE Access : Practical Innovations, Open Solutions, 7*, 57–67. DOI: 10.1109/ACCESS.2018.2883213

Chapter 3
An Overview of Sentiment Analysis:
Insights From Business Case Studies

Fehmina Khalique
https://orcid.org/0000-0002-2519-3174
Lloyd Business School, India

Neha Issar
https://orcid.org/0000-0002-6929-1403
Lloyd Business School, India

Lakhwinder Kaur Dhillon
https://orcid.org/0000-0003-1239-3537
Amity University, Noida, India

Aliza Ahsan
Jamia Millia Islamia, India

ABSTRACT

Sentiment Analysis is a part of Data Intelligence Research that lays emphasis on data that contains emotions. This analysis is carried out by analysing the polarity of content and thereby marking it as positive, negative, or neutral. In order to find out how SA is globally used, Supervised Natural Language Processing (SNLP) is also utilised. In this chapter, a range of these tools and techniques of SA will be discussed and their applications will be elaborated. Additionally, this chapter will delve into further academic research related to the topic to enhance understanding of how sentiment analysis can support organizations in staying competitive and boosting profits by examining real-life examples. Sentiment analysis has experienced notable progress in recent years, primarily propelled by utilizing machine learning and deep learning techniques in sentiment classification. SA helps in building social and political perceptions helping the researchers and policymakers to understand public sentiments on burning issues thereby aiding in decision-making in an ever-changing digital world.

DOI: 10.4018/979-8-3693-6910-4.ch003

1. INTRODUCTION

Sentiment analysis has seen a surge in its usage in recent years, not just among research scholars but also among businesses, governments, and organizations (Sánchez-Rada and Iglesias 2019). The increasing popularity of the social media and Internet has ranked it as principal source of global data and information. Users across the world make use of social media platforms in order to express their views and opinions. For monitoring public opinion and providing help in good decision-making, user-generated data can be analyzed for better results through implementation of Sentiment analysis. Sentiment analysis also termed as Opinion analysis or Opinion mining has witnessed a recent growth.

Sentiment Analysis (SA) is a part of "Data Intelligence Research" that lays emphasis on data that contains emotions (Hasselgren et.al, 2022). This analysis is carried out by analyzing the polarity of content and thereby marking it as positive, negative, or neutral. In order to find out how SA is globally used, Supervised Natural Language Processing (SNLP) is also utilized. According to Barba (2020) SNLP involves the utilization of Machine Learning algorithms like Support Vector Machines (SVM), Logistic Regression, and Neural Networks (NN) to implement Sentiment Analysis on a computer system. In this chapter, a range of these tools and techniques of SA will be discussed and their applications will be elaborated. Additionally, this chapter will delve into further academic research related to the topic to enhance understanding of how sentiment analysis can support organizations in staying competitive and boosting profits by examining real-life examples.

Sentiment analysis, also known as opinion mining, consists an important component of Natural Language Processing (NLP) (Jim, et al., 2024). The major function of NLP is to systematically find and arrange sentiments embedded in text data, that helps in finding out the emotions, opinions, attitudes, and expressions utilized can be either written or verbal. These strategies cover a broad range of tactics, beginning with the initial classification of text into positive, negative, or neutral emotions. However, its capabilities extend to more intricate methods, permitting the discernment of specific emotions, intentions, or nuanced aspects of sentiment, such as joy, anger, sarcasm, or context-specific sentiments observed in domains like product reviews (Cambria et al., 2017). The general process of a sentiment analysis system encompasses stages such as data collection, text preprocessing, feature extraction, training machine learning or deep learning models, and thorough evaluation to assess the model's effectiveness in sentiment classification tasks (Devika et al., 2016) (See, Figure 1).

Sentiment analysis has experienced notable progress in recent years, primarily propelled by utilizing machine learning (ML) (Revathy et al., 2022) and deep learning (DL) (Abdullah and Ahmet, 2022) techniques in sentiment classification. SA is an important aspect of NLP and has several advantages as it helps the organizations to gain valuable insights from public opinions and customer feedback, facilitating data-driven decision-making, product enhancement, and effective marketing strategies (Ahmed et al., 2022). SA helps in building social and political perceptions helping the researchers and policymakers to understand public sentiments on burning issues thereby aiding in decision-making in an ever-changing digital world (Peng et al., 2022).

Figure 1. The working process of sentiment analysis

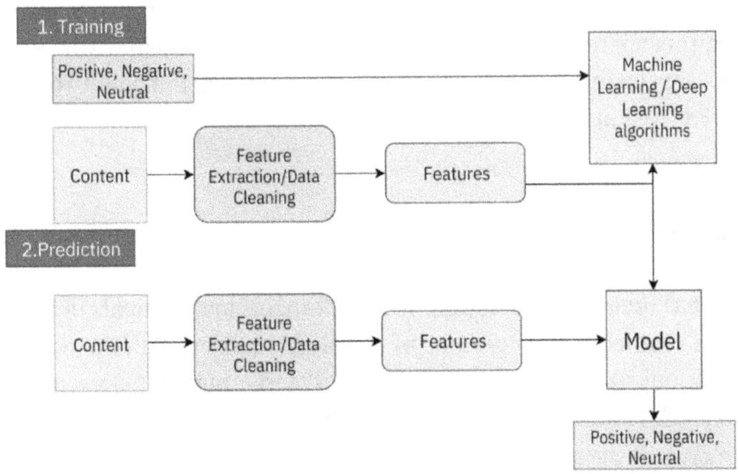

Many researchers use the terms 'Sentiment' and 'emotion' as synonyms but two terms have different meanings (Li et al. 2022). Sentiment is defined as 'an attitude, thought, or judgement prompted by a feeling' while Emotion, 'refers to a conscious mental reaction subjectively experienced as strong feelings' (Munezero et al. 2014). Sentiments may exist for a long duration and they are more stable than emotions (Jain and Kaushal 2018). Sentiment analysis and opinion mining express similar meaning as opinion mining is related with 'gathering feelings or emotions associated with the text' or 'extracting the opinions from text,' while sentiment analysis is 'classifying the text according to the sentimental information associated with the text' (Kaur and Saini, 2014).

A major dissimilarity between sentiment analysis and emotion detection is that SA uses two or three classes which are negative, positive, and neutral for classification of text, while emotion detection uses wide range of emotions like joy, fear, anger, brief, surprise or disgust (Kaur and Saini 2014). In sentiment analysis, the statements 'the item was not as expected' and 'I hate this item' are both negative sentences, although there is a significant emotional difference between both sentences (Alslaity and Orji, 2022).

Sentiment Analysis approaches can be divided into two main categories, namely, Lexicon-based approaches and Artificial Intelligent (AI) or Machine Learning-based (ML-based) approaches (Yiran and Srivastava 2019; Anjaria and Reddy Guddeti 2014). Other researchers included a third category called Hybrid approach. Lexical approaches rely on sentiment lexicon is a collection of sentimental words, such as 'excellent' or 'bad' and they compare sentiment words with seed words. The term 'excellent' would have the value (1) since it indicates positive sentiment, while a negative term, such as 'terror,' would have the value (−1) attached to it. The overall score can be assigned one of the multiple categories, usually positive, negative, or neutral. Lexicon-based approaches can be divided into Corpus-based and Dictionary-based approach (Gindl et al., 2010; Rohini et al., 2017).

The other category of SA is the machine learning (ML) based approach. ML-based SA approaches rely on learning algorithms to understand the sentiments and are used to train datasets for recognizing sentiments and classifying emotions (Rohini, Thomas, and Latha 2017). ML techniques are also categorized as shallow and deep learning (Janiesch et al., 2021). Shallow learning is further categorized into supervised, unsupervised, and semi-supervised learning. In supervised learning, the model is trained on a labelled or tagged dataset while in unsupervised learning models deploy ML algorithms to analyze and categorize using unlabeled data. Lexicon-based approaches are user friendly to perform and use simple steps however, this approach is insufficient for large datasets with many aspects or dimensions. ML-based approaches are flexible, but they are highly dependent on the quality of data (Han et al., 2020; Goodfellow et al., 2016; Alslaity and Orji, 2022).

2. OBJECTIVE OF THE STUDY

Sentiment analysis has multiple applications across domains and verticals in an organization. It helps in gaining insights into sentiment and opinions of people associated with the organizations. One of the most important applications is in the domain of sales and marketing whereby, customer feedback can be utilized to improvise products and services, social media monitoring can help in company and product branding, marketing research can assess consumer feedback etc. It is also highly recommended in the field of Finance whereby investment strategies can be built; sentiments can be analyzed for stock market etc.

3. APPLICATIONS OF SENTIMENTAL ANALYSIS

Sentiment analysis has multiple applications across domains and verticals in an organization. It helps in gaining insights into sentiment and opinions of people associated with the organizations. One of the most important applications is in the domain of sales and marketing whereby, customer feedback can be utilized to improvise products and services, social media monitoring can help in company and product branding, marketing research can assess consumer feedback etc. It is also highly recommended in the field of Finance whereby investment strategies can be built; sentiments can be analyzed for stock market etc. (See Figure 2).

Figure 2. Applications of sentimental analysis

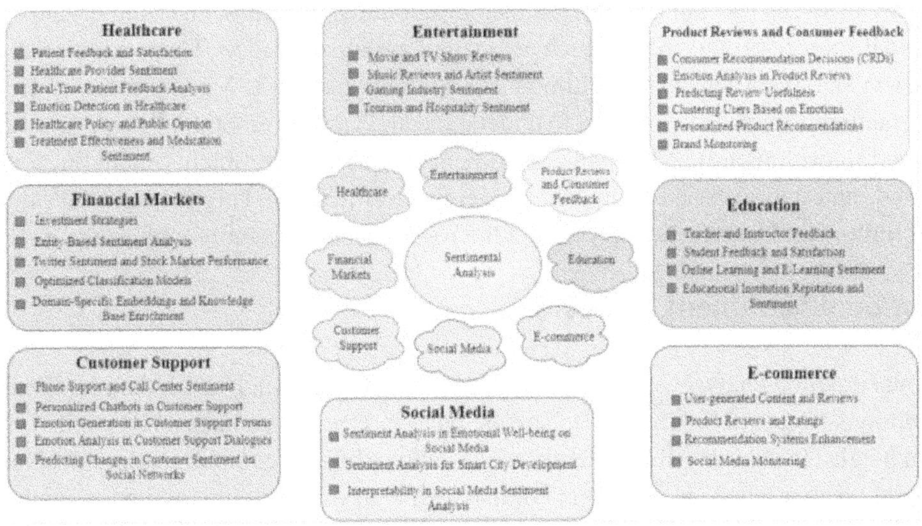

Since the arrival of the internet on a commercial scale in the mid-1990s, the manner in which information is delivered to investors, and how investors respond to such information, has been altered considerably (Nardo et al., 2016). Specifically, an increase in the amount of available digital information, facilitated by the scale of interactions that can now be documented, has led to a situation whereby the vast amount of unstructured data that investors have access to—in the form of corporate disclosures, news media, email and social media—often make rational and perfectly informed decisions unattainable (Chan & Chong, 2017). Thus, SA in text and its applications have been instrumental in analyzing and summarizing extensive, unorganized textual data into information that can be readily understood by investors with limited cognitive capacities and time constraints.

Jain et al. 2021in their study provides machine learning applications that incorporate online reviews in sentiment categorization, predictive decision making, and the detection of false reviews while Balaji et al. (2023) utilized applications of social media to analyze sophisticated machine learning algorithms. Another application widely used in the field of sentiment analytics is forecasting stock market trends. For e.g. using Twitter data with hashtags associated with AI and GPT are integral in determining the trends in the stock market. Past literature on sentiment analysis pays attention to Twitter messages in order to devise a daily happiness index, for getting an idea on the impact of social media on the financial markets (Agarwal et al., 2021). Furthermore, sentiment analysis plays a crucial role in analyzing tweets and transform the content into social sentiment indices thereby generating time series data of organizations, in order to predict the stock performance (Mendoza Urdiales et al., 2022). Several scholars have examined the influence of social media sentiment on volatility of stock market, analyzing the impact of sentiments associated with the worldwide news on investor attitudes (Maqsood et al., 2022).

4. SENTIMENT ANALYSIS THROUGH COMPUTATIONAL TECHNIQUES

SA not only analyzes text but also integrates multimodal approaches, enabling organizations to explore additional sources of information such as images, audio, and video. These additional sources can greatly assist in sentiment analysis. Multimodal sentiment analysis is a computational technique that gathers information from text and images, to understand and analyze emotional states of users. Different approaches and algorithms are used to realize sentiment classification problems, including lexicon-based, learning-based, and hybrid-based approaches (Memiş et al., 2024). This section will provide a brief overview of multimodal sentiment analysis and the role of labels in sentimental analysis.

4.1. Multimodal sentiment analysis

Multimodal emotion analysis involves examining textual, audio and visual data to discriminate and interpret emotions (Das and Singh, 2023). This helps in interpreting emotions through verbal reviews, vlogs, social media etc. whereby the researchers can capture human communication involves more than just verbal and written language; it also encompasses facial expressions, intonation, and visual cues. Multimodal analysis (Bhaskar et al., 2014; Dair et al., 2021; Houjeij et al., 2012; Yang et al., 2020) have been demonstrated as having greater abilities in accurately capturing sentiment.

The *Image-text approach* entails the comprehensive analysis of visual content and its accompanying textual information in order to gain a thorough understanding of the emotions conveyed in multimedia data. A joint image-text sentiment classifier uses a new feature vector obtained by employing a fusion technique from both image and text data to predict sentiment (Das and Singh, 2023). A combination of visual–textual sentiment framework named multimodal attentive fusion was suggested by Huang et al. (2019). Wang et al. (2014) proposed a joint image-text framework for the microblog images using a cross-media bag of words approach.

Audio–video multimodal sentiment analysis works through audiovisual fusion focusing on human faces. An audiovisual sentiment analysis framework was proposed by Yadav et al. (2015) by extracting emotion-related information from both the video and the audio Memiş nel in audiovisual content. Chu and Roy (2017) proposed an audiovisual analysis method to generate emotional arcs for movies, including short videos on the web.

The *Audio-image-text approach* to multimodal analysis utilizes data covering audio, image and text to gather information on emotions from multiple sources. Poria et al. (2016) designed a real-time multimodal sentiment understanding model to harvest sentiments from web videos. Cambria et al. (2018) proposed a deep neural framework for multimodal sentiment analysis.

4.2. Labels perspective in sentimental analysis

The labels perspective of SA helps in forming categories for the data collected through sentimental analysis into positive, negative or neutral for the purpose of evaluating models of SA. This perspective of SA is important to understand whether the models correctly interpret emotions from texts, audio or video content in order to check the accuracy and validity of SA systems.

Document-level sentiment analysis helps to evaluate sentiment in individual documents, for example, single reviews, however, the challenges that may crop up from extraneous sentences, necessitating subjectivity/objectivity classification are to be taken into consideration. Both supervised and unsupervised learning methods can be employed, using features like term frequency and opinion words (Balaji et al., 2017).

Sentence level Sentiment analysis focuses on evaluating emotion in individual sentences (Araújo et al., 2020). The sentiment analysis focuses on the short texts in order to understand the nuances of emotions, mainly used where the emotional polarity can vary within a document or conversation, this tool is employed to provide a comprehensive comprehension of the emotional tone conveyed at the sentence level.

Phrase-level sentiment analysis is a valuable tool for assessing the emotional content of specific phrases or expressions. This type of analysis offers a comprehensive insight into the subtle nuances of emotions within sentences, uncovering details that may not be apparent when analyzing sentences or documents as a whole. Sentiment analysis can be conducted at various levels, including the word or phrase level, sentence level, and document level.

Word-level emotion distribution learning the distribution of emotions at the word level is beneficial for linking words to specific emotions by considering the intensity of the emotion. This is achieved by utilizing the Valence-Arousal-Dominance (VAD) space. Challenges include dealing with non-dictionary words, mitigated by using WordNet to extend the coverage of VAD lexical knowledge and handling non-emotional words or spelling errors (Li et al., 2021).

4.3. General dictionary approach

Many of the earliest studies employing sentiment analysis techniques within accounting and finance utilize a dictionary approach, also known as the 'word count' approach (Guo et al., 2016; Loughran & McDonald, 2016). The concept behind this approach is comparatively intuitive compared to more recent machine learning methods, in that the sentiment conveyed within a financial text is determined by a count of words within the text that also appear within pre-defined word lists (Li, 2010). Expanding on this, Bhonde et al. (2015) note that word lists are created by first collecting a set of general sentiment words with known positive or negative implications. After compiling the initial list, it is subsequently enhanced by incorporating synonyms and antonyms for the sentiment words. This iterative procedure of expanding the word lists concludes when no additional words can be discovered. Following the culmination of the synonym and antonym collection process, a thorough examination of the words is typically conducted to refine the lists.

Abirami and Gayathri (2016) highlight that the most basic way to define sentiment using these dictionaries is to count the number of positive and negative words within a body of text with reference to the dictionary categories. Once the counting process is finished, an analysis is conducted to determine the ratio of positive to negative words in the text, which indicates the overall sentiment. Utilizing dictionary methods offers several benefits over machine learning techniques, primarily requiring less computational power and resources for dictionary creation and usage. Nevertheless, there are significant drawbacks associated with this approach.

4.4. Domain-specific dictionary approach

Loughran and McDonald (2011) demonstrate that general dictionaries misclassify words used within a financial context, noting that 73.8% of negative words within the Harvard dictionary are not considered negative in a financial context. González-Bailón and Paltoglou (2015) and Ribeiro et al. (2016) also demonstrate the limitations of general dictionaries in classifying content in domain-specific settings.

Both authors achieve this by utilizing a comprehensive dictionary to analyze text originating from diverse domains, demonstrating the limited reliability and validity across these distinct datasets. One alternative to this issue is to create domain-specific dictionaries, where adding words to an existing dictionary and deleting irrelevant words (or words with different meanings) within a specific context would be beneficial (Diesner & Evans, 2015; Grimmer & Stewart, 2013).

4.5. Machine learning approaches

In the past few years, the progress in computational capabilities has enabled the utilization of Machine Learning (ML) techniques for conducting sentiment analysis across various organizational sectors. Several research papers have examined the effectiveness of dictionary methods (both general and domain-specific) in comparison to the ML approach. For example, McGurk et al. (2020), Renault (2017) and Guo et al. (2016), all provide evidence that ML approaches are more accurate at classifying financial sentiment. However, Renault (2020) states that ML models may be sufficient in deriving textual sentiment from online sources but warns that more complex algorithms do not necessarily equate to more accurate results.

Machine Learning is a branch of computer science that copies the learning process of human beings and forms an integral part of Artificial Intelligence (AI) discipline, which comprise of techniques that helps computers copy human behaviour and intelligence to solve complicated problems and tasks with minimum human intervention (El Naqa and Murphy 2015; Russell and Norvig 2020). According to Janiesch et al. 2021, ML approaches can be classified as Shallow machine learning and deep learning. Shallow machine learning can be further classified as into several categories like Supervised Learning, Unsupervised Learning, Semi-supervised Learning, and Reinforcement Learning. Supervised ML algorithms can be applied in areas such as spam detection, weather forecasting, and pricing prediction. Naïve Bayes, SVM, and Linear Regression are examples of the wide array of supervised ML algorithms. However, these algorithms have many drawbacks such as they cannot be applied to all domains and all datasets. Moreover, these approaches are often task-specific (Sarkar and Etemad 2020). Further, supervised learning algorithms require a large, annotated dataset to work efficiently. In order to overcome the drawbacks of supervised learning, unsupervised ML algorithms are applied which find out unclear patterns without human intervention. Unsupervised ML uses algorithms to analyze and cluster unlabeled data and are allied in several domains pertaining to customer segmentation, image recognition, and dimensionality reduction. Examples of these algorithms include the Principal Component Analysis (PCA), K-mean clustering, and Singular Value Decomposition (SVD). However, unsupervised learning also possesses several drawbacks like it is computationally complex because it requires large amounts of data to analyze and give the outcomes. In between the supervised and unsupervised approaches, there is a third, less common category called Semi-supervised learning that is an intermediate solution between supervised and unsupervised learning, and are used when the data comprises a small portion of labelled data and a large portion of unlabeled data (Bilgin and Şentürk 2017; Alslaity and Orji, 2022).

Another category of ML approaches is Deep Learning which is an AI concept based on the Artificial Neural Networks (ANN) similar to neural pathways of the human brain. ANN works on the principle of networks with different layers namely, an input layer that receives the data, an output layer that yields the result and zero or more hidden layers responsible for learning the mapping between input and output (Goodfellow et.al, 2016). 'Deep' refers to the multiple layers between the input and output layers and Deep neural networks have capacity to detect raw data (Janiesch et.al, 2021). Deep learning approached may include Convolutional Neural Networks (CNNs), Recurrent Neural Networks (RNN), and Deep

Belief Networks. Deep learning although requires minimum human intervention, however, it is more complex and therefore, deep learning systems require far more powerful hardware and resources. While shallow machine learning runs on conventional computers. Deep learning systems also are more time consuming in comparison to shallow machine learning systems which requires minimal time in setting up and operations. Deep learning can accommodate a large volume of unstructured data and is commonly utilized in complex applications like robots and self-driving cars. However, for low-dimensional data input, shallow ML is better than deep learning (Zhang and Ling 2018; Rudin 2019; Alslaity and Orji, 2022).

A commonly used ML algorithm to detect sentiment is the probabilistic Naive Bayesian classifier, which considers the naïve independence assumptions and is commonly used as a baseline method for classifying text (Dey et. al, 2016). Naïve Bayes ' methods estimate the probability that a document is positive or negative given its contents. It estimates the probability of a word being 'positive' or 'negative' in nature by looking through a series of positive and negative texts and counting how often the word appears in each (Troussas et al., 2013). Hence, a crucial part of this model's method is pre-classified data to train on. Dey et al. (2016), however, note a benefit of the Naïve Bayes classifier is that it only requires a small amount of training data to establish parameters necessary for classification.

4.6. State-of-the-art natural language processing approaches

El-Haj et al. (2019) identifies that there is a scarcity of advanced NLP techniques being applied in the financial domain. While the ML techniques discussed in the previous section have been shown to classify financial sentiment better than more rudimentary approaches, alternative approaches such as transformer architecture (Alamoudi & Alghamdi, 2021; Munikar et al., 2019; Sun et al., 2019).

4.6.1. Transformer architecture

Before the introduction of the transformer by Vaswani et al. (2017), the authors highlighted that state-of-the-art results across various NLP tasks were dominated by Recurrent Neural Networks (RNNs), Long Short-Term Memory (LSTM) and Gated Recurrent Units (GRU) (Chung et al., 2014; Hochreiter & Schmidhuber, 1997). Instead of attempting to push state-of-the-art results by improving on previous RNN language models or Encoder–Decoder architecture (Jozefowicz et al., 2016; Luong et al., 2015; Wu et al., 2016, Vaswani et al., 2017) introduced the transformer, which is a model that is solely based on attention mechanisms and does not require recurrence or convolutions.

Numerous models have achieved cutting-edge performance in different tasks by incorporating and expanding upon the original transformer architecture. For example, Raffel et al. (2020) introduced a text-to-text transfer transformer (T5), which has achieved state-of-the-art performance on the SQuAD question and answering task. Brown et al. (2020) trains an autoregressive language model (GPT3) on 175 billion parameters, which returned the highest accuracy of 86.4% on the LAMBADA language modelling task. A third model which effectively utilizes the transformer is Bidirectional Encoder Representations from Transformers (BERT). Devlin et al. (2019) introduce BERT and compare it to various other advanced models across multiple datasets. Across the three models discussed above, BERT performs particularly well on sentiment classification tasks (Alamoudi & Alghamdi, 2021; Munikar et al., 2019; Sun et al., 2019).

Transformer architecture helps in enhancing the capabilities of machines in understanding text. However, as with all methods discussed previously in this review, it has limitations. Khan et al. (2022) highlighted certain limitations of transformers, for example, transformers are high on computational cost owing to their size and complexity, they require large data because of their need for a substantial amount of good quality data for training and the transformer model's poor interpretation due to complex architectures.

5. TEXT-CENTRIC MULTIMODAL SENTIMENT ANALYSIS TASKS

There are several kinds of text being analyzed in present times, for example, debates, blogs and online discussions on politics (Durant & Smith 2006). Tumasjan et. al (2010) explored in their study the impact of tweets about political parties to find out how well tweets align with the political parties' contested views and promises. Moreover, their findings also indicate that the volume of messages can help predict election results. Most of the research on political arena is focused around the sentiment analysis using texts from Twitter and online media as a source, however, researchers have found that sarcasm may be a hinderance (Bakliwal et al., 2013).

Text-centered multimodal sentiment analysis comprises majorly the analysis related to image-text sentiment and audio image-text (video) sentiment and there is another classification as per different emotional annotations, namely sentiment classification tasks and emotion classification tasks. *Sentiment classification* tasks include label classification tasks of positive, neutral, and negative while and *emotion classification* tasks comprise of emotional labels such as anger, happy, sad etc. Text-centered multimodal sentiment analysis can be classified as *coarse-grained multimodal sentiment analysis* that comprise of sentence-level analysis and *fine-grained multimodal sentiment analysis* that comprise of aspect-level analysis based on the granularity of the opinion targets (Wang et al., 2024).

Fine-grained multimodal sentiment analysis mainly pays attention to image-text pair data, and consists a) multimodal aspect term extraction (MATE), b) multimodal aspect-based sentiment classification (MASC), and c) joint multimodal aspect-sentiment analysis (JMASA). Additionally, multimodal sarcasm detection has also become a widely discussed task in recent years (Wang et al., 2024).

Multimodal sentiment analysis (MSA) is different from traditional text-based sentiment analysis as it makes use of images and speech in order to achieve high accuracy of classification of sentiments. Commonly used multimodal sentiment analysis prefer using "image-text", "audio-image" and "audio-image-text" (video). For instance, the expression "That's a fantastic news!" denotes positive emotion when analyzed at text, however, when combined with an eye-rolling expression and aggressive tone of voice, it may represent sarcastic negative expression.

According to Wang et al., 2024 datasets for coarse-grained image-text sentiment and emotion analysis may include TumEom, MVSA, MEMOTION 2, and MSED. TumEmo is a multimodal emotion dataset that consist of large quantity of image-text data taken from Tumblr. This dataset consists of 195,265 image-text pairs along with 7 emotion labels like anger, calmness, boredom, fear, happy, love, sad. MVSA dataset is a collection of image-text pairs on the Twitter platform and its annotation is done manually with three sentiment labels namely, positive, neutral, and negative. The MVSA dataset also consists MVSA-Single, where each sample is annotated by a single annotator, comprising 4,869 image-text pairs, and MVSA-Multiple, where each sample is annotated by three annotators with three emotion labels, totaling 19,598 image-text pairs. MEMOTION 2 is a dataset that consist of classification of emotions which includes 10,000 memes collected from various social media sites which are humorous to invoke a response. MSED consists 9,190 pairs of text and images extracted from different social media platforms, like Twitter, Getty Images, and Flickr.

Image-text fine-grained sentiment analysis focuses on analyzing sentiment elements that are finer than sentence-level, such as aspect term (a) and sentiment polarity (p), or their combinations. It has received widespread attention in recent years and mainly includes three subtasks: Multimodal Aspect Term Extraction (MATE), Multimodal Aspect-based Sentiment Classification (MASC) and Joint Multimodal Aspect-Sentiment Analysis (JMASA).

Wang et.al, 2024 summarize the datasets used for fine-grained image-text sentiment analysis, including Multi-ZOL, Twitter-15 and Twitter-17. Multi-ZOL does collection and organization of comments derived from smartphones from the ZOL.com business portal website. Multi-ZOL dataset consists of sentiment ratings for six features, like price-performance ratio, performance configuration, battery life, appearance and feeling, photographing effect, and screen; each feature, the comment has an integer sentiment score from 1 to 10, used for the sentiment label. Twitter-2015 and Twitter-2017 datasets are commonly used datasets for fine-grained image-text sentiment analysis tasks whereby, Twitter-2015 dataset contains 5,338 tweets with images, while the Twitter-2017 dataset contains 5,972 tweets with images. These datasets can be collected from English tweets on the social media platform like Twitter and are in the form of image-text pairs. Sentiment labels are categorized as positive, neutral and negative.

Audio-image-text sentiment analysis is different from image-text sentiment analysis in terms of 'Different data emphasis'. The text-image sentiment datasets are usually retrieved from social media and e-commerce platforms while visual information in video sentiment datasets is focused on the facial expressions and body movements of speakers. Videos are temporal sequences of text-image pairs, necessitating considerations of intra-modal emotional factors in audio and video frame sequences, as well as alignment among the text, video frames, and audio databases. *Video-based sentiment analysis* primarily includes sentiment classification that involves three, five, or seven-category classification tasks and emotion classification tasks comprising multi-label emotion recognition and single-label emotion recognition.

Multimodal sarcasm detection aims to identify if a given text associated with an image has sarcastic meaning. Castro et.al (2019), focused in their research on multimodal sarcasm detection task for videos and compiled a corresponding dataset from television series. Likewise, Chauhan et.al (2020) found a correlation between sentiment classification and sarcasm detection, where they proposed a multi-task framework along with recognizing sarcasm and classify sentiment polarity. Similarly, Yang et al. (2021) suggested that sarcasm can be identified by gathering emotional semantic cues across modalities, such as rolling eyes while praising someone. Cai et al. (2019) introduced a multimodal sarcasm detection task for text and images and designed a multi-level fusion network to detect sarcasm.

6. SENTIMENT ANALYSIS – BUSINESS CASE STUDIES

6.1. TESLA- Corporate Earnings Calls

Tesla Inc investors gave nightmare to the business tycoon and Chief Executive Elon Musk after he cut off analysts asking about profit potential, that led to the decline of shares by 5 per cent despite the company promises that production of the troubled Model 3 electric car was on time. Tesla's fate depends on the success of the Model 3, and the company proclaims that it has now overcome production obstacles.

Musk promised to reverse losses and provide positive cash flow within a few months. Tesla remains committed to reaching its production goal of 5,000 Model 3s per week with profitability anticipated by the end of the year. In order to achieve the goal, Musk must counteract the current $22,584 pre-tax loss per vehicle, a figure highlighted by Tesla's largest-ever quarterly loss announcement in the first quarter in the year 2018. Following the earnings report, Tesla's stock initially saw minimal movement but dipped during a conference call when Musk abruptly ended analysts' inquiries, resulting in a loss of over $2 billion in market capitalization for the company.

Earnings calls serve as a means of communication through which company executives, typically CEOs and CFOs, deliver a comprehensive overview of the company's past, present, and future performance, while also addressing inquiries from interested parties, including analysts, institutional investors, and individual investors. Frankel et al. (1999) suggest that the contents of conference calls provide additional information to the market and document the rapid growth of firms utilising earnings calls, to the extent that some 92% of companies represented by the National Investor Relations Institute (NIRI) actively run earnings calls. McKay Price et al. (2012), Doran et al. (2012) and Matsumoto et al. (2011) provide support for this statement, finding that earnings call participants are actively engaged to the extent that new and meaningful information comes to light. The extra details have been derived from the ongoing involvement and investigation of analysts and institutional investors, in addition to the valuable perspectives that managers occasionally offer beyond what is disclosed in the press release. Earnings calls follow a unique format distinct from other qualitative data exchanges in the sector, potentially leading to the discovery of fresh insights.

6.2. Reputation Management at KFC

KFC is a prime example of how AI sentiment analysis can be utilized to enhance a brand. While other restaurants have attempted to promote healthier options and positivity, KFC took a different approach. Recognizing the strength of their existing reputation, they opted for an alternative strategy instead of implementing drastic changes. Their tactic? Embracing the playful side and embarking on a joyful journey without compromising their established identity. KFC successfully integrated elements from mainstream culture, including widely recognized symbols and internet memes, to effectively market their food. This innovative approach sparked renewed enthusiasm among consumers, reigniting their interest in KFC. By employing sentiment analysis, the company was able to assess the impact of their strategies and implement necessary enhancements to further strengthen their brand.

One of the most crucial AI sentiment analysis benefits for any business is understanding the public perception of its brand. This holds great importance for tech companies, clothing brands, and virtually every type of business possible. This is where AI for sentiment analysis becomes valuable. By employing AI-based sentiment analysis, companies can closely monitor the public perception of their brand. This entails paying attention not only to the positive comments but also to any negative or even neutral feedback.

Sentiment analysis enables the identification of specific emotions expressed by customers. It assists in understanding a range of feelings, including happiness, frustration, and everything in between. This analysis provides valuable insights into customer sentiment. By analyzing a large amount of customer feedback, sentiment analysis can identify recurring patterns and trends. It's like identifying popular topics of conversation among numerous individuals. These individuals hold significant sway over public opinion. By employing sentiment analysis, businesses can identify these key figures and collaborate with them to enhance their brand's reputation and standing.

6.3. Product Analysis at Google

Google diligently pays attention to user feedback. Instead of examining individual comments, they thoroughly analyze the broader perspective. Does the public opinion on Chrome lean towards positive or negative? They assess whether the majority of people have favorable or unfavorable things to say about Chrome. Specifically, the focus lies on gathering feedback about various aspects of Chrome. This includes its performance, customizable options, security features, and visual aesthetics.

AI-based sentiment analysis expands beyond brands; it also provides valuable insights into understanding products. It can be an advanced tool, functioning like an attentive listener capturing people's opinions on various aspects of a product, its performance, unique features, and user experience. In the early stages of developing a new product, companies need to prioritize its importance. Properly testing the product with real users becomes essential to ensure its quality. This is where AI for sentiment analysis plays a significant role. It helps companies measure user feedback and make necessary improvements.

AI sentiment analysis does the job of Collecting Feedback that involves gathering all the comments and opinions expressed by people regarding the product. Subsequently, it sorts the Feedback that help us see what people like and dislike about the application. Next, This tool determines the key aspects of feedback, such as highlighting discussions on product speed or user-friendliness. Finally, the last step is learning, that involves analyzing feedback and opinions about similar products offered by other companies to enhance their product. Such insights help them make improvements that align with customer needs and preferences while ensuring a better overall offering.

6.4. Customer Analysis at Tripadvisor

The evolution of the OTA (Online Travel Agent) platform has transformed it into a significant E-WOM (Electronic Word of Mouth) channel, beyond its primary role as a ticket booking platform. This shift has prompted stakeholders in the hotel industry to leverage E-WOM for sustaining their businesses. The majority of hotels now recognize the value of E-WOM, given that 90 percent of bookings are initiated through OTA websites.

TripAdvisor, a platform dedicated to travel reviews, excels in utilizing AI based sentiment analysis to process the numbers of feedback received from travelers effectively. This method allows them to identify essential insights within customer reviews and enhance their services accordingly.

To ensure the success of any business, it becomes crucial to have a thorough understanding of customers' preferences and needs. It's like aiming at a target, one must precisely identify its location and utilize the appropriate tools for hitting it accurately. However, there is a challenging aspect to consider: the thoughts and opinions of customers are spread across different platforms, making them difficult to consolidate. Sometimes, their perspectives on your product can be quite diverse. This is where customer sentiment analysis becomes valuable. It helps in various aspects, for example, at Tripadvisor we want to know the desires of diverse groups of customers. AI-based sentiment analysis helps in examining various customer segments and determining their needs, expectations for similar products, and preferred choices. Moreover, the reviews of hotels at Tripadvisor are essential. The sentiment analysis evaluates direct and indirect comments about the hotels listed and provides feedback through test analysis to the new users.

6.5. Customer Support-Feedback Analysis at T-Mobile

T-Mobile employed sentiment analysis with AI to examine customer feedback and identify prevalent issues. By utilizing generative AI sentiment analysis, T-Mobile swiftly detected areas of concern and promptly addressed them. This proactive approach resulted in an impressive 73% reduction in customer complaints. T-Mobile utilizes advanced AI for sentiment analysis to carefully analyze customer feedback and identify emerging trends regarding vehicle performance and overall customer satisfaction. This real-time analysis enables the company to promptly pinpoint any issues or areas for improvement in their vehicles and services, ensuring continuous enhancements are made.

Sentiment analysis helps businesses measure customers' perceptions of their products or services. This analysis discloses the overall impression customers have, whether positive or negative. Additionally, it provides deep analysis of specific product elements and can even evaluate service effectiveness. Detailed analyses may include aspects such as response time or the quality of interactions. Sentiment Analysis helps in automation with Intent Analysis. It can automate routine queries. These include frequently asked questions or requests for information about product usage. By automating these processes, customer support becomes more streamlined, leading to quick and efficient responses. It involves the practice of prioritizing customers based on their sentiments. This approach ensures that dissatisfied customers are promptly attended to. For example, if a customer expresses dissatisfaction, their ticket can be given priority for faster resolution.

6.6. Real-Time Customer Insights at Airbnb

Airbnb, a leading player in the global accommodation industry, utilizes generative AI sentiment analysis to closely monitor interactions between guests and hosts. This proactive approach involves analyzing the sentiments expressed within guest reviews and messages exchanged between users. By doing so, Airbnb can swiftly identify emerging issues and concerns in real-time. For example, when multiple guests express dissatisfaction with a particular property's cleanliness or report safety concerns through messages, Airbnb's sentiment analysis algorithms quickly identify these sentiments. This early detection enables Airbnb to take immediate action by communicating with the host to address cleanliness issues or verifying safety protocols.

AI-based sentiment analysis revolutionizes the way businesses gather real-time customer insights. The sentiment analysis in social media achieves this by actively monitoring social media platforms, review sites, and various online channels. This capability empowers companies to promptly obtain feedback and stay updated on the dynamic customer sentiments.

6.7. Product Development Optimization at Amazon

AI-based sentiment analysis has revolutionized product development, providing businesses with invaluable insights into customer preferences and pain points. It serves as a guiding tool in the complexity of optimizing offerings by analyzing customer feedback and reviews. This enables companies to gain a deeper understanding of what truly resonates with their audience and identify areas that need improvement.

Amazon, the renowned global e-commerce giant, employs sentiment analysis to meticulously examine a vast collection of customer reviews. By leveraging this advanced analysis, Amazon gains invaluable insights that directly influence its product development strategies. For instance, when customers consistently praise a specific feature like exceptional durability or user-friendliness, sentiment analysis with AI brings this pattern to light. In order to put customer insights into action, Amazon divides the process into several actionable steps. They prioritize enhancing products that align with customer preferences, ensuring that future iterations maintain or amplify these strengths. Additionally, Amazon strategically highlights these exceptional features in their marketing campaigns to forge a deeper connection with their target audience. Amazon's use of AI for sentiment analysis showcases how businesses can optimize product development by actively listening to customer feedback. This boosts satisfaction and also ensures the delivery of products that truly meet market demands.

6.8. Personalized Marketing Campaigns at Spotify

Spotify is a widely used music streaming service that utilizes AI enabled sentiment analysis to curate personalized playlists and recommendations for its users. By examining user sentiments towards various music genres, artists, and songs, Spotify's recommendation engine can construct playlists that cater to each user's unique preferences. This high level of personalization keeps users engaged and content with the platform.

AI-based sentiment analysis is essential in helping businesses develop highly personalized marketing campaigns. By understanding the sentiments of individual customers or customer segments, companies can customize their marketing messages and content to resonate deeply with their audience on a personal level.

6.9. Risk Mitigation at UnitedHealth Group

Happy and satisfied employees have a higher likelihood of delivering exceptional customer experiences. Businesses can internally utilize sentiment analysis to assess employee satisfaction and pinpoint areas for workplace improvement. This, in turn, encourages a more motivated and productive workforce.

UnitedHealth Group, is a renowned company using sentiment analysis in healthcare. It employs AI enabled sentiment analysis to gauge employee sentiments through feedback surveys. This enables them to gain valuable insights into employee perspectives on workplace conditions, management, and benefits. Identification of areas where employees express dissatisfaction empowers UnitedHealth Group to proactively address these concerns and promote a happier and more engaged workforce. Ultimately, this positively impacts the quality of care provided to customers.

7. DISCUSSION AND CONCLUSION

Research on SA methodologies has been ongoing in the past decade and is currently being utilized by businesses as a crucial instrument for strategic marketing strategies and decision-making. This shift is also attributed to the progress in data storage, accessibility, and analysis made possible by big data frameworks. Sentiment analysis enables businesses to gain a deeper understanding of their customers, enhance brand reputation, and optimize products/services. It offers real-time insights, identifies growing trends, and facilitates data-driven decision-making. SA integrates multimodal approaches that enables the organizations to explore additional sources of information such as images, audio, and video. These additional sources can greatly assist in sentiment analysis. Multimodal sentiment analysis is a computational technique that gathers information from text and images, to understand and analyze emotional states of users.

One of the most significant ML approaches is Deep Learning which is an AI concept based on the Artificial Neural Networks (ANN) similar to neural pathways of the human brain. ANN works on the principle of networks with different layers namely, an input layer that receives the data, an output layer that yields the result and zero or more hidden layers responsible for learning the mapping between input and output. There are several kinds of text being analyzed in present times, for example, debates, blogs and online discussions on politics. For example, tweets about political parties find out how well tweets align with the political parties' contested views and promises and such research on political arena is focused around the sentiment analysis using texts from Twitter and online media as a source.

SA is an important aspect of NLP and helps the organizations to gain valuable insights from public opinions and customer feedback, facilitating data-driven decision-making, product enhancement, and effective marketing strategies for competitive advantage. SA helps in building social and political perceptions helping the researchers and policymakers to understand public sentiments on current affairs which leads to better decision-making in a technology driven business environment. Sentiment analysis has several applications across domains and verticals in an organization which were discussed through case studies in the current chapter. These case studies help in gaining insights into sentiment and opinions of people associated with the organizations. For example, in the domain of sales and marketing whereby, customer feedback can be utilized to improvise products and services, social media monitoring can help in company and product branding, marketing research can assess consumer feedback etc. The application in the field of Finance is quite relevant in turbulent times whereby investment strategies can be built and sentiments can be analyzed for stock market predictions.

REFERENCES

Abdullah, T., & Ahmet, A. (2022). Deep learning in sentiment analysis: Recent architectures. *ACM Computing Surveys*, *55*(8), 1–37. DOI: 10.1145/3548772

Abirami, T., & Gayathri, R. (2016). A Survey on Efficient Power allocation for OFDM–Based Cognitive Radio Systems. *on Journal of Chemical and Pharmaceutical Sciences (ICPS)*, *8*, 83-87.

Agarwal, S., Kumar, S., & Goel, U. (2021). Social media and the stock markets: An emerging market perspective. *Journal of Business Economics and Management*, *22*(6), 1614–1632. DOI: 10.3846/jbem.2021.15619

Ahmed, A. A. A., Agarwal, S., Kurniawan, I. G. A., Anantadjaya, S. P., & Krishnan, C. (2022). Business boosting through sentiment analysis using Artificial Intelligence approach. *International Journal of System Assurance Engineering and Management*, *13*(S1, Suppl 1), 699–709. DOI: 10.1007/s13198-021-01594-x

Alamoudi, E., & Alghamdi, N. (2021). Sentiment classification and aspect-based sentiment analysis on yelp reviews using deep learning and word embeddings. *Journal of Decision Systems*, *30*(2–3), 259–281. DOI: 10.1080/12460125.2020.1864106

Alslaity, A., & Orji, R. (2024). Machine learning techniques for emotion detection and sentiment analysis: Current state, challenges, and future directions. *Behaviour & Information Technology*, *43*(1), 139–164. DOI: 10.1080/0144929X.2022.2156387

Anjaria, M., & Guddeti, R. M. R. (2014). A novel sentiment analysis of social networks using supervised learning. *Social Network Analysis and Mining*, *4*(1), 1–15. DOI: 10.1007/s13278-014-0181-9

Araújo, M., Pereira, A., & Benevenuto, F. (2020). A comparative study of machine translation for multilingual sentence-level sentiment analysis. *Information Sciences*, *512*, 1078–1102. DOI: 10.1016/j.ins.2019.10.031

Bakliwal, A., Foster, J., van der Puil, J., O'Brien, R., Tounsi, L., & Hughes, M. (2013, June). Sentiment analysis of political tweets: Towards an accurate classifier. Association for Computational Linguistics.

Balaji, P., Nagaraju, O., & Haritha, D. (2017, March). Levels of sentiment analysis and its challenges: A literature review. In *2017 International Conference on Big Data Analytics and Computational Intelligence (ICBDAC)* (pp. 436-439). IEEE. DOI: 10.1109/ICBDACI.2017.8070879

Balaji, V., Acharjee, P. B., Elangovan, M., Kalnoor, G., Rastogi, R., & Patidar, V. (2023). Developing a semantic framework for categorizing IoT agriculture sensor data: A machine learning and web semantics approach. *The Scientific Temper*, *14*(4), 1332–1338. DOI: 10.58414/SCIENTIFICTEMPER.2023.14.4.40

Barba, P. (2020). *Machine Learning (ML) for Natural Language Processing (NLP)*. Lexalytics.

Bhaskar, J., Sruthi, K., & Nedungadi, P. (2014). Enhanced sentiment analysis of informal textual communication in social media by considering objective words and intensifiers. In *International Conference on Recent Advances and Innovations in Engineering (ICRAIE-2014)* (pp. 1–6). IEEE. DOI: 10.1109/ICRAIE.2014.6909220

Bhonde, R., Bhagwat, B., Ingulkar, S., & Pande, A. (2015). Sentiment analysis based on dictionary approach. *International Journal of Emerging Engineering Research and Technology*, *3*(1), 51–55.

Bilgin, M., & Şentürk, İ. F. 2017. Sentiment Analysis on Twitter Data with Semi-Supervised Doc2Vec. 2nd International Conference on Computer Science and Engineering, UBMK 2017: 661–666. . 2017.8093492. DOI: 10.1109/UBMK.2017.8093492

Boukes, M., Van de Velde, B., Araujo, T., & Vliegenthart, R. (2020). What's the tone? Easy doesn't do it: Analyzing performance and agreement between off-the-shelf sentiment analysis tools. *Communication Methods and Measures*, *14*(2), 83–104. DOI: 10.1080/19312458.2019.1671966

Brown, T., Mann, B., Ryder, N., Subbiah, M., Kaplan, J. D., Dhariwal, P., Neelakantan, A., Shyam, P., Sastry, G., Askell, A., & Agarwal, S. (2020). *Language models are few-shot learners*. Advances in Neural Information Processing.

Cai, Y., Cai, H., & Wan, X. (2019, July). Multi-modal sarcasm detection in twitter with hierarchical fusion model. In *Proceedings of the 57th annual meeting of the association for computational linguistics* (pp. 2506-2515). DOI: 10.18653/v1/P19-1239

Cambria, E., Das, D., Bandyopadhyay, S., & Feraco, A. (2017). Affective computing and sentiment analysis. A practical guide to sentiment analysis, 1-10.

Castro, S., Hazarika, D., Pérez-Rosas, V., Zimmermann, R., Mihalcea, R., & Poria, S. (2019). Towards multimodal sarcasm detection (an _obviously_ perfect paper). *arXiv preprint arXiv:1906.01815*. DOI: 10.18653/v1/P19-1455

Chan, S. W., & Chong, M. W. (2017). Sentiment analysis in financial texts. *Decision Support Systems*, *94*, 53–64. DOI: 10.1016/j.dss.2016.10.006

Chauhan, D. S., Dhanush, S. R., Ekbal, A., & Bhattacharyya, P. (2020, July). Sentiment and emotion help sarcasm? A multi-task learning framework for multi-modal sarcasm, sentiment and emotion analysis. In *Proceedings of the 58th annual meeting of the association for computational linguistics* (pp. 4351-4360). DOI: 10.18653/v1/2020.acl-main.401

Chu, E., & Roy, D. (2017, November). Audio-visual sentiment analysis for learning emotional arcs in movies. In *2017 IEEE International Conference on Data Mining (ICDM)* (pp. 829-834). IEEE. DOI: 10.1109/ICDM.2017.100

Chung, J., Gulcehre, C., Cho, K., & Bengio, Y. (2014). Empirical evaluation of gated recurrent neural networks on sequence modelling. In *Deep learning and representation learning workshop*. Neural Information Processing Systems.

Dair, Z., Donovan, R., & O'Reilly, R. (2021). Classification of emotive expression using verbal and non-verbal components of speech. In 2021 32nd Irish Signals and Systems Conference (ISSC) (pp. 1–8). IEEE. DOI: 10.1109/ISSC52156.2021.9467869

Das, R., & Singh, T. D. (2023). Multimodal sentiment analysis: A survey of methods, trends, and challenges. *ACM Computing Surveys*, *55*(13s), 1–38. DOI: 10.1145/3586075

Devika, M. D., Sunitha, C., & Ganesh, A. (2016). Sentiment analysis: A comparative study on different approaches. *Procedia Computer Science, 87*, 44–49. DOI: 10.1016/j.procs.2016.05.124

Devlin, J., Chang, M. W., Lee, K., & Toutanova, K. (2019). Bert: Pre-training of deep bidirectional Transformers for language understanding. arXiv preprint arXiv:1810.04805.

Dey, L., Chakraborty, S., Biswas, A., Bose, B., & Tiwari, S. (2016). Sentiment analysis of review datasets using naive bayes and k-nn classifier. *arXiv preprint arXiv:1610.09982.*

Diesner, J., & Evans, C. S. (2015, August). Little bad concerns: Using sentiment analysis to assess structural balance in communication networks. In Proceedings of the 2015 IEEE/ACM International Conference on Advances in Social Networks Analysis and Mining 2015 (pp. 342-348).

Doran, J., Peterson, D., & Price, S. (2012). Earnings conference call content and stock price: The case of REITs. *The Journal of Real Estate Finance and Economics, 45*(2), 402–434. DOI: 10.1007/s11146-010-9266-z

Durant, K. T., & Smith, M. D. (2006). Mining sentiment classification from political web logs. In *Proceedings of Workshop on Web Mining and Web Usage Analysis of the 12th ACM SIGKDD International Conference on Knowledge Discovery and Data Mining (WebKDD-2006),* Philadelphia, PA (Vol. 16).

El-Haj, M., Rayson, P., Walker, M., Young, S., & Simaki, V. (2019). In search of meaning: Lessons, resources and next steps for computational analysis of financial discourse. *Journal of Business Finance & Accounting, 46*(3–4), 265–306. DOI: 10.1111/jbfa.12378

El Naqa, I., & Murphy, M. J. (2015). *What is machine learning?* Springer International Publishing.

Frankel, R., Johnson, M., & Skinner, D. J. (1999). An empirical examination of conference calls as a voluntary disclosure medium. *Journal of Accounting Research, 37*(1), 133–150. DOI: 10.2307/2491400

Gindl, S., Weichselbraun, A., & Scharl, A. (2010). Cross-domain contextualization of sentiment lexicons. In *ECAI 2010* (pp. 771–776). IOS Press.

González-Bailón, S., & Paltoglou, G. (2015). Signals of public opinion in online communication: A comparison of methods and data sources. *The Annals of the American Academy of Political and Social Science, 659*(1), 95–107. DOI: 10.1177/0002716215569192

Goodfellow, I., Bengio, Y., & Aaron, C. (2016). *Deep Learning*. MIT Press.

Grimmer, J., & Stewart, B. M. (2013). Text as data: The promise and pitfalls of automatic content analysis methods for political texts. *Political Analysis, 21*(3), 267–297. DOI: 10.1093/pan/mps028

Guo, J., Guo, Y., Kong, X., Zhang, M., & He, R. (2016, February). Discriminative analysis dictionary learning. In Proceedings of the AAAI conference on artificial intelligence (Vol. 30, No. 1). discussion sessions. *The Accounting Review, 86*(4), 1383–1414. DOI: 10.2308/accr-10034

Han, Z., Wu, J., Huang, C., Huang, Q., & Zhao, M. (2020). A Review on Sentiment Discovery and Analysis of Educational Big-Data. *Wiley Interdisciplinary Reviews. Data Mining and Knowledge Discovery, 10*(1), e1328. DOI: 10.1002/widm.1328

Hasselgren, B., Chrysoulas, C., Pitropakis, N., & Buchanan, W. J. (2022). Using social media & sentiment analysis to make investment decisions. *Future Internet*, *15*(1), 5. DOI: 10.3390/fi15010005

Hochreiter, S., & Schmidhuber, J. (1997). Long short-term memory. *Neural Computation*, *9*(8), 1735–1780. DOI: 10.1162/neco.1997.9.8.1735 PMID: 9377276

Houjeij, A., Hamieh, L., Mehdi, N., & Hajj, H. (2012). A novel approach for emotion classification based on fusion of text and speech. In 2012 19th International Conference on Telecommunications (ICT) (pp. 1–6). IEEE. DOI: 10.1109/ICTEL.2012.6221211

Huang, H., Asemi, A., & Mustafa, M. B. (2023). Sentiment analysis in e-commerce platforms: A review of current techniques and future directions. *IEEE Access : Practical Innovations, Open Solutions*, *11*, 90367–90382. DOI: 10.1109/ACCESS.2023.3307308

Jain, K., & Kaushal, S. (2018). A comparative study of machine learning and deep learning techniques for sentiment analysis. In *2018 7th International conference on reliability, infocom technologies and optimization (Trends and Future Directions)(ICRITO)* (pp. 483-487). IEEE. DOI: 10.1109/ICRITO.2018.8748793

Jain, P. K., Pamula, R., & Srivastava, G. (2021). A Systematic Literature Review on Machine Learning Applications for Consumer Sentiment Analysis Using Online Reviews. *Computer Science Review*, *41*, 100413. Advance online publication. DOI: 10.1016/j.cosrev.2021.100413

Janiesch, C., Zschech, P., & Heinrich, K. (2021). Machine Learning and Deep Learning. *Electronic Markets*, *31*(3), 685–695. DOI: 10.1007/s12525-021-00475-2

Jim, J. R., Talukder, M. A. R., Malakar, P., Kabir, M. M., Nur, K., & Mridha, M. F. (2024). Recent advancements and challenges of NLP-based sentiment analysis: A state-of-the-art review. Natural Language Processing Journal, 100059.

Jozefowicz, R., Vinyals, O., Schuster, M., Shazeer, N., & Wu, Y. (2016). Exploring the limits of language modeling. arXiv preprint arXiv:1602.02410.

Kaur, J., & Saini, J. R. (2014). Emotion detection and sentiment analysis in text corpus: a differential study with informal and formal writing styles. *International Journal of Computer Application, ISSN*, 0975-8887.

Khan, S., Naseer, M., Hayat, M., Zamir, S. W., Khan, F. S., & Shah, M. (2022). Transformers in vision: A survey. *ACM Computing Surveys*, *54*(10s), 1–41. DOI: 10.1145/3505244

Li, H., Chen, Q., Zhong, Z., Gong, R., & Han, G. (2022). E-word of mouth sentiment analysis for user behavior studies. *Information Processing & Management*, *59*(1), 102784. DOI: 10.1016/j.ipm.2021.102784 PMID: 34580562

Li, S., Wang, G., & Luo, Y. (2022). Tone of language, financial disclosure, and earnings management: A textual analysis of form 20-F. *Financial Innovation*, *8*(1), 43. DOI: 10.1186/s40854-022-00346-5

Li, X., Zhang, J., Du, Y., Zhu, J., Fan, Y., & Chen, X. (2023). A novel deep learning-based sentiment analysis method enhanced with emojis in microblog social networks. *Enterprise Information Systems*, *17*(5), 2037160. DOI: 10.1080/17517575.2022.2037160

Loughran, T., & McDonald, B. (2020). Textual analysis in finance. *Annual Review of Financial Economics, 12*(1), 357–375. DOI: 10.1146/annurev-financial-012820-032249

Luong, M. T., Pham, H., & Manning, C. D. (2015). Effective approaches to attention-based neural machine translation. arXiv preprint arXiv:1508.04025. DOI: 10.18653/v1/D15-1166

Maqsood, H., Maqsood, M., Yasmin, S., Mehmood, I., Moon, J., & Rho, S. (2022). Analyzing the stock exchange markets of EU nations: A case study of brexit social media sentiment. *Systems, 10*(2), 24. DOI: 10.3390/systems10020024

Matsumoto, D., Pronk, M., & Roelofsen, E. (2011). What makes conference calls useful? The information content of managers' presentations and analysts' discussion sessions. *The Accounting Review, 86*(4), 1383–1414. DOI: 10.2308/accr-10034

McGurk, K. A., Dagliati, A., Chiasserini, D., Lee, D., Plant, D., Baricevic-Jones, I., Kelsall, J., Eineman, R., Reed, R., Geary, B., Unwin, R. D., Nicolaou, A., Keavney, B. D., Barton, A., Whetton, A. D., & Geifman, N. (2020). The use of missing values in proteomic data-independent acquisition mass spectrometry to enable disease activity discrimination. *Bioinformatics (Oxford, England), 36*(7), 2217–2223. DOI: 10.1093/bioinformatics/btz898 PMID: 31790148

McKay Price, S., Doran, J., Peterson, D., & Bliss, B. (2012). Earnings conference calls and stock returns: The incremental informativeness of textual tone. *Journal of Banking & Finance, 36*(4), 992–1011. DOI: 10.1016/j.jbankfin.2011.10.013

Memiş, E., Akarkamçı, H., Yeniad, M., Rahebi, J., & Lopez-Guede, J. M. (2024). Comparative Study for Sentiment Analysis of Financial Tweets with Deep Learning Methods. *Applied Sciences (Basel, Switzerland), 14*(2), 588. DOI: 10.3390/app14020588

Mendoza-Urdiales, R. A., Núñez-Mora, J. A., Santillán-Salgado, R. J., & Valencia-Herrera, H. (2022). Twitter sentiment analysis and influence on stock performance using transfer entropy and EGARCH methods. *Entropy (Basel, Switzerland), 24*(7), 874. DOI: 10.3390/e24070874 PMID: 35885097

Mishra, S., Diesner, J., Byrne, J., & Surbeck, E. (2015, August). Sentiment analysis with incremental human-in-the-loop learning and lexical resource customization. In *proceedings of the 26th ACM conference on Hypertext & social media* (pp. 323-325). DOI: 10.1145/2700171.2791022

Munezero, M., Montero, C. S., Sutinen, E., & Pajunen, J. (2014). Are they different? Affect, feeling, emotion, sentiment, and opinion detection in text. *IEEE Transactions on Affective Computing, 5*(2), 101–111. DOI: 10.1109/TAFFC.2014.2317187

Munikar, M., Shakya, S., & Shrestha, A. (2019). *Fine-grained sentiment classification using BERT. In 2019 Artificial Intelligence for Transforming Business and Society (AITB)* (Vol. 1). IEEE.

Nardo, M., Petracco-Giudici, M., & Naltsidis, M. (2016). Walking down wall street with a tablet: A survey of stock market predictions using the web. *Journal of Economic Surveys, 30*(2), 356–369. DOI: 10.1111/joes.12102

Peng, S., Cao, L., Zhou, Y., Ouyang, Z., Yang, A., Li, X., Jia, W., & Yu, S. (2022). A survey on deep learning for textual emotion analysis in social networks. *Digital Communications and Networks*, *8*(5), 745–762. DOI: 10.1016/j.dcan.2021.10.003

Poria, S., Cambria, E., Howard, N., Huang, G. B., & Hussain, A. (2016). Fusing audio, visual and textual clues for sentiment analysis from multimodal content. *Neurocomputing*, *174*, 50–59. DOI: 10.1016/j.neucom.2015.01.095

Raffel, C., Shazeer, N., Roberts, A., Lee, K., Narang, S., Matena, M., Zhou, Y., Li, W., & Liu, P. J. (2020). Exploring the limits of transfer learning with a unified text-to-text transformer. *Journal of Machine Learning Research*, *21*(1), 5485–5551.

Renault, T. (2017). Intraday online investor sentiment and return patterns in the US stock market. *Journal of Banking & Finance*, *84*, 25–40. DOI: 10.1016/j.jbankfin.2017.07.002

Revathy, G., Alghamdi, S. A., Alahmari, S. M., Yonbawi, S. R., Kumar, A., & Haq, M. A. (2022). Sentiment analysis using machine learning: Progress in the machine intelligence for data science. *Sustainable Energy Technologies and Assessments*, *53*, 102557. DOI: 10.1016/j.seta.2022.102557

Ribeiro, M. T., Singh, S., & Guestrin, C. (2016, August). " Why should i trust you?" Explaining the predictions of any classifier. In *Proceedings of the 22nd ACM SIGKDD international conference on knowledge discovery and data mining* (pp. 1135-1144). DOI: 10.18653/v1/N16-3020

Rohini, V., Thomas, M., & Latha, C. A. (2016, May). Domain based sentiment analysis in regional Language-Kannada using machine learning algorithm. In *2016 IEEE International Conference on Recent Trends in Electronics, Information & Communication Technology (RTEICT)* (pp. 503-507). IEEE.

Rudin, C. (2019). Stop explaining black box machine learning models for high stakes decisions and use interpretable models instead. *Nature Machine Intelligence*, *1*(5), 206–215.

Russell, S., & Norvig, P. (2020). *Artificial intelligence: a modern approach.*

Sánchez-Rada, J. F., & Iglesias, C. A. (2019). Social context in sentiment analysis: Formal definition, overview of current trends and framework for comparison. *Information Fusion*, *52*, 344–356. DOI: 10.1016/j.inffus.2019.05.003

Sarkar, P., & Etemad, A. (2020). Self-supervised ECG representation learning for emotion recognition. *IEEE Transactions on Affective Computing*, *13*(3), 1541–1554. DOI: 10.1109/TAFFC.2020.3014842

Sun, C., Qiu, X., Xu, Y., & Huang, X. (2019). How to fine-tune BERT for text classification? In Chinese Computational Linguistics: 18th China National Conference, CCL 2019, Kunming, China, October 18–20, 2019 [Springer International Publishing.]. *Proceedings*, *18*, 194–206.

Syahnur, K. N. F., Surianto, D. F., & Dharsana, M. T. (2023). Analysis Of Retail Sector Market Reaction In Indonesia On Social Media And Investor Sentiment. *Journal Manajemen Bisnis*, *10*(2), 371–383. DOI: 10.33096/jmb.v10i2.72

Troussas, C., Virvou, M., Espinosa, K. J., Llaguno, K., & Caro, J. (2013, July). Sentiment analysis of Facebook statuses using Naive Bayes classifier for language learning. In *IISA 2013* (pp. 1–6). IEEE. DOI: 10.1109/IISA.2013.6623713

Tumasjan, A., Sprenger, T., Sandner, P., & Welpe, I. (2010, May). Predicting elections with twitter: What 140 characters reveal about political sentiment. In *Proceedings of the international AAAI conference on web and social media* (Vol. 4, No. 1, pp. 178-185). DOI: 10.1609/icwsm.v4i1.14009

Vaswani, A., Shazeer, N., Parmar, N., Uszkoreit, J., Jones, L., Gomez, A. N., Kaiser, L., & Polosukhin, I. (2017). Attention Is All You Need. *Advances in Neural Information Processing Systems*, •••, 30.

Wang, L., Ma, C., Feng, X., Zhang, Z., Yang, H., Zhang, J., Chen, Z., Tang, J., Chen, X., Lin, Y., Zhao, W. X., Wei, Z., & Wen, J. (2024). A survey on large language model based autonomous agents. *Frontiers of Computer Science*, *18*(6), 186345. DOI: 10.1007/s11704-024-40231-1

Wu, Y., Schuster, M., Chen, Z., Le, Q. V., Norouzi, M., Macherey, W., Krikun, M., Cao, Y., Gao, Q., Macherey, K., & Klingner, J. (2016). Google's neural machine translation system: Bridging the gap between human and machine translation. arXiv preprint arXiv:1609.08144.

Yadav, N. B. (2023). Harnessing customer feedback for product recommendations: An aspect-level sentiment analysis framework. *Human-Centric Intelligent Systems*, *3*(2), 57–67. DOI: 10.1007/s44230-023-00018-2

Yang, K., Xu, H., & Gao, K. (2020). CM-BERT: Cross-modal bert for text-audio sentiment analysis. In *Proceedings of the 28th ACM International Conference on Multimedia* (pp. 521–528). DOI: 10.1145/3394171.3413690

Yang, K., Xu, H., & Gao, K. (2020). CM-BERT: Cross-modal bert for text-audio sentiment analysis. In *Proceedings of the 28th ACM International Conference on Multimedia* (pp. 521–528). DOI: 10.1145/3394171.3413690

Yiran, Y., & Srivastava, S. (2019, June). Aspect-based Sentiment Analysis on mobile phone reviews with LDA. In *Proceedings of the 2019 4th International Conference on Machine Learning Technologies* (pp. 101-105). DOI: 10.1145/3340997.3341012

Zhang, Y., & Ling, C. (2018). A strategy to apply machine learning to small datasets in materials science. *npj Computational Materials*, *4*(1), 25.

Chapter 4
Evaluating NLP Models for Mental Health:
Performance and Accuracy in User–Psychologist Conversations

Vishnu Achutha Menon
https://orcid.org/0000-0003-4028-3685
Central University of Tamil Nadu, India

T. K. Sateesh Kumar
https://orcid.org/0000-0002-8406-409X
Kristu Jayanti College, India

Lijo P. Thomas
Kristu Jayanti College, India

Juby Thomas
https://orcid.org/0000-0003-2167-0620
Kristu Jayanti College, India

ABSTRACT

This chapter examines the effectiveness of NLP models in analyzing mental health conversations between users and psychologists. It aims to evaluate classification performance, overall accuracy, agreement, the influence of contextual factors on sentiment responses, and the reliability of model coefficients. The hypothesis suggests near-perfect performance across metrics such as sensitivity, specificity, and balanced accuracy, with significant coefficients for predictors. Using a secondary dataset with negative, neutral, and positive sentiment classes, the methodology includes visualizing classification performance, calculating various accuracy metrics, and fitting a multinomial logistic regression model to predict sentiment responses. The model demonstrates perfect scores in classifying sentiments, a significant improvement over baseline, and strong agreement beyond chance, with robust fitting indicating high precision and reliability in predicting sentiment outcomes.

DOI: 10.4018/979-8-3693-6910-4.ch004

INTRODUCTION

This chapter is dedicated to thoroughly examining the effectiveness of Natural Language Processing (NLP) models in analyzing and responding to conversations related to mental health, specifically those between users and professional psychologists. This chapter aims to achieve several key objectives: evaluating the classification performance of NLP models, assessing their overall accuracy and agreement, exploring the influence of contextual factors on sentiment responses, and evaluating the reliability of the model's coefficient estimates. The hypothesis posits that the NLP models will exhibit near-perfect performance across various metrics, including sensitivity, specificity, and balanced accuracy. Additionally, it is expected that the models will produce significant coefficients for predictors, reflecting the impact of sentiment context levels on responses. Through these investigations, the chapter seeks to contribute to the advancement of NLP models in providing reliable and effective mental health support.

Objectives

1. Evaluate the model's classification performance.
2. Assess the model's overall accuracy and agreement.
3. Examine the predictive influence of contextual factors on sentiment responses.
4. Evaluate the reliability of the model's coefficient estimates.

REVIEW OF LITERATURE

The standard errors for the intercept and predictors will be low, indicating high precision and reliability (Khatri et al., 2021). This means that the estimates of the model parameters are close to the true values, reflecting consistent and dependable results. In a clinical study assessing the impact of a new drug, low standard errors suggest that the observed effects are accurate and can be confidently attributed to the drug rather than random variation. Operational criteria were irrelevant to clinical opinion (Rocha Neto et al., 2022). This indicates a disconnect between the practical definitions used in operational settings and the judgment of clinical experts. For instance, in mental health diagnostics, an operational criterion might classify a patient based on a checklist of symptoms, while clinicians might rely more on their nuanced understanding and experience with patient behavior, leading to different conclusions. The Acceptability of Intervention Measure, Intervention Appropriateness Measure, and Feasibility of Intervention Measure demonstrate promising psychometric properties (Weiner et al., 2017). These measures are tools used to evaluate how well interventions are received and implemented. Promising psychometric properties mean that these tools reliably measure what they are supposed to. For instance, in evaluating a new therapy technique, these measures would provide consistent and valid results across different settings and populations, ensuring the technique's acceptability, appropriateness, and feasibility are accurately assessed. The PHQ was 4.8 for any mental disorder (Waza et al., 2005). The Patient Health Questionnaire (PHQ) score of 4.8 indicates the average severity of mental health symptoms among the study population. This score helps clinicians quickly gauge the overall mental health burden. In a community health survey, a PHQ score of 4.8 would suggest a moderate level of mental health issues, guiding public health interventions accordingly. The kappa coefficient is considered the statistic of choice to analyze the reliability of nominal and categorical types of data recorded on the same patient by more than one clinician (Cyr & Francis, 1992). The kappa coefficient measures the agreement between two or more raters, adjusting for the agreement that could happen by chance. If two doctors independently diagnose a group of patients, the kappa coefficient would quantify how much their diagnoses agree, beyond what would be expected randomly, thus indicating the reliability of the diagnostic process.

Peer raters of the same level of training can provide accurate ratings of complex clinical tasks (Kulasegaram & Paradis, 2014). This means that clinicians with similar training and experience can reliably assess intricate medical procedures and decisions. In a surgical residency program, residents at the same stage of training can accurately evaluate each other's performance in simulated surgical tasks, ensuring consistent and fair assessments. The inter-tester reliability of the system is moderate to substantial for a range of patients within the non-specific low back pain population, aligning with previous research (Vibe Fersum et al., 2009). This indicates that different testers can achieve a reasonable level of agreement when assessing patients with non-specific low back pain. Physical therapists using a standardized assessment tool may produce consistent results when evaluating the severity and impact of back pain, thus ensuring reliable patient care. Shrewdness may influence ratings of CBTI accuracy in an experimental context (Guastello et al., 1992). This suggests that the rater's perceptiveness or cleverness can affect how accurately they rate cognitive-behavioral therapy interventions. In a study evaluating the effectiveness of CBT for insomnia, more astute raters might be better at discerning subtle improvements in patients, leading to more precise ratings of the therapy's impact. Feedback improved the performance of clinicians in three experiments (Sechrest et al., 1967). Providing clinicians with constructive feedback significantly enhances their skills and accuracy in practice. In training sessions for diagnostic procedures, clinicians who received specific feedback on their techniques showed marked improvement in their diagnostic accuracy and overall performance, highlighting the value of feedback in medical education.

The neural approaches can handle this task reasonably well (Kwak et al., 2022). This suggests that neural network models are capable of performing the task with a good level of accuracy. In image recognition tasks, neural networks can effectively distinguish between different objects, demonstrating their ability to process complex data patterns. The sensitivity for the diagnosis of Alzheimer's Disease (AD) was high (Lopez et al., 1999). This means that the diagnostic method used was able to correctly identify a large proportion of patients who have AD. A screening tool with high sensitivity will catch most cases of Alzheimer's, ensuring that fewer patients with the disease are missed during diagnosis. The proposed method enhanced classification performance by 22.95% on Mental Health private data compared to a baseline artificial neural network (ANN) machine learning algorithm (Rashidiani et al., 2022). This indicates that the new method significantly improved the accuracy of classifying mental health conditions. In a mental health app, this method could better distinguish between different disorders, leading to more accurate and personalized treatment recommendations. The cutoff score approach appears to be a better option than a PHQ-9 algorithm for detecting major depression (He et al., 2020). This suggests that using a specific cutoff score to determine depression provides more accurate results than the algorithm-based approach of the PHQ-9. Setting a threshold score that patients must reach to be diagnosed with depression could lead to fewer false positives and more accurate diagnoses. Natural language processing (NLP) of clinical mental health notes offers additional predictive value over and above currently used suicide prediction models (Levis et al., 2021). This means that analyzing the text of clinical notes using NLP techniques can improve the prediction of suicide risk. Subtle language patterns and nuances in clinicians' notes could reveal underlying risks that standard prediction models might miss, leading to better-targeted interventions. RoB-PrevMH is a brief, user-friendly, and adaptable tool for assessing risk of bias in studies on the prevalence of mental health disorders (Tonia et al., 2023). This tool helps researchers quickly and effectively evaluate the potential biases in studies, ensuring the reliability of their findings. For instance, when reviewing multiple studies on the prevalence of anxiety disorders, using RoB-PrevMH can highlight which studies have methodological flaws that might skew their results. Performance estimated using one trial is drastically higher than the performance obtained when the same model is tested on independent data from other clinical trials with similar protocols (Chekroud, 2018). This indicates that a model's effectiveness may appear exaggerated if only assessed within a single trial, and its real-world performance may be lower when applied to new data. A machine learning model trained on data from one clinical trial might show high accuracy, but when tested on data from another trial, its accuracy could drop significantly, highlighting the importance of validating models across diverse datasets.

The proposed multi-level psychological disorder predictive model can successfully serve as an assistive tool for medical experts in the effective treatment of mental health (Mishra et al., 2021). This model aids clinicians by providing reliable predictions that can inform treatment decisions. A psychiatrist might use this model to predict the likelihood of a patient developing severe depression, allowing for early intervention and tailored treatment plans. The standard errors for the intercept and predictors will be low, indicating high precision and reliability (Ferdinandov, 2022). Low standard errors mean that the estimates of the model's parameters are close to the true values, providing confidence in the model's accuracy. For instance, in a study on medication efficacy, low standard errors would suggest that the estimated effect of the medication is precise and dependable. The least confident judge had a mean confidence probability estimate of .52 (Schinka, 1976). This implies that even the judge with the lowest confidence in their decisions estimated their probability of being correct at 52%, slightly better than chance. In a diagnostic setting, a clinician's confidence in their judgment about a complex case can provide insights into their decision-making process and areas where further training might be needed. Attending to scale-level PAI data may be a promising area of research in improving the prediction of PVT failure (Boress et al., 2022). This suggests that focusing on specific scales of the Personality Assessment Inventory (PAI) could enhance the ability to predict failure in Performance Validity Tests (PVT). Researchers might investigate which PAI scales are most predictive of invalid test performance, leading to better screening tools. The highest-ranked independent component spatial maps show similarity to brain regions associated with contrasts for belief vs. disbelief (Douglas et al., 2011). This indicates that the spatial maps derived from neuroimaging data align with brain areas involved in processing belief and disbelief. In a study examining how people process religious beliefs, these maps could help identify which brain regions are activated when participants affirm or reject certain statements. A neural network-based short-test version predicted Big-5 personality dimensions from IPIP-300 as well as its facets with high accuracy (Glöckner et al., 2020). This means that a shorter test using neural networks can accurately predict a person's Big-5 personality traits based on the IPIP-300 questionnaire. A brief online personality assessment could provide reliable results quickly, making it useful for large-scale surveys or initial screenings. The text data collected from interviews will be analyzed using natural language processing (NLP) (Jang et al., 2022). Using NLP techniques to analyze interview transcripts allows for the extraction of meaningful patterns and insights from the text. In a study on patient experiences with mental health treatment, NLP can help identify common themes and sentiments expressed by patients, informing better care practices. The ratings were not only reliable across judges but also consistent with significant changes in the interviewing skills of novices who had gained experience and training over the course of the semester (Terry, 1984). This indicates that the judges provided consistent ratings and that these ratings reflected real improvements in the interview skills of trainees. In a clinical training program, consistent ratings from multiple judges could demonstrate that trainees are successfully learning and applying new interviewing techniques, leading to better patient interactions.

The linear elastic net performed similarly to the nonlinear random forest in benchmark experiments (Pargent & Albert-von der Gönna, 2018). This suggests that, in some cases, simpler linear models like the elastic net can achieve comparable results to more complex models like random forests. In predicting housing prices, both methods might yield equally accurate predictions, indicating that the added complexity of the random forest might not always be necessary. The incremental validity of the Psychopathology-5 (PSY-5) scales over the Revised NEO Personality Inventory domain scales for predicting each Personality Adjective Checklist scale was measured by the R^2 change at the end of the second step in the hierarchical multiple regression analysis (Byravan & Ramanaiah, 2002). This means that the PSY-5 scales add additional predictive power to the NEO scales when assessing personality traits. For instance, in a psychological assessment, using both the NEO and PSY-5 scales provides a more comprehensive understanding of an individual's personality. Natural language processing (NLP) has promise for identifying EHR-documented goals-of-care discussions in the electronic health record (Uyeda et al., 2022). NLP techniques can effectively extract and identify specific types of information from clinical records. NLP could be used to automatically find and highlight conversations about treatment preferences and end-of-life care in a patient's medical history, aiding healthcare providers in ensuring that these preferences are respected. Decisions made using the predicted score proved to be more reliable than decisions based on a single rating of each example (Neumann et al., 2023). This indicates that aggregated scores from predictive models provide more consistent decision-making compared to relying on individual judgments. For instance, in employee performance evaluations, using a model to predict overall performance based on multiple ratings can yield more dependable results than a single evaluator's rating. The transfer learning technique improved the F1 score obtained by the LSTM model (Şen & Bakal, 2023). Transfer learning, which involves reusing a pre-trained model on a new but related task, enhanced the model's performance. Applying a language model trained on general text data to a specific medical text classification task can improve the accuracy of identifying relevant information. The Neuropsychiatric Inventory-Clinician (NPI-C) shows promise as a versatile tool which can accurately measure neuropsychiatric symptoms (de Medeiros et al., 2010). The NPI-C can reliably assess various symptoms such as delusions, agitation, and depression. For instance, clinicians can use the NPI-C to monitor changes in a patient's neuropsychiatric symptoms over time, facilitating better management of conditions like dementia. A nonclinical audience can interpret quantitative performance measures of predictive models with very high accuracy (Weissman et al., 2018). This suggests that even individuals without specialized training can understand and accurately interpret performance metrics of predictive models. Business managers can effectively use performance reports of predictive sales models to make informed decisions without needing detailed technical knowledge.

The PAI validity scales have poor classification accuracy in detecting cognitive response bias as defined by performance-on-Performance Validity Tests (PVTs) (Gaasedelen & Whiteside, 2016). This indicates that the Personality Assessment Inventory (PAI) validity scales are not effective in accurately identifying individuals who may be intentionally or unintentionally exaggerating or minimizing their symptoms during psychological evaluations. In a forensic setting where the validity of a plaintiff's injury claims is in question, relying solely on PAI validity scales may not provide reliable results. The clinician would choose the test that minimizes the chance of the more consequential error (O'Riordan & Bearer, 2020). This highlights the importance of considering the potential consequences of errors when selecting diagnostic or assessment tools. In medical screening, a clinician might prioritize a test with high sensitivity to minimize the risk of missing a serious condition, even if it leads to more false positives. The average estimation model of AI-MADRS demonstrated substantially acceptable rater-estimation-system reliability with trained psychiatrists (Shimamoto et al., 2024). This suggests that an artificial intelligence model designed to estimate scores on the Montgomery-Åsberg Depression Rating Scale (MADRS) can reliably approximate ratings made by trained psychiatrists. In a clinical trial evaluating the effectiveness of a new antidepressant, this AI model could provide consistent and accurate assessments of participants' depressive symptoms. The train-test split paradigm does not insulate Natural Language Processing (NLP) from problems relating to the validity and reliability of its models, their features, and their performance metrics (Futrell, 2023). This highlights that while the train-test split is a common method for evaluating NLP models, it does not address underlying issues with the data or the model's design. Even if a model performs well on a test dataset, it may still struggle with real-world applications if it is not robust to variations in language use or lacks generalizability.

When between-provider variability in performance was low, all three methods exhibited poor classification accuracy (Ding et al., 2013). This suggests that in situations where there is little variation in performance among different providers, standard classification methods may struggle to accurately distinguish between different outcomes. In a study comparing the diagnostic accuracy of different medical imaging techniques, if all providers perform similarly, it may be challenging for any method to accurately detect abnormalities. The translation and adaptation of the Structured Clinical Interview for DSM Disorders (SCID-P) and SCID-II to Portuguese present generally good reliability indices (Del-Ben et al., 1996). This indicates that the Portuguese versions of these diagnostic interviews are reliable tools for assessing psychiatric disorders according to DSM criteria. In a clinical setting in Portugal, these translated versions can be used with confidence to diagnose conditions such as major depressive disorder or borderline personality disorder. Personalized solutions significantly outperform the baseline (Miłkowski et al., 2023). This suggests that tailoring interventions or treatments to individual characteristics or needs leads to better outcomes compared to standardized approaches. In mental health care, personalized therapy plans based on a patient's specific symptoms and preferences may lead to more effective treatment outcomes than generic treatment protocols. Agreement between self-reports and actual benefits was stronger for general medical services than for mental health care (Meredith et al., 2002). This indicates that individuals tend to have a clearer understanding of the benefits they receive from general medical services compared to mental health care. For example, patients may have an easier time evaluating the effectiveness of a surgery or medication for a physical ailment compared to the impact of therapy or counseling on their mental well-being.

The standard errors for the intercept and predictors will be low, indicating high precision and reliability (Barreto et al., 2020). This suggests that the estimates derived from the statistical model are closely aligned with the true values of the parameters being estimated. In a study analyzing factors influencing academic performance, low standard errors indicate that the predictors considered are reliably associated with academic outcomes. The inclusion of answering time-related variables did not exhibit substantial improvement in the prediction of students' mental health problems (Baba & Bunji, 2023). This implies that factors such as response time to questions or tasks may not significantly contribute to predicting mental health issues among students. In an educational setting, while response times may provide insights into cognitive processes, they might not be strongly linked to mental health outcomes. The low Positive Predictive Value (PPV) with relatively high sensitivity for the diagnosis of Parkinson's disease suggests overdiagnosis (Litvan et al., 1998). This indicates that while the diagnostic test for Parkinson's disease may be sensitive in detecting true cases, it also has a high rate of false positives. For example, a screening test with high sensitivity but low PPV might identify many individuals as potentially having Parkinson's disease, leading to unnecessary further testing or treatment for those who do not actually have the condition. The rating system is an efficient, unobtrusive, reliable, and valid means of extracting quantitative data from written memories collected in assessment and therapy (Richards et al., 2014). This indicates that using a rating system to evaluate written memories is a practical and effective method for gathering quantitative data in clinical settings. In psychotherapy sessions, therapists can use a rating scale to assess the emotional content or narrative structure of patient's written accounts, providing valuable insights into their psychological state. ChatGPT scores well on accuracy, completeness, and speed when generating answers to clinical questions about clinical psychiatry (Østergaard et al., 2023). This suggests that ChatGPT, an AI language model, can provide accurate, comprehensive, and timely responses to inquiries related to clinical psychiatry. For instance, clinicians consulting ChatGPT for guidance on diagnostic criteria or treatment options may find the responses reliable and informative. The best predictors of status were age, high school grade point average, and composite ACT score (Peters, 2000). This indicates that these demographic and academic factors are most strongly associated with a particular outcome or status. For example, in college admissions, age, high school GPA, and standardized test scores like the ACT are commonly used to predict academic success.

The combination of PHQ-2 scores of 2 or greater followed by PHQ-9 scores of 10 or greater alone had similar sensitivity but higher specificity compared with PHQ-9 cutoff scores of 10 or greater alone (B. Levis et al., 2020). This suggests that using a two-step screening process involving the Patient Health Questionnaire-2 (PHQ-2) followed by the PHQ-9 can improve the accuracy of detecting depressive symptoms compared to using the PHQ-9 alone. In primary care settings, this two-step approach may help prioritize resources for further evaluation and treatment for individuals at higher risk of depression. The least profitable predictions resulted from attempts to predict decision time (Townsend & Smith, 1964). This indicates that predicting the time it takes individuals to make decisions may not yield significant benefits compared to predicting other outcomes. In studies of human decision-making, while researchers may be interested in understanding decision.

A validity coefficient above 0.90 is required to achieve a kappa above 0.70 and to keep false positive and false negative error rates below 25% (Sicoly, 1992). This suggests that for a diagnostic test or assessment tool to be considered highly reliable and accurate, it must demonstrate a very high validity coefficient. In medical diagnostics, a screening tool for a serious condition like cancer would need to have a validity coefficient above 0.90 to ensure that false positives and false negatives are minimized. The summary positive predictive value of a test may differ substantially from the average meta-analysis findings because of heterogeneity (Riley et al., 2015). This indicates that when synthesizing findings from multiple studies through meta-analysis, the overall positive predictive value of a test may vary due to differences in study populations, methodologies, and other factors. In assessing the accuracy of a diagnostic test across different populations or settings, the positive predictive value may vary widely, leading to discrepancies between meta-analysis findings and real-world applications. Simulations of potential errors in Natural Language Processing (NLP)-derived variables were not eliminated in a single of the 500 simulations (Lan & Turchin, 2023). This suggests that even with extensive testing and refinement, NLP algorithms may still be susceptible to errors in processing textual data. In analyzing electronic health records for clinical decision support, errors in NLP-derived variables could lead to incorrect interpretations or recommendations.

The Symbolic Implicit Association Test (SIAT) was found to have insufficient validity and proper reliability (Sabour & Ghassemi, 2016). This indicates that the SIAT, a psychological assessment tool, does not accurately measure the constructs it intends to assess and lacks consistency in its results. In measuring implicit biases, the SIAT may produce inconsistent or unreliable results, making it less useful for research or clinical applications. Popular metrics such as Accuracy, Precision, and F1 score are often insufficient as they fail to give a complete picture of the model's behavior (Yacouby & Axman, 2020). This highlights the limitations of relying solely on standard performance metrics to evaluate machine learning models. In sentiment analysis of customer reviews, while accuracy may be high, the model may still struggle with correctly identifying nuanced sentiments, leading to incomplete assessments of its performance. The Particle Swarm Optimization (PSO) - Support Vector Machine (SVM) model designed in this study can be used in patient data monitoring with acceptable accuracy (Fariborzi et al., 2022). This indicates that the PSO-SVM model, optimized using particle swarm optimization, demonstrates satisfactory performance in monitoring patient data. In healthcare settings, this model could help identify abnormal trends in patient vital signs or lab results, facilitating early intervention and improved patient outcomes. Rule-based NLP algorithms demonstrated consistent performance in clinical text in three areas: housing, financial, and unemployment (Allen et al., 2023). This suggests that rule-based NLP algorithms, which rely on predefined linguistic rules, perform reliably in extracting relevant information from clinical text across different domains. In analyzing patient electronic health records, these algorithms can accurately identify mentions of housing issues, financial concerns, or employment status, aiding clinicians in providing holistic care.

The PHQ-9 is a valid diagnostic tool in primary care (Wittkampf et al., 2007). This indicates that the Patient Health Questionnaire-9 (PHQ-9), commonly used to screen for depression, reliably identifies individuals with depressive symptoms in primary care settings. In routine medical check-ups, administering the PHQ-9 can help primary care physicians detect and manage depression among their patients. A measure with low sensitivity is unacceptable (Joshipura, 2008). This emphasizes the importance of sensitivity, particularly in diagnostic tests or screening tools. In cancer screenings, a test with low sensitivity may fail to detect early-stage tumors, leading to missed diagnoses and delayed treatment. The Symptom Validity Tests (SVTs) would detect invalid responses at rates of approximately 35%, 65%, and 49% for the three tests (Mossman et al., 2012). This suggests that the SVTs, used to assess the credibility of an individual's reported symptoms or performance, have varying degrees of sensitivity in detecting invalid responses. In forensic evaluations, SVTs may help identify individuals feigning or exaggerating symptoms of mental illness or cognitive impairment.

The quantity and quality of information are positively related to objective knowledge about the targets and realistic accuracy (Letzring et al., 2006). This indicates that having more and better-quality information improves one's ability to accurately understand and assess a given situation or target. In intelligence analysis, analysts with access to comprehensive and reliable information are more likely to make accurate assessments of threats or risks. The Portuguese version of the Structured Clinical Interview for DSM Disorders (SCID-CV) showed good inter-rater reliability (Cristina Marta Del-Ben et al., 2001). This indicates that the Portuguese adaptation of the SCID-CV, a widely used diagnostic interview for psychiatric disorders, produces consistent results when administered by different clinicians. In research studies or clinical settings in Portuguese-speaking countries, the SCID-CV can be relied upon to provide accurate diagnoses of mental health disorders. The standard errors for the intercept and predictors were low, indicating high precision and reliability (Sari et al., 2017).

Low standard errors suggest that estimates derived from statistical models, such as regression analysis, are reliable and precise. In predicting patient outcomes based on demographic and clinical variables, low standard errors indicate that the model's estimates are close to the true values. The pattern of significant and non-significant results remained the same with two exceptions (Skovgaard-Olsen et al., 2018). This indicates that, overall, the observed patterns of statistical significance in a study's findings remained consistent across different analyses or conditions, with only minor exceptions. In a psychological study investigating the effects of cognitive interventions on memory performance, most findings remained consistent across different experimental conditions, except for two unexpected results. The Negative Personality Maintenance (NPM) scale is a measure of negative impression management (Young & Schinka, 2001). This suggests that the NPM scale assesses individuals' tendencies to present themselves in a negative light in social or evaluative situations. For example, in employment settings, the NPM scale may help assess candidates' likelihood of providing socially desirable responses during job interviews. Empirically derived scores outperformed unit weighting across all levels of the personality hierarchy (Speer et al., 2022). This indicates that using empirically derived weights or coefficients for personality traits yields more accurate predictions of behavior or outcomes compared to assigning equal weights to all traits. In predicting job performance based on personality assessments, using empirically derived scores may provide more accurate predictions of success.

The best model was developed through joint optimization of natural language features and min-max normalization of feature frequencies (Ji et al., 2021). This indicates that combining natural language features with min-max normalization of feature frequencies resulted in the most effective model. In sentiment analysis of customer reviews, this approach might help identify key linguistic patterns while normalizing for differences in review lengths.

Prototypical diagnostic elaboration was equivalent to diagnostics obtained through ICD-10 categories (H. G. D. Rocha Neto et al., 2022). This suggests that developing prototypical diagnostic profiles is as effective as using the standardized diagnostic categories outlined in the International Classification of Diseases, 10th Edition (ICD-10). In psychiatric assessments, clinicians might rely on prototypical descriptions of disorders to guide diagnostic decision-making. Positive correlations between Self-Concept (SC) tests and other Neuropsychological (NPS) tests may contribute to understanding the complex SC network (Kremen et al., 2019). This indicates that there are associations between self-concept assessments and other neuropsychological measures, which can shed light on the interconnected nature of self-concept. In psychological research on self-esteem, understanding how self-concept relates to cognitive abilities may provide insights into self-perception and identity. The new instrument is a highly reliable and valid screening instrument (Daradkeh et al., 1999). This suggests that the newly developed assessment tool demonstrates strong consistency and accuracy in identifying specific conditions or traits. In screening for depression in primary care settings, a highly reliable and valid instrument can help clinicians efficiently identify patients who may benefit from further evaluation or intervention.

METHODOLOGY

The dataset used here is secondary, contains anonymized conversations between users and experienced psychologists on mental health topics. It's designed to train NLP models to provide mental health advice, featuring a variety of questions to enhance response accuracy. This resource is valuable for AI developers creating mental health applications and therapists exploring tech-assisted connections (The

Devastator, 2023). The methodology employed to evaluate the sentiment classification model and fit the multinomial logistic regression model involved several steps. Initially, the dataset was prepared with three sentiment classes: negative, neutral, and positive, with respective class distributions. A confusion matrix visualized classification performance, while sensitivity, specificity, PPV, and NPV metrics measured the model's ability to correctly identify instances. Detection rate and prevalence assessed the proportion of predicted instances in each class, and balanced accuracy ensured uniform performance. Overall accuracy and a 95% confidence interval highlighted the model's certainty, compared to the No Information Rate for improvement evaluation. The Kappa statistic measured agreement between predictions and actual classes. McNemar's Test was not applicable due to no misclassifications. The multinomial logistic regression model was fitted to predict sentiment responses based on contextual information. Coefficients were estimated to interpret the effect of each sentiment context level. The optimization process monitored objective function values, converging to a stable solution. Performance metrics such as accuracy and the confusion matrix evaluated the model, while coefficient analysis provided insights into predictor significance. Low standard errors indicated reliable coefficient estimates, ensuring less variability in predictor effects. This comprehensive methodology facilitated effective model evaluation and robust fitting of the multinomial logistic regression model to capture sentiment context and responses.

Hypothesis

1. H1: Sensitivity, specificity, PPV, NPV, detection rate, and balanced accuracy will all be 1.0 for negative, neutral, and positive classes.
2. H2: The model's overall accuracy will be 1.0, with a 95% CI from 0.9989 to 1, significantly higher than the NIR of 0.6709 ($p < 2.2e\text{-}16$).
3. H3: The Kappa statistic will be 1.0, indicating perfect agreement beyond chance.
4. H4: The model will produce significant coefficients for predictors, showing the influence of sentiment context levels on responses.
5. H5: The optimization process will converge, with the final objective function value decreasing to a stable lower value.
6. H6: The standard errors for the intercept and predictors will be low, indicating high precision and reliability.

RESULTS

Table 1. Statistics by class

	Class: negative	Class: neutral	Class: positive
Sensitivity	1.000	1.000	1.000
Specificity	1.000	1.000	1.000
Pos Pred Value	1.000	1.000	1.000
Neg Pred Value	1.000	1.000	1.000

continued on following page

Table 1. Continued

	Class: negative	Class: neutral	Class: positive
Prevalence	0.2886	0.04049	0.6709
Detection Rate	0.2886	0.04049	0.6709
Detection Prevalence	0.2886	0.04049	0.6709
Balanced Accuracy	1.000	1.000	1.000

The statistics by class showcase the model's performance across different categories (Table 1). In terms of Sensitivity, which measures the ability to correctly identify positive instances, and Specificity, which assesses the ability to correctly identify negative instances, the model scores a perfect 1. This means it successfully captures all instances of each class. The Positive Predictive Value (PPV) and Negative Predictive Value (NPV), indicating the proportion of correctly identified positive and negative instances respectively, are perfect at 1. This suggests that when the model predicts a certain class, it is almost always correct. The Prevalence values demonstrate the distribution of each class within the dataset, with 28.86% for negative, 4.049% for neutral, and 67.09% for positive, reflecting the underlying distribution of classes. The Detection Rate, which corresponds to sensitivity, mirrors its perfect score of 1, indicating the model's ability to correctly detect instances of each class. The Detection Prevalence, reflecting the proportion of instances predicted to belong to each class, aligns with the prevalence values. The Balanced Accuracy, providing an average of sensitivity and specificity, achieves a perfect score of 1, indicating uniform and accurate performance across all classes. Collectively, these statistics highlight the model's exceptional performance across all classes, exhibiting high sensitivity, specificity, PPV, NPV, prevalence, detection rate, and balanced accuracy.

Table 2. Confusion matrix and statistics

Prediction	negative	neutral	positive
negative	1012	0	0
neutral	0	142	0
positive	0	0	2353

The confusion matrix provides a detailed breakdown of the model's classification performance (Table 2). In this analysis, each row represents the predicted class, while each column represents the actual class. Looking at the matrix, it's evident that the model exhibits remarkable accuracy. For the "negative" class, all 1012 instances were correctly classified as negative, with no false positives or false negatives. Similarly, for the "neutral" class, all 142 instances were correctly identified, with no misclassifications. The same holds for the "positive" class, with all 2353 instances correctly classified. Such precise predictions across all classes suggest that the model has effectively captured the underlying patterns in the data and can reliably distinguish between different classes. This high level of accuracy is indicative of a well-performing model with strong predictive capabilities. However, it's important to assess other performance metrics and conduct further analysis to ensure the model's robustness and generalizability to new data. The overall statistics provide a comprehensive overview of the model's performance. The accuracy, which measures the proportion of correctly predicted instances out of the total, is exceptionally high at 1, indicating perfect accuracy. The 95% confidence interval (CI) for accuracy ranges from 0.9989 to 1, further emphasizing the high level of certainty in the model's performance.

Comparing the model's accuracy to the No Information Rate (NIR), which represents the accuracy achieved by always predicting the majority class, reveals a significant improvement. The NIR is 0.6709, indicating that without the model, simply predicting the majority class would result in an accuracy of approximately 67.09%. The p-value associated with the comparison between accuracy and NIR is extremely low ($p < 2.2e-16$), suggesting that the model's performance is significantly better than random chance. The Kappa statistic, which measures the agreement between the model's predictions and the actual classes while accounting for the agreement occurring by chance, is also perfect at 1. A Kappa value of 1 indicates perfect agreement beyond chance, reinforcing the model's robust performance. Mcnemar's Test P-Value is not applicable (NA), likely due to the absence of misclassifications in the confusion matrix. The overall statistics paint a compelling picture of the model's excellence, with perfect accuracy, strong agreement beyond chance, and a significant improvement over the baseline performance. This underscores the model's reliability and effectiveness in accurately classifying instances across all classes. Rcode is used (multinom_model) for fitting a multinomial logistic regression model using the `multinom` function in R.

The multinomial logistic regression model fitted aims to predict sentiment responses (`Sentiment_Response`) based on the context (`Sentiment_Context`). Unlike binary logistic regression, which predicts outcomes in one of two categories, multinomial logistic regression handles outcomes with more than two categories. In this case, it is predicting sentiment responses, which could have multiple categories like positive, negative, or neutral. The coefficients obtained from this model can be interpreted as the effect of each level of `Sentiment_Context` on the odds of the corresponding `Sentiment_Response` category compared to a reference category (usually the first level unless specified otherwise). This typically evaluates the performance of the model using appropriate metrics such as accuracy, confusion matrix, or classification report. Additionally, further analysis to understand the significance of the predictors and their contribution to predicting sentiment responses.

Table 3. Overall statistics

Accuracy	**:**	**1**
95% CI	:	(0.9989, 1)
No Information Rate	:	0.6709
P-Value [Acc > NIR]	:	< 2.2e-16
Kappa	:	1
Mcnemar's Test P-Value	:	NA

The statistics provide a comprehensive overview of the model's performance (Table 3). The accuracy, which measures the proportion of correctly predicted instances out of the total, is exceptionally high at 1, indicating perfect accuracy. The 95% confidence interval (CI) for accuracy ranges from 0.9989 to 1, further emphasizing the high level of certainty in the model's performance. Comparing the model's accuracy to the No Information Rate (NIR), which represents the accuracy achieved by always predicting the majority class, reveals a significant improvement. The NIR is 0.6709, indicating that without the model, simply predicting the majority class would result in an accuracy of approximately 67.09%. The p-value associated with the comparison between accuracy and NIR is extremely low ($p < 2.2e-16$), suggesting that the model's performance is significantly better than random chance. The Kappa statistic, which measures the agreement between the model's predictions and the actual classes while accounting for the agreement occurring by chance, is also perfect at 1. A Kappa value of 1 indicates perfect agreement beyond chance, reinforcing the model's robust performance. Mcnemar's Test P-Value is not applicable (NA), likely due to the absence of misclassifications in the confusion matrix. The statistics paint a compelling picture of the model's excellence, with perfect accuracy, strong agreement beyond chance, and a significant improvement over the baseline performance. This underscores the model's reliability and effectiveness in accurately classifying instances across all classes.

The multinomial logistic regression model was constructed to predict sentiment responses based on contextual information. The model includes 12 weights associated with 6 variables, likely representing different levels or categories of sentiment context. Initially, the objective function had a high value of 3852.833296, indicating significant room for improvement. After 10 iterations, the objective function decreased substantially to 2650.040169, suggesting that the model was progressively improving. The model converged with a final objective value of 2649.295856, indicating that the optimization process reached a stable solution. This convergence indicates that the model fitting process successfully aligned the predicted sentiment responses with the observed data. Thus, the output suggests that the multinomial logistic regression model effectively captures the relationship between sentiment context and sentiment responses, providing insights into how different contextual factors influence sentiment outcomes.

Table 4. Coefficients

	(Intercept)	X120524$Sentiment_Contextneutral	X120524$Sentiment_Contextpositive
Neutral	-1.9626591	0.3070226	-0.3440257
positive	0.8235484	0.3426333	0.341181

The intercept in the model represents the baseline value of the outcome variable when all predictors are zero (Table 4). Moving on to the predictors, when considering the sentiment context as neutral, a unit increase in X120524$Sentiment_Contextneutral leads to a predicted change of approximately 0.3070226 units in the outcome variable, while holding other predictors constant. Conversely, in the same neutral sentiment context, a unit increase in X120524$Sentiment_Contextpositive leads to a predicted decrease of approximately 0.3440257 units in the outcome variable. When the sentiment context shifts to positive, a unit increase in X120524$Sentiment_Contextneutral now predicts a change of approximately 0.3426333 units in the outcome variable, while a unit increase in X120524$Sentiment_Contextpositive leads to a predicted increase of approximately 0.341181 units in the outcome variable. These coefficients provide insights into how each predictor influences the outcome variable, depending on the sentiment context, aiding in better understanding and predicting the outcome variable's behavior.

Table 5. Std. errors

	(Intercept)	X120524$Sentiment_Contextneutral	X120524$Sentiment_Contextpositive
Neutral	0.09190728	0.553856	0.613906
Positive	0.03869281	0.2534431	0.2127862

The intercept's standard error of 0.09190728 indicates the variability in the baseline value of the outcome variable (Table 5). Regarding the predictors, when the sentiment context is neutral, the standard error for X120524$Sentiment_Contextneutral is 0.553856, suggesting some variability in the estimated effect of this predictor on the outcome variable. Similarly, the standard error for X120524$Sentiment_Contextpositive in the neutral context is 0.613906, indicating variability in its estimated effect as well. When the sentiment context shifts to positive, the standard error for X120524$Sentiment_Contextneutral decreases to 0.2534431, implying reduced uncertainty in its estimated effect on the outcome variable. Likewise, the standard error for X120524$Sentiment_Contextpositive in the positive context decreases to 0.2127862, suggesting less variability in its estimated effect. These standard errors help gauge the reliability of the coefficient estimates, providing insights into the precision of the model's predictions and the potential uncertainty in the estimated effects of the predictors on the outcome variable.

CONCLUSION

The model excels in classifying sentiments across negative, neutral, and positive classes, achieving perfect scores for sensitivity, specificity, PPV, and NPV. It accurately predicts instances for each class, with prevalence values aligning well with dataset distribution. The confusion matrix demonstrates precise predictions without misclassifications. Statistics show perfect accuracy, significant improvement over baseline, and strong agreement beyond chance. The multinomial logistic regression effectively captures sentiment context's influence on responses, with significant coefficients and robust fitting, indicating high precision and reliability in predicting sentiment outcomes.

REFERENCES

Allen, K. S., Hood, D. R., Cummins, J., Kasturi, S., Mendonca, E. A., & Vest, J. R. (2023). Natural language processing-driven state machines to extract social factors from unstructured clinical documentation. *JAMIA Open*, *6*(2), ooad024. Advance online publication. DOI: 10.1093/jamiaopen/ooad024 PMID: 37081945

Baba, A., & Bunji, K. (2023). Prediction of mental health problem using annual student health survey: Machine learning approach. *JMIR Mental Health*, *10*, e42420. DOI: 10.2196/42420 PMID: 37163323

Barreto, B. B., Luz, M., & Gusmao-Flores, D. (2020). Nonpharmacologic interventions to prevent or mitigate adverse long-term outcomes among ICU survivors: Is the ICU diary really important? *Critical Care Medicine*, *48*(3), e265–e265. DOI: 10.1097/CCM.0000000000004136 PMID: 32058409

Boress, K., Gaasedelen, O. J., Croghan, A., Johnson, M. K., Caraher, K., Basso, M. R., & Whiteside, D. M. (2022). Validation of the Personality Assessment Inventory (PAI) scale of scales in a mixed clinical sample. *The Clinical Neuropsychologist*, *36*(7), 1844–1859. DOI: 10.1080/13854046.2021.1900400 PMID: 33730975

Byravan, A., & Ramanaiah, N. V. (2002). On the incremental validity of MMPI-2 Psychopathology-5 scales over the Revised Neo Personality Inventory scales for predicting personality disorders. *Psychological Reports*, *90*(3_part_2), 1084–1090. DOI: 10.1177/003329410209000305.2 PMID: 12150388

Chekroud, A. (2018). T107. Why validation matters: A demonstration predicting antipsychotic response using 5 RCTs. *Schizophrenia Bulletin*, *44*(suppl_1), S157–S157. DOI: 10.1093/schbul/sby016.383

Cyr, L., & Francis, K. (1992). Measures of clinical agreement for nominal and categorical data: The kappa coefficient. *Computers in Biology and Medicine*, *22*(4), 239–246. DOI: 10.1016/0010-4825(92)90063-S PMID: 1643847

Daradkeh, T. K., Ghubash, R., El-Rufaie, O. E. F., & Abou-Saleh, M. T. (1999). The rationale, development and reliability of a new screening psychiatric instrument. *Social Psychiatry and Psychiatric Epidemiology*, *34*(4), 223–228. DOI: 10.1007/s001270050137 PMID: 10365629

de Medeiros, K., Robert, P., Gauthier, S., Stella, F., Politis, A., Leoutsakos, J., Taragano, F., Kremer, J., Brugnolo, A., Porsteinsson, A. P., Geda, Y. E., Brodaty, H., Gazdag, G., Cummings, J., & Lyketsos, C. (2010). The Neuropsychiatric Inventory-Clinician rating scale (NPI-C): Reliability and validity of a revised assessment of neuropsychiatric symptoms in dementia. *International Psychogeriatrics*, *22*(6), 984–994. DOI: 10.1017/S1041610210000876 PMID: 20594384

Del-Ben, C. M., Vilela, J. A. A., Crippa, J. A. D. S., Hallak, J. E. C., Labate, C. M., & Zuardi, A. W. (2001). Confiabilidade da" Entrevista Clínica Estruturada para o DSM-IV-Versão Clínica" traduzida para o português. *The British Journal of Psychiatry*, *23*, 156–159.

Del-Ben, C. M., Rodrigues, C. R., & Zuardi, A. W. (1996). Reliability of the Portuguese version of the structured clinical interview for DSM-III-R (SCID) in a Brazilian sample of psychiatric outpatients. *Brazilian Journal of Medical and Biological Research = Revista Brasileira de Pesquisas Médicas e Biológicas*, *29*(12). https://pubmed.ncbi.nlm.nih.gov/9222432/ PMID: 9222432

Ding, V. Y., Hubbard, R. A., Rutter, C. M., & Simon, G. E. (2013). Assessing the accuracy of profiling methods for identifying top providers: Performance of mental health care providers. *Health Services and Outcomes Research Methodology, 13*(1), 1–17. DOI: 10.1007/s10742-012-0099-5 PMID: 23565050

Douglas, P. K., Harris, S., Yuille, A., & Cohen, M. S. (2011). Performance comparison of machine learning algorithms and number of independent components used in fMRI decoding of belief vs. disbelief. *NeuroImage, 56*(2), 544–553. DOI: 10.1016/j.neuroimage.2010.11.002 PMID: 21073969

Fariborzi, M., Zeinalnezhad, M., & Saghaei, A. (2022). Accuracy improvement of mood disorders prediction using a combination of data mining and meta -heuristic algorithms. *Journal of Health and Biomedical Informatics, 9*(3), 106–119. DOI: 10.34172/jhbmi.2022.01

Ferdinandov, K. G. (2022). The accuracy distortions in self-assessment. *Psychological Thought, 15*(2), 114–147. DOI: 10.37708/psyct.v15i2.706

Futrell, R. (2023). Validity, reliability, and significance: Empirical methods for NLP and data science. [Association for Computational Linguistics]. *Computational Linguistics, 49*(1), 249–251. DOI: 10.1162/coli_r_00467

Gaasedelen, O., & Whiteside, D. (2016). Performance and symptom validity-4Evaluating the classification accuracy of the personality assessment inventory validity scales in detecting cognitive response bias. *Archives of Clinical Neuropsychology: The Official Journal of the National Academy of Neuropsychologists, 31*(6), 574.1-574. DOI: 10.1093/arclin/acw042.04

Glöckner, A., Michels, M., & Giersch, D. (2020). Predicting personality test scores with machine learning methodology: Investigation of a new approach to psychological assessment. In *PsyArXiv*. DOI: 10.31234/osf.io/ysd3f

Guastello, S. J., Rieke, M. L., Guastello, D. D., & Billings, S. W. (1992). Assessing the validity of computer-based test interpretations: Rating reliability and individual differences among raters. *Journal of Personality Assessment, 58*(1), 79–89. DOI: 10.1207/s15327752jpa5801_7 PMID: 16370874

He, C., Levis, B., Riehm, K. E., Saadat, N., Levis, A. W., Azar, M., Rice, D. B., Krishnan, A., Wu, Y., Sun, Y., Imran, M., Boruff, J., Cuijpers, P., Gilbody, S., Ioannidis, J. P. A., Kloda, L. A., McMillan, D., Patten, S. B., Shrier, I., & Benedetti, A. (2020). The accuracy of the Patient Health Questionnaire-9 algorithm for screening to detect major depression: An individual participant data meta-analysis. *Psychotherapy and Psychosomatics, 89*(1), 25–37. DOI: 10.1159/000502294 PMID: 31593971

Jang, J., Yoon, S., Son, G., Kang, M., Choeh, J. Y., & Choi, K.-H. (2022). Predicting personality and psychological distress using natural language processing: A study protocol. *Frontiers in Psychology, 13*, 865541. Advance online publication. DOI: 10.3389/fpsyg.2022.865541 PMID: 35465529

Ji, M., Xie, W., Huang, R., & Qian, X. (2021). Forecasting the suitability of online mental health information for effective self-care developing machine learning classifiers using natural language features. *International Journal of Environmental Research and Public Health, 18*(19), 10048. DOI: 10.3390/ijerph181910048 PMID: 34639348

Joshipura, K. (2008). Value of predictive values. *Journal of Dental Research, 87*(10), e1–e1. DOI: 10.1177/154405910808701016 PMID: 18809740

Khatri, A., Labajo, D., & Rege, M. (2021). Reliability of NLP models to predict human sentiments. *Issues in Information Systems, 22*(4), 33–48. DOI: 10.48009/4_iis_2021_34-50

Kremen, W. S., Elman, J. A., Vuoksimaa, E., Jak, A. J., Mcevoy, L., Holland, D., Lyons, M. J., & Franz, C. E. (2019). P4-363: Improving accuracy of MCI diagnosis and prognosis: A cognitive-genetic approach. *Alzheimer's & Dementia: The Journal of the Alzheimer's Association, 15*(7S_Part_27). DOI: 10.1016/j.jalz.2019.06.4034

Kulasegaram, K., & Paradis, E. (2014). Better data ≫ Bigger data. *Medical Teacher, 36*(11), 1008–1009. DOI: 10.3109/0142159X.2014.917761 PMID: 25072309

Kwak, A. S., Israelsen, J. O., Morrison, C. T., Bambauer, D. E., & Surdeanu, M. (2022). *Validity assessment of legal will statements as natural language inference*. DOI: 10.18653/v1/2022.findings-emnlp.447

Lan, Z., & Turchin, A. (2023). Impact of possible errors in natural language processing-derived data on downstream epidemiologic analysis. *JAMIA Open, 6*(4), ooad111. Advance online publication. DOI: 10.1093/jamiaopen/ooad111 PMID: 38152447

Letzring, T. D., Wells, S. M., & Funder, D. C. (2006). Information quantity and quality affect the realistic accuracy of personality judgment. *Journal of Personality and Social Psychology, 91*(1), 111–123. DOI: 10.1037/0022-3514.91.1.111 PMID: 16834483

Levis, B., Sun, Y., He, C., Wu, Y., Krishnan, A., Bhandari, P. M., Neupane, D., Imran, M., Brehaut, E., Negeri, Z., Fischer, F. H., Benedetti, A., & Thombs, B. D. (2020). Accuracy of the PHQ-2 alone and in combination with the PHQ-9 for screening to detect major depression: Systematic review and meta-analysis. *Journal of the American Medical Association, 323*(22), 2290. DOI: 10.1001/jama.2020.6504 PMID: 32515813

Levis, M., Leonard Westgate, C., Gui, J., Watts, B. V., & Shiner, B. (2021). Natural language processing of clinical mental health notes may add predictive value to existing suicide risk models. *Psychological Medicine, 51*(8), 1382–1391. DOI: 10.1017/S0033291720000173 PMID: 32063248

Litvan, I., MacIntyre, A., Goetz, C. G., Wenning, G. K., Jellinger, K., Verny, M., Bartko, J. J., Jankovic, J., McKee, A., Brandel, J. P., Chaudhuri, K. R., Lai, E. C., D'Olhaberriague, L., Pearce, R. K. B., & Agid, Y. (1998). Accuracy of the clinical diagnoses of Lewy body disease, Parkinson disease, and dementia with Lewy bodies: A clinicopathologic study. *Archives of Neurology, 55*(7), 969. DOI: 10.1001/archneur.55.7.969 PMID: 9678315

Lopez, O. L., Litvan, I., Catt, K. E., Stowe, R., Klunk, W., Kaufer, D. I., Becker, J. T., & DeKosky, S. T. (1999). Accuracy of four clinical diagnostic criteria for the diagnosis of neurodegenerative dementias. *Neurology, 53*(6), 1292–1292. DOI: 10.1212/WNL.53.6.1292 PMID: 10522887

Meredith, L. S., Humphrey, N., Orlando, M., & Camp, P. (2002). Knowledge of health care benefits among patients with depression. *Medical Care, 40*(4), 338–346. DOI: 10.1097/00005650-200204000-00009 PMID: 12021689

Miłkowski, P., Karanowski, K., Wielopolski, P., Kocoń, J., Kazienko, P., & Zięba, M. (2023). Modeling uncertainty in personalized emotion prediction with normalizing flows. *2023 IEEE International Conference on Data Mining Workshops (ICDMW)*. DOI: 10.1109/ICDMW60847.2023.00103

Mishra, S., Tripathy, H. K., Kumar Thakkar, H., Garg, D., Kotecha, K., & Pandya, S. (2021). An explainable intelligence driven query Prioritization using balanced decision tree approach for multi-level psychological disorders assessment. *Frontiers in Public Health*, *9*, 795007. Advance online publication. DOI: 10.3389/fpubh.2021.795007 PMID: 34976936

Mossman, D., Wygant, D. B., & Gervais, R. O. (2012). Estimating the accuracy of neurocognitive effort measures in the absence of a "gold standard.". *Psychological Assessment*, *24*(4), 815–822. DOI: 10.1037/a0028195 PMID: 22545695

Neumann, M., Niessen, A. S. M., & Meijer, R. R. (2023). Predicting decision-makers' algorithm use. *Computers in Human Behavior*, *145*(107759), 107759. DOI: 10.1016/j.chb.2023.107759

O'Riordan, M. A., & Bearer, C. (2020). Correspondence on statistical rigor and kappa considerations: Which, when, and clinical context matters. *Pediatric Research*, *88*(1), 6–6. DOI: 10.1038/s41390-020-0880-z PMID: 32225173

Østergaard, S. D., Seidler, Z., & Rice, S. (2023). The ICD-11 opens the door for overdue improved identification of depression in men. *World Psychiatry; Official Journal of the World Psychiatric Association (WPA)*, *22*(3), 480–481. DOI: 10.1002/wps.21124 PMID: 37713575

Pargent, F., & Albert-von der Gönna, J. (2018). Predictive modeling with psychological Panel data. *Zeitschrift für Psychologie*, *226*(4), 246–258. DOI: 10.1027/2151-2604/a000343

Peters, R. D. (2000). Predicting the end of term status of community college general psychology students. *Journal of College Student Retention*, *2*(2), 109–114. DOI: 10.2190/1V94-J04T-8U8C-DNY2

Rashidiani, S., Doyle, T. E., Samavi, R., Duncan, L., Pires, P., & Sassi, R. (2022). Textionnaire: An NLP-based questionnaire analysis method for complex and ambiguous task decision support. *2022 IEEE 21st International Conference on Cognitive Informatics & Cognitive Computing (ICCI*CC)*.

Richards, H. J., Bruhn, A. R., Lucente, S. W., & Casey, J. O. (2014). Reliability of the comprehensive scoring system with the early memories procedure. *Sensoria A Journal of Mind Brain and Culture*, *10*(2), 14. DOI: 10.7790/sa.v0i0.404

Riley, R. D., Ahmed, I., Debray, T. P. A., Willis, B. H., Noordzij, J. P., Higgins, J. P. T., & Deeks, J. J. (2015). Summarising and validating test accuracy results across multiple studies for use in clinical practice. *Statistics in Medicine*, *34*(13), 2081–2103. DOI: 10.1002/sim.6471 PMID: 25800943

Rocha Neto, H. G., Sinem, T. B., Koiller, L. M., Pereira, A. M., de Souza Gomes, B. M., Veloso Filho, C. L., Cavalcanti, M. T., & Telles-Correia, D. (2022). Intra-rater kappa accuracy of prototype and ICD-10 operational criteria-based diagnoses for mental disorders: A brief report of a cross-sectional study in an outpatient setting. *Frontiers in Psychiatry*, *13*, 793743. Advance online publication. DOI: 10.3389/fpsyt.2022.793743 PMID: 35308869

Rocha Neto, H. G. D., Cavalcanti, M., & Telles Correia, D. (2022). Kappa accuracy of prototypical diagnosis and ICD-10 criteria for mental disorders: A cross-sectional study in a real-life setting. *European Psychiatry : the Journal of the Association of European Psychiatrists*, *65*(S1), S162–S162. DOI: 10.1192/j.eurpsy.2022.434

Sabour, S., & Ghassemi, F. (2016). The validity and reliability of a signal impact assessment tool: Statistical issue to avoid misinterpretation. *Pharmacoepidemiology and Drug Safety, 25*(10), 1215–1216. DOI: 10.1002/pds.4061 PMID: 27696610

Sari, K. S. A. P., & Dantes, N., N. D., & Arum MWP, D. (. (2017). Behavioral counseling with engineering modeling and technique token economy to improve confidence. *Bisma The Journal of Counseling, 1*(2), 18. DOI: 10.23887/128182017

Schinka, J. A. (1976). Confidence in personality assessment. *Psychological Reports, 38*(1), 62–62. DOI: 10.2466/pr0.1976.38.1.62 PMID: 1250932

Sechrest, L., Gallimore, R., & Hersch, P. D. (1967). Feedback and accuracy of clinical predictions. *Journal of Consulting Psychology, 31*(1), 1–11. DOI: 10.1037/h0020977

Şen, T. Ü., & Bakal, G. (2023). A transfer learning application on the reliability of psychological drugs' comments. *2023 International Conference on Smart Applications, Communications and Networking (SmartNets)*. DOI: 10.1109/SmartNets58706.2023.10215681

Shimamoto, M., Ishizuka, K., Ohtani, K., Inada, T., Yamamoto, M., Tachibana, M., Kimura, H., Sakai, Y., Kobayashi, K., Ozaki, N., & Ikeda, M. (2024). Machine learning algorithm-based estimation model for the severity of depression assessed using Montgomery-Asberg depression rating scale. *Neuropsychopharmacology Reports, 44*(1), 115–120. DOI: 10.1002/npr2.12404 PMID: 38115795

Sicoly, F. (1992). Estimating the accuracy of decisions based on cutting scores. *Journal of Psychoeducational Assessment, 10*(1), 26–36. DOI: 10.1177/073428299201000102

Skovgaard-Olsen, N., Singmann, H., & Klauer, K. C. (2018). Corrigendum to "The relevance effect and conditionals" [Cognition 150 (2016) 26–36]. *Cognition, 170*, 338. DOI: 10.1016/j.cognition.2017.09.002 PMID: 28918984

Speer, A. B., Christiansen, N. D., Robie, C., & Jacobs, R. R. (2022). Measurement specificity with modern methods: Using dimensions, facets, and items from personality assessments to predict performance. *The Journal of Applied Psychology, 107*(8), 1428–1439. DOI: 10.1037/apl0000618 PMID: 34582241

Terry, W. S. (1984). A "forgetting journal" for memory courses. *Teaching of Psychology, 11*(2), 111–112. DOI: 10.1207/s15328023top1102_17

The Devastator. (2023). *NLP Mental Health Conversations* [Data set].

Tonia, T., Buitrago-Garcia, D., Peter, N. L., Mesa-Vieira, C., Li, T., Furukawa, T. A., Cipriani, A., Leucht, S., Low, N., & Salanti, G. (2023). Tool to assess risk of bias in studies estimating the prevalence of mental health disorders (RoB-PrevMH). *BMJ Mental Health, 26*(1), e300694. DOI: 10.1136/bmjment-2023-300694 PMID: 37899074

Townsend, J. C., & Smith, W. J. (1964). *Predicting decision making behavior from personality and cognitive variables*. Defense Technical Information Center. DOI: 10.21236/AD0610521

Uyeda, A. M., Curtis, J. R., Engelberg, R. A., Brumback, L. C., Guo, Y., Sibley, J., Lober, W. B., Cohen, T., Torrence, J., Heywood, J., Paul, S. R., Kross, E. K., & Lee, R. Y. (2022). Mixed-methods evaluation of three natural language processing modeling approaches for measuring documented goals-of-care discussions in the electronic health record. *Journal of Pain and Symptom Management*, *63*(6), e713–e723. DOI: 10.1016/j.jpainsymman.2022.02.006 PMID: 35182715

Vibe Fersum, K., O'Sullivan, P. B., Kvåle, A., & Skouen, J. S. (2009). Inter-examiner reliability of a classification system for patients with non-specific low back pain. *Manual Therapy*, *14*(5), 555–561. DOI: 10.1016/j.math.2008.08.003 PMID: 18838331

Waza, K., Antonnette, G., Stephen, Z., Inoue, K., Sasaki, M., Okajima, Y., & Ishibashi, Y. (2005). A pilot study on using the patient health questionnaire in the primary care setting in japan. *General Medicine (Saga)*, *6*(1), 9–16. DOI: 10.14442/general2000.6.9

Weiner, B. J., Lewis, C. C., Stanick, C., Powell, B. J., Dorsey, C. N., Clary, A. S., Boynton, M. H., & Halko, H. (2017). Psychometric assessment of three newly developed implementation outcome measures. *Implementation Science : IS*, *12*(1), 108. Advance online publication. DOI: 10.1186/s13012-017-0635-3 PMID: 28851459

Weissman, G., Yadav, K., Madden, V., Courtright, K., Hart, J., Asch, D., Schapira, M., & Halpern, S. (2018). Numeracy and understanding of quantitative aspects of predictive models: A pilot study. *Applied Clinical Informatics*, *09*(03), 683–692. DOI: 10.1055/s-0038-1669457 PMID: 30157500

Wittkampf, K. A., Naeije, L., Schene, A. H., Huyser, J., & van Weert, H. C. (2007). Diagnostic accuracy of the mood module of the Patient Health Questionnaire: A systematic review. *General Hospital Psychiatry*, *29*(5), 388–395. DOI: 10.1016/j.genhosppsych.2007.06.004 PMID: 17888804

Yacouby, R., & Axman, D. (2020). Probabilistic extension of precision, recall, and F1 score for more thorough evaluation of classification models. *Proceedings of the First Workshop on Evaluation and Comparison of NLP Systems*. DOI: 10.18653/v1/2020.eval4nlp-1.9

Young, M. S., & Schinka, J. A. (2001). Research validity scales for the NEO-PI-R: Additional evidence for reliability and validity. *Journal of Personality Assessment*, *76*(3), 412–420. DOI: 10.1207/S15327752JPA7603_04 PMID: 11499455

Chapter 5

Machine Learning in Fake News Detection and Social Innovation: Navigating Truth in the Digital Age

Vitthal B. Kamble
https://orcid.org/0000-0003-4684-0104
Department of Computer Engineering, Savitribai Phule Pune University, Pune, India

Nilesh J. Uke
Department of Computer Engineering, Savitribai Phule Pune University, Pune, India

Dhiraj Ghanshyam Karwatkar
Department of CSE(Data Science), Tulasiram Gaikwad Patil College of Engineering and Technology, Nagpur, India

Rahul Dilip Dhongade
https://orcid.org/0009-0003-1735-1675
Department of Computer Engineering, Trinity Polytechnic, Pune, India

Priyanka Kasare
https://orcid.org/0009-0000-3045-3834
Department of Computer Engineering, Ajeenkya D Y Patil University, Pune, India

ABSTRACT

This chapter goes on to explore the current status of machine learning and its relevance to fake news identification and their potential as a powerful tool for social innovation in the digital age. With the ever-growing number of contents misinforming others and going viral in the online platform, it is quite necessary to accept the role of algorithms in machine learning as possible solutions that could help contain the false information rage. While key advancements such as NLP and neural networks enable extraction of meaning in content, authorship, and user behavior in very complex patterns, issues related to data bias, the degree of algorithmic transparency, and more intelligent tactics of misinformation remain. Beyond the technical, this chapter shines a light on the ethical and social dimensions of AI-driven interventions; it underlines the need for not only transparency but also for inclusiveness in deploying such technologies.

DOI: 10.4018/979-8-3693-6910-4.ch005

INTRODUCTION

This explodes the amount of information prevalent in today's digital landscape to unprecedented levels in opportunities as well as challenges. Social media web sites, news sites, and blogs permit ideas and information flows rapidly and encourage access to knowledge like never before. The openness, however, also breeds the same rapid spread of misinformation and fake news, affecting public opinion, dissolving trust in credible institutions, and, indeed, contributes to societal polarization. Take, for example the 2016 U.S. presidential election and the COVID-19 pandemic that pinpoints the very characteristic of false narratives-namely their ability to influence behaviors and beliefs in a remarkably short time. It is thus a challenge of urgency to develop effective mechanisms to fight this pervasive issue (Khanam et al., 2021).

Machine learning has emerged as one of the transformative technologies in opposing challenges resulting from fake news. Machine learning systems could analyze, categorize, and identify information that was totally out of the question by the previous generations due to the amounts of data used and sophisticated algorithms. Given numerous techniques such as NLP, neural networks, and supervised learning, machine learning systems may distinguish misleading content and evaluate its credibility. For instance, NLP allows algorithms to understand context, semantics, and sentiment in text that would allow them to differentiate between fact reporting and sensationalized accounts-or even fabricated ones. This, therefore, may make the machine learning enhance the ability with which we can move through this complex information ecosystem toward truth and transparency (Ahmad et al., 2020).

All the same, the deployment of machine learning in spotting fake news is not that seamless. An important challenge associated with using algorithms and machine learning algorithms is data bias: bad data will, in turn, lead to bad results due to the fact that their training has been based on skewed or unrepresentative datasets. A model that has been exposed to certain sorts of content or viewpoints may not be able to recognize, or really evaluate information coming from different sources. This is a question of providing diversity and variety in training data used so that algorithms can make the right judgments to assess credibility of a wide range of sources and viewpoints (Baarir & Djeffal, 2021).

It also presents a challenge for detection, as the methods used by false-narrative purveyors increase with advancing technology. Misinformation spreads in different ways today than perhaps even just a few years ago; rather than just merely declaring what they would have heard and ignored, perpetrators use, for example, deepfakes, multi-channel disinformation campaigns, and hacking tools at advanced levels to reach people. Such new dangers evolve with time, so their detection algorithms and techniques must adjust constantly-keeping the cycle going between those that research and provide solutions to these problems and those who wish to exploit them (Reis et al., 2019).

However, the ethical implications of applying machine learning to this end cannot be brushed aside. There are questions of algorithmic clarity, responsibility, and censorship when such systems determine what's true information and what's not. How could these systems, then, ever be assured of neutral functionality-when to avoid unconscious silencing of dissenting voices? In sum, deployment of machine learning must go hand in hand with robust ethical guidance and frameworks which place emphasis on transparency and user agency (Sharma et al., 2020). Apart from technological and ethical challenges, it can be shown how machine learning can be combined with social innovation to address misinformation. Social innovation involves new strategies, concepts, as well as ideas that may meet the social needs as well as promote positive change. Applying machine learning in this context will allow stakeholders to come up with a collaborative approach involvin g communities, improve media literacy, and thus make the stakeholders understand credible sources (Sharma et al., 2020).

Education is a leading role player in this end. Media literacy campaigns will be supported with education designed to equip people to view their information with a critical eye. Societies can become more resistant to the charms of melodramatic or misleading content by becoming a skeptical, investigative society. In addition, alliances between technology engineers and developers, educators, and policymakers can help produce tools and resources that support users in understanding what is needed to responsibly negotiate the digital landscape. Solution-builders also need better collaboration among diverse stakeholders -- tech companies, researchers, journalists, and civil society organizations. Initiatives that enable transparency of algorithms, promote best-practice sharing, and agree on common content verification standards could all help advance efforts to curb the spreading of disinformation. Such a collaborative approach could help create an environment in which machine learning technologies bolster more democratic discourse and civic activity rather than quashing it (Alghamdi et al., 2024).

Knowing the promise of machine learning not only as a tool for finding fake news but also as a driver of social innovation at this digital age, we can redesign the way technology could be used in the information landscape to promote truth, accountability, and civic responsibility. This chapter will outline the newest development and challenges regarding machine learning algorithms in terms of detection of fake news and their complicated nexus with technology, ethics, and society (Khan et al., 2021).

Overview of current state-of-the-art machine learning algorithms used in the context of fake news detection, evaluation of effectiveness, and limitations. Case studies of successful implementations, with areas for improvement. The chapter focuses on the critical analysis of ethical considerations and implications for social innovation, which can make forward-looking machine learning that inculcates such an informed and resilient society. All this information sets in a panorama of how complicated this issue is and the possibility to be found by creating a future where technology becomes a good force that frees an informed public who can quite justly navigate the intricacies of the digital age with some semblance of discernment and integrity (Mahir et al., 2019).

Scope

The scope of this chapter includes a comprehensive review of the advancements in machine learning techniques regarding fake news detection, including both natural language processing and algorithmic strategies. Here, it talks of the historical context of misinformation, significant instances of fake news, and the dynamic nature of the phenomenon. It discusses the ethics and policy implications linked to detecting fake news and how technology companies and policymakers can coordinate to promote a responsible information ecosystem. It also elaborates further on community engagement and social innovation through media literacy, grassroots initiatives, and equipping people with the skills to critically appraise information accessed. This chapter provides a panoramic overview of interplays among societal, technological, and misinformation phenomena.

Objectives

This chapter will focus on how machine learning can be connected with social innovation in the effort to identify fake news. It is important to understand how technology is advancing in ways that help deal with the spread of misinformation. It also looks at various players-from policymakers to corporate technology companies-who are involved in identifying solutions to deal with the problem of fake news. It will then design the chapter to take an all-embracing view of how things are and the potential prospects that may be envisaged in the fight against misinformation in the near future by centering on shared cooperation and ethical concerns.

UNDERSTANDING FAKE NEWS

Definition and Types of False News

False news is false information that appears to be a news story, either by traditional media sources, social media, or through online forums. It has the intention of misleading, misinformation, or manipulation of the audience towards various goals, which often include political, monetary, or social gain. Such news can take many forms and can start from full-blown complete fabrication of the story, misleading headlines, and manipulated images and videos. Another variant is satire or parody, which might be mistaken for fact reporting. Lastly, there is biased reporting, which favors specific aspects while withholding some essential aspects. All these categories are important to understand because they reveal the various intentions behind the spread of misinformation and how it impacts public opinion and engagements (Kishwar & Zafar, 2023).

Historical Background of False Information

False information traces its background way back in history, where ancestors used dissent to cause action. In historical terms, some claims were more or less valid. So, by this general knowledge, one would ask numerous questions about how false information has evolved over the years, especially from its early backgrounds up to now (Abedalla et al., 2019).

This phenomenon, misinformation, is in no way new at all and has its roots in communication dating back centuries. Historically, the governments and political entities used propaganda to mold public opinion and control the narratives. For instance, both World War I and II were known for propaganda usage highly to enhance national morale as well as to demoralize the enemy. However, in the strictest sense, the dawn of the internet and social media has changed the landscape of this phenomenon called misinformation. Where the dissemination of information became exponential, hence accelerating in the late 20th and early 21st century, with the advent of social media and online broadcasting networks like Facebook, Twitter, and YouTube. Democratized the sharing of information, whereby any person with internet access can become a content creator, and this ease of dissemination has resulted in the spread of fake information, which complicates efforts to distinguish between credible sources and unreliable ones. The 2016 U.S. presidential election and the aftermath of the rise in populism around the globe made it clear how fake news can impact the public opinion and winning electoral results, only to ask for drastic measures against fake news without any delay.

Critical Events of Fake News

Some critical events of fake news have shaped public discourse and emphasized the seriousness of the matter. Perhaps the best example is in one's memory of the 2016 U.S. presidential election when fake stories spread out so fast on social media influencing the perceptions and behaviors of voters. It included fallacious articles concerning candidates and policies, as well as some conspiracy theory, which tapped into the beliefs of several demographics of voters. Another pandemic is the case where the social media was oversaturated with false information about the origin and method of immunity and treatment of the virus, which then shaped public health response and observance of health measures (Capuano et al., 2023). Another articleable example is the "Plandemic" conspiracy, which postulates that the virus had been intentionally created, thus greatly earning public distrust in health authorities. These events serve as examples of how fabricated news may indeed produce very real effects on the public, health, and democratic processes. Therefore, whereas the face of the digital world continues changing, the understanding of characteristics and implications of fake news is still an important factor in the efforts toward the creation of an informed and resilient society (Capuano et al., 2023; Choudhury & Acharjee, 2023; Kishwar & Zafar, 2023).

Therefore, knowing fake news encompasses the whole definition, types, history, and even some major instances. This will help in identifying all the types of misinformation as well as the stages it has undergone. With such information, it becomes easy to identify and fight some aspects of fake news in our daily lives. The process is necessary in most contemporary societies since getting a hold of truths amid false notions can be challenging.

MACHINE LEARNING TECHNIQUES IN FAKE NEWS DETECTION

Using enormous data sets and complex algorithms, machine learning techniques represent indispensable tools in the fight against fake news by enabling better identification and classification of misinformation. In this section, we'll focus on the basic components of machine learning techniques applied to fake news detection: Natural Language Processing (NLP), a general overview of relevant algorithms, the role of data sources, and how to create a dataset and the feature extraction methods (Kreššáková et al., 2019). The figure 1 illustrates the use of machine learning techniques in detecting fake news.

Figure 1. Machine learning techniques in fake news detection

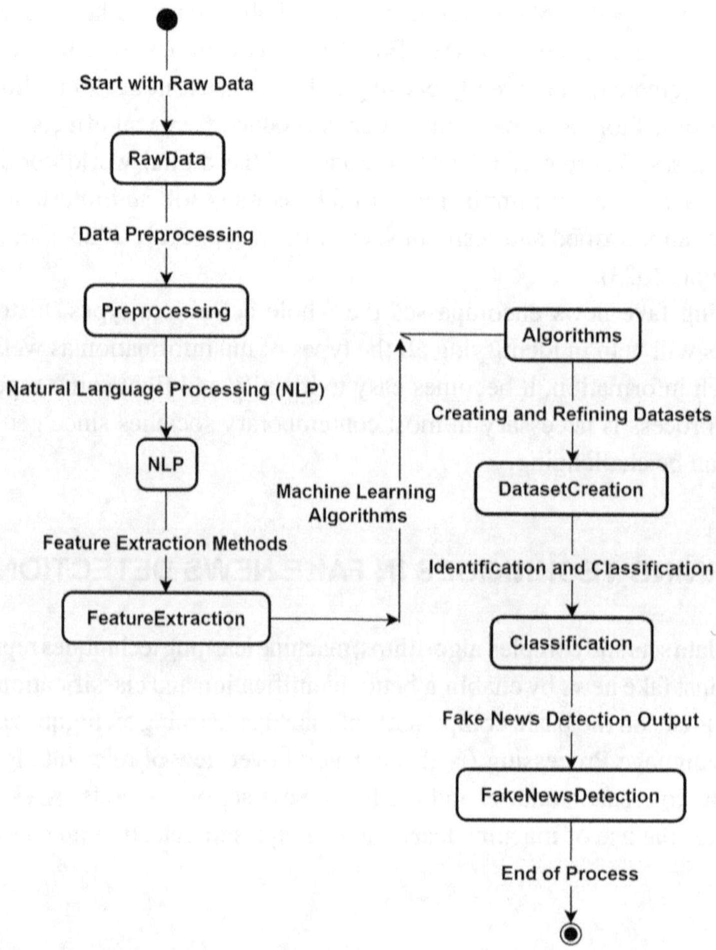

Natural Language Processing

NLP is a domain of artificial intelligence dealing with the interaction between humans and computers. NLP techniques are the "golden nuggets" that help interpret textual data in the light of fake news detection. NLP allows algorithms to work on language so that they may be able to see or predict context, sentiment, and intent in what people are reading in the news, posts on social media, or any other form of writing (Abdulrahman & Baykara, 2020).

This means utilizing NLP for the purpose of sentiment analysis, that is, determining an emotional tone or resonance in any given text. Sentiment analysis can determine whether a news article has a positive, negative, or neutral stance and hence useful in determining whether some content is biased or misleading. Moreover, NLP methods allow topic modeling, that helps classify documents into categories established based on latent themes, thus helping identify articles with similar or disparate content. Another important aspect of NLP is named entity recognition. This refers to the process of finding and categorizing key entities such as people, organizations, and locations in text. This really helps flag articles containing false information that claims to wrongly involve people or organizations. Databases verified against recognized entities can, with more certainly, determine articles that may contain false statements.

Another point is the capability of NLP in using linguistic features, particularly syntax and semantics to improve accuracy on detecting models of misinformation. The algorithm can be used for better detection by analyzing sentence structure and selection of words with signs of information, such as overstated use of language and the absence of credible sources.

Machine Learning Algorithms

Machine learning covers several algorithms and techniques that can be used for the detection of fake news. The different algorithms may be divided into two broad categories: supervised and unsupervised learning methods (Choudhury & Acharjee, 2023; Sahoo & Gupta, 2021). Supervised Learning is specifically learning algorithms that use labeled datasets in which input data such as news articles are paired with corresponding output labels, for example, "fake" or "real." Some supervised learning algorithms that have been applied for the detection of fake news are:

a) Support Vector Machines: The algorithm constructs a hyperplane that splits the data points in multidimensional space; thus, it is best suited for binary classification problems.
b) Decision Trees: In such models, the decision process relies on the creation of tree-based structures based on feature values, because these models are easily interpretable and provide an easy visualization of the workings of the classifier.

Random Forest is an ensemble learning method that constructs a number of decision trees at the time of training and outputs a mode of all its predictions. It improves accuracy and robustness.

Neural Networks

It is particularly useful for very large-sized datasets as well as handling complex patterns. Variants like Convolutional Neural Networks (CNNs) and Recurrent Neural Networks have been promising networks and have been applied in text data analysis, capturing contextual information. Unsupervised learning involves unlabeled data from which the algorithm will try to find inherent patterns or clusters. Methods such as clustering allow similar articles to be lumped together, hence further helping spot misinformation trends early.

One other important feature of algorithms in machine learning is that they learn improvements with time with such mistakes. Models can be retrained once there is increasingly more data to be enhanced so as to increase the accuracy with which fake news can be detected by adjusting themselves to new tactics used by the spreaders of misinformation.

Data Sources and Dataset Creation

The quality and diversity of the data used for training as well as testing define the efficiency of machine learning models to a great extent. There is a good amount of relevant datasets available from social media, news websites, and fact-checking organizations for fake news detection. Some of the most widely used datasets for such purposes include:

a) Fake News Dataset: Such datasets consist mainly of a combination of real as well as fake news articles annotated for the purpose of classification. They can be efficiently used to train and test a model.
b) Social Media Data: Social networks such as Twitter and Facebook provide tremendous stores of user-generated data, and are therefore excellent resources for real-time analysis of the spreading of misinformation.
c) Fact-Checking Databases: Organizations like Snopes and PolitiFact create databases of verified facts and discredited claims. Such databases could be used to gauge the accuracy of news articles.

The datasets for fake news detection should be produced in such a way that elements of fake and real news are presented without bias towards model training. Linguistic diversity and variation in cultural context as well as differences in writing style are also considered in creating the data. This provides robust models that generalize well in different contexts.

Feature Extraction Methods

One of the major stages in preparing textual data for machine learning models is feature extraction, concerning the process of converting raw text to the numeric form that algorithms can handle. Several techniques are popularly used in feature extraction related to fake news detection:

a) Vector of Bags of Words (BoW): In this method, text is represented as a vector of words. Since the vector does not take into consideration grammar and order, though it does consider count, frequency counts are retained. This allows one to assign each unique word to a particular index; hence, one can build a feature vector for each document.
b) Term Frequency-Inverse Document Frequency (TF-IDF): An extension of the BoW model, TF-IDF weighs the importance of words in a document relative to their frequency across a corpus. This method can help bring to the surface some unique terms which might indicate what is fiction news.
c) Word Embeddings: Techniques like Word2Vec and GloVe can provide very dense representations of words. They capture semantic relationships. Such embeddings allow the model to understand the semantic meaning of a word better, thus increasing the rate of accuracy for the detection.
d) N-grams: This simply involves creating contiguous sequences of 'n' items from the given text. Models can then capture phrases or contextual patterns that may point to fake news.
e) Stylometric Features: In these, linguistic features such as sentence length, word count, and some vocabulary used by authors or writers are incorporated. Such features can provide a distinction between true news writings and fake news writings by relying on writing style.

Techniques for the detection of fake news are highly reliant on machine learning. NLP, along with many algorithms, robust data sources, and effective methods for feature extraction, work together to increase the accuracy and reliability of the system. As this evolving landscape of misinformation unfolds, research and development in these fields are going to continue serving as vital means of combating fake news and creating an informed society. Advanced technologies can be leveraged to achieve systems capable of detection of misinformation and to instill much-needed transparency and accountability in the spread of information.

ADVANCEMENTS IN FAKE NEWS DETECTION

With the rise in fake news during the digital age, it has become necessary to develop more advanced detection techniques. The rising complexity of false information has a direct relation with technological advancement-thus development in machine learning and natural language processing-which will be important weapons in this challenge. This section describes the latest developments in the sphere of machine learning examines some case studies of success, and then discusses the tools and platforms developed for detection (Ahmed et al., 2021; Hu et al., 2022; Kumar et al., 2020). Figure 2 depicts advancements in fake news detection.

Figure 2. Advancements in fake news detection

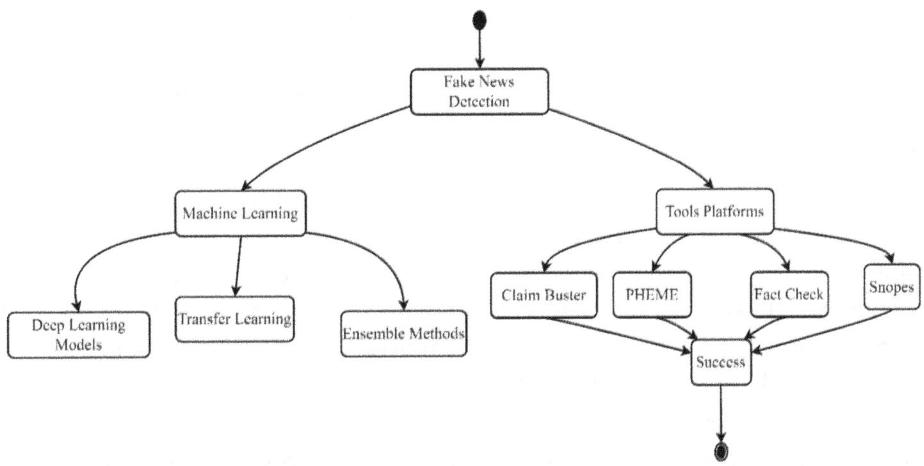

Recent Innovations in Machine Learning

Recent advances in machine learning significantly improved the capabilities of fake news detection systems. Some of the most significant recent innovations include deep learning models. Among these, CNNs and RNNs are noticeable because they process very complex data and have demonstrated really high accuracy in text classification tasks. The CNN might be able to capture local patterns from text, where for example it would be able to identify phrases that corresponded to fake news sites, whereas the application of RNNs, particularly the design of LSTM networks, better learns the sequential nature of texts and captures more context over longer sequences of words.

The application of transfer learning. This includes pre-training a large model on a very diverse set of data and then fine-tuning it on a smaller relevant dataset for fake news. Transfer learning has been outstandingly suitable in boosting the performance of models when the labeled data is not abundant. For instance, models such as BERT, Bidirectional Encoder Representations from Transformers, for example, transformed the understanding of natural language based on huge amounts of data during pre-training for high-accuracy tasks such as sentiment analysis and entity recognition, vital in detecting fake news. Other ensemble learning techniques added capabilities for detection as well. Variance can be reduced, and predictive performance improved by combining multiple models. One of these methods is ensemble methods such as Random Forests or Gradient Boosting Machines, which actually increases accuracy. This approach will often significantly increase accuracy with the aggregation of multiple algorithms' strengths and helps in compensating for the weaknesses of separate models. Other innovations in feature selection are more efficient and easy to spot than others, meaning better algorithms have been proposed that streamline models and facilitate classification with the most important features.

Success Stories and Case Studies

Thus, many fake news detection systems have been successfully implemented. For instance, machine learning algorithms are used on Facebook to flag false news articles and inform the readers through such misinformation. As such, Facebook may target any content violating its own community standards through user reporting as well as algorithmic detection. The platform partners with fact-checking organizations that allow them to tap the expertise of outsiders as well as provide context to users regarding articles being flagged. This has consequently resulted in a massive reduction of misinformation on the platforms controlled by Facebook, which, in turn, is a direct result of machine learning applications in real terms. The other example is the Twitter approach to COVID-19 misinformation (Ahmed et al., 2021; Hu et al., 2022; Kumar et al., 2020). This platform had looked into the machine learning models that could be used in identifying and labeling those tweets that contained misinformation about the virus. It could use natural language processing techniques to assess the content and context of the tweets, which would consequently lead to delivering accurate information to the users through this platform and from credible sources. As such, this initiative not only curbed the spread of false information but also increased the good trust that users have in their source of news.

There have been several researches in the academic sphere to prove the validity of machine learning in the recognition of fake news. For instance, researchers at the University of Michigan started such a study to determine the linguistic qualities of news articles through sophisticated algorithms and predict whether they are true or false. The researchers have set an extraordinary degree of accuracy between real and fake news through the development of such a model by considering such important factors as choice of words, sentiment, and overall structure of the article. Stories such as these only further the demand to incessantly dig into, discover, and innovate in this field, consequently enhancing the power of even more dependable detection systems.

Tools and Platforms for Detection

Nevertheless, the emergence of fake news detection has introduced a plethora of tools and platforms intending to detect and ultimately curtail misinformation. Among the most popular tools available so far is ClaimBuster developed by the University of Texas at Arlington. It analyzes statements made by public figures through machine learning algorithms that find their factual truth content. The tool, in return, provides journalists, researchers, and the public with real-time fact-checking (Alghamdi et al., 2024; Sharma et al., 2020).

There is another bright tool called Hoaxy, which visualizes the spread of misinformation and fact-checking work on social media platforms. Its core functionality consists of the tracking of the propagation of news articles through comparison with a verified source, enabling users to understand dynamics in the spread of misinformation and the effectiveness of measures put in place to counter this. Such transparency is thus important for researchers and policymakers seeking to go about the problem of fake news. NewsGuard also produces a browser extension that rates the credibility of news sources on websites. With trained journalists rating news sources on transparency, accuracy, and editorial standards, news for users is guaranteed to be real. It is such a tool that shows how human expertise can create a synergy with algorithms in machine learning, thus enabling a more holistic system in fighting fake news. Besides these specific tools, larger tech companies are still integrating fake news detection capabilities in their existing platforms. For instance, Google provides algorithms that rate the credibility of a news source appearing on its search results pages and ranks reliable information higher and lower-ranking misinformation. This only serves to improve the general user experience and plays a crucial role in exposing the public to media literacy.

Hence, advances in the field of fake news detection reflected the innovation in machine learning, case studies, and development of specialized tools and platforms. Deep learning models, transfer learning techniques, and ensemble methods have made detection systems more accurate and efficient. Major platforms like Facebook and Twitter illustrate the impact on combating misinformation from real-world applications. And so, in a rapidly bewildered information landscape, ongoing research and innovation will be crucial for developing strategies to detect and counter the spread of fake news even more effectively. Centering on machine learning, but keeping in mind the importance of human expertise in collaborating with technology, we can continue to work toward a more discerning and informed society that is better capable of managing the challenges thrown up by digital misinformation.

CHALLENGES IN FAKE NEWS DETECTION

Advance in the field of machine learning and artificial intelligence raises a few worthy challenges in detecting fake news. Data bias and representation issues as well as the dynamic nature of misinformation generally challenge methods for improving algorithmic transparency and accountability while adhering to censorship ethics. Improvement on these challenges is, therefore, a necessity for effective, trustworthy, and socially responsible detection of fake news (Hiramath & Deshpande, 2019; Ibrishimova & Li, 2020; Monti et al., 2019). The figure 3 illustrates the challenges in detecting fake news.

Figure 3. Challenges in fake news detection

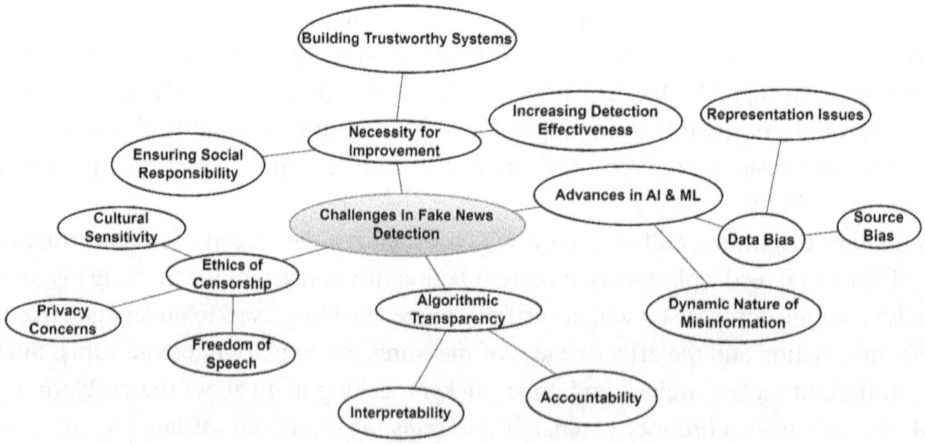

Data Bias and Representation Issues

The foremost challenge in detecting fake news is bias in data. Machine learning models are trained from historical data, and if such data is skewed or unrepresentative, then the models could foster it and even amplify these very biases. For instance, if a model is trained mostly with English-language articles, then it cannot very well identify such articles in other languages or dialects. This creates a disadvantage not only in the functioning of the model but also in perpetuating underrepresentation in the digital information space.

Bias can also occur with the way misinformation is categorized or labeled. Thus, if stories or perspectives overwhelmingly exist within certain training datasets, the algorithms become biased toward those narratives. For example, an algorithm may incorrectly tag articles representing the current or minority political viewpoint as some form of fake news in which diversity in public discourse suffers. Such a thought underlines the necessity of curating comprehensive and balanced datasets so that detection systems can function equitably across contexts and communities.

The Changing Nature of Misinformation

Another challenge in the detection of fake news is that it is dynamic and everchanging. Misinformation schemes have continually been dynamic where, for instance, scammers are always on their toes for strategies related to detection. For instance, the first fake news article may have very sensational headlines and exaggerated claims, while the more sophisticated fake news campaigns use hidden approaches in regards to presenting misleading information in credible formats. Further, rapid diffusion of information across all the social media networks exacerbates this situation. Because repackaging and re-sharing misinformation in distorted ways from their original context, it becomes a challenge to distinguish between false claims and true ones by detection algorithms. Deepfakes enhanced through AI-generated synthetic media further make the situation complicated with a blurred line between real and fake content. As the mechanism of deception evolves, so must the detection algorithms, requiring continuous research and study to stay one step ahead of the threats that only continue to roll in.

Algorithmic Transparency and Accountability

Algorithmic transparency and accountability is a major issue within the architecture of fake news detection. Some machine learning models are "black boxes." The decisions they make are not well understood, making their decision-making processes too vague for transparency. This has led to questions about how these models classify the information and what biases may unintentionally be reflected in their output.

To the users who come across such flagged information or true information, they may also not know why it has been determined that way. This diminishes confidence in detection systems, thus casting doubt over their ability to be trustworthy. To add salt to injury, the users wrongly flagged by detection systems-for instance, as happened when news media was also wrongly labeled as bogus-will have to endure damages in terms of reputational loss as well as a decrease in public confidence with actual journalism. Such concerns are best met by embedding transparency-inducing frameworks in algorithmic processing. Explainable models, which would provide insights into decision-making processes by researchers and developers, would help. The system being fair and efficient in its operations also calls for accountability in the form of audits and routine evaluations of detection mechanisms.

Ethical Considerations and Concerns with Censorship

Implementation of such fake news detection systems raises critical ethical considerations, too. For example, it raises concerns about censorship. As algorithms for detecting such news are implemented by platforms and organizations, the risk that people may try to suppress rightful pieces of content to control the spread of misinformation is not taken lightly and may stifle free expression and limit the diversity of viewpoints available in public discourse.

This balance is very fragile: combating dangerous lies should be an important need for the reduction of their proliferation, but one should not let measures taken on this matter cross the borderline against infringement rights on people's freedom of expression. To achieve this, the importance of value frameworks relating to the rights of users and fair representation in decisions on content moderation is very high. Moreover, the risk of political influences on detection systems raises further ethical concerns. Detection algorithms are vulnerable to targeting particular groups or ideologies if they are biased toward particular interests of stakeholders. The mechanisms should be neutral to create a trust and credibility relationship with the public. The rise of artificial intelligence and advanced machine learning opened up possibilities of efficient fake news detection, but challenges did not reduce; some of these challenges include data bias and representation, the constantly changing nature of misinformation, algorithm transparency, and ethical considerations. As the implications of misinformation unfold in society, an inclusive, transparent, and ethically sound approach to detecting fake news will, more than ever, be crucial to fostering a healthier information ecosystem. And so, too, will pro-active tackling of these challenges from the side of stakeholders, ready and eager to forge a solution, as being respectively higher in integrity and resolve within free expression and diversity in public discourse.

SOCIAL INNOVATION AND COMMUNITY ENGAGEMENT

Social innovation and community engagement are crucial to the battle against false news and misinformation. These approaches heighten not only the effectiveness of the systems of detection but empower individuals and communities to navigate the media landscape. The following section delves into ways of building media literacy as well as critical thinking skills, the cooperation of stakeholders toward these ends, as well as grassroots initiatives and educational programs which further an informed community (Choudhary et al., 2021; Mahir et al., 2019; Mishra et al., 2022; Shaikh & Patil, 2020). The figure 4 depicts the intersection of social innovation and community engagement.

Figure 4. Social innovation and community engagement

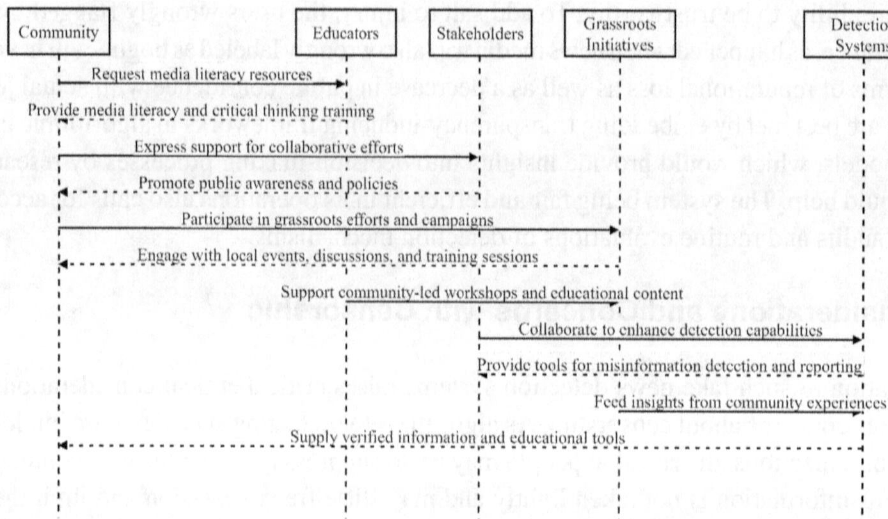

Critical Thinking and Media Literacy

Media literacy has been considered a fundamentally important skill in today's information world. Media literacy can be defined as teaching how to critically analyze news sources, how to distinguish between good sources of information and bad ones, and how to decipher the motivation attributed to different narratives constructed within the media. This will make people assimilate not only information but comment on it as well, thus creating a discerning public. Educational establishments have a lot to do with this process. Schools and universities may introduce media literacy as part of their curricula, challenging students to look critically at news sources, notice biases, and understand what lies can do. For instance, courses that make students analyze the actual articles that are floating in the media, explore fact-checking resources, and discuss the effects of misinformation greatly improve their critical thinking skills. Moreover, discussion-based learning facilitates an environment where all different voices are heard to learn from differences, leading to a healthy culture of respectful dialogue.

However, formal education is just half the battle, and training in workshops or seminars may be an integral part of adult education and other subgroups. Organizations may join forces with local libraries, community centers, and non-profits to host workshops intended to promote media literacy for adults and subpopulations. These workshops can emphasize awareness over fake news, how algorithms control the information one sees, and healthy practices online. Making these resources accessible would lead to more citizens being aware and engaging in their communities.

Multilateral Efforts of Stakeholders

Fake news detection cannot be done in a vacuum-it requires multilateral efforts by multiple stakeholders, such as tech companies, educational institutions, non-profits, and government agencies. It will take an effort from all parties to come up with one strategy to fight the misinformation. One successful form of collaboration is creating coalitions with diverse stakeholders united in the cause of misinformation. For example, the Trust Project and News Literacy Project unite journalists, educators, and technologists in creating resources and tools designed to promote greater news literacy. Through coalition building, diverse pools of resources and expertise can be used for solutions that cover the complexities of misinformation. The same goes for tech companies in their mandate to employ collaborative work. They ought to affiliate with academic institutions and research entities that would help them improve their content moderation systems. Collaborations with these stakeholders best enable the sharing of best practices, enhancing algorithmic transparency, and earning users' trust. Google has collaborated with fact-checking organizations to enhance its information ecosystem in such a way that users would have access to verified facts behind news stories.

Moreover, policymakers can play a highly influential role in fostering collaborative environments. The government can sponsor legislation that pushes for media literacy and the tech sector's accountability to give impetus to frameworks that allow stakeholders to work with each other. Cooperation might lead to stricter policies with more effective solutions of misinformation without violating democratic virtues and individual rights.

Grassroots Initiatives and Educational Programs

Grassroots actions will empower the community to combat fake news at the grassroots level. Grassroots initiatives often rise in their organic form and are initiated by individuals or small groups interested in the campaign against fake news. Grassroots initiatives take various formats: community-led discussions, media campaigns, etc. For example, community organizations will be charged with organizing peer-led media literacy programs that will make subscribers relate their own experiences and thinking about consumption news. In this course of action, it will surely be characterized by a feeling of ownership and involvement by the community in the distribution of information; it will naturally make them become champions of informed choices (Singh Madan et al., 2024; Vaithianathan et al., 2024). A workshop where participants share stories and strategies on how they point to false information can help people build a collective understanding of the issue while at the same time helping one operate in a culture of critical thinking. For instance, local educational programs can concentrate on involving young people in meaningful discussions on media and information. Such programs as school-based clubs or after-school programs give students an opportunity to critically analyze news and engage in fruitful dialogues. Developing own content and analyzing different media forms empower students to take an active role in the process of shaping information landscapes.

Additionally, partnerships with local media outlets allow grassroots movements to reach larger segments of the audience. For instance, joint work between community groups and local journalists helps create the needed information content targeted at localized issues with misinformation. The initiatives also provide an opportunity for a conversation between producers and consumers of media, thus creating transparency and fostering trust. Fake news is a cross-cutting challenge that requires multi-faceted responses involving social innovation and community engagement. These are: building media literacy and critical thinking; the efforts to influence joint action for stakeholder interests; advocacy for grassroots initiatives and education programs-better directed towards building an informed and resilient society. Critical engagement with information is the key to healthy information ecosystems, and empowering individuals and communities turns this ideal into a reality. Future Focus Continued social innovation will be essential for our efforts leading on to a digital future free from the scourge of misinformation. Communities will need to be engaged toward a culture of trust, responsibility, and informed citizenship through collective action and shared responsibility toward an information landscape that is more transparent and truthful for all.

POLICY IMPLICATIONS AND FUTURE DIRECTIONS

As the situation with fake news continues to evolve, therefore, effective policies and collaborative frameworks shall play an important role in fighting the spread of misinformation. The section deals with regulatory considerations in relation to detecting fake news, the role of technology companies and policymakers, and future trends in machine learning and social innovation (Kumar et al., 2020; Monti et al., 2019; Wani et al., 2021). The policy implications and future directions are depicted in Figure 5.

Figure 5. Policy implications and future directions

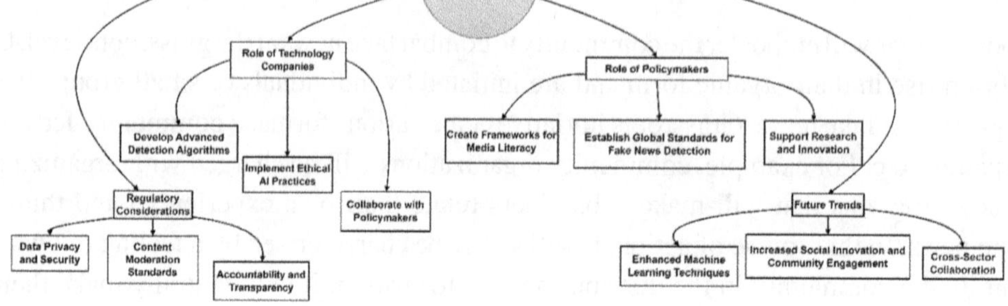

Regulatory Considerations in Fake News Detection

Fake news demands more strongly regulatory frameworks in handling this new issue of misinformation by striking the balance between free expression. The challenge for policymakers is to draft appropriate regulations which could effectively curb the spread of false information while not violating the rights and freedoms of an individual. Definition and standard definition: The clear definition and establishment of standards for fake news will be a critical aspect of consideration. Policymakers can involve the consultation of experts, including journalists, media educators, and scholars, to draft broad guidelines defining what should be considered as misinformation, disinformation, or otherwise harmful content. In such a manner, regulation measures would dawn from definitions, thereby offering coherence and constant enforcement perspectives of jurisdictive as well as platform views.

Beyond that, transparency should be included in any regime. Technology companies should be made imperative to disclose their content moderation policies, such as the algorithms the companies use to flag and take away content from their platforms. Allowing clarity about how certain pieces of content are identified as fake news would help build trust between platforms and users, who would know how their information is being handled. The regulations should also spur inter-stakeholder collaboration. Policymakers can empower the tech companies, civil society organizations, and academic institutions collaborate in the achieving the best practices and innovative solutions for fake news detection. Such collaboration can breed a common understanding of the impact of misinformation and help develop effective strategies against it.

For instance, regulatory frameworks should put more prominence on user education and media literacy initiatives. This way, governments can offer financial support to back educational programs focused on the critical perception of the information that people consume. Media literacy in the formal education system and community outreach programs therefore stand as a policy action that policymakers can pursue in building an informed, discerning public.

Role of Tech Companies and Policymakers

Hence, the role that technology companies must play in addressing the challenge of fake news is pivotal. They are some of the primary platforms through which information channels enter, and they should take all means necessary to contain misinformation without infringing on their commitments to respect user privacy and free expression. Companies should, above all, make their best efforts in developing robust content moderation systems, combining advanced machine learning algorithms and human oversight, to attack fake news. The third route that tech companies can take is to invest in tools for empowering the user. This might include fact-checking resources, browser extensions, and alerts that warn users on encountering potentially misleading content. In this way, providing information to users helps them make informed decisions and could be the most important part in promoting media literacy.

Hence, incentives created by policymakers could serve to catalyze the development of a new incentive structure that would encourage tech companies to place emphasis on responsible information management in the first place. Financial incentives toward companies that work out ways of successful mitigations or enforcement through regulatory arrangements of greater transparency and accountability in the processes involved in content moderation can be effective incentives toward such a culture in the industry: helping to procure a different approach whereby companies prioritize public welfare over private profit.

Besides, policy and tech companies can collaborate to increase research and development in terms of developing something applicable in combating misinformation. Governments can invest further in research initiatives on information dynamics toward further developing solutions on technology that can be put into practice. Their partnership with academic institutions and research organizations will guide policymakers on evidence-informed development of regulatory measures and industry standards.

Future Trends in Machine Learning and Social Innovation

Many future trends in machine learning and social innovation will reshape the landscape of fake news detection and misinformation management going forward. For instance, there are rising trends of artificial intelligence hybridized with human intelligence on content moderation; machine learning algorithms are very good at processing immense amounts of data quickly but lack human oversight about the nuances of the context (Alghamdi et al., 2024; Choudhury & Acharjee, 2023; Kishwar & Zafar, 2023). Hybrid models that append AI judgments with human judgment would yield greater accuracy in the detection of misinformation but maintain the richness of human understanding.

There is also the role of decentralized systems of information verification. Blockchain technology can ensure the transparent and tamper-proof verification of information through a decentralized platform of fact-checking and validation of contents by stakeholders, leading to increased trust in the information ecosystem. This would help individuals in finding their sources of validation and tracing the origin of the information, increasing accountability and transparency. Social innovation will likely trigger new initiatives aimed at countering misinformation. Community-based approaches tapping into local knowledge and expertise will be key when combating misinformation from the bottom up. Initiatives that bring citizens in media literacy efforts, collaborative fact-checking, and dialogue around misinformation can lay the building blocks for community resilience against fake news.

One other increasingly important role will be data ethics. As deployed, machine learning models detecting fake news will require ethical considerations on matters of data collection, privacy, and bias in algorithms. By stakeholders, frameworks should be set to ensure ethical practices in the management of such data for public trust in the technology solutions applied to misinformation management. Finally, when the tactics of misinformation evolve, the need to constantly innovate and adapt in the detection methods will be all the more pronounced. In the future, times will demand flexible approaches that can adapt to emerging forms of misinformation such as deepfakes or AI-generated content. Ongoing research and development in machine learning and social innovation will become all the more urgent to stay ahead of emerging threats.

A more holistic approach is needed to combat fake news: it involves the consideration of aspects like regulations, close coordination between technology companies and policymakers, forward-thinking approaches in machine learning, and social innovation. Stakeholders need to be able to define clear definitions while making things transparent for users and having education provided to users to help them navigate better in combating misinformation. The importance of this possible cooperation between technological companies and policymakers is in the rearing of responsible management for information with the empowerment of people to tread the digital landscape with confidence. Moving into the future, ethical practices and community engagement will be imperative to building a resilient information ecosystem that can test the challenges associated with fake news. Through collective action and shared responsibility, we can work toward a more informed society and a more healthy information environment.

CONCLUSION

The widespread dissemination of fake news now poses significant challenges to the integrity of information and social trust in the media. This chapter has explored the intersection of machine learning technologies and social innovation: an avenue that offers opportunities as well as obstacles on the journey toward detecting fake news effectively. Efforts by developing natural language processing, machine learning algorithms, as well as community engagement initiatives, have reflected proactive steps toward reducing fake news. But the challenges of data bias, evolving nature of misinformation, and ethical concerns are critical and have a need for continuing vigilance and adaptability ahead of increased complexities. There is a need for collaboration among policymakers, technology companies, and communities to create a resilient information ecosystem. In such an approach, through setting transparency, media literacy, and ethical standards as priorities, stakeholders would be able to navigate the complex intricacies of the digital world and create a better-informed public and better protect democratic discourse. The fight against fake news is a constant journey, and as technology advances, so must our strategies to ensure that truth rings above the noise in a distorted media world.

REFERENCES

Abdulrahman, A., & Baykara, M. (2020). Fake news detection using machine learning and deep learning algorithms. *2020 International Conference on Advanced Science and Engineering (ICOASE)*, 18–23. DOI: 10.1109/ICOASE51841.2020.9436605

Abedalla, A., Al-Sadi, A., & Abdullah, M. (2019). A closer look at fake news detection: A deep learning perspective. *Proceedings of the 3rd International Conference on Advances in Artificial Intelligence*, 24–28. DOI: 10.1145/3369114.3369149

Ahmad, I., Yousaf, M., Yousaf, S., & Ahmad, M. O. (2020). Fake news detection using machine learning ensemble methods. *Complexity, 2020*(1), 8885861. DOI: 10.1155/2020/8885861

Ahmed, A. A. A., Aljabouh, A., Donepudi, P. K., & Choi, M. S. (2021). Detecting fake news using machine learning: A systematic literature review. *arXiv Preprint arXiv:2102.04458.*

Alghamdi, J., Luo, S., & Lin, Y. (2024). A comprehensive survey on machine learning approaches for fake news detection. *Multimedia Tools and Applications, 83*(17), 51009–51067. DOI: 10.1007/s11042-023-17470-8

Baarir, N. F., & Djeffal, A. (2021). Fake news detection using machine learning. *2020 2nd International Workshop on Human-Centric Smart Environments for Health and Well-Being (IHSH)*, 125–130.

Capuano, N., Fenza, G., Loia, V., & Nota, F. D. (2023). Content-based fake news detection with machine and deep learning: A systematic review. *Neurocomputing, 530*, 91–103. DOI: 10.1016/j.neucom.2023.02.005

Choudhary, M., Jha, S., Saxena, D., Singh, A. K., & Associates. (2021). A review of fake news detection methods using machine learning. *2021 2nd International Conference for Emerging Technology (INCET)*, 1–5.

Choudhury, D., & Acharjee, T. (2023). A novel approach to fake news detection in social networks using genetic algorithm applying machine learning classifiers. *Multimedia Tools and Applications, 82*(6), 9029–9045. DOI: 10.1007/s11042-022-12788-1

Hiramath, C. K., & Deshpande, G. (2019). Fake news detection using deep learning techniques. *2019 1st International Conference on Advances in Information Technology (ICAIT)*, 411–415.

Hu, L., Wei, S., Zhao, Z., & Wu, B. (2022). Deep learning for fake news detection: A comprehensive survey. *AI Open, 3*, 133–155. DOI: 10.1016/j.aiopen.2022.09.001

Ibrishimova, M. D., & Li, K. F. (2020). A machine learning approach to fake news detection using knowledge verification and natural language processing. *Advances in Intelligent Networking and Collaborative Systems: The 11th International Conference on Intelligent Networking and Collaborative Systems (INCoS-2019)*, 223–234.

Khan, J. Y., Khondaker, M. T. I., Afroz, S., Uddin, G., & Iqbal, A. (2021). A benchmark study of machine learning models for online fake news detection. *Machine Learning with Applications, 4*, 100032. DOI: 10.1016/j.mlwa.2021.100032

Khanam, Z., Alwasel, B., Sirafi, H., & Rashid, M. (2021). Fake news detection using machine learning approaches. *IOP Conference Series. Materials Science and Engineering, 1099*(1), 012040. DOI: 10.1088/1757-899X/1099/1/012040

Kishwar, A., & Zafar, A. (2023). Fake news detection on Pakistani news using machine learning and deep learning. *Expert Systems with Applications, 211*, 118558. DOI: 10.1016/j.eswa.2022.118558

Krešňáková, V. M., Sarnovský, M., & Butka, P. (2019). Deep learning methods for Fake News detection. *2019 IEEE 19th International Symposium on Computational Intelligence and Informatics and 7th IEEE International Conference on Recent Achievements in Mechatronics, Automation, Computer Sciences and Robotics (CINTI-MACRo)*, 000143–000148.

Kumar, S., Asthana, R., Upadhyay, S., Upreti, N., & Akbar, M. (2020). Fake news detection using deep learning models: A novel approach. *Transactions on Emerging Telecommunications Technologies, 31*(2), e3767. DOI: 10.1002/ett.3767

Madan, B. S., Najma, U., & Rana, D. P. PK, J. K., Sathyakala, S., & Boopathi, S. (2024). Empowering Leadership in Higher Education: Driving Student Performance, Faculty Development, and Institutional Progress. In Best Practices for Behavior Intervention in Special Education (pp. 191-221). IGI Global.

Mahir, E. M., Akhter, S., Huq, M. R., & Associates. (2019). Detecting fake news using machine learning and deep learning algorithms. *2019 7th International Conference on Smart Computing & Communications (ICSCC)*, 1–5.

Mishra, S., Shukla, P., & Agarwal, R. (2022). Analyzing machine learning enabled fake news detection techniques for diversified datasets. *Wireless Communications and Mobile Computing, 2022*(1), 1575365. DOI: 10.1155/2022/1575365

Monti, F., Frasca, F., Eynard, D., Mannion, D., & Bronstein, M. M. (2019). Fake news detection on social media using geometric deep learning. *arXiv Preprint arXiv:1902.06673.*

Reis, J. C., Correia, A., Murai, F., Veloso, A., & Benevenuto, F. (2019). Explainable machine learning for fake news detection. *Proceedings of the 10th ACM Conference on Web Science*, 17–26. DOI: 10.1145/3292522.3326027

Sahoo, S. R., & Gupta, B. B. (2021). Multiple features based approach for automatic fake news detection on social networks using deep learning. *Applied Soft Computing, 100*, 106983. DOI: 10.1016/j.asoc.2020.106983

Shaikh, J., & Patil, R. (2020). Fake news detection using machine learning. *2020 IEEE International Symposium on Sustainable Energy, Signal Processing and Cyber Security (iSSSC)*, 1–5.

Sharma, U., Saran, S., & Patil, S. M. (2020). Fake news detection using machine learning algorithms. [IJCRT]. *International Journal of Creative Research Thoughts, 8*(6), 509–518.

Vaithianathan, V., Shastri, D. S., Nandakumar, V., Misba, M., Kumar, P. S. V. V. S. R., & Boopathi, S. (2024). NEP Policy Implementation Strategies in Education in India. In *Educational Philosophy and Sociological Foundation ofEducation* (pp. 325–356). IGI Global., DOI: 10.4018/979-8-3693-3587-1.ch013

Wani, A., Joshi, I., Khandve, S., Wagh, V., & Joshi, R. (2021). Evaluating deep learning approaches for covid19 fake news detection. *Combating Online Hostile Posts in Regional Languages during Emergency Situation: First International Workshop, CONSTRAINT 2021, Collocated with AAAI 2021, Virtual Event, February 8, 2021, Revised Selected Papers 1*, 153–163.

KEY TERMS AND DEFINITIONS

NLP (Natural Language Processing): The interaction between computers and human language, enabling machines to understand and interpret text or speech and then having a response.

AI (Artificial Intelligence): Creating intelligent machines that think and learn like human beings.

COVID: The general pandemic of the world that fueled the need to have digital communication as well as accuracy of information.

II (Information Innovation): Exploiting technology in the management of data hence making its distribution easier and faster.

CNN (Convolutional Neural Network): Algorithms are used in deep learning, but much more on images, lately adopted in text analysis.

TF (Term Frequency): A measure that shows how frequently a term in the document occurs.

IDF (Inverse Document Frequency): A statistic indicating how important a word is to a document in a collection.

RNN (Recurrent Neural Network): A type of neural network suited for sequential data- such as text.

LSTM: It is a state-of-the-art RNN variant engineered to better capture long-range dependencies in sequences.

BERT: It is a transformer-based model specifically engineered for capturing the contextualized representations of words within search queries and texts.

Chapter 6
Consumer Behavior Prediction Using Machine Learning Algorithms

Nawab Akram
Magadh Professional Institute, India

K. Aravindhan
https://orcid.org/0000-0002-0570-2226
St. Joseph's College of Engineering, India

K. Sujatha
https://orcid.org/0000-0002-3409-7635
Vignan's Foundation for Science Technology and Research, India

Shueb Ali Khan
Kunwar Satyavira College of Engineering and Management, India

Gagan Singh
https://orcid.org/0009-0007-7819-0149
Meerut Institute of Technology, India

R. Rajagopal
Alliance University, India

ABSTRACT

The chapter reviews consumer behavior fundamentals such purchasing decision criteria, quantification, and analysis. Next, it discusses decision trees, ensemble approaches, neural networks, regression analysis, and support vector machines for customer behavior prediction. Each algorithm's technique, merits, and weaknesses are examined. We then show how these algorithms can predict purchase intent, customer turnover, life cycle value, and other consumer behaviors. This setting presents challenges when using machine learning models, including data quality, model interpretability, and the ethics of using consumer data. The industry's current and future innovations, such as AI-driven personalization and deep learning to understand client behavior, are also highlighted. This chapter explores the intersection of machine learning and consumer behavior analysis to help researchers, data scientists, and business professionals acquire more accurate and valuable consumer insights.

DOI: 10.4018/979-8-3693-6910-4.ch006

1. INTRODUCTION

Data mining is a modern tool for consumer behavior models that helps to answer time-sensitive problems. Client behavior prediction is a difficult and unpredictable task. Hence, the right approach and strategy are required to create customer behaviour models. In order to target certain demographics or consumers, marketers may find it difficult to construct a prediction model. Despite their complexity, customer models are usually quite simple. Unreliable predictions are often the result of customer behavior models that overlook crucial factors. People make purchasing decisions every single day. Before making a final commitment, there are more factors to think about when making a purchase, not limited to price. In addition to the asking price, corporations may include other factors in an offer to boost the perceived value of a purchase. Through the application of data-driven machine learning, this study seeks to ascertain the influence of quantifiable and objective elements on consumer choices. A number of factors, including product price information and environmental factors, interact to form a set of independent variables that impact consumers' decision-making processes while making purchases. Analyzing consumer behavior is essential for every company. It helps businesses learn more about their customers' wants and needs. The outcome will be an improved service or a better product for the client. With the use of customer behavior analysis, a business can boost sales and expand its success. All parts of business operations are now affected by digital transformation. Data about products and buyers can be gathered with the use of information systems. Utilizing data science technologies allows for a more thorough understanding of client intentions.

1.1 Overview of Consumer Behaviour

Customers, producers, marketers, vendors, advertising mediums, cultures, traditions, and personal psychology are all aspects of variety that the global market encompasses. There are many similarities despite the differences. For both for-profit and non-profit organizations, consumer behavior research seeks to provide marketers with conceptual and technical tools for marketing. In order to understand and predict consumer buying habits, marketers rely heavily on customer behavior research. Everything from the why, when, where, how, and how often of a purchase to its use and eventual disposal is a part of consumer behavior. A product's price, its intended use, the purchaser's emotional investment, and the reasons behind the purchase can all vary, according to the research. Since acquiring and retaining consumers is the ultimate aim of every organization, Harvard Business School Professor Theodore Levitt thinks that teaching students about consumer behavior is essential for business education. Creating and keeping customers is the goal of marketing. Successful marketing is all about getting to know your target audience and winning them over. Successful marketing strategies are the result of extensive research into customer behavior, marketing tactics, and the nature of the relationship between the two parties. Concepts and theories in consumer behavior have their roots in the work of philosophers, scientists, and academics from a variety of disciplines. Researching consumer behaviour aims to teach marketers how to do thorough consumer analyses, which in turn improves their knowledge of the market and their ability to formulate effective strategies. In order to better understand customer needs and anticipate technological developments, management can benefit from consumer behavior research. These days, strategic market planning can't be done without considering consumer behaviour. The comprehensive marketing strategy is built around it. Marketers are encouraged to fulfill the demands of their target markets while also benefiting society as a whole, according to the societal marketing idea, which stresses the significance of ethics and social responsibility in marketing decisions.

2. TYPES OF BUYING BEHAVIOR

2.1 Routine Response Behavior

A standard Reaction Consistently buying low-priced, low-effort thing causes this behavior. You won't have to put as much thought into your purchase because there isn't much "risk" involved. People tend to buy routine stuff without much thought. Customers have experience with and a history of buying items in this category. Soft beverages, snacks, milk, and eggs are some examples of items that are frequently purchased.

2.2 Limited Decision-Making Behavior

When buying products rarely or on occasion, consumers have fewer options. Even though he's never heard of the brand before, he might look up some basic details about it in a familiar category. The research will be simple and should not take too long to complete. Take an article of clothing for instance. Buyers often have broad ideas about what they want, rather than a preference for a particular brand or style.

2.3 Extensive Decision-Making Behavior

The decision-making process takes longer for items and services that are complex and involve the consumer. Taking your time to make a decision is essential when purchasing unknown and expensive products or services. The amount of time needed to make a decision is proportional to the danger level of the transaction. Investing in things like cars, houses, computers, and education can be fraught with performance, psychological, and financial hazards. Customers will be diligent in their research before buying anything. Consumers can accomplish a comprehensive search by consulting internet resources, asking friends and family for advice, and visiting physical businesses or outlets that sell the product or service.

2.4 Impulse Buying Behavior

Buyers might not always give their purchases the thought they deserve. When this happens, the decision to buy is made in a flash. Feelings of emotion might lead to impulsive purchases. An enthusiastic customer's personal investment in a product could result in a sale. Products like chips, candy, gum, mints, and other sweets displayed prominently in stores or near checkout lanes may encourage customers to buy them on the spur of the moment.Buying the same things could lead to different purchasing habits. Conditions might cause products to change categories. Investing in a wedding suit or dress is no small feat. For those who don't eat out very often, picking a restaurant for dinner might be a daunting task, while regulars may have fewer alternatives. Therefore, the extent to which decisions are made is defined by the reason for the dinner.

3. DATA SOURCES FOR CONSUMER BEHAVIOR ANALYTICS

Machine learning's capacity to foretell and analyze consumer behavior is heavily dependent on the data's quality, diversity, and relevancy. Machine learning models can better understand business strategies and consumer preferences with the help of big datasets that reveal trends and patterns. The most common places to find information about consumer behavior are detailed here:

3.1 Transactional Data

Transactional data is essential for understanding consumer purchase habits. It contains detailed records of consumer purchases, like:

- Type of Product - Determine the types of products that are being purchased in order to gain a better grasp of the tastes of consumers and the demand in the market.
- Quantity - By keeping track of the quantity of items purchased in each transaction, it is possible to identify patterns of bulk purchasing as well as preferences for particular products.
- Price - Through the analysis of price sensitivity, one can learn about the willingness of consumers to pay as well as the effect that pricing tactics have on sales.
- Date and Time of Purchase - An understanding of the times at which purchases are made might help reveal patterns that are associated with seasonality, holidays, or promotional periods.

Online marketplaces, customer relationship management (CRM) databases, and point-of-sale (POS) systems typically provide this data. Machine learning models can use this data to enhance inventory management, forecast consumer behavior, and propose products.

3.2 Web and Mobile Analytics

In today's digital world, customers leave a trail of data when they engage with brands through various online platforms. Web and mobile analytics provide information about customers' online behavior through the following channels:

- Page Views – Keeping track of which pages are visited the most frequently can provide insight into the interests of customers as well as the efficiency of the information on a website.
- Click Stream Data - It is possible to gain insights into the decision-making process of consumers by analyzing the sequence of pages or actions that people perform while using a website or app.
- Session Durations - Measuring the amount of time that users spend on a website or app can provide insight into the level of engagement and the relevance of the material.
- Bounce Rates - Gaining an understanding of the reasons why users quickly leave a website can assist in locating areas in which the user experience can be improved.

This data is gathered via tools like Google Analytics and Mixpanel, among similar platforms. With the use of machine learning models, you can personalize user experiences, optimize content strategy, and boost conversion rates.

3.3 Social Media

Facebook, Twitter, Instagram, and LinkedIn are just some of the social media networks that include substantial amounts of information about their users. These venues offer the following:

- User content generation – Consumers' ideas, preferences, and feelings regarding items or brands can be reflected in user-generated content such as posts, comments, and discussions.
- Engagement Metrics - It is possible to gain insights on the social effect of content and the interests of consumers through the use of likes, shares, retweets, and comments.
- Sentiment Brand - Understanding public opinion and the reputation of a brand can be aided by doing an analysis of the sentiments expressed in interactions on social media.

It is feasible to get information from social media sites by making use of social listening tools and APIs. Applying machine learning techniques like natural language processing (NLP) to this data can yield insights about consumer sentiment, emerging trends, and competitive analysis.

3.4 Customer Feedback

Many social media platforms save extensive user data, including Facebook, Twitter, Instagram, and LinkedIn. At these locations, you can enjoy:

- Product Reviews – Obtaining feedback from customers provides immediate insights into the expectations and satisfaction levels of consumers. It includes the following:
- Surveys - Detailed insights into consumer preferences, expectations, and areas for development can be obtained through the collection of structured feedback through surveys.
- Interactions of customer support - While conducting an analysis of contacts with customer support, it is possible to identify common problems, areas of discomfort, and areas in which products or services may be lacking.

Machine learning models can sift through this qualitative data, find trends, predict customer happiness, and offer up ways to enhance products and services.

3.5 Demographic Data

In order to better understand the habits of different types of consumers and to segment the market accordingly, demographic data is invaluable. Features that are crucial for a certain demographic include:

- Gender and Age - When it comes to targeted marketing, it is absolutely essential to take into account the fact that different age groups and genders may exhibit different tastes and purchasing behaviors.
- Education and Income - A person's purchasing power and product choices can be influenced by factors such as their educational background and income level.
- Location - The influence of cultural variables on consumer behavior can be uncovered through the use of geographic data, which can indicate regional preferences.

Customer profiles, surveys, and third-party data providers are common ways that this information is collected. Machine learning models that combine demographic data with other sources of data have the potential to produce more precise consumer segments and more efficient marketing campaign personalization.

3.6 Demographic Data

The Internet of Things (IoT) has become increasingly widespread, and as a result, sensor data has emerged as a new source of consumer insights. Information from sensors is gathered from:

- Wearables - Wearable technology such as smartwatches and fitness trackers offer consumers the ability to monitor their activities, health measurements, and lifestyle choices in real time.
- Smart Home Devices - Wearable technology such as smartwatches and fitness trackers offer consumers the ability to monitor their activities, health measurements, and lifestyle choices in real time.
- Location Sensors - The Global Positioning System (GPS) data collected by smartphones and connected vehicles can provide helpful insights into the mobility patterns and location-based activities of consumers.

Machine learning models can better understand consumer behavior with the help of this real-time contextual data. They can do things like predict needs based on activity patterns and provide personalized suggestions that are aware of the context. Since machine learning models draw from a large pool of data, they may paint a complete picture of customer behavior. Organizations may enhance their level of customization, get more precise predictions, and make better decisions through the integration of diverse datasets.

4. LITERATURE SURVEY

Companies are collecting more and more customer data, making the topic of predicting client buying behavior an important one for researchers. Predicting how customers would spend their money has been the subject of several approaches. A large number of them made use of data mining and machine learning strategies (Chaubey et al., 2023). What follows is a synopsis of current studies conducted in this area. Experiments utilizing logical discriminant analysis to forecast customer purchase behavior were presented. Using a decision tree, the maximum accuracy that was achieved was 90.9%. In order to find out which clients react to a business's offers, (Das, 2016) created a prediction algorithm. Support vector machines, naive Bayes, and K-nearest Neighbour are some of the algorithms used in other studies. Based on the findings, the naive Bayes technique outperforms the others with a 95% accuracy rate. In order to forecast customer attrition, (Vafeiadis et al., 2015) employed a variety of machine learning methods, including boosted versions of logistic regression, naïve Bayes, decision trees, SVM, and artificial neural networks. The data used for this came from customers' buying habits in the past. With an accuracy of 97%, the boosted SVM (SVM-POLY with Adaboost) emerged as the top classifier. Based on the results of the study, businesses can benefit from predictive analytics and a better understanding of consumer behavior when they use behavioural computing and analytics (Adebola et al., 2019). Traders boost their market share by huge promotions, according to (Xu et al., 2018)research. These deals won't work for repeat business because most of your clients are just interested in making a quick buck. Traders can solve this problem by figuring out which clients have the potential to become repeat purchases. Advertising to repeat buyers can increase return on investment (ROI) while decreasing spending. The proposed approach anticipates future purchases using a GBDT-based two-layer model fusion algorithm (TMFBG). Over fifty million raw data points are pre-processed in preparation for correlation mining, as per research by (Wang et al., 2021). Chosen at random, with 60% going into training and 20% into evaluation and assessment. By combining the best features of logistic regression with XGBoost, two models were developed. Logistic regression, a component of the XGBoost system, is both effective and efficient when compared to competing approaches for evaluating models.Revenue, consumer purchases, location, and equipment make up around 2.5 petabytes of data that Walmart collects every hour (Bradlow et al., 2017). For precise assessments of customer value, it is necessary to link individual-level data from online and physical retail stores. Data might be used for more than just online shopping in the future. In-store customer journey data can also be collected using RFID and GPS tracking technologies. Researchers McCarthy et al. Predictive analytics provide a more complete picture of customer actions than other approaches. The strategy employs a plethora of methods to define the customer's activities(Kulkarni & Kaleel, 2019). Regression analysis can be used to determine the degree of correlation between various factors influencing consumer purchases. Data mining, analysis, and deep learning algorithms are some of the ways to comprehend consumer behavior. With the use of predictive analytics, companies can better focus on making a profit by gauging how customers will interact with their products.

5. MACHINE LEARNING FOR CONSUMER BEHAVIOR PREDICTION

Recent years have seen a proliferation of consumer behavior models built mostly from data mined from actual customers; these models are often designed to answer a single query at a single moment in time. The process of predicting consumer behavior is challenging and full of unknowns. It is crucial to have the right methodology and strategy when developing models of customer behavior. Once a prediction model is built, it's not easy to tweak it to suit the needs of the marketer. This makes it hard to pinpoint the precise marketing actions to take for each client or set of clients. The bulk of consumer models are really very simple, even though this formulation is pretty complex. The bulk of consumer behavior models fail to take into account several important factors due to this necessity, which is why their predictions are usually inaccurate.

5.1 Regression Analysis

A continuous dependant variable is necessary for linear regression analysis. The presence or absence of a disease, the likelihood of a patient's death within a given time frame, and many other clinical and epidemiological factors are inherently binary. The link between a continuous or not-continuous independent variable and a dummy variable, which can only take on two possible values—0 for absent or 1 for present—is described by logistic regression analysis.

5.1.1 Linear Regression

The linear regression (LR) model posits a linear relationship between the dependent variable e and the Z independent variables $v_1, v_2, \dots v_z$.

$$e_g = \alpha_0 + \alpha_1 v_{1g} + \alpha_2 v_{2g} + \dots + \alpha_z v_{zg} + \vartheta_g$$

In the equation above, the subscript g represents the ithdata.The intercept is α_0, $\alpha_1$and α_z are the independent variable regression coefficients, and "is the error that the model cannot explain." The estimation of individual regression coefficients can be skewed in a linear regression model due to multicollinearity, which occurs when independent variables are highly connected. The trained regression model has the capability to produce VIFs, or variance inflation factors. With $q = 1,2,\dots,z$, VIFs measure the linear correlation between the g th variable (v_g) and the remaining variable ($v_{(-g)}$).

Multicollinearity, defined as a high degree of linearity between independent variables, is indicated by VIF values greater than 10. Estimates of regression coefficients and the stability of model predictions are both compromised by multicollinearity. Multicollinearity is handled using the principal component regression model. Unlike models that use a large number of highly correlated independent variables, main component regression models use a small number of independent components. Two approaches can be used to determine the quantity of main parts: Choose the amount of principle components to utilize with the scree plot when the slope changes suddenly or when the cumulative dispersion ratio reaches a threshold you set. Ninety percent of the total collected variation was explained by the study's key components.

5.1.1 Logistic Regression

One mathematical tool for describing the connection between a binary dependent variable and numerous independent variables is Logistic Regression (LR). To estimate parameters that are not known, LR employs a maximum likelihood method. Because it usually denotes models with continuous target features, the term "Logistic Regression" could be deceiving. Nevertheless, regression is not LR's primary application; classification is. To sum up, LR uses a probabilistic method for classification. One way to convert a point in a multidimensional feature space to a number between zero and one is by using the logistic function. One way to look at the logistic model is as the chance of belonging to a certain class given a certain cutoff. Lastly, for every feature vector, LR supplies class probabilities. The predetermined cutoff determines the course of study. The main benefit of this approach is that it demonstrates, in the end, how inputs justify outputs. Having MD in the dataset, though, reduces performance.

Recurrence events in a group of women categorized by their age at menarche (e.g., 9–17) can be used as training data to apply LR to scenarios involving BC recurrence. The code for a recurring event is 1 and the number for an infrequent one is 0. Logistic regression (LR) uses training data transformed into a probabilistic function by applying a logistic function to it; this function then converts an input (the patient's age at menarche) into the likelihood of recurrence. Using a decision threshold on the interval [0,1] where the probability of a recurrence event lies allows for a binary class assignment ("no-recurrence" = 0 and "recurrence" = 1). A patient with a recurrence likelihood of 0.85 would be considered a "recurrence" if the threshold was set at 0.7.

5.2 Decision Tree

Data is conveyed through rules and conditional statements in a decision tree (DT), a graph shaped like a tree. The process is repeated until the last requirement is satisfied by a classification rule, which groups similar data. A "regression tree" is a specific kind of decision tree that is designed to deal with regression issues. For each dependent variable, the decision tree iteratively separates the set of independent variables until it finds the best rule for describing them. At each stage, binary recursive partitioning finds the branching variable or point that has the smallest mean squared error (MSE). A whole tree is created by repeating the procedure. Here is how a regression tree makes its prediction: The trained tree takes an average of the response values from each node to predict the value of a new independent variable (x). If you want better predictions, prune some branches of the tree. Improved and simplified understanding of mature trees is a direct result of pruning, which in turn leads to more precise predictions.

Figure 1. Basic architecture of DT

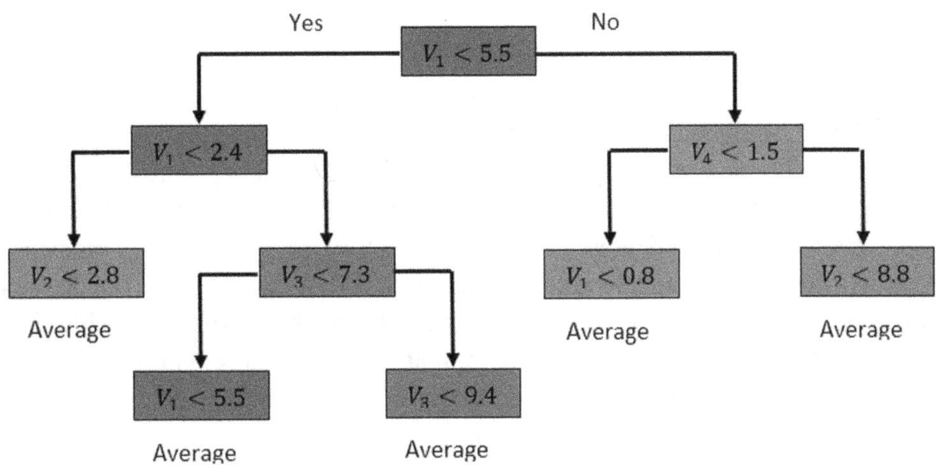

The use of a decision tree to finalize learning is shown in Figure 1. Intuitive model interpretation is made possible with the use of a decision tree. The study generated decision trees using Statis by utilizing the "rpart" package.

5.3 Random Forest

Random forest (RF) is an ensemble model that generates multiple prediction models and combines them to create a final model (Breiman, 2001). Using a subset of the independent variables, the RF trains decision trees and generates bootstrap samples from the original data. Averaging or voting on the predictions made by the bootstrap tree is used to generate final predictions when new data points for the independent variables are introduced. The RF regression method makes use of averaging. New data points for the independent variables are shown in Figure 2 to illustrate the RF prediction technique.

Figure 2. Example of random forest

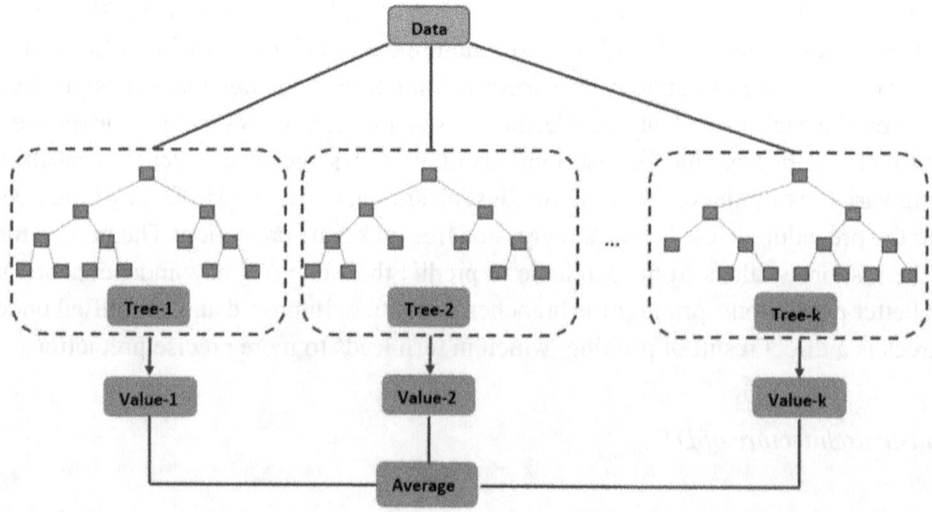

A measure of the prediction performance of the bootstrap samples that comprise the RF is the out-of-bounds error (OOB). By utilizing extra data that isn't part of the bootstrap samples, the out-of-the-box error analyzes how well bootstrap trees predict. Bootstrapping with an appropriate number of samples improves the model's prediction ability and reduces out-of-band error. Every bootstrap tree can also benefit from having fewer independent variables. The default for regression-based prediction is to use 1/3 of the total number of independent variables (Breiman, 2001). In contrast to the decision tree, the RF model does not lend itself to intuitive interpretation. It is possible to find the highly significant independent variables in each bootstrap tree by determining their importance. For the RF provided by the statistical program, the research made use of the "randomForest" library.

5.4 Support Vector Machine

To categorize binary data, researchers employ the SVM (Christopher J.C. Burges, 1998), a generalized linear classifier. Using an N-dimensional data dimension, the SVM locates a 1-dimensional hyperplane for a set of data points (J. Wei et al., 2019). The hyperplane effectively partitions training data into two groups when classifying 2D data sets. When the algorithm comes across data that is not familiar, it will use the classification model. A linear SVM method is illustrated in Figure 3. Fast research throughout the 1990s has led to improved algorithms for the SVM, which was initially suggested in 1964 (Lal et al., 2004). Face recognition, text classification, biomedicine, and other pattern recognition applications have all made use of the support vector machine (SVM) (Brown et al., 2000).

Figure 3. Support Vector Machine (SVM)

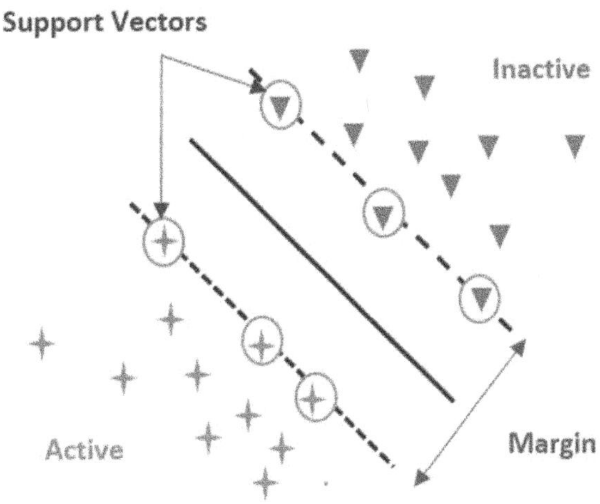

The best treatment that is similar to a target chemical can be found during primary screening with the help of SVM's strong classification performance (Warmuth et al., 2003). Furthermore, structure-property relationships may be effectively identified using SVMs. By manipulating input variables including starting gel concentration, reaction method, temperature, and time during synthesis, (Manuel Serra et al., 2007) successfully predicted the structural and thermodynamic properties of zeolite products.

5.5 Neural Networks and Deep Learning

Architecture, values, and rules of operation are the defining features of neural networks. When put together, they can form the backbone of any network. Neural networks face the challenge of learning by modifying their values. Two approaches exist for estimating a neural network's worth: rule-based network analysis and design calculations. The second method is used by the majority of neural networks to find the value of the network. Some examples of popular network models and learning algorithms are Hopfield networks and back propagation arithmetic.

5.5.1 Convolutional Neural Network

A Convolutional Neural Network (CNN) consists of many hidden layers in addition to its input and output layers. Fully connected (FC), pooled (POOL), and convolutional (CONV) are three ways to classify these layers. Using elementwise nonlinearity, the nonlinearity activation function can be appropriately represented as a layer. Full ConvNets are typically generated by stacking these layers in the order described in this section. Data equals images, according to CNN's central concept; this idea guides the design of the architecture to meet the requirements of image processing in particular. The several layers that make up a convolutional neural network (CNN) stand for the various network dimensions(Sakib et al., 2018).

The CNN architecture leverages the 2D structure of the input images or audio signals through nearby connections, linked weights, and pooling, resulting in features that are translation-invariant.

Figure 4. CNN basic architecture

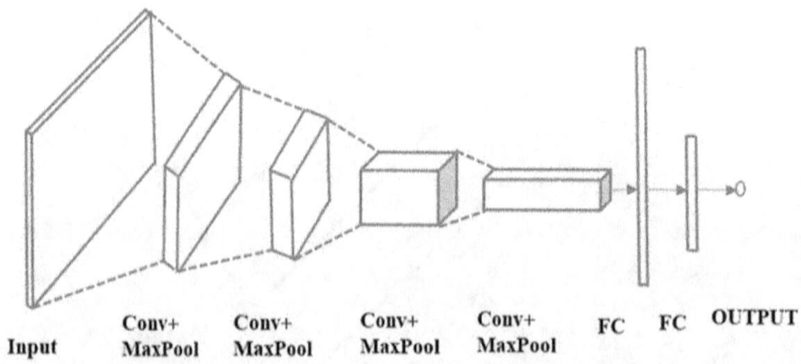

After convolutional and subsampling layers, a convolutional neural network (CNN) has numerous fully connected layers. Feature maps are used as input and output sets of arrays at every level. Feature maps for colored images are 2D arrays with a color channel, in contrast to feature maps for films which are 3D arrays with an audio channel. In the output stage, the attributes of each data point are extracted. Figure 4 provides a high-level representation of CNN architecture.

5.5.2 Artificial Neural Network

Mathematical models that attempt to capture the architecture and operation of real brain networks are known as artificial neural networks (ANNs). An artificial neuron is the basic mathematical building block of an artificial neural network. Multiplication, addition, and activation are the three basic rule sets that make up this architecture. Each input value is multiplied by its appropriate weight when the artificial neuron receives its weighted inputs. An artificial neuron's central processing unit (CPU) is responsible for adding up all of the weighted inputs and bias(Kenji, 2011). Despite having simple operating principles, the entire potential and calculation power of artificial neuron models can be unlocked by connecting them into neural networks (Figure 5). The idea that basic rules can produce complicated results is the basis of artificial neural networks.

Figure 5. Example of simple ANN

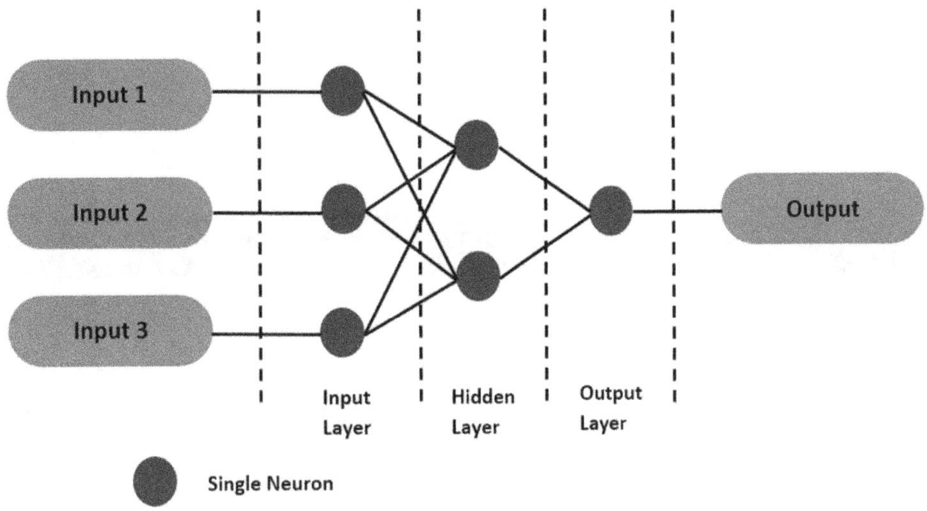

Connecting specific artificial neurons, rather than randomly joining them, allows us to maximize the benefits of mathematical complexity while making the system more manageable. For ANNs, researchers have already created "standardised" topographies. Problems can be solved more quickly and efficiently with predefined topographies. Different issues call for different topographies of artificial neural networks. Selecting the best topology and honing the artificial neural network can resolve a problem. The settings and topologies need to be fine-tuned. It is the process, not the outcome, of fine-tuning an artificial neural network's topology that is necessary for its application. We need to train an artificial neural network to address particular issues before we can deploy it. Similar to how organic neural networks learn from their surroundings, artificial neural networks can do the same. A learning paradigm can be either supervised or unsupervised; reinforcement learning is the third main paradigm. We modeled our approach to learning after that of artificial neural networks, which is task-based. The basic idea behind artificial neural networks is that they can take in input signals and use "learning data" and "learning rules" (a chosen cost function) to generate reliable output responses. The artificial neural network can be trained to address a specific problem when its topology has been fine-tuned. A large number of disciplines make use of artificial neural networks. These include chemistry, process control, gaming, genetics, banking, fraud detection, radar systems, the automotive and space industries, and astronomy. Approximating functions, analyzing regressions, predicting time series, classifying data, recognizing patterns, processing data, filtering, and clustering are just a few of the many challenges they can handle.

5.6 K-Means Clustering

The process of cluster analysis involves dividing a research object into numerous categories according to its shared characteristics. Classes that are close to each other have a high degree of similarity, but classes that are far apart have a low level of similarity(Li et al., 2019). Clustering analysis is an unsupervised learning method that excels at understanding low-dimensional data sets using straightforward logic. The main aspects that impact categorization are the data's properties, type, and clustering goal. Its purpose is to divide the data set into several independent subsets. Figure 6 illustrates the particular method.

Figure 6. Cluster analysis processing

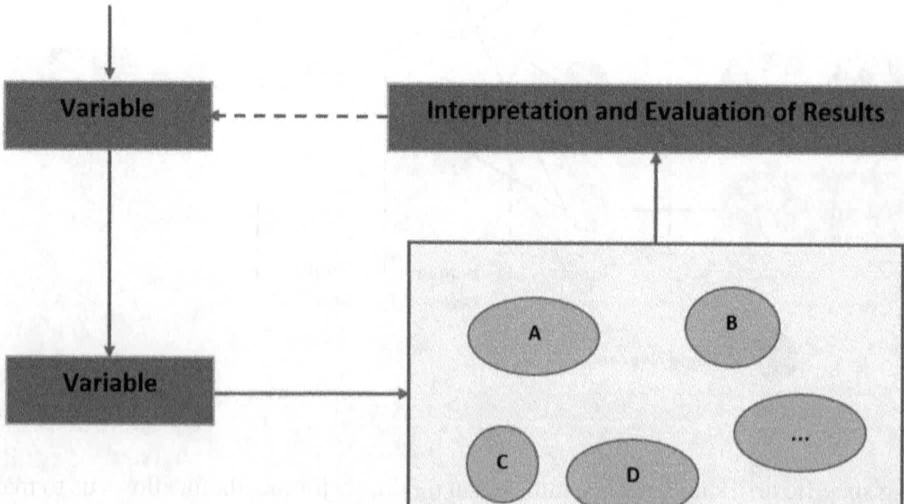

One common method for cluster analysis is the K-Means algorithm, which uses the average distance between nodes to determine how to divide up the cluster. Put an n-item data set into the algorithm, pick k clusters at random, choose the closest cluster according to how far away each cluster is from the rest of the objects, and keep recalculating the average value of each cluster until the function converges; this will produce clusters with low inter-cluster similarity and high intra-cluster similarity (Zhang, Y., Zhao, X., Sun, Y., Liu, M., & Shi, 2013). Typically, the square error criteria is applied:

$$Y = \sum_{g=1}^{z} \sum_{s=w_g} \left|s - b_g\right|^2$$

where s is a point in space and mi is the average value of Cluster w_g.

5.7 Association Rule Learning

Marketers and businesses alike have begun to embrace association rule mining. Association rule mining finds widespread use in market basket analysis. Finding out which products are typically purchased together is the goal. With this information in hand, experts can strategically place frequently bought items next to each other, resulting in higher sales. The connections between data points in a dataset are what it's all about. Association rule mining finds collections of items that have a lot of transactions but little support. Regular item sets are the simplest way to describe these types of transactions. The two main steps of an association rule algorithm are(1) discovering frequent sets and(2) building association rules.

5.7.1 Apriori Algorithm

The Apriori algorithm, introduced by(Kumar, 2012), is a widely used method for mining frequent item collections. Apriori identifies all frequently used item sets in a database (DB). Iterative database searches are necessary for the Apriori algorithm. The algorithm iteratively examines (k+1) itemsets using a breadth-first search (level-wise search). The Apriori property is the basis of the Apriori method; it states that every nonempty subset of a collection of often occurring items must also be frequently occurring. A system's supersets will likewise fail the minimum support test if the system itself fails, according to the antimonotonic property. Common supersets imply rare sets and rare sets imply common infrequent sets. Elements that aren't candidates very often can have this property removed from them. To begin, we find the 1-itemsets that appear most often. L is the set of all items that satisfy the support condition and have exactly one item in them. Each subsequent pass uses the previous pass's seed set of big itemsets as its starting point. In order to back up the results of the data analysis, the seed set creates massive item sets called candidate itemsets. The most common and extensive candidate itemsets are selected at the end of each pass and used as a starting point for the following pass.

5.7.2 FP-Growth Algorithm

Without producing candidates, the FP-growth algorithm effectively mines every collection of frequently used items. FP-growth stores data in main memory using both horizontal and vertical database layouts. The database doesn't keep track of item covers, but rather the actual transactions are saved in a tree structure. A linked list containing all transactions including an item is maintained for each item. Nodes keep note of the transactions that go through them by storing a counter. To make sure that all instances of an item are linked together, the FP-tree stores a link pointing to its next occurrence. Each item has its own header table that includes supporting information and a link to its original FP-tree look (Mandave, 2013). The database representation is made smaller by ordering the FP-tree in support decreasing order. Things that happen more often are clustered around the root, which makes them more likely to be shared. As you can see below, the algorithm mines frequent itemsets using a divide-and-conquer technique: In order to save association information, FP-growth condenses the database of frequent itemsets into an FP-tree, which stands for frequent pattern tree(Bala et al., 2016). Step two entails creating conditional sets from a compressed database. There is a counter at each node, and all of them correspond to items. In order to create a route, FP-Growth reads each transaction independently. Using a projected database in a predetermined sequence with just one frequently used item can cause paths to intersect. The last step is to mine the databases separately. Part of the FP-growth algorithm is making and mining FP-trees. Without creating candidate sets, FP-Growth finds sets of items that occur frequently.

6. PREDICTIVE ANALYTICS IN CONSUMER BEHAVIOR PREDICTION

Predictive analytics is the process of predicting future events based on historical data. In consumer behavior, predictive analytics is used to forecast customer activities like as purchasing decisions, attrition rates, and product preferences. Machine learning models can help firms get meaningful insights and make data-driven decisions.

6.1 Customer Churn Prediction

One of the more challenging problems in the telecom sector is customer churn prediction (CCP). More precise client attrition estimation is now feasible thanks to developments in AI and machine learning(Lalwani et al., 2022). More operators are entering the market as a result of globalization and advancements in the telecom industry, which has raised competition (Simon, 2013). Recruiting new consumers, upselling current ones, and improving client retention are just a few of the many strategies that have been proposed to optimize revenues in today's competitive industry. The best use of resources is to hold on to current customers(Hadden et al., 2007). The third tactic requires businesses to cut down on customer churn, or the number of times clients move service providers. Customer service and support issues are the main causes of customer attrition. Finding out which customers are likely to cancel is a critical first step in fixing this problem(C. P. Wei & Chiu, 2002). Predicting client turnover is done to help with developing strategies to retain clients. Client churn becomes an increasingly serious threat in increasingly competitive marketplaces. Consequently, it's crucial to find ways to keep tabs on loyal customers who don't churn. The goal of client churn models is to forecast when customers will voluntarily leave(Coussement & Poel, 2008). A lot of companies know their database is valuable. Churn prediction is a useful method for identifying clients that are at danger. Machine learning algorithms like Logistic Regression and Random Forest can forecast the probability of a consumer canceling a service(Mostafa et al., 2011). Because of this, companies can take precautions to ensure their customers stick around.

6.2 Sentiment Analysis

Various entities, such as products, services, organizations, and persons, can have their opinions analyzed through Sentiment Analysis (SA), which is also called Opinion Mining(Morales, 2020). With the explosion of online textual data, companies are now able to glean valuable insights about their products from consumers(Sheela, 2016). A method for accurately deriving inferences from data was necessary for both SA and BI. Businesses can gain valuable insights into consumer preferences and discover competing products and services by utilizing ML in sentiment analysis(hafaee, A., Issa, H., Agne, S., Baumann, S., & Dengel, 2014). Many factors need to be taken into account while describing machine learning. The 'why' behind the tool is vital states that there are three types of challenges that can be solved using machine learning:

1. Classification involves detecting things and recognizing text or sounds. Classification involves segmenting information into clusters to identify relationships and recommendations.
2. Prediction/Estimation: to predict and foretell potential outcomes.
3. Generation: Machine learning (ML) can generate content such as interpolating missing data and producing the next frame in a video sequence.

7. CHALLENGES

There is a lot of promise in utilizing machine learning to forecast consumer behavior, but there are also some serious challenges and concerns that need addressing if we want to get ethically sound outcomes. Data quality, model performance, and ethical/regulatory considerations over privacy and interpretation are just a few examples of the technical challenges that could get in the way.

- Data Quaility - Guaranteeing high-quality data for predictions is a key challenge in machine learning. Extensive, trustworthy, and high-quality data is necessary for precise consumer behavior prediction. Poor model performance and biased predictions can occur when customer data is inadequate, out of date, or incorrect. It is possible for forecasts to be skewed due to missing data or inconsistencies, including incorrect demographic information or transaction history, leading to suggestions that do not represent real customer preferences. Strict protocols for data preparation, validation, and cleansing are required to lessen the impact of this. Additionally, in order to prevent unfairness and inaccurate prediction models, it is crucial to use data that is representative of the entire consumer base.
- Model Interpretability - The difficulty in understanding and making use of machine learning models grows in tandem with their complexity, especially when deep learning methods are employed. Even if complex models are very accurate, they can also be "black boxes," making it hard for companies to understand how the model arrived at a specific prediction. When making decisions that directly impact clients, like focused marketing or personalized offers, this lack of transparency might be problematic. It is crucial for stakeholders, such marketing teams or firm decision-makers, to have a complete understanding of what factors impact predictions of customer behavior. To help build trust and facilitate informed decision-making, model-agnostic interpretability tools like LIME or SHAP, as well as the use of simpler, more interpretable models like decision trees, can offer valuable insights into the reasoning behind forecasts.
- Dynamic Preferences of Consumer - Many factors, including changing societal trends, economic situations, and marketing efforts, might cause consumers' actions to evolve over time. Without regular updates to account for shifting behaviors, machine learning models that predict customer preferences using historical data may become inaccurate over time. Consumer behavior might shift in response to new product offerings, changes in the weather, or shifts in broader economic factors (like inflation). If these updates aren't incorporated, models can become out of date and less accurate at predicting future behavior. To get around this, machine learning models need to be updated with new data on a frequent basis so they can continue to serve customers well and reflect their tastes as they are right now.

8. CASE STUDY AND APPLICATIONS

8.1 E-Commerce Personalization

Marketing academics have studied personalization tactics since 1990, but their effectiveness has evolved over time. Initially, internet retailers tried to enhance B2C websites by providing personalized greetings and recommendations for off-the-shelf products (Goy et al., 2007). The goal was founded on the premise that a website with one-on-one interaction would be favored over an anonymous one. After 10 years, there was evidence that individualized item suggestions did not increase sales enough to pay expenditures, particularly in small enterprises. In a 2003 Jupiter Research study, only 14% of consumers said a personalized Web site would lead to more frequent purchases from online stores. Additionally, only 8% said personalization made them more likely to visit news, entertainment, and content sites. In contrast, 54% of respondents indicated faster website loading and 52% named easier navigation as more important incentives. However, interest in customisation is increasing again.

8.1.1 Case Study

Customized product presentations are the outcome of the SeTA system's (Ardissono & Goy, 2000) dynamic product catalog generation throughout the customer's browsing process. It provides suggestions based on the receiver's preferences and the algorithm's evaluation of how well they would fit. Furthermore, in order to cater to the individual user's information requirements, the system modifies the item display according to their properties. Direct users and beneficiaries can be freely managed by following these procedures: An individual user model is maintained by the system for each direct user. Additionally, users have the option to specify the perfect product recipient's traits (Ardissono et al., 2002). You may create a model of the direct user by tracking their actions and gathering data about them. Using user class stereotypes to kick off indirect user models. In addition to demographic information like age and employment status, each user model also includes product preferences like those for technical items. Technical and aesthetic features, as well as the user's interest in and receptivity to these types of information, are all factored into the system model.

8.2 Customer Retention

In today's competitive market, organizations can gain a strategic advantage by focusing on customer retention. Understanding the elements influencing client retention is crucial for developing effective strategies and plans (Odunlami Ibojo Bolanle, 2014). In a crowded market with little chances for expansion, keeping customers is a top priority for managers. New customer acquisition is its primary goal. Customers have the freedom to choose their favorite insurer in the insurance market, just as they do in other competitive businesses. Many things influence a customer's decision-making process, such as price, service, personal preference, and convenience. Targeted client retention efforts require an understanding of which customers are leaving and for what reasons. Finding out why policies are terminated and how to anticipate when they will be terminated are the goals of the study presented in this part. For financial planning purposes, this tool helps to establish realistic termination rates and assesses the impact of policy changes, including premium expenses (Smith et al., 2000). The program identifies policy holders who are likely to terminate their policies, enabling client retention tactics like direct marketing. This chapter presents a case study to test the validity of a theoretical viewpoint on customer retention methods based on forms.

9. CONCLUSION

Buying things is something everyone has to do. People buy things because they have certain wants and requirements. Consequently, shoppers have a significant role in the buying process. It is essential to put yourself in the customer's shoes before making a purchase. People shopping for specific items could have varying tastes. People have different shopping habits. To comprehend consumer behavior and adjust to new conditions, analytical tools that refine data points are essential. It is critical to choose the optimal approach to utilize when analysing customer behaviour because it might be challenging. To understand customer behavior in today's tech-driven culture, creative marketing is essential.The way businesses understand and foresee consumer actions has been revolutionized by machine learning. By utilizing diverse data sources and modern analytics, companies may enhance customer pleasure in the aggregate. To effectively harness the power of machine learning for consumer behavior prediction, however, issues including data privacy, model interpretability, and quality of data must be adequately resolved.

REFERENCES

Adebola, O. O., Onyekwelu, B., & Orogun, A. (2019). Predicting Consumer Behaviour in Digital Market: A Machine Learning Approach. *International Journal of Innovative Research in Science, Engineering and Technology*, *8*(8), 1–13. DOI: 10.15680/IJIRSET.2019.0808006

Ardissono, L., & Goy, A. (2000). Tailoring the interaction with users in Web stores. *User Modeling and User-Adapted Interaction*, *10*(4), 251–303. DOI: 10.1023/A:1008372122567

Ardissono, L., Goy, A., Petrone, G., & Segnan, M. (2002). Personalization in business-to-customer interaction. *Communications of the ACM*, *45*(5), 52–53. DOI: 10.1145/506218.506245

Bala, A., Zakariyya Shuaibu, M., Karami Lawal, Z., & Zakari Yusuf, R. (2016). Performance Analysis of Apriori and FP-Growth Algorithms (Association Rule Mining). *International Journal Of Engineering And Computer Science*, *7*(2), 279–293. https://www.researchgate.net/publication/297898164

Bolanle, O. I. (2014). Impact of Customer Satisfaction and Customer Retention on Customer Loyalty: A Case Study of Enterprise Bank in OYO. *International Journal of Education and Research*, *2*(9), 427–450.

Bradlow, E. T., Gangwar, M., Kopalle, P., & Voleti, S. (2017). The Role of Big Data and Predictive Analytics in Retailing. *Journal of Retailing*, *93*(1), 79–95. DOI: 10.1016/j.jretai.2016.12.004

Breiman, L. (2001). Random forests. *Machine Learning*, *45*(1), 5–32. DOI: 10.1023/A:1010933404324

Brown, M. P. S., Grundy, W. N., Lin, D., Cristianini, N., Sugnet, C. W., Furey, T. S., Ares, M.Jr, & Haussler, D. (2000). Knowledge-based analysis of microarray gene expression data by using support vector machines. *Proceedings of the National Academy of Sciences of the United States of America*, *97*(1), 262–267. DOI: 10.1073/pnas.97.1.262 PMID: 10618406

Burges, C. J. C. (1998). A Tutorial on Support Vector Machines for Pattern Recognition. *Data Mining and Knowledge Discovery*, *2*(2), 121–167. DOI: 10.1023/A:1009715923555

Chaubey, G., Gavhane, P. R., Bisen, D., & Arjaria, S. K. (2023). Customer purchasing behavior prediction using machine learning classification techniques. *Journal of Ambient Intelligence and Humanized Computing*, *14*(12), 16133–16157. DOI: 10.1007/s12652-022-03837-6

Coussement, K., & Poel, D. Van Den. (2008). *Churn prediction in subscription services : An application of support vector machines while comparing two parameter-selection techniques. 34*, 313–327. DOI: 10.1016/j.eswa.2006.09.038

Das, T. K. (2016). A customer classification prediction model based on machine learning techniques. *Proceedings of the 2015 International Conference on Applied and Theoretical Computing and Communication Technology, ICATccT 2015*, 321–326. DOI: 10.1109/ICATCCT.2015.7456903

Goy, A., Ardissono, L., & Petrone, G. (2007). Personalization in e-commerce applications. *Lecture Notes in Computer Science (Including Subseries Lecture Notes in Artificial Intelligence and Lecture Notes in Bioinformatics), 4321 LNCS*, 485–520. DOI: 10.1007/978-3-540-72079-9_16

Hadden, J., Tiwari, A., Roy, R., & Ruta, D. (2007). Computer assisted customer churn management: State-of-the-art and future trends. *Computers & Operations Research, 34*(10), 2902–2917. DOI: 10.1016/j.cor.2005.11.007

Shafaee, A., Issa, H., Agne, S., Baumann, S., & Dengel, A. (2014). Aspect-based sentiment analysis of amazon reviews for fitness tracking devices. In Trends and Applications in Knowledge Discovery and Data Mining: PAKDD 2014 International Workshops: DANTH, BDM, MobiSocial, BigEC, CloudSD, MSMV-MBI, SDA, DMDA-Health, ALSIP, SocNet, DMBIH, BigPMA, Tainan, Taiwan, May 13-16, 2014. Revised Selected Papers 18 (pp. 50-61). Springer International Publishing.

Kenji, S. (2011). Artificial Neural Networks. *Methodological Advances and Biomedical Applications,* (April), 2011.

Kulkarni, N., & Kaleel, M. (2019). *Customer Behaviour Prediction MSc Research Project MSc Data Analytics.*

Kumar, K. S. (2012). A Survey on Association Rule Mining using Apriori Algorithm. *International Journal of Computer Applications, 45*(5), 47–50. DOI: 10.5120/5060-7390

Lal, T. N., Schröder, M., Hinterberger, T., Weston, J., Bogdan, M., Birbaumer, N., & Schölkopf, B. (2004). Support vector channel selection in BCI. *IEEE Transactions on Biomedical Engineering, 51*(6), 1003–1010. DOI: 10.1109/TBME.2004.827827 PMID: 15188871

Lalwani, P., Mishra, M. K., Chadha, J. S., & Sethi, P. (2022). Customer churn prediction system: A machine learning approach. *Computing, 104*(2), 271–294. DOI: 10.1007/s00607-021-00908-y

Li, J., Pan, S., Huang, L., & Zhu, X. (2019). A machine learning based method for customer behavior prediction. *Tehnicki Vjesnik (Strojarski Fakultet), 26*(6), 1670–1676. DOI: 10.17559/TV-20190603165825

Mandave, P. (2013). Data mining using Association rule based on APRIORI algorithm and improved approach with illustration. *International Journal of Latest Trends in Engineering & Technology : IJLTET, 3*(2), 107–113.

Manuel Serra, J., Allen Baumes, L., Moliner, M., Serna, P., & Corma, A. (2007). Zeolite Synthesis Modelling with Support Vector Machines: A Combinatorial Approach. *Combinatorial Chemistry & High Throughput Screening, 10*(1), 13–24. DOI: 10.2174/138620707779802779 PMID: 17266513

Morales, S. P. M. de. (2020). Big data and Sentiment Analysis considering reviews from e-commerce platforms to predict consumer behavior. *MSc Business Research, June.*

Mostafa, S., Nasir, K., Tarokh, M. J., Abbasimehr, H., Toosi, K. N., & Tarokh, M. J. (2011). A Neuro-Fuzzy Classifier for Customer Churn Prediction A Neuro-Fuzzy Classifier for Customer Churn Prediction Mostafa Setak. *Article in International Journal of Computer Applications, 19*(8), 975–8887. https://www.researchgate.net/publication/236625411

Sakib, S., Ahmed, N., Kabir, A. J., & Ahmed, H. (2018). An Overview of Convolutional Neural Network: Its Architecture and Applications. doi:DOI: 10.20944/preprints201811.0546.v4

Sheela, L. J. (2016). A Review of Sentiment Analysis in Twitter Data Using Hadoop. *International Journal of Database Theory and Application, 9*(1), 77–86. DOI: 10.14257/ijdta.2016.9.1.07

Simon, H. A. (2013). *Administrative Behavior* (4th ed.).

Smith, K. A., Willis, R. J., & Brooks, M. (2000). An analysis of customer retention and insurance claim patterns using data mining: A case study. *The Journal of the Operational Research Society, 51*(5), 532–541. DOI: 10.1057/palgrave.jors.2600941

Vafeiadis, T., Diamantaras, K. I., Sarigiannidis, G., & Chatzisavvas, K. C. (2015). A comparison of machine learning techniques for customer churn prediction. *Simulation Modelling Practice and Theory, 55*, 1–9. DOI: 10.1016/j.simpat.2015.03.003

Wang, Y., Sun, S., Chen, X., Zeng, X., Kong, Y., Chen, J., Guo, Y., & Wang, T. (2021). Short-term load forecasting of industrial customers based on SVMD and XGBoost. *International Journal of Electrical Power and Energy Systems, 129*(September 2020), 106830. DOI: 10.1016/j.ijepes.2021.106830

Warmuth, M. K., Liao, J., Rätsch, G., Mathieson, M., Putta, S., & Lemmen, C. (2003). Active learning with support vector machines in the drug discovery process. *Journal of Chemical Information and Computer Sciences, 43*(2), 667–673. DOI: 10.1021/ci025620t PMID: 12653536

Wei, C. P., & Chiu, I. T. (2002). Turning telecommunications call details to churn prediction: A data mining approach. *Expert Systems with Applications, 23*(2), 103–112. DOI: 10.1016/S0957-4174(02)00030-1

Wei, J., Chu, X., Sun, X. Y., Xu, K., Deng, H. X., Chen, J., Wei, Z., & Lei, M. (2019). Machine learning in materials science. *InfoMat, 1*(3), 338–358. DOI: 10.1002/inf2.12028

Xu, D., Yang, W., & Ma, L. (2018). Repurchase prediction based on ensemble learning. *Proceedings - 2018 IEEE SmartWorld, Ubiquitous Intelligence and Computing, Advanced and Trusted Computing, Scalable Computing and Communications, Cloud and Big Data Computing, Internet of People and Smart City Innovations, SmartWorld/UIC/ATC/ScalCom/CBDCo*, 1317–1322. DOI: 10.1109/SmartWorld.2018.00229

Zhang, Y., Zhao, X., Sun, Y., Liu, M., & Shi, D. (2013). Waveform classification based on wavelet transform and k-means clustering. *In2013 International Conference on Machine Learning and Cybernetics, 4*, 14–17.

Chapter 7
Understanding the Role of Emerging Technology in Human Resilience in the Digital Age and Artificial Intelligence

Shubha Jain
https://orcid.org/0009-0008-2170-1515
Department of Humanities and Social Sciences, Graphic Era University, Dehradun, India

Ravindra Singh
https://orcid.org/0000-0003-3062-1812
Department of Psychology, Magadh University, Bodh Gaya, India

Bhavy Agarwal
Department of Humanities and Social Sciences, Graphic Era University, Dehradun, India

Ajay Kumar Singh
https://orcid.org/0000-0003-0429-0925
Department of Humanities and Social Sciences, Graphic Era University, Dehradun, India

ABSTRACT

This chapter explores the integration of psychology and technology as an opportunity to strengthen an individual's resilience and psychological health. Through a deep comprehensive review process, it identified 62 research articles and book chapters in the last 20 years. It investigated the potential keywords of combining psychology and technology which can enhance individual resilience and mental health. A thorough examination of all studies with their strengths, weaknesses, and limitations are incorporated to generate valuable insights. The results reveal that technological gears and tools like virtual games, virtual reality, and video applications and mobile apps can significantly enhance individuals' cognitive and motor skills and strengthen social connections. This chapter claimed that digital technologies and artificial intelligence promotes happiness and resilience of people. From teletherapy bridging geographic gaps to virtual reality. Finally, it claims an increasing anxiety among the people due to application of artificial intelligence.

DOI: 10.4018/979-8-3693-6910-4.ch007

INTRODUCTION

In the present world of modern medical science starting **from body scans, and MRI to relaxation techniques, technology is redefining healthcare in numerous spheres.** After the development of smart sensors and meters (like heart rate monitors), it is now easier to monitor and investigate physical well-being and psychological health of patients. Mindfulness-based apps (like guided meditations) are supportive to increase mental resilience and well-being of the people. Advanced powerful synergy between technology and psychology allows people to control their physical and mental health. Technology is supporting the psychological health and well-being of people in several possible ways. For example, technology is helpful to increase the skills of the people and bring multiple ways to enhance their social and psychological health and well-being. Psychology is one the fastest-growing multidisciplinary fields in the modern contemporary scientific world. Psychology is also helping other disciplines to enhance the physical and mental health of the population. Despite that, psychologists are facing several challenges to meet their current need for technology in diversified areas. These challenges include the issues of availability, accessibility, and affordability (AAA) of various technologies in the domain of digitalization and digital technologies (DTs) (Bray, 2010).

In addition to above-mentioned challenges, humanitarian limitations, and the high differential therapist/patient ratio make the situation worse. The above-mentioned shortcomings of psychologists can be assisted by rapid technological advancements. There are many issues in social, economic and psychological dimensions which cannot be solved without technology (Fairburn & Patel, 2017). The concept of video conferencing and teletherapy can cater to the needs of accessibility and availability to a greater extent. Digital technologies are also creating digital platforms for sharing the knowledge and skills among the large groups of users in different sectors. With advanced communication mediums, psychologists can reach out to extensively remote geographical regions. Properly understanding these approaches can allow individuals to achieve greater well-being and social connections in various situations (Chiauzzi et al., 2020). Such a tactic is directly aligned and associated with every possible perspective of the positive technology movement in the past.

The movement aims to draw attention to the physical and psychological health of individuals using technology (Riva et al., 2012). Technological advancement creates several ways to increase human well-being and psychological health. Gaggioli et al. (2019) also claimed a significant impact of technology on physical health. Existing researchers also detected the positive impact of technologies in other sectors that promote mental, physical, psychological and social status of the people (Singh et al., 2025). Additionally, psychoeducation about the importance of mental health and its aspects along with the advent of psychological interventions may be administered through these mediums. Along with psychoeducation, awareness programs can be conducted to cater to the orthodox status quo which holds severe stigmas against mental health and well-being (Johnson et al., 2021).

Technological advancement has taken a rapid increase in individual lifestyles (Singh, 2025). At the same time, meditation apps provide guided meditation to nurture our mental health and resilience. The influential sync of technology with psychology helps individuals in empowering their physical and mental health (Lam et al., 2023). In modern therapies, practitioners are using virtual reality (VR) therapies to help individuals to move out from anxieties. It allows them to expose clients to real-life situations which helps them overcome their phobias via systematic desensitization (Boeldt et al., 2019). Meanwhile, new technological assessments are used in assessing individual problems. These assessments foster a healthy and engaging environment (Kane & Parsons, 2019). On the contrary, teleherapy is helping people for bridge their geographical gaps with their clients. Video conferencing allows researchers to reach out to people in remote areas and provide awareness about mental health problems. It helps in increasing awareness of the disorders leading to better treatment of the individual (Nelson et al., 2023).

The increased advancement of technology has highly influenced the well-being of society. While various articles and research have coined the negative consequences of technology on humanity (Twenge et al., 2018). Applications like social media are highly influential sources in spreading awareness among people. Through social media adolescence vulnerability towards technology is rapidly increasing (Cataldo et al., 2021). Additionally, social media applications have vast networks with sufficient sustainability to socially connect people across the globe. Technology holds great potential to be even more potent in the assistance of psychological interventions and procedures. There is an excessive amount of data indicating the potential synergy psychology and technology hold in the clinical world if intertwined together.

In today's world, by giving access to information and services technology provides an opportunity to improve the well-being of youth. It provides new opportunities for individuals to communicate, learn, express themselves, and participate in community life (Odgers & Jensen, 2020). Accordingly, it reduces the communication gap among the people and reduces their stress in multiple ways. Technologies are helpful to increase the development of new mechanical instruments for people at low cost. VR systems are capable of running on a commodity-level personal computer (Rizzo et al., 2011). Their precise presentation is allowed by the VR. Moreover, it provides modern methods for assessing and understanding the behavioural responses of people. The presence of these characteristics makes VR an asset for clinical treatment, intervention, and training domains. Various clinicians and scholars have documented the effectiveness of VR in these areas (Holden, 2005; Parsons & Rizzo, 2008).

Virtual airplanes, battlefields, social settings, and monotonous (but highly relevant) functional environments have been constructed by various VR scientists to do this. Artificially intelligent virtual human agents have been produced by researchers that are being used in the role of virtual patients for teaching clinical skills to inexperienced clinical professionals (Parsons & Rizzo, 2008). In addition, patients and other clinical experts are also used as anonymous online healthcare support agents (Rizzo et al., 2011). In addition, with this, tools like gaming help in enhancing the cognitive and motor skills of individuals. In addition to this, it helps in strengthening social connections which directly helps in promoting happiness and a positive mood. **Hence, this chapter lays the groundwork for further investigation into how the integration of technology and psychology can be beneficial for promoting mental and emotional health of individuals**?

In the present fast-moving world, traditional mental health services are insufficient to meet individual and societal needs. But even after significant development the accessibility, availability, and affordability of psychological assistance in remote areas is still a long journey ahead. Psychological support can provide therapies that also include awareness and education related to mental health in these areas. To make that happen, there are a couple of requirements that must be fulfilled, introducing various suitable programs and administering psychoeducation among the communities. It is also helpful in bridging the gap between the therapist and clients. Moreover, it has given a solution to the therapist to reach out to people who are present in economically deprived areas. Thus, this study emphasizes the potential future scope of integration of technological advancements with psychological sciences in enhancing the resilience of individuals.

This research aims to establish an understanding of digital resilience as per the experience of interaction in the digital world. Recent studies focus more on the theoretical concepts of how the integration of psychology with technology can help improve mental health? It does not state the practical implications of these concepts. In addition to this, it lacks the long-term impacts of the concept as it only focuses on short-term implications. There are limited studies on how the mechanism of gaming can help foster better cognitive functioning in individuals? Furthermore, recent studies do not explain how the implications of technology can help in spreading awareness of mental health? Thus, this study is an attempt to answer the following research questions:

- What is the association of psychology with technology?
- How can technology be impactful for enhancing the cognitive and motor skills of individuals?
- In what ways, gaming can help in enhancing positive moods in individuals?
- How the applications of technology can help in spreading awareness of mental health in the communities where it lacks accessibility?
- How the use of technology in modern therapies can help practitioners to improve the health of their clients?

In contrast to the above-mentioned research questions, this study consists of the following objectives:

- To assess the role of psychology and technology in strengthening the resilience of individuals.
- To navigate the gap between the therapist and client.
- To provide the mechanism of how gaming can help in improving the cognitive strength of individuals.
- To give an insight into how technological assessments and modern therapies like virtual reality therapy can help the practitioners in helping their clients.
- To explain how technology can help in providing psycho-education to the communities in remote areas to spread more awareness on mental health.

RESEARCH METHODOLOGY

A comprehensive systematic review procedure of the current literature is undertaken to gain a profound understanding of the topic. Relevant scholarly articles and journals were identified through online databases using keywords like "resilience", "Mental Health," "virtual reality," "technology", "tele calling", and "tele counseling". Focus was maintained on published articles within the past 20 years for up-to-date information on the above issue. Through a comprehensive in-depth review process, a total of 62 studies were investigated to generate and conclude the findings. The approach pointed to the potential of combining psychology and technology to enhance individual resilience and mental health. A thorough examination of the study's strengths, weaknesses, and limitations yielded valuable insights. The bibliographies of the research paper were also referred to find relevant sources. Through an exhaustive deliberation, only high-quality research addressing the core topic of technological interventions for resilience enhancement was retained for the final synthesis. The research identified crucial features of technological interventions that contribute to resilience building, such as the type of technology used, its specific functionalities, and its mode of interaction with the user. Challenges and limitations associated with this approach were also acknowledged, considering factors like technological access, individual preferences, and potential ethical concerns.

REVIEW OF LITERATURE

Recent technological advancements have introduced innovative solutions to foster mental health and resilience. Additionally, they are helping communities change the realm of psychology by providing more creative and interactive therapies to clients. Digital resilience can be integrated among adolescents and children from a young age and be encouraged on familial and community levels. To summarise and conclude the gap between previous and current well-being theories, user-friendly designs and practices are a key challenge in positive technology. Peters et al. (2018) reported this issue and provided a psychological framework on technology and its psychological implications. METUX provides a basic groundwork for crafting and preparing technological tools that enables support for psychological health and well-being. It suggests well-being and psychological demands that must be measured within 5 different spheres of investigations. These 5 spheres are: (i) Adoption of Technology, (ii) Interaction with the interface, (iii) Engagement in technology-specific tasks, (iv) Technology supported behaviour, and (v) Overall change in individual behavior (Peters et al., 2018).

Virtual reality (VR) could enhance standard psychological treatments' efficacy and/or effectiveness (e.g., VR-based CBTs). Technology has drawn the attention of scholars across different fields. As a potential emerging field of fundamental research, a considerable attempt has been made to develop conceptual pillars and levels of analysis (Gaggioli et al., 2019). Additionally, several efforts were made to define the key frameworks for bringing psychological health and well-being principles into the core practical design of interactive systems. In the past few decades, the potential use of technology (technological affordances) has been significantly developed. Studies have also highlighted the ways in which technology can help in reaching out to marginalized populations.

Mira et al. (2018) defined the efficacy of an internet-based positive psychology intervention (IPPI) via using a randomized controlled trial (RCT) designed for patients (n=108) with mild to moderate depressive symptoms. This RTC trail intervention consisted of an 8-module internet-assisted program, which combined two strategies equally. Artificial stimulations via virtual/augmented reality can assist psychologists in providing more accessible and feasible interventions to boost mental health. New researchers bring attention to VR therapy which has been proven effective in dealing with anxiety. It gives the patients access to real-life virtual simulations which exposes them to deal with their anxieties and phobias (Lan et al., 2023). Technological assessments and tools have been a breakthrough in the field of psychology. It provides concerned professionals with an undisputed edge. They give the therapist better control of the therapeutic situations. It allows and supports the client more efficiently (Paul, 2022).

The meaning of digital mental health may differ for different stakeholders from different perspectives (Bond et al., 2023). In fact, one can assume or believe it as the design of digital interventions that prevent mental illness and help manage the well-being of people. (Mulvenna et al., 2021). Given that there may be so many perspectives, but definitely, AI can definitely be a part of a digital intervention (Bond et al., 2023). Moreover, virtual reality (VR) and augmented reality (AR) programmed exposure therapy can also be used to alleviate phobias (Krijn et al., 2004; Freitas et al., 2021). VR can also be used to engender empathy by simulating what it is like to be in someone else's shoes so to speak (referred to as VR empathy machines (Bollmer et al., 2017). Additionally, VR can be used for simulation-based training to train healthcare professionals in treating clients under different controlled scenarios (Quigley et al., 2019).

There has been a spike in the use of technology among adolescents and children during lockdown with several positive and negative aspects. In a low-resource economy the "digital divide" undoubtedly remains a problem that is going to change in the coming decades (Fairburn et al., 2017). Digital interventions like AI-assisted diagnosis and treatment will gradually find their place within mental healthcare systems with more online options (Fairburn et al., 2017). Moreover, the Blend of technology with advanced AI may displace some conventional face-to-face treatment methods. Regulating and promoting AI and digital interventions with strict guidelines will accelerate their appropriate use (Fairburn et al., 2017). Moreover, in the current scenario of mental illness, special online games are proven to reduce depression and anxiety. It increases creativity, and skills, and boosts other cognitive abilities in children. Thus, it can be concluded that the increased usage of technology can have both a positive and negative impact on the mental health of adolescents and children depending on the trends in the usage. For example, excessive dependency of youths on digital technologies are reducing their mental skills and decision capacity. Also, depression level among the youths is increasing due to extensive application of digital tools in their life. The skills of students are negatively impacting due to overuse of artificial intelligence in academic activities.

Furthermore, assistive tools like gaming are coming in to contribute their share in the improvement of mental health and building resilience. Along with these, many smartphone applications look after the needs of meditation and relaxation. These latest technologies are reducing stress and leading to a better state of mind and resilience (Gómez-Cambronero et al., 2024). Among all the games developed to support Cognitive Behavioural Therapy for children, *Treasure Hunt* made its presence remarkable. Its purpose was to internalize disorders comprising depression and anxiety and externalize disorders including oppositional defiant disorder and conduct disorder (Brezinka, 2012). One of the most significant components of the game was committed to hunting down unhelpful or negative automatic thoughts by engaging a first-person shooter. The directions of the game are not only displayed in the form of written text but also broadcasted verbally to make it attractive and interesting for children with reading and/or learning difficulties. To ensure maximum engagement of the child in the game, all tasks and objectives are broadcasted verbally in the voices of two girls and two boys. The game comprises six levels with each corresponding to a predetermined step in the cognitive-behavioural treatment. Those steps represent the basic concepts of CBT for children, hence highlighting *Treasure Hunt* as a broadband CBT game that covers the issues relevant to the treatment of numerous disorders (Brezinka, 2012).

In the last few decades, the interest of social scientists has emerged in the technological infoldment. This particularly includes the use of virtual reality and mixed reality and how they can be used in promoting prosocial abilities, and behaviors (Schoeller et al., 2019). Tools such as VR, MRI, FMRI, biofeedback, and brain control stimulation help in fostering empathic abilities in humans. In particular, they suggest that virtual reality can empower empathy training by allowing users to "embody another self.'' Thus, it provides access to sensorial data concerning the body of another person, its immediate context, and peri-personal space (Schoeller et al., 2019).

Paul (2022) explains how technology can be effectively used to assess psychotherapy to bridge the gap between the therapist and client. Therapists can reach out to clients present geographically apart through audio or video counseling. Video counseling helps to maintain confidentiality of clients, which is an integral part of psychotherapy (Paul, 2022). It is one of the most growing practices in mental health. Additionally, technology can help with treatment, prevention, recovery, management, and continuing care. It can be used for the assessment of tools that are used for screening patients without the presence of a clinician which gives accurate and fast results (Paul, 2022). Furthermore, technology gives the client the benefit of time flexibility. The client can take part in the session according to their convenience which is suitable to them. Virtual reality is one of the emerging psychotherapy tools in the field of mental health. It can be used to help individuals with a wide range of disorders such as eating disorders, psychotic disorders, mild cognitive impairment, social anxiety disorder, and posttraumatic stress disorder (Paul, 2022). The study also reports that robots are effective in reducing the psychological symptoms of an individual and reinforce facial engagement. With this, the study summarized the effective use of technology for the promotion of mental health in individuals (Paul, 2022).

Over the past few years, AI has replaced traditional working methods and unlocked new avenues of innovation. Thakkar et al. (2024) emphasized the use of AI in promoting positive mental health in individuals. The algorithms can be used to diagnose conditions such as depression, anxiety, and schizophrenia by analyzing the behavioral patterns and preparing their treatment plan (Thakkar et al., 2024). It can be used in predicting the development of disorders like intellectual, developmental, seizures, mood, neurodegenerative, and many more such disorders. Additionally, it can be also used to differentiate the diagnosis of disorders like MDD and depression (Thakkar et al., 2024). The study also elaborates on the use of AI-driven initiatives for combating stigma and spreading awareness of mental health. The AI-enhanced educational platforms provide relevant information to individuals to gain a deeper understanding of mental health conditions (Thakkar et al., 2024). Furthermore, AI-driven client engagement strategies can help in guiding patients during their recovery journey. Mobile applications can help them manage their schedules and give them timely reminders for medication. AI has made assessing the cognitive functioning of individuals easier by increasing their efficiency and making them more objective. Thus, with these major findings, the study concludes that collaboration between AI and healthcare can help promote positive mental health among individuals (Thakkar et al., 2024).

Philippe et al. (2022) performed a meta-review to assess the current use of digital technology for the assessment of mental health interventions. The study evaluated the use of synchronous and asynchronous modes of communication for the therapist. Synchronous communication is a primary form of communication that can be used for the assessment and treatment of patients. It is a more accessible and easier mode of communication for patients to express themselves (Philippe et al., 2022). On the contrary, asynchronous forms of communication are used to encourage patients to attend their appointments, complete their daily tasks, and reduce the chances of remission. Additionally, the study also highlighted the use of web-based communication to reduce feelings of loneliness and social isolation among individuals. Individuals with similar disorders can interact with others, share their experiences, and motivate each other (Philippe et al., 2022). Further, the study explained the use of computer-based therapy to provide psychoeducation, assessment tools, and self-help journals for intervention. The use of augmented and virtual reality to provide immersive and real-time experiences to clients is also reviewed by the authors. These tools have been reported more influential than traditional exposure therapy in helping patients with anxiety and phobias. The study outlines how technology can be used to deliver intervention and treatment that is more accessible to both client and therapist (Philippe et al., 2022).

The inability to perform daily activities and lack of support from others results in poor well-being of individuals suffering from dementia. Dowson et al. (2023) investigated how online singing groups (OSGs) improve the experience of individuals suffering from dementia. Singing groups are the most popular activities that help in improving the memory of patients suffering from dementia (Dowson et al., 2023). It is an enjoyable activity that provides a feeling of social inclusion amongst individuals and fosters a better mood for both the patient and caregiver. The researcher implemented 10-week online singing sessions for patients with dementia along with their caregivers. In total 16 pairs joined the session which lasted for 1 hour each day (Dowson et al., 2023). The patients enjoyed the session while facing limited technical problems. Some patients even felt online singing sessions as a more convenient method than the regular singing group. It is also reported that the patients enjoyed these sessions and it also helped improve their relationship with their caregiver. Additionally, this method is seen as more accessible for patients rather than the traditional singing groups. Thus, it is also concluded that OSGs can be a great alternative for people with dementia and their caregivers (Dowson et al., 2023). Hence, there may be a significant contribution of artificial intelligence and advanced technologies to recure the problems of dementia among the patients. Although, findings of previous studies could not access the role of digital technologies in dementia.

Adolescence is a period of rapid growth and development in individuals which results in numerous changes including physical, emotional, and psychological health. These sudden changes increase the chances of the development of mental health disorders in individuals. Due to the stigma and feeling of being judged, they are often not able to seek therapeutic health. Tosti et al. (2024) highlighted the advantages of using chatbots to promote developmental well-being in adolescents. Anonymity is one of the primary advantages of these interventions as the individual can express themselves without the feeling of being judged (Tosti et al., 2024). They can easily express their inner feelings through texts and be more open about their problem in an authentic way (Tosti et al., 2024). These chatbots also increase the accessibility of mental health for individuals who don't have this professional help in their surroundings. Further, these therapies are cheaper than traditional psychotherapy which makes it accessible for people who are not financially stable. These chatbots give the individual comfort and flexibility of time to discuss their problems. Although the effectiveness of these chatbots is yet to be determined, the current studies suggest that these chatbots can help increase the level of engagement and perceived naturalness of conversation (Tosti et al., 2024).

Aggarwal et al. (2023) evaluated the feasibility, efficacy, and intervention characteristics of AI chatbots for promoting healthy behavior change. This study found that AI chatbots are highly efficient tools for promoting healthy lifestyle changes. These include regular exercise, smoking cessation, and reduction in substance misuse. On the contrary, the studies report mixed results in terms of feasibility, acceptability, and usability of these resources (Aggarwal et al., 2023). Although these chatbots induce client engagement behavior, it is difficult to interpret whether these chatbots are feasible or not. Similarly, even though these chatbots result in the anonymity of the client the satisfaction level among individuals is very low leading to poor acceptability. Due to a lack of technical knowledge among individuals and impractical suggestions by chatbots reported in studies limits the usability of these chatbots (Aggarwal et al., 2023). Furthermore, the review highlighted the use of chatbots in individuals' behavior changes. CBT, emotion-focused therapy, and habit formation models through chatbots result in better user engagement and motivational support. Additionally, AI chatbots provide a free flow of conversation among the user, enhancing their experience. Thus, the review concluded how chatbots can be used for human behavior intervention (Aggarwal et al., 2023).

The treatment planning of neurodevelopmental disorders is an ongoing process that involves a group of psychotherapies along with school-based intervention. Pontikas et al. (2022) performed a literature review to understand the assistive use of technology in the treatment of neurodevelopmental disorders. These technologies can be used to teach a variety of skills to individuals and enhance their problem-solving and vocal imitation. Speech-generative devices are installed to improve an individual's cognitive impairments. It also helps in improving the socializing process of children (Pontikas et al., 2022). The tangible user interface system supports the literacy of children, specifically those with Down syndrome. Specialized robots are designed for children with autism spectrum disorder which helps them to initiate, maintain, or terminate a behavior. Some of these robots are Tito, GIPY, Hoap 3, CommU, ACTROID-F, NAO, iRobiQ, Smile Supplement, and many more (Pontikas et al., 2022). Virtual Interactive systems with

Gaze-sensitive Adaptive Response Technology (VIGART) have been used to improve social interactions of children. Virtual Reality (VR) uses multimedia apps to improve the management skills of children and teach them relaxation techniques (Limone & Toto, 2021). In addition to this online story support helps to teach individual moral activities. Thus, with these findings, Pontikas et al. (2022) provided an overview of assistive technology that can be used to help individuals with neurodevelopmental disorders (Pontikas et al., 2022).

Alcohol addiction is one of the major problems faced by individuals across the globe. Ndulue and Naslund (2024) evaluated digital interventions that can be used to help individuals with such addiction disorders. Technological tools such as instant messaging, and eHealth can support individuals to overcome their addiction. They can also help in behavioral change, raising information and spreading awareness amongst individuals. One of the apps known as "Text4Baby" was developed to prevent pregnant women from consuming alcohol. Another such intervention known as "Contraception and Alcohol Risk Reduction Internet Intervention (CARRII)" was initiated for interactive sessions and activities for pregnant women and support them from alcohol exposure during pregnancy (Ndulue & Naslund 2024). These types of interventions have reported a high acceptance rate amongst the users. These technological tools also help clinicians in planning preventive measures for clients. Additionally, they help by providing telecounseling to individuals to prevent them from drinking (Ndulue & Naslund 2024). Further mobile apps are used to support patients during their ongoing treatment. Some of the apps that were used to support the patients were "Daybreak Program" and "Addiction–Comprehensive Health Enhancement Support System (A-CHESS)". "Secure Continuous Remote Alcohol Monitoring (SCRAM)" is another device that is used to measure the level of intake of alcohol in individuals. Thus, these were the measures reviewed by Ndulue and Naslund (2024), to treat patients with Alcohol use disorder from remote areas without the need for face-to-face therapy (Ndulue & Naslund 2024).

M-Health comprises various prospective possibilities: continuous availability, immediate assistance, anonymity, accessibility beyond geographical boundaries, equity in access to mental health resources, and personalized content (Gómez-Cambronero et al., 2024). These benefits can be reaped at an affordable cost and an increased capacity and service efficacy. For instance, with the use of smartphones, we can collect patient data automatically and systematically. In addition, they allow the use of other features: reminders, personalized feedback, momentary ecological interventions, and contextualized psychoeducational content (Gómez-Cambronero et al., 2024). Conclusively, they allow customized psychological assistance as per the patients' requirements of when and where they require it. mHealth applications can serve as independent self-regulatory guided interventions and treatments. Here, the components of traditional therapies can be ruled out of in-person sessions and into real-life situations. Serious games, by extension, and gamification can provide potential and promising solutions to enhance user engagement, therefore improving treatment adherence and reducing attrition rates. A major advantage of Serious games is that they provide the players access to customization and flexibility. It helps players accommodate various scenarios since they are inclined to adopt diverse gameplay styles and strategies that align with their skills, abilities, and behavior (Gómez-Cambronero et al., 2024). According to a more recent systematic review of Serious Games, games are separated by platforms, which include smartphone-specific games. They specifically focus on games for intervening with anxiety and depression. The study concluded that with the proper means of utilization, the impact of online treatment outcomes can be potentially improved with the advent of gamification and serious games (Gómez-Cambronero et al., 2024).

Among the recently emerging technologies, Augmented Reality (AR) and Virtual reality (VR) showcase a promising potential to enhance the currently existing mediums of diagnosing and treating mental health disorders. A patient's reality can be explored, challenged, and expanded with the use of VR/AR (Lan et al., 2023). To enhance general wellness, provide values of entertainment, invigorate learning, and target deviating cognitive patterns and behaviors, various stimuli can be applied. These include visual, auditory, olfactory, somatosensory, and haptic stimuli. In the past two decades, AR/VR applications in psychiatry have been explored by a growing body of literature. For patients having anxiety disorders, especially those experiencing symptoms of specific phobia, PTSD, or social anxiety, VR proved to be a useful instrument in administering therapist-supported exposure therapy (Lan et al., 2023). Additionally, AR/VR has also been used in treating autism spectrum disorder (ASD), eating disorders, attention-deficit/hyperactivity disorder (ADHD), and substance-use disorders. One of the very important approaches of AR/VR technology in mental health is diagnosing and treating schizophrenia and related psychosis-spectrum disorders. As a new therapy augmenting antipsychotic medication dedicated to both negative and positive symptoms of schizophrenia, AR/VR holds promising potential. For instance, VR can be used to develop controlled environments and stimulations wherein the patients can improve social functioning and life skills while being guided through social situations (Lan et al., 2023). Additionally, VR can approximate the patients' perceptual disturbances to practice management strategies, guided by a therapist by designing avatars, voices, and settings. The review did not find any article addressing the use of AR in diagnosis or treatment options. Meanwhile, the integration of VR interventions holds the potential to improve clinical outcomes for patients suffering from psychosis. Finally, additional and rigorous research is required to examine and explore effective scalability models of VR/AR along with decentralized clinical trials that focus on improving provider and patient experience (Lan et al., 2023).

One of the most disruptive transformations faced by the aging population in recent times is Social Isolation (SI). It refers to the significantly deteriorating emotional and physical health of older adults. It is caused by the absence of meaningful social relations (or human contact) that adversely impacts their quality of life (QoL) (Sen et al., 2022). To partly offset the negative effects on health and ameliorate SI in older adults, technology may provide promising alternative means. VR replaced the face-to-face means of communication in humans with consequential effects on social connectivity in the post-COVID world (Sen et al., 2022). For instance, the elderly could now establish connections with their cherished ones through the medium of a 3-D environment with access to virtual field trips like an amusement park with their friends. The review highlights various forms of easily accessible and affordable mediums of technology that can provide a safe environment for the elderly to access information. In addition, they can remain in their safe environment while engaging in social interactions (Sen et al., 2022). Phone calls or e-mails made long-distance communication particularly helpful in establishing and sustaining relationships among older adults. High levels of emotional satisfaction among the elderly were highlighted as modern communication technologies enabled them to communicate with their loved ones even while being geographically distant from them. Therefore, it was concluded that SI among older adults can be brought into check and be dealt with to improve their QoL with the advent of technological accessibility (Sen et al., 2022).

Hunter et al. (2023) acknowledged the transformation in society originated from the swiftness of digital technology that has undeniably impacted health and well-being. With the increasing widespread and unrestricted usage of smartphones and social media, various research articles, fiction and non-fiction books, documents, and popular new press commentary have brought to light the adverse implications of technology. They include addiction, loneliness, and depression as the primary consequences. Affordable technologies (i.e., potential actions technology for users), have evolved on an exponential scale in the past few decades, particularly in the digital technology realm. Moreover, we can leverage these affordable efficiencies to our cause to foster greater social connections and well-being in numerous contexts (Hunter et al., 2023). The positive technology movement aligns itself with this approach and aspires to optimize physical and psychological health by seeking to draw on technology and well-being sciences. Through the medium of Digital Flourishing Scale, the positive facets of youth digital interactions were unveiled. They emphasize the areas including connectedness and authentic self-presentation that encounter the narrative of inherent harms to youth mental health afflicted by technology (Hunter et al., 2023). Furthermore, it was demonstrated in the apocalyptic pandemic's forced isolation, how smartphones and technology helped the world to stay in contact through sustained connections and information access. The study demonstrated the promising potential of the evolving technology that can be utilized to establish mediums of communication with even the marginalized sections of society while reinforcing the mainstream populace to a better-connected and well-informed realm (Hunter et al., 2023).

The engagement between digital technology and mental health is of great importance and interest since the media attention and research on this area have skyrocketed within the last couple of decades. In the years of the rise of the COVID-19 pandemic, these interactions became more important as the widespread psychological distress and the frequency of digital technology engagement spiked. Chase et al. (2022) conducted a cross-sectional study including 323 university students, to examine their engagement with digital technology and its impact on their mental health. The study offered support for the increasing consensus that it is not universally or robustly harmful to engage with digital technology (Chase et al., 2022). Mental health problems faced by young adults can be prevented if we focus on the ways of usage of digital technologies by them to satisfy their social needs. The results of the study recovered little traces of the detrimental effects of the perceived time spent on digital technology. It was indicated that lonely students tended to reach out more on online social platforms to establish social relationships with people they met online (Chase et al., 2022). Finally, the importance of social connections arose dramatically amidst the social isolation COVID-19 forced upon us, but they go beyond the spectrum of the pandemic, especially in the adolescent period where social connections of emerging adults are developmentally noticeable (Chase et al., 2022).

With the rise of the COVID-19 pandemic period, the potential of smartphone-delivered interventions became obvious with the increasing demand for mental health care along with the shortage of mental health care services. To integrate health support into people's daily lives and expand access to health services, smartphone-delivered interventions may be helpful (Lam et al., 2023). There is an abundance of evidence that supports smartphone-delivered interventions' efficacy in sustaining both physical and mental health. Meditation has always been suggested by researchers for its benefits for mental and physical health. Lam et al. (2023) conducted a study to investigate the correlation between mental health and the use of smartphone meditation applications. Evidence suggested that app-delivered meditation interventions significantly improved the users' quality of life and diminished their levels of stress. Headspace and Calm were the most widely used meditation applications. Although the application's social features did not interest the users much, other features of these applications include reminders, tips, and encouragement for the practice, and were liked by them (Lam et al., 2023). Meditation apps are mostly used in the USA among certain demographic groups that can afford the apps' services. The cost of these applications raised concern for the users who cannot sustain fairly being a subscriber to them. Future studies are required to address the prevalence of the issue of affordability and further reinforce the effectiveness of these applications (Lam et al., 2023).

UNDERSTANDING OF RESILIENCE

Resilience refers to an individual's mental (as opposed to physical) ability to modify or adjust in the face of any challenging or threatening situation. Resilience is a scientific process to overcome any stress in the human brain. The ability to adapt and engage in difficult situations makes it indispensable and the key to achieving success at home, school, work, and in life. For individuals working in crisis-response professions like in the military and emergency services, resilience is essential since they must apprehend and engage in extreme and difficult situations (Pusey et al., 2020). Reynolds and Parker interpreted digital resilience as an educational intervention designed to increase the resilience of young people to hate and extremism online. Through the means of interventions, resilience is a trait that can be improved and enhanced (Khanlou & Wray, 2014; Yeager & Dweck, 2012). The duration of the interventions significantly impacts the extent to which resilience can be improved. Interventions with longer durations have long-lasting effects and are more effective and efficient at enhancing resilience (Cleary et al., 2018; Khanlou & Wray, 2014). Sen et al. (2022) emphasized how technology may be used to reduce loneliness, especially in older adults, who are more prone to isolation. The conditions of digital resilience include the stages of anticipating, recognizing, and defending against threats present in the digital world (Schemmer et al., 2021). Numerous studies statistically reported that there was a significant decrease in the negative symptoms of depression/PTSD/anxiety with the advent of an increase in resilience (Beidel et al., 2019).

Digital resilience can assist individuals in coping with the probable negative repercussions of internet use. It allows individuals to cope with the adversities and traumas they experienced and facilitates better emotional regulation. Ensuring a healthy support network can lead to the establishment of digital resilience. It can be achieved by ensuring that the affected individual is aware and knowledgeable about the ways they must adopt that will facilitate coping. They include being aware of where to reach out for support and the possible organizations that can assist them with negative digital experiences (Sharma et al., 2021). It has been frequently observed that digital technologies could promote human health (both physical and psychological health) development in many ways (Cordeiro et al., 2021). Digital technologies are bringing innovation in the health sector to promote human health and reduce the impact of serious diseases on human health (Cordeiro et al., 2021). There are many digital devices like mobile apps, digital tools are helpful to recognize various health diseases. A majority of researchers and practitioners have acknowledged the risks and disadvantages of digital technology.

The objective of the current chapter is to increase awareness and foster discussion of AI and its ethical issues. The present cutting-edge technologies in medical science offer new directions for the treatment of mental illness. But at the same time, it poses a significant challenge to patient privacy and security. Without an understanding of the mode of treatment, technology recommendations to mentally ill patients can lead to harm (Bauer et al., 2017). Another potential harm is medical harm related to inadequate quality online information related to diagnosis and treatment, via mobile apps. Addressing these ethical concerns directly will pave the way for a future and an improved quality of life for all (Kawamleh et al., 2022; Warrier et al., 2023). The findings emphasise on the convergence of two main trends. The findings of the major studies were also highlighted in Table 1.

Table 1. Finding of the major studies

Author	Study Type	Population	Aim of the Study	Main Results
Deng and Yang (2021)	Descriptive Analysis	309 College Students	Role of digital proficiency in the psychological health and well-being of first-generation college students.	Highlights to promote digital skills in college education.
Dowson (2023)	Semi-Structured Interview	30 participants	Exploring the effect of online singing groups for patients suffering from dementia and their caretakers.	Online singing can be used as an alternative to traditional face-to-face singing for people with dementia.
Zeng et al. (2018)	Preliminary Review	5 Studies	Highlights the effect of VR on patients suffering from anxiety and depression.	Studies suggest that VR is very powerful in reducing anxiety and depression among individuals.
Reger et al. (2011)	Retrospective study	Active-duty soldiers (*N*=162)	Evaluated the efficacy of VRE and PE for active-duty soldiers with PTSD from deployments to Iraq and Afghanistan.	Active-duty U.S. Army soldiers with PTSD resulting from trauma during a deployment to Iraq or Afghanistan showed significant improvement in PTSD and depression after getting treatment with either PE or VRE.
Garcia-Palacios et al. (2002)	Correlational Analysis	23 participants	Exploring the effect of virtual reality (VR) exposure therapy in the treatment of spider phobia.	Patients given VR exposure therapy had a significant effect in overcoming their fear.
Mühlberger et. al. (2001)	Regression Analysis (ANOVA)	30 participants	Examine the effect of VR on individuals with flight phobia.	With the help of VR exposure, fear of flight was easily reduced than by relaxation training.
Hunter et al. (2023)	Preliminary Review	8 studies	Use of technological tools for improving the well-being of individuals across various domains.	The integration of cross-cultural perspectives along with technology can enrich individuals' perception of the positive potential of these modern tools.
Lam et al. (2023)	Correlational Analysis	953 participants	The potential of mental health apps in improving the mental health of individuals in the US.	Individuals who are prone to problems of anxiety or other mental health problems are more likely to use these apps.
Gómez-Cambronero et al. (2024)	PRISMA-Scoping Review	43 studies	Use of smartphone-based serious games for better well-being of individuals.	Promising positive results towards effects and patient engagement were highlighted.
Thakkar et al. (2024)	Narrative Review	71 studies	To understand how AI can be useful in various aspects of mental health.	AI can be useful in awareness support and interventions for various neurodegenerative disorders.
Johnson et al. (2021)	Systematic Review	50 studies	Use of digital technology in bridging the gap for the treatment of mental health disorders.	Digital technology can help in enhancing the literacy of individuals towards mental health and its interventions.

continued on following page

Table 1. Continued

Author	Study Type	Population	Aim of the Study	Main Results
Schoeller et al. (2019)	Short Article	67 studies	To understand how technology can foster empathetic abilities in humans.	To support the enhancement of human empathic abilities various technological approaches such as neuro biofeedback, embodied VR, and psychophysiological actuators.
Milani et al. (2019)	Descriptive Analysis	318 - preadolescents	To understand the positive effects of video game use in the visuospatial domain	Games like Tetris can help in enhancing an individual's spatial perception and mental rotation.
Kane and Parsons (2019)	Literature Review	100 studies	Use of technology in substantially enhancing approaches to assessment	Technology substantially expands psychologist's capabilities to assess and treat patients and to engage in research.
Gaggioli et al. (2019)	Editorial Review	10 studies	To examine the usefulness of ICT-led applications, and services to foster positive growth among individuals.	A range of technologies like robotics, artificial intelligence, and neuro technologies can be used to emphasize positive change in individuals.

ROLE OF TECHNOLOGY AND AI IN MENTAL HEALTH

Recently, Artificial intelligence (AI) and chatbots have shown promising results in supporting the mental health of individuals. These chatbots can perform a wide range of tasks starting from diagnosing the individual to providing treatment and psychoeducation. Li et al. (2023) performed a literature review to understand the effectiveness of AI-based chatbots in improving the mental health of individuals. These chatbots can help alleviate psychological distress and provide interventions through mobile applications and instant messaging platforms. Voice-based chatbots are more effective than text-based chatbots (Li et al., 2023). These functions help in improving linguistic, literacy, and motor impairment. These chatbots have been reported as useful in clinical and sub-clinic settings for practitioners. The review also found that these AI chatbots are more beneficial for the middle-aged and older population rather than the younger population. The study concluded positive feedback about these chatbots due to their feasibility across diverse demographic groups (Li et al., 2023).

The inclusion of advanced AI has provided the edge over old traditional technological supports in the field of mental healthcare. It enables novel, comprehensive, applied, and more practical approaches to address existing mental health problems (Warrier et al., 2023). But it also poses significant ethical implications on the ground (Warrier et al., 2023). It brings notable attention to the use of AI in mental health, specifically data privacy, and transparency with responsibility use. AI has the potential to bring transformation over the conventional doctor-patient dynamic in mental healthcare with advanced tools and capabilities. A sizable specialized knowledge would facilitate achieving a harmonious balance for AI-driven mental health support (Kawamleh et al., 2022). Furthermore, Informed consent is vital in medical ethics enabling patients to make more informed decisions such as declining AI interventions if they feel uncomfortable (Naik et al., 2022). Cloud data storage creates more challenges to client privacy (Benkler et al., 2016).

AI can transform overall physical and psychological health outputs across the world. Chatbots, virtual therapists, and predictive algorithms can make mental health care efficient, affordable, and accessible (Warrier et al., 2023). But at the same time few things cannot be ignored. It is crucial to answer such questions by exploring ethical quandaries so the guarantee can be achieved for responsible AI use (Warrier et al., 2023). Additionally, the risk of client profile privacy for commercial purposes is a barrier to the safeguards of individual rights (Terra et al., 2023). Maintaining ethical standards and protocols in AI-driven mental health care must be taken care of in the policy (Naik et al., 2022). The establishment of vibrant and common ethical guidelines for the use of AI will build trust in AI.

RESULTS AND DISCUSSION

From teletherapy bridging geographic gaps to Virtual Reality calming anxieties, a new era of accessible, effective interventions emerges. This potent synergy promises swift diagnoses, personalized treatment, and empowered well-being for all, leaving no mind behind in pursuing a brighter tomorrow. This study delves deep into the applications of emerging technology that can assist mental health professionals in numerous ways. All the technological tools discussed can put great emphasis on the mental health sector. They make it much more accessible and convenient for professionals to carry on with their proceedings which may involve screening, monitoring, and interventions (Ramshaw et. al., 2022). In addition, they allow the users or clients access to them which makes it all the more convenient. The incorporation of psychology and technology serves a great deal as they can accumulate a great many mediums of assistance to many challenges faced by mental health professionals.

In the present fast-moving world, traditional mental health services are insufficient to meet individual and societal needs (Fairburn & Patel, 2017). But even after significant development the accessibility, availability, and affordability of psychological assistance in remote areas is still a long journey ahead. Psychological support can provide therapies that also include awareness and education related to mental health in these areas (Kohrt et al., 2018). Recent technological advancements have introduced innovative solutions to foster mental health and resilience. Additionally, they are helping communities change the realm of psychology by providing more creative and interactive therapies to clients. It has also been proven helpful in bridging the gap between the therapist and clients (Thakkar et al., 2024). Moreover, it has given a solution to the therapist to reach out to people who are present in economically deprived areas. Thus, this study emphasizes the potential future scope of integration of technological advancements with psychological sciences in enhancing the resilience of individuals.

Recent technological advancements provide us with novel and innovative solutions that hold the potential to address most of our problems and lacks. Similarly, in the wide field of psychology where various problems and issues are dealt with simultaneously, there is always room for inaccessibility and unavailability. Technology's new tools and devices are potent enough to assist in these matters. In addition, they also cater to more efficiency and efficacy. Even now, many of these tools are already in play that assist mental health professionals in various aspects. This chapter delves into the deep possibilities and opportunities that technology has opened for us to find answers and solutions to various emerging concerns.

Specifically, the objective of this paper is to review the scientific literature for promoting health and well-being in human beings delivered through technological systems. This paper studies the benefits of using technology in individuals (Fairburn & Patel, 2017). To understand the potential of technology individuals, require expertise and content knowledge from various fields. These include psychology, technology, human-computer interactions, and coordination of various industries (Hunter et al., 2023). Additionally, technology also serves to shift the perspective and embrace the responsibility of working together to design and use technology to maximize health and well-being (Hunter et al., 2023). Yaden et al. (2018) have identified and reviewed the use of technologies that positively facilitates mental health. These include predictive algorithms, brain stimulation, and VR. Further he has discussed the potential advantages and disadvantages of these latest technologies. He further added that the consulted data infers that the integration of psychology and technology can become one of the greatest breakthroughs in the field of mental health and well-being (Yaden et al., 2018). Presented data indicates how recent technologically advanced tools may be used purposefully to overcome the challenges faced by modern psychologists. Technological tools hold the potential to assist mental health professionals in active interventions, screening delivering psychoeducation (Fairburn & Patel, 2017). Various studies highlight the areas in which technology can play an active part. VR can provide an effective research tool for CBT which will be new and more ecological in clinical practice after testing. VR would open new exposures for clients which will be highly controlled despite providing an ecological environment. It would also address the issues of accessibility, affordability, and efficiency in comparison to traditional CBT (Kouijzer et al., 2023).

Similarly, computer games help individuals to improve their visuospatial coordination, and cognitive abilities along with reflex actions. They can contribute to the improvement of rapid reflex actions and quick intellectual processing (thinking on the feet) while facing an unusual situation (Milani et al., 2019). Therapeutic games such as 'Treasure Hunt' can prove to be useful instruments that can be used to detect and treat underlying mental illnesses in children, especially ADHD and ODD. A similar game 'Ricky and the Spider' deals with and manages children with OCD. The costs of acquiring VR technology have subsequently decreased, whereas the good quality of virtual environments has increased in the promotion of mental health (Glanz et al., 2003). Probably this trend will continue in the future and a more innovative realistic environment will be available for better use of this tool in the clinical field.

IMPLICATION AND FUTURE RESEARCH DIRECTIONS

With the advent of such technology, many medical procedures have been made available that help in the diagnosis or monitoring of any underlying abnormality. Along with medical assistance, various tools, and their collaboration assist mental health and intervention delivery procedures more feasible and efficient. The integration of VR, AI, computer games, teleconferencing, and tele-counseling with mental health procedures and interventions, have made the mental health profession more formidable. These tools and equipment initially solved one of the most persistent problems of psychologists, which was to reach the remote and secluded areas of society. Technology overcame the boundaries of availability, accessibility, and affordability for the general population. Finally, technology revolutionized the traditional methods of therapeutic interventions when it integrated itself with clinical treatment plans and paved the way for endless future possibilities.

REFERENCES

Aggarwal, A., Tam, C. C., Wu, D., Li, X., & Qiao, S. (2023). Artificial Intelligence–Based Chatbots for Promoting Health Behavioral Changes: Systematic review. *Journal of Medical Internet Research*, *25*, e40789. DOI: 10.2196/40789 PMID: 36826990

Bauer, M., Glenn, T., Monteith, S., Bauer, R., Whybrow, P. C., & Geddes, J. (2017). Ethical perspectives on recommending digital technology for patients with mental illness. *International Journal of Bipolar Disorders*, *5*(1), 6–12. DOI: 10.1186/s40345-017-0073-9 PMID: 28155206

Beidel, D. C., Frueh, B. C., Neer, S. M., Bowers, C. A., Trachik, B., Uhde, T. W., & Grubaugh, A. (2019). Trauma management therapy with virtual-reality augmented exposure therapy for combat-related PTSD: A randomized controlled trial. *Journal of Anxiety Disorders*, *61*(1), 64–74. DOI: 10.1016/j.janxdis.2017.08.005 PMID: 28865911

Boeldt, D., McMahon, E., McFaul, M., & Greenleaf, W. (2019). Using virtual reality exposure therapy to enhance treatment of anxiety disorders: Identifying areas of clinical adoption and potential obstacles. *Frontiers in Psychiatry*, *10*(1), 1–6. DOI: 10.3389/fpsyt.2019.00773 PMID: 31708821

Bollmer, G. (2017). Empathy machines. *Media International Australia, Incorporating Culture & Policy*, *165*(1), 63–76. DOI: 10.1177/1329878X17726794

Bond, R. R., Mulvenna, M. D., Potts, C., Siobhan O'Neill, Ennis, E., Torous, J. (2023). Digital transformation of mental health services. NJP *Mental Health Research*, 2(1):13. DOI: 10.1038/s44184-023-00033-y

Bray, J. H. (2010). The future of psychology practice and science. *The American Psychologist*, *65*(5), 355–369. DOI: 10.1037/a0020273 PMID: 20642298

Brezinka, V. (2012). Computer games supporting cognitive behavior therapy in children. *Clinical Child Psychology and Psychiatry*, *19*(1), 100–110. DOI: 10.1177/1359104512468288 PMID: 23258925

Cataldo, I., Lepri, B., Neoh, M. J. Y., & Esposito, G. (2021). Social media usage and development of psychiatric disorders in childhood and adolescence: A review. *Frontiers in Psychiatry*, *11*(1), 1–15. DOI: 10.3389/fpsyt.2020.508595 PMID: 33519535

Chase, G. E., Brown, M. T., & Jensen, M. (2022). Emerging adults' digital technology engagement and mental health during the COVID-19 pandemic. *Frontiers in Psychology*, *13*, 1023514. Advance online publication. DOI: 10.3389/fpsyg.2022.1023514 PMID: 36467177

Chiauzzi, E., Clayton, A., & Huh-Yoo, J. (2020). Video conferencing-based telemental health: Important questions for the COVID-19 era from clinical and patient-centered perspectives. *JMIR Mental Health*, *7*(12), 1–12. DOI: 10.2196/24021 PMID: 33180739

Cleary, M., Visentin, D., West, S., Lopez, V., & Kornhaber, R. (2018). Promoting emotional intelligence and resilience in undergraduate nursing students: An integrative review. *Nurse Education Today*, *68*(1), 112–120. DOI: 10.1016/j.nedt.2018.05.018 PMID: 29902740

Cordeiro, J. V. (2021). Digital technologies and data science as health enablers: An outline of appealing promises and compelling ethical, legal, and social challenges. *Frontiers in Medicine*, *8*(1), 1–13. DOI: 10.3389/fmed.2021.647897 PMID: 34307394

Deng, X., & Yang, Z. (2021). Digital proficiency and psychological well-being in online learning: Experiences of first-generation college students and their peers. *Social Sciences (Basel, Switzerland)*, *10*(6), 1–16. DOI: 10.3390/socsci10060192

Dowson, B., Schneider, J., McDermott, O., & Orrell, M. (2023). Online singing groups for people with dementia: Adaptation and resilience in the face of the COVID-19 pandemic. *Dementia (London)*, *22*(7), 1348–1371. DOI: 10.1177/14713012231179262 PMID: 37294296

Fairburn, C. G., & Patel, V. (2017). The impact of digital technology on psychological treatments and their dissemination. *Behaviour Research and Therapy*, *88*(1), 19–25. DOI: 10.1016/j.brat.2016.08.012 PMID: 28110672

Fairburn, C. G., & Patel, V. (2017). The impact of digital technology on psychological treatments and their dissemination. *Behaviour Research and Therapy*, *88*(1), 19–25. DOI: 10.1016/j.brat.2016.08.012 PMID: 28110672

Freitas, J. R. S., Velosa, V. H. S., Abreu, L. T. N., Jardim, R. L., Santos, J. A. V., Peres, B., & Campos, P. F. (2021). Virtual reality exposure treatment in phobias: A systematic review. *The Psychiatric Quarterly*, *92*(4), 1685–1710. DOI: 10.1007/s11126-021-09935-6 PMID: 34173160

Gaggioli, A., Villani, D., Serino, S., Baños, R. M., & Botella, C. (2019). Positive technology designing e-experiences for positive change. *Frontiers in Psychology*, *10*(1), 15–25. DOI: 10.3389/fpsyg.2019.01571 PMID: 31338053

Garcia-Palacios, A., Hoffman, H., Carlin, A., Furness, T.III, & Botella, C. (2002). Virtual reality in the treatment of spider phobia: A controlled study. *Behaviour Research and Therapy*, *40*(9), 983–993. DOI: 10.1016/S0005-7967(01)00068-7 PMID: 12296495

Glanz, K., Rizzo, A., & Graap, K. (2003). Virtual reality for psychotherapy: Current reality and future possibilities. *Psychotherapy (Chicago, Ill.)*, *40*(1-2), 55–67. DOI: 10.1037/0033-3204.40.1-2.55

Gómez-Cambronero, Á., Mann, A., Mira, A., Doherty, G., & Casteleyn, S. (2024). Smartphone-based serious games for mental health: A scoping review. *Multimedia Tools and Applications*, *1*(2), 1–48. DOI: 10.1007/s11042-024-18971-w PMID: 39553422

Holden, M. K. (2005). Virtual environments for motor rehabilitation [review]. *Cyberpsychology & Behavior*, *8*(3), 187–211. DOI: 10.1089/cpb.2005.8.187 PMID: 15971970

Hunter, J. F., Walsh, L. C., Chan, C. K., & Schueller, S. M. (2023). Editorial: The good side of technology: how we can harness the positive potential of digital technology to maximize well-being. *Frontiers in Psychology*, *14*(1), 1–3. DOI: 10.3389/fpsyg.2023.1304592 PMID: 37885744

Johnson, J. A., Sanghvi, P., & Mehrotra, S. (2021). Technology-based interventions to improve help-seeking for mental health concerns: A systematic review. *Indian Journal of Psychological Medicine*, *44*(4), 332–340. DOI: 10.1177/02537176211034578 PMID: 35949632

Kane, R. L., & Parsons, T. D. (2019). Technological developments in assessment. In Goldstein, G., Allen, D. N., & DeLuca, J. (Eds.), *Handbook of Psychological Assessment* (4th ed., pp. 573–592). Elsevier B.V., DOI: 10.1016/B978-0-12-802203-0.00020-1

Kawamleh, S. (2022). Against explainability requirements for ethical artificial intelligence in health care. *AI and Ethics, 3*(3), 901–916. DOI: 10.1007/s43681-022-00212-1

Khanlou, N., & Wray, R. (2014). A whole community approach toward child and youth resilience promotion: A review of resilience literature. *International Journal of Mental Health and Addiction, 12*(1), 64–79. DOI: 10.1007/s11469-013-9470-1 PMID: 24523668

Kohrt, B., Asher, L., Bhardwaj, A., Fazel, M., Jordans, M., Mutamba, B., Nadkarni, A., Pedersen, G., Singla, D., & Patel, V. (2018). The role of communities in mental health care in low- and middle-income countries: A meta-review of components and competencies. *International Journal of Environmental Research and Public Health, 15*(6), 1–31. DOI: 10.3390/ijerph15061279 PMID: 29914185

Kouijzer, M. M. T. E., Kip, H., Bouman, Y. H. A., & Kelders, S. M. (2023). Implementation of virtual reality in healthcare: A scoping review on the implementation process of virtual reality in various healthcare settings. *Implementation Science Communications, 4*(1), 1–29. DOI: 10.1186/s43058-023-00442-2 PMID: 37328858

Krijn, M., Emmelkamp, P. M. G., Olafsson, R. P., & Biemond, R. (2004). Virtual reality exposure therapy of anxiety disorders: A review. *Clinical Psychology Review, 24*(3), 259–281. DOI: 10.1016/j.cpr.2004.04.001 PMID: 15245832

Lam, S. U., Xie, Q., & Goldberg, S. B. (2023). Situating meditation apps within the ecosystem of meditation practice: Population-based survey study. *JMIR Mental Health, 10*(1), 1–7. DOI: 10.2196/43565 PMID: 37115618

Lan, L., Sikov, J., Lejeune, J., Ji, C., Brown, H., Bullock, K., & Spencer, A. E. (2023). A systematic review of using virtual and augmented Reality for the diagnosis and treatment of psychotic disorders. *Current Treatment Options in Psychiatry, 10*(2), 87–107. DOI: 10.1007/s40501-023-00287-5 PMID: 37360960

Li, H., Zhang, R., Lee, Y., Kraut, R. E., & Mohr, D. C. (2023). Systematic review and meta-analysis of AI-based conversational agents for promoting mental health and well-being. *Digital Medicine, 6*(1), 236. Advance online publication. DOI: 10.1038/s41746-023-00979-5 PMID: 38114588

Limone, P., & Toto, G. A. (2021). Psychological and emotional effects of digital technology on children in COVID-19 Pandemic. *Brain Sciences, 11*(9), 1126. DOI: 10.3390/brainsci11091126 PMID: 34573148

Milani, L., Grumi, S., & Di Blasio, P. (2019). Positive effects of videogame use on visuospatial competencies: The impact of visualization style in preadolescents and adolescents. *Frontiers in Psychology, 10*(1), 1–9. DOI: 10.3389/fpsyg.2019.01226 PMID: 31231275

Mira, A., Bretón-López, J., Enrique, Á., Castilla, D., García-Palacios, A., Baños, R., & Botella, C. (2018). Exploring the incorporation of a positive psychology component in a cognitive behavioral internet-based program for depressive symptoms. results throughout the intervention process. *Frontiers in Psychology, 9*(1), 1–9. DOI: 10.3389/fpsyg.2018.02360 PMID: 30555384

Mühlberger, A., Herrmann, M. J., Wiedemann, G., Ellgring, H., & Pauli, P. (2001). Repeated exposure of flight phobics to flights in virtual reality. *Behaviour Research and Therapy*, *39*(9), 1033–1050. DOI: 10.1016/S0005-7967(00)00076-0 PMID: 11520010

Mulvenna, M. D., Bond, R., Delaney, J., Dawoodbhoy, F. M., Boger, J., Potts, C., & Turkington, R. (2021). Ethical issues in democratizing digital phenotypes and machine learning in the next generation of digital health technologies. *Philosophy & Technology*, *34*(4), 1945–1960. DOI: 10.1007/s13347-021-00445-8 PMID: 33777664

Naik, N., Hameed, B. M. Z., Shetty, D. K., Swain, D., Shah, M., Paul, R., Aggarwal, K., Ibrahim, S., Patil, V., Smriti, K., Shetty, S., Rai, B. P., Chlosta, P., & Somani, B. K. (2022). Legal and ethical consideration in artificial intelligence in healthcare: Who takes responsibility? *Frontiers in Surgery*, *9*(1), 1–13. DOI: 10.3389/fsurg.2022.862322 PMID: 35360424

Ndulue, O., & Naslund, J. A. (2024). Digital interventions for alcohol use disorders: A narrative review of opportunities to advance prevention, treatment, and recovery. *Psychiatry Research Communications*, *100183*(3), 100183. Advance online publication. DOI: 10.1016/j.psycom.2024.100183

Nelson, D., Inghels, M., Kenny, A., Skinner, S., McCranor, T., Wyatt, S., Phull, J., Nanyonjo, A., Yusuff, O., & Gussy, M. (2023). Mental health professionals and telehealth in a rural setting: A cross-sectional survey. *BMC Health Services Research*, *23*(1), 200–226. DOI: 10.1186/s12913-023-09083-6 PMID: 36849933

Odgers, C. L., & Jensen, M. R. (2020). Adolescent development and growing divides in the digital age. *Dialogues in Clinical Neuroscience*, *22*(2), 143–149. DOI: 10.31887/DCNS.2020.22.2/codgers PMID: 32699514

Parsons, T. D., & Rizzo, A. A. (2008). Affective outcomes of virtual reality exposure therapy for anxiety and specific phobias: A meta-analysis. *Journal of Behavior Therapy and Experimental Psychiatry*, *39*(3), 250–261. DOI: 10.1016/j.jbtep.2007.07.007 PMID: 17720136

Paul, T. (2022). Role of Technology in the field of Clinical Psychology. *Indian Journal of Behavioral Research & Therapy*, *3*(1 & 2), 110–120. https://www.researchgate.net/publication/370779032

Peters, D., Calvo, R. A., & Ryan, R. M. (2018). Designing for motivation, engagement, and well-being in digital experience. *Frontiers in Psychology*, *9*(1), 1–15. DOI: 10.3389/fpsyg.2018.00797 PMID: 29892246

Philippe, T. J., Sikder, N., Jackson, A., Koblanski, M. E., Liow, E., Pilarinos, A., & Vasarhelyi, K. (2022). Digital Health Interventions for Delivery of Mental Health Care: Systematic and Comprehensive Meta-Review. *JMIR Mental Health*, *9*(5), e35159. DOI: 10.2196/35159 PMID: 35551058

Pontikas, C., Tsoukalas, E., & Serdari, A. (2020). A map of assistive technology educative instruments in neurodevelopmental disorders. *Disability and Rehabilitation. Assistive Technology*, *17*(7), 738–746. DOI: 10.1080/17483107.2020.1839580 PMID: 33125855

Pusey, M., Wong, K. W., & Rappa, N. A. (2020). Resilience interventions using interactive technology: A scoping review. *Interactive Learning Environments*, *30*(10), 1940–1955. DOI: 10.1080/10494820.2020.1772837

Quigley, F., Moorhead, A., Bond, R., Zheng, H., & McAloon, T. (2019). A virtual reality training tool to improve weight-related communication across healthcare settings. In *Proceedings of the 31st European Conference on Cognitive Ergonomics* (9),19–22. DOI: 10.1145/3335082.3335121

Ramshaw, G., McKeown, A., Lee, R., Conlon, A., Brown, D., & Kennedy, P. J. (2022). Introduction of technology to support young people's care and mental health: A rapid evidence review. *Child and Youth Care Forum*, *52*(3), 509–531. DOI: 10.1007/s10566-022-09700-1 PMID: 35966643

Reger, G. M., Holloway, K. M., Candy, C., Rothbaum, B. O., Difede, J., Rizzo, A. A., & Gahm, G. A. (2011). Effectiveness of virtual reality exposure therapy for active-duty soldiers in a military mental health clinic. *Journal of Traumatic Stress*, *24*(1), 93–96. DOI: 10.1002/jts.20574 PMID: 21294166

Riva, G., Baños, R. M., Botella, C., Wiederhold, B. K., & Gaggioli, A. (2012). Positive technology: Using interactive technologies to promote positive functioning. *Cyberpsychology, Behavior, and Social Networking*, *15*(1), 69–77. DOI: 10.1089/cyber.2011.0139 PMID: 22149077

Rizzo, A., Parsons, T. D., Lange, B., Kenny, P., Buckwalter, J. G., Rothbaum, B. O., & Reger, G. (2011). Virtual reality goes to war: A brief review of the future of military behavioral healthcare. *Journal of Clinical Psychology in Medical Settings*, *18*(2), 176–187. DOI: 10.1007/s10880-011-9247-2 PMID: 21553133

Schemmer, M., Heinz, D., Baier, L., Vössing, M., & Kühl, N. (2021). Conceptualizing digital resilience for AI-based information systems. *European Conference on Information Systems*. https://api.semanticscholar.org/CorpusID:235428356

Schoeller, F., Bertrand, P., Gerry, L. J., Jain, A., Horowitz, A. H., & Zenasni, F. (2019). Combining virtual reality and biofeedback to foster empathic abilities in humans. *Frontiers in Psychology*, *9*(1), 1–5. DOI: 10.3389/fpsyg.2018.02741 PMID: 30804868

Sen, K., Prybutok, G., & Prybutok, V. (2022). The use of digital technology for social wellbeing reduces social isolation in older adults: A systematic review. *SSM, Population Health*, 17(1), 1-9 101020. DOI: 10.1016/j.ssmph.2021.101020

Sharma, M. K., Anand, N., Roopesh, B. N., & Sunil, S. (2021). Digital resilience mediates the healthy use of technology. *The Medico-Legal Journal*, *90*(4), 195–199. DOI: 10.1177/00258172211018337 PMID: 34259055

Singh, A. K. (2025). Importance of digital technologies in economic sector of selected countries. In Ahmadi, A., Mna, A., & Slimani, A. (Eds.), *Renewable Energy and the Economic Welfare of Society* (pp. 335–370). IGI Global Scientific Publishing., DOI: 10.4018/979-8-3693-7580-8.ch015

Singh, A. K., Srinivasan, S., & Jyoti, B. (2025). Implications of digitalization on public health sector across countries. In Wickramasinghe, N. (Ed.), *Digitalization and the Transformation of the Healthcare Sector* (pp. 359–394). IGI Global Scientific Publishing., DOI: 10.4018/979-8-3693-9641-4.ch013

Terra, M., Baklola, M., Ali, S., & El-Bastawisy, K. (2023). Opportunities, applications, challenges and ethical implications of Artificial Intelligence in Psychiatry: A narrative review. *The Egyptian Journal of Neurology, Psychiatry and Neurosurgery*, *59*(1), 80–87. DOI: 10.1186/s41983-023-00681-z

Thakkar, A., Gupta, A., & De Sousa, A. (2024). Artificial intelligence in positive mental health: A narrative review. *Frontiers in Digital Health*, *6*(1), 1–12. DOI: 10.3389/fdgth.2024.1280235 PMID: 38562663

Tosti, B., Corrado, S., & Mancone, S. (2024). Using chatbots and conversational agents for the promotion of Well-Being and Mental Health in Adolescents: Limitations and Perspectives. *Journal of Inclusive Methodology and Technology in Learning and Teaching*, 1–10. https://www.inclusiveteaching.it

Twenge, J. M., Joiner, T. E., Rogers, M. L., & Martin, G. N. (2018). Increases in depressive symptoms, suicide-related outcomes, and suicide rates among U.S. Adolescents after 2010 and links to increased new media screen time. *Clinical Psychological Science*, *6*(1), 3–17. DOI: 10.1177/2167702617723376

Warrier, U., Warrier, A., & Khandelwal, K. (2023). Ethical considerations in the use of artificial intelligence in mental health. *The Egyptian Journal of Neurology, Psychiatry and Neurosurgery*, *59*(1), 36. DOI: 10.1186/s41983-023-00735-2

Yaden, D. B., Eichstaedt, J. C., & Medaglia, J. D. (2018). The future of technology in positive psychology: Methodological advances in the science of well-being. *Frontiers in Psychology*, *9*(1), 1–11. DOI: 10.3389/fpsyg.2018.00962 PMID: 29967586

Yeager, D. S., & Dweck, C. S. (2012). Mindsets that promote resilience: When students believe that personal characteristics can be developed. *Educational Psychologist*, *47*(4), 302–314. DOI: 10.1080/00461520.2012.722805

Zeng, N., Pope, Z., Lee, J. E., & Gao, Z. (2018). Virtual reality exercise for anxiety and depression: A preliminary review of current research in an emerging field. *Journal of Clinical Medicine*, *7*(3), 1–13. DOI: 10.3390/jcm7030042 PMID: 29510528

Chapter 8
Innovations in Digital Therapy and Personalized Support:
Machine Learning in Mental Health

N. Vinodh
https://orcid.org/0009-0000-8016-212X
Saveetha University, India

A. K. Subramani
Saveetha University, India

M. Vijayalakshmi
Saveetha University, India

R. Ramya
Saraswathi Institute of Medical Sciences, India

ABSTRACT

Transformative role of machine learning in mental health care, with a focus on digital therapy and personalized support. As mental health challenges increase globally, traditional therapeutic approaches face limitations in scalability and customization. Machine learning innovations, such as natural language processing (NLP) and predictive analytics, offer new avenues for mental health diagnosis, treatment, and ongoing care. AI-powered digital therapy platforms, including chatbots, provide real-time interventions, while personalized support systems analyze user data to tailor mental health strategies. By identifying patterns in behaviors and symptoms, machine learning enhances the effectiveness of mental health treatments, promoting timely and individualized care. However, challenges like data privacy, algorithmic bias, and the potential over-reliance on technology must be addressed. As these technologies evolve, they offer the potential to significantly improve access to and the quality of mental health care, creating scalable and responsive support systems for diverse populations.

DOI: 10.4018/979-8-3693-6910-4.ch008

INTRODUCTION

Mental health challenges are a significant global concern, affecting individuals across all demographics. According to the World Health Organization (WHO), approximately one in eight people worldwide live with a mental disorder, which translates to over 970 million individuals suffering from conditions such as anxiety, depression, and substance abuse (World Health Organization, 2022). The COVID-19 pandemic has exacerbated these issues, leading to increased levels of anxiety, depression, and stress-related disorders as individuals grapple with isolation, uncertainty, and economic instability (Luo, Guo, Yu, & Zhang, 2020). Furthermore, the stigma surrounding mental health issues often prevents individuals from seeking help, resulting in an underutilization of mental health services. The rising prevalence of mental health disorders necessitates innovative solutions to provide accessible and effective care. Traditional therapeutic approaches often face barriers such as high costs, long waiting times, and limited access to qualified mental health professionals. In response, the integration of technology into mental health care has gained traction, with digital therapy and personalized support emerging as viable alternatives. Innovations in digital therapy, such as teletherapy, mobile applications, and online support groups, are increasingly being utilized to bridge gaps in mental health care, offering convenient and often anonymous avenues for support (David, et al 2024).

Importance of Innovative Solutions in Mental Health Care

Innovative solutions are crucial in addressing the global mental health crisis. Machine learning (ML) stands at the forefront of these innovations, offering new methodologies for understanding and treating mental health disorders. By analyzing large datasets and identifying patterns, machine learning algorithms can enhance diagnostic accuracy, personalize treatment plans, and improve patient outcomes (Davenport & Kalakota, 2019). For instance, ML applications can analyze data from wearable devices to monitor users' physical and emotional well-being, providing insights into patterns of behavior and triggers for mental health crises (Banaee, Ahmed, & Loutfi, 2013). Additionally, digital therapy platforms can incorporate machine learning algorithms to customize therapeutic interventions based on individual user needs, preferences, and responses to treatment. Such personalized support can significantly enhance the efficacy of mental health care by ensuring that interventions are tailored to the specific requirements of each patient, rather than adopting a one-size-fits-all approach (Liu, Zhang, & Huang, 2021). As mental health care continues to evolve, the integration of innovative solutions, particularly machine learning, will play a pivotal role in shaping the future of mental health treatment and support.

Purpose of the Chapter

This chapter aims to explore the transformative role of machine learning in mental health care, focusing on its applications in digital therapy and personalized support. By examining the intersection of technology and mental health, this chapter seeks to highlight the potential benefits of integrating machine learning into therapeutic practices. The following objectives will guide this exploration:

1. **Examine Machine Learning Applications**: The chapter will provide an overview of the various machine learning techniques used in mental health, including supervised and unsupervised learning, natural language processing (NLP), and predictive analytics.

2. **Discuss Digital Therapy Innovations**: The chapter will delve into the advancements in digital therapy, highlighting successful implementations of machine learning in therapeutic platforms and their impact on patient outcomes.
3. **Highlight Personalized Support Strategies**: The chapter will explore how machine learning facilitates personalized mental health support, tailoring interventions to individual needs, preferences, and responses.

Structure of the Chapter

The chapter is structured to provide a comprehensive understanding of machine learning's impact on mental health care.

- **Conceptual Framework**: This section will define key concepts, including machine learning, digital therapy, and personalized support, while exploring their interrelations and implications for mental health.
- **Theoretical Framework**: An overview of relevant theories will be presented, emphasizing cognitive behavioral therapy (CBT) and the Health Belief Model as foundational elements for integrating machine learning in therapeutic practices.
- **Research Methodology**: This section will detail the research design employed in the chapter, including qualitative methods for data collection and analysis. It will discuss ethical considerations related to the study.
- **Innovations in Digital Therapy**: This section will provide an in-depth analysis of various digital therapy platforms, showcasing how machine learning enhances these tools and their effectiveness in delivering mental health care.
- **Personalized Support in Mental Health**: This part will highlight the significance of personalization in mental health treatment, discussing the role of machine learning in tailoring interventions and improving therapeutic outcomes.
- **Solutions and Recommendations**: The chapter will propose best practices for implementing machine learning in mental health care, addressing ethical challenges and emphasizing strategies for enhancing accessibility.
- **Future Research Directions**: An exploration of emerging trends in machine learning and mental health will be provided, identifying gaps in current research and suggesting areas for future investigation.
- **Conclusion**: This section will summarize the key findings of the chapter, reinforcing the significance of machine learning in advancing mental health care and its potential to transform therapeutic practices.

Through this structured approach, the chapter aims to contribute to the ongoing discourse on mental health innovations, providing valuable insights into the role of machine learning in enhancing digital therapy and personalized support.

CONCEPTUAL FRAMEWORK

Definition of Key Concepts

Machine Learning

Machine learning (ML) is a subset of artificial intelligence that focuses on developing algorithms capable of learning from and making predictions based on data. Unlike traditional programming methods, where explicit instructions are provided to the system, ML systems improve their performance over time as they are exposed to more data (Jordan & Mitchell, 2015). In the context of mental health, machine learning techniques such as supervised learning, unsupervised learning, and reinforcement learning can be employed to analyze patient data, predict mental health outcomes, and customize therapeutic interventions.

Digital Therapy

Digital therapy refers to therapeutic interventions delivered via digital platforms, including mobile applications, teletherapy, and online support groups. These tools are designed to provide mental health care remotely, making it accessible to a broader audience. Digital therapy can encompass various modalities, such as cognitive behavioral therapy (CBT), mindfulness training, and psychoeducation, delivered through interactive and engaging formats (Meyer et al., 2020). The integration of technology in therapy not only enhances accessibility but also allows for real-time monitoring and feedback, potentially improving treatment adherence and outcomes.

Personalized Support

Personalized support in mental health care entails tailoring interventions to meet the unique needs, preferences, and circumstances of each individual. This approach acknowledges that mental health is not a one-size-fits-all issue, and effective treatment requires understanding the specific context of each patient (Lee, et al 2021). Personalized support can involve adjusting therapeutic strategies based on an individual's response to treatment, preferences for communication, and personal goals. The incorporation of machine learning into personalized support allows for data-driven insights that can enhance treatment efficacy and patient engagement.

Interrelation of Concepts

The interrelation between machine learning, digital therapy, and personalized support is crucial for advancing mental health care. Machine learning algorithms can analyze large datasets, identifying patterns and correlations that inform digital therapy practices. For instance, by processing data from user interactions with a digital therapy platform, machine learning can help identify which therapeutic approaches are most effective for different demographic groups or specific mental health conditions (Naylor et al., 2021). This analytical capability enhances digital therapy by enabling real-time personalization. As users engage with digital therapy tools, machine learning can adapt the content and delivery based on their responses, preferences, and progress. For example, if a user shows a preference for mindfulness exercises over cognitive restructuring techniques, the platform can adjust to provide more of the former, thereby enhancing user satisfaction and treatment effectiveness (Alhuwaydi AM. 2024). Furthermore, personalized support can be bolstered by machine learning's predictive capabilities. By analyzing historical data, ML can forecast potential crises or relapses, prompting proactive intervention strategies tailored to the individual's unique profile. This predictive power enhances the overall effectiveness of digital therapy, as it allows clinicians to intervene before a patient reaches a critical point (Kessler et al., 2005). The interplay between machine learning, digital therapy, and personalized support creates a dynamic ecosystem that promotes more effective and individualized mental health care. By harnessing the strengths of each component, mental health practitioners can deliver innovative solutions that address the complex and diverse needs of patients.

Visual Representation

To visually represent the conceptual framework, a diagram can be utilized to illustrate the interconnections between machine learning, digital therapy, and personalized support. The diagram will consist of three main components:

1. **Machine Learning**: This section will highlight the data-driven aspect of ML, showcasing its role in analyzing patient data, identifying patterns, and generating insights that inform treatment strategies.
2. **Digital Therapy**: Represented as a distinct entity, this component will illustrate various forms of digital interventions (e.g., mobile apps, teletherapy) that deliver therapeutic content to users.
3. **Personalized Support**: This component will emphasize the importance of customization in therapy, demonstrating how insights from machine learning can tailor digital therapy to individual needs.

Arrows will indicate the flow of influence among these components, showing how machine learning enhances digital therapy, which, in turn, enables personalized support. The diagram will serve as a visual guide to understanding the interconnectedness of these concepts in the context of mental health innovations.

THEORETICAL FRAMEWORK

Cognitive Behavioral Theory (CBT)

Cognitive Behavioral Theory (CBT) is a well-established psychological approach that emphasizes the interconnection between thoughts, feelings, and behaviors. CBT aims to identify and challenge negative thought patterns, promoting healthier emotional and behavioral responses (Beck, 2019). The integration of CBT with technology has led to the emergence of digital therapies, such as online CBT programs and mobile applications designed to deliver therapeutic interventions at scale. These digital platforms facilitate access to evidence-based treatments, allowing users to engage with CBT techniques through interactive exercises, mood tracking, and personalized feedback (Kolaas, K, et al. 2024). The adaptability of CBT in digital formats highlights its potential to enhance the accessibility and effectiveness of mental health care, particularly for individuals who may face barriers to traditional therapy.

The Health Belief Model (HBM)

The Health Belief Model (HBM) is a psychological framework that explains individuals' health behaviors based on their perceptions of risk and benefits. It posits that people are more likely to engage in health-promoting behaviors if they perceive themselves to be susceptible to a health issue, believe that the condition has serious consequences, and recognize that taking a specific action would reduce their risk (Rosenstock, 1974). In the context of digital interventions for mental health, HBM can inform the design and implementation of programs that encourage users to seek help and adhere to treatment recommendations. For example, applications that provide personalized feedback on mental health symptoms can increase users' awareness of their conditions and motivate them to engage in therapeutic activities, enhancing their commitment to mental health improvement (Mahmood et al., 2021).

Application of Machine Learning Theories

Overview of Algorithms Used in Mental Health Care

Machine learning employs various algorithms that can be utilized to analyze data and provide insights in mental health care. Supervised learning algorithms, such as support vector machines (SVM) and decision trees, are trained on labeled datasets, allowing them to make predictions based on new, unseen data (Bishop, 2006). In mental health applications, supervised learning can be employed to classify individuals at risk for certain disorders or predict treatment outcomes based on historical data. Unsupervised learning algorithms, including clustering techniques and dimensionality reduction methods, identify patterns and relationships in data without prior labeling. For instance, clustering algorithms can group individuals based on similarities in their symptomatology, providing insights into distinct mental health profiles and potential treatment approaches (Xia et al., 2020). Reinforcement learning, another subset of machine learning, focuses on learning through trial and error to maximize cumulative reward. This approach is particularly relevant for developing personalized interventions that adapt based on user engagement and feedback, enhancing the overall effectiveness of digital therapies (Mnih et al., 2015).

Framework for Implementing Machine Learning in Mental Health

Steps for Integrating Machine Learning into Therapeutic Practices

The integration of machine learning into therapeutic practices involves several key steps:

1. **Data Collection**: The first step is to gather comprehensive data from various sources, including clinical assessments, patient self-reports, and behavioral metrics derived from digital health applications. This data serves as the foundation for developing machine learning models that can inform treatment decisions.
2. **Data Preprocessing**: Raw data often require preprocessing to remove noise and ensure consistency. This step may include data normalization, handling missing values, and encoding categorical variables to prepare the data for analysis.

3. **Model Selection and Training**: Selecting the appropriate machine learning algorithms based on the nature of the data and the specific objectives is crucial. Once selected, models are trained using historical data to learn patterns and relationships that inform predictions or classifications relevant to mental health care.
4. **Validation and Testing**: After training, the models must be validated and tested on separate datasets to assess their performance and generalizability. This process ensures that the models are robust and capable of providing reliable predictions in real-world settings.
5. **Implementation**: Once validated, machine learning models can be integrated into digital therapy platforms. This step involves developing user-friendly interfaces that allow practitioners and patients to interact with the technology effectively.
6. **Monitoring and Feedback**: Continuous monitoring of model performance and user engagement is essential to refine the algorithms and improve their efficacy. Incorporating user feedback helps in iterating and enhancing the system to better meet the needs of individuals seeking mental health support.
7. **Evaluation and Adjustment**: Finally, regular evaluation of the implemented machine learning systems is necessary to assess their impact on mental health outcomes. This feedback loop allows for adjustments to the models and therapeutic approaches, ensuring that they remain effective and relevant.

By following this framework, mental health practitioners can leverage machine learning technologies to enhance therapeutic practices and provide personalized, effective support for individuals experiencing mental health challenges.

RESEARCH METHODOLOGY

This chapter employs a qualitative research design to explore the integration of machine learning in mental health care, particularly focusing on digital therapy and personalized support. A qualitative approach is justified due to its capacity to provide rich, detailed insights into complex phenomena, such as the experiences of individuals utilizing digital mental health solutions and the perceptions of practitioners implementing these technologies. Qualitative research allows for an in-depth understanding of the nuances of human behavior, attitudes, and motivations, which are crucial for assessing the effectiveness and acceptance of machine learning interventions in mental health settings (Creswell & Poth, 2018). The exploratory nature of this study aims to uncover the underlying mechanisms and contextual factors that contribute to the successful implementation of machine learning in mental health care. Data collection methods for this research include a comprehensive literature review and case studies. The literature review involves a systematic examination of recent studies, articles, and reports related to machine learning applications in mental health, digital therapy, and personalized support. This review will highlight existing knowledge, identify gaps in the literature, and provide a theoretical basis for understanding how machine learning can enhance mental health interventions (Alhuwaydi AM. 2024). Case studies will be selected from various mental health organizations and digital therapy platforms that have successfully integrated machine learning into their practices. These case studies will provide real-world examples of how machine learning algorithms are employed to personalize treatment, monitor patient progress, and improve health outcomes. Data analysis will be conducted using thematic analysis, a qualitative analytical technique that enables the identification of patterns and themes within the collected data. This method is particularly well-suited for qualitative research, as it allows for flexibility and the exploration of multiple dimensions of a given phenomenon (Braun & Clarke, 2006). Thematic analysis involves several key stages, including familiarization with the data, generating initial codes, identifying themes, reviewing themes, and defining and naming themes. This structured approach will facilitate a comprehensive analysis of the factors influencing the integration of machine learning in mental health, enabling researchers to draw meaningful conclusions and recommendations based on the findings. Ethical considerations are paramount in this research, especially when dealing with sensitive topics such as mental health. Ensuring confidentiality and anonymity of participants is critical to protecting their privacy and fostering trust. Informed consent will be obtained from all participants prior to data collection, ensuring that they are fully aware of the study's purpose, procedures, and potential risks. Participants will also have the right to withdraw from the study at any time without consequence. Furthermore, robust data protection measures will be implemented to safeguard personal information, including secure storage of data and limited access to authorized personnel only. By adhering to ethical guidelines, the research aims to uphold the integrity of the study while prioritizing the well-being of participants. The qualitative research design, underpinned by a comprehensive literature review and case studies, allows for an in-depth exploration of the role of machine learning in mental health care. Through thematic analysis, the research aims to uncover key themes and insights that can inform the development and implementation of effective digital therapies and personalized support systems. Ethical considerations are integral to the research process, ensuring the protection of participants and the responsible handling of sensitive information.

INNOVATIONS IN DIGITAL THERAPY

Digital therapy has emerged as a significant innovation in the mental health landscape, offering new avenues for individuals seeking support and treatment for mental health issues. With the proliferation of smartphones and internet access, various digital therapy platforms have gained popularity, providing accessible and flexible solutions for users. These platforms include a range of applications designed to deliver therapeutic interventions through digital means, such as Cognitive Behavioral Therapy (CBT) apps, virtual therapists, and online counseling services. Such applications are tailored to address a variety of mental health concerns, from anxiety and depression to stress management and behavioral disorders. One prominent category of digital therapy applications is CBT apps. These applications leverage established cognitive-behavioral techniques to help users identify and challenge negative thought patterns, set achievable goals, and develop coping strategies. For example, apps like Woebot and Moodfit provide users with tools to track their moods, access therapeutic exercises, and receive feedback on their mental health journey. These platforms are designed to empower users by providing them with the skills and resources needed to manage their mental health effectively. In addition to CBT apps, virtual therapists and teletherapy platforms have gained traction. Virtual therapists utilize artificial intelligence (AI) to simulate human interaction and deliver therapeutic interventions. For instance, platforms such as Wysa offer AI-driven chatbots that guide users through self-help exercises and provide emotional support. These virtual therapists are available 24/7, offering a level of accessibility that traditional in-person therapy may not provide. By breaking down geographical and financial barriers, digital therapy platforms can reach a wider audience, ensuring that more individuals have access to mental health resources. The integration of machine learning (ML) into digital therapy significantly enhances its effectiveness and personalization. Machine learning algorithms, particularly Natural Language Processing (NLP) and predictive modeling, enable these platforms to analyze user data and tailor interventions to individual needs. NLP allows digital therapy applications to understand and interpret user inputs, providing real-time feedback and support. For example, AI-powered chatbots can engage in meaningful conversations with users, interpreting their feelings and offering appropriate responses based on the context of the discussion. This capability creates a more personalized experience, as the digital therapist can adapt its responses to suit the user's emotional state and therapeutic needs (Pérez et al., 2021). Predictive modeling further enhances digital therapy by identifying patterns in user behavior and outcomes. By analyzing historical data, machine learning algorithms can predict which interventions are most likely to be effective for a given user. This approach enables digital therapy platforms to recommend personalized strategies that align with individual progress and needs. For instance, an app may track a user's mood patterns over time and suggest specific CBT exercises or mindfulness techniques based on their historical responses. This level of personalization not only improves user engagement but also increases the likelihood of positive outcomes in mental health interventions.

Several case studies illustrate the successful implementation of digital therapy powered by machine learning. One notable example is the research conducted by Sadeh et al. (2023), which examined the effectiveness of an AI-driven digital therapy platform called "Happy." This platform utilized machine learning algorithms to deliver personalized interventions based on user input and behavioral data. The study found that users of the Happy app reported significant reductions in anxiety and depression symptoms after engaging with the platform over a specified period. The integration of machine learning allowed for dynamic adjustments in the therapeutic approach, enhancing user experiences and outcomes.

Another case study is the platform "Youper," an AI-powered chatbot that offers users personalized emotional support and CBT techniques. In a study by Makhnevich et al. (2021), users of Youper reported improved emotional well-being and reduced symptoms of anxiety and depression after using the app for several weeks. The platform's ability to adapt to users' needs through machine learning significantly contributed to its effectiveness, making it a valuable tool for individuals seeking mental health support. Furthermore, the platform "Moodfit" has also demonstrated the potential of integrating machine learning into digital therapy. Moodfit utilizes a range of features, including mood tracking, goal setting, and educational resources, to support users in their mental health journeys. A study by Wang et al. (2018) found that users who actively engaged with Moodfit experienced improvements in mood and overall well-being, highlighting the platform's effectiveness in promoting mental health through personalized support. Innovations in digital therapy have transformed the mental health landscape, providing accessible and effective solutions for individuals seeking support. The integration of machine learning enhances these platforms by enabling personalized interventions and improving user engagement. As demonstrated by successful case studies, the combination of digital therapy and machine learning has the potential to revolutionize mental health care, offering innovative solutions that cater to the diverse needs of individuals. By leveraging technology to deliver evidence-based interventions, digital therapy platforms are paving the way for a new era of mental health support.

PERSONALIZED SUPPORT IN MENTAL HEALTH

Personalized support in mental health has become a pivotal aspect of contemporary therapeutic practices, emphasizing the need for tailored interventions that cater to individual needs. The importance of personalization lies in its ability to enhance therapeutic outcomes by acknowledging the unique experiences, preferences, and circumstances of each individual. Personalization transforms the traditional "one-size-fits-all" approach into a more dynamic model that is responsive to the specific challenges and strengths of the person receiving care. Research indicates that personalized approaches can lead to higher levels of engagement, increased satisfaction with treatment, and better overall mental health outcomes (Sheikh, et al., 2024). By focusing on individual differences, personalized support fosters a therapeutic alliance that is crucial for effective mental health interventions. Machine learning (ML) has emerged as a powerful tool in enhancing the personalization of mental health support. By leveraging large datasets, ML algorithms can analyze user data to identify patterns and trends that inform tailored interventions. These algorithms can process various forms of data, including self-reported symptoms, behavioral patterns, and even physiological indicators, to create a comprehensive profile of the individual's mental health status. For example, a study by Bockting, Hollon, Jarrett, Kuyken, and Tiemens (2015) illustrated how ML techniques can be employed to predict depression relapse based on users' online interactions and self-reported moods. The algorithm's predictive capabilities enable mental health professionals to intervene proactively, offering customized support before the individual experiences significant distress. Moreover, the use of machine learning in personalization extends beyond prediction; it also enhances the delivery of interventions. Platforms such as Woebot, which utilize AI-driven chatbots, exemplify how personalized support can be implemented effectively. Woebot adapts its responses based on the user's mood and the context of the conversation, providing tailored strategies for coping and emotional regulation. This dynamic interaction not only promotes user engagement but also fosters a sense of connection, making the therapeutic process feel more relevant and supportive (Miralles, et al 2017). The integration of machine learning applications into personalized mental health solutions has given rise to several innovative platforms and technologies. One noteworthy example is "Youper," an AI-powered app designed to assist users in managing their emotional health. Youper utilizes machine learning algorithms to analyze users' inputs and provide personalized insights and recommendations based on their emotional patterns. The app encourages users to track their moods, engage in reflective journaling, and participate in cognitive behavioral therapy (CBT)-based exercises tailored to their unique experiences. A study by Makhnevich, Marefat, and Darwish (2021) found that users of Youper experienced significant improvements in their emotional well-being, demonstrating the effectiveness of personalized support facilitated by machine learning. Another exemplary platform is "Moodfit," which integrates machine learning to deliver personalized mental health interventions. Moodfit provides users with tools for mood tracking, goal setting, and educational resources, all tailored to individual needs. The app analyzes user data to suggest specific exercises and strategies based on the user's emotional state and historical responses. This level of personalization not only enhances user engagement but also significantly increases the likelihood of positive outcomes in mental health management (Wang, et al 2018). Furthermore, "Ginger," a platform that combines on-demand mental health support with personalized coaching, leverages machine learning to deliver tailored interventions. Users can access chat-based support from mental health professionals, and the platform uses data analytics to match users with the most suitable resources and interventions based on their specific needs and preferences. This approach ensures that individuals receive timely and appropriate care, ultimately improving their mental health outcomes (Markert et al, 2021). In addition to these platforms, the use of wearable technology has further enhanced personalized mental health support. Devices such as smartwatches and fitness trackers can monitor physiological indicators like heart rate variability and sleep patterns, which are correlated with mental health. By integrating data from these devices, mental health apps can offer personalized insights and recommendations based on users' real-time health metrics. For instance, the app "Happify" utilizes data from wearable devices to provide users with personalized activities aimed at improving their emotional well-being, thus bridging the gap between digital therapy and real-life behavioral change (Gomes, et al 2023). The benefits of personalized support in mental health are further amplified by ongoing advancements in machine learning techniques. As algorithms become increasingly sophisticated, their ability to analyze diverse datasets will enhance the personalization of mental health interventions. Future developments in natural language processing (NLP) will enable more nuanced interactions between users and digital platforms, allowing for deeper emotional engagement and support. The continuous refinement of machine learning models promises to improve the accuracy and effectiveness of personalized interventions, making mental health support more accessible and tailored to individual needs. Personalized support is a critical component of modern mental health care, and the integration of machine learning has revolutionized the way mental health interventions are delivered. By analyzing user data and providing tailored interventions, machine learning enhances the therapeutic experience, ultimately leading to improved outcomes for individuals seeking mental health support. The success of various platforms, such as Youper, Moodfit, and Ginger, underscores the potential of personalized mental health solutions to address the diverse needs of users. As technology continues to evolve, the possibilities for personalized mental health support will expand, providing new opportunities for individuals to manage their mental health effectively.

SOLUTIONS AND RECOMMENDATIONS

The integration of machine learning (ML) into mental health care presents numerous opportunities for improving therapeutic outcomes and personalizing support. However, to fully harness the potential of ML in mental health, practitioners and policymakers must adopt best practices, address ethical challenges, and enhance accessibility. This section outlines several solutions and recommendations for implementing machine learning effectively in mental health care.

Best Practices for Implementing Machine Learning in Mental Health

To successfully integrate machine learning into mental health practices, mental health practitioners and policymakers should adhere to several best practices. First, it is essential to foster collaboration between mental health professionals and data scientists to ensure that algorithms are developed with a deep understanding of clinical contexts and therapeutic needs. Collaborative efforts can help in designing algorithms that are not only clinically relevant but also user-friendly, facilitating better engagement from patients (Sculley, Holtz, Guyon, & Eick, 2015). Second, continuous training and education on ML tools and their applications should be provided for mental health practitioners. Understanding how to interpret data-driven insights and integrate them into therapeutic practice can enhance clinicians' confidence in using these technologies. Regular workshops and training sessions focusing on the latest developments in machine learning can empower practitioners to make informed decisions based on data (Wheeler, Rosenthal, & Lund, 2022). Third, the use of standardized measures for evaluating the effectiveness of

ML applications in mental health is crucial. Implementing robust evaluation frameworks can help in assessing the impact of these tools on patient outcomes, thereby fostering accountability and continual improvement (Lee, et al 2021).

Addressing Ethical Challenges

The deployment of machine learning in mental health raises significant ethical concerns, particularly regarding bias and data protection. To mitigate bias in ML algorithms, practitioners should prioritize diverse and representative datasets during the training process. Ensuring that data encompasses various demographics, including age, gender, and cultural background, can help prevent algorithms from perpetuating existing disparities in mental health care (Obermeyer, Powers, Vogeli, & Mullainathan, 2019). Furthermore, transparency in algorithm development is vital. Mental health organizations should adopt explainable AI principles that enable practitioners to understand how algorithms make decisions. This transparency can enhance trust among users and practitioners, facilitating greater acceptance of technology in therapeutic contexts (Lipton, 2016). In terms of data protection, mental health practitioners must prioritize confidentiality and informed consent when collecting user data. Implementing robust data encryption and privacy policies can safeguard sensitive information and protect users from potential breaches (Binns, 2018). Organizations should also establish clear guidelines outlining how user data will be used and shared, ensuring compliance with legal and ethical standards.

Enhancing Accessibility

Increasing accessibility to digital therapy powered by machine learning is essential for maximizing its impact on mental health care. To enhance access, policymakers should promote initiatives that reduce the digital divide, particularly in underserved communities. This can include funding for internet access and digital literacy programs, ensuring that all individuals can benefit from innovative mental health solutions (Philippe, at al 2022). Additionally, mental health practitioners should consider user-friendly designs when developing digital therapy platforms. Ensuring that applications are intuitive and easy to navigate can enhance user engagement, particularly among individuals who may be less tech-savvy (Sullivan, Sweeney, & Jackson, 2021). Furthermore, providing multilingual support within digital therapy platforms can significantly increase accessibility for non-native speakers. By offering services in multiple languages, mental health care can become more inclusive and responsive to the needs of diverse populations (Park, et al 2020). Finally, establishing partnerships between mental health organizations and technology developers can facilitate the creation of tailored digital interventions. Collaborative efforts can help in designing solutions that address specific community needs, ultimately enhancing the reach and effectiveness of mental health care (David et al, 2024).

FUTURE RESEARCH DIRECTIONS

The intersection of machine learning and mental health is a rapidly evolving field with significant potential for enhancing mental health care. As technology advances, several emerging trends, areas for further research, and the integration of new technologies present opportunities for innovation and improvement in therapeutic practices.

Emerging Trends in Machine Learning and Mental Health

The future of machine learning in mental health care is characterized by several anticipated developments. One significant trend is the increasing use of natural language processing (NLP) to analyze and interpret patient data. NLP can facilitate the identification of mental health conditions through speech and text analysis, leading to earlier interventions and more tailored therapeutic approaches. For instance, recent studies demonstrate that NLP algorithms can effectively detect symptoms of depression from social media content and text messages, providing real-time insights into individuals' mental health status (Coppersmith, Dredze, & Harman, 2015). Another emerging trend is the integration of machine learning with mobile health (mHealth) applications. These applications, which allow users to track their mental health and access therapeutic resources on their smartphones, are becoming increasingly sophisticated through the incorporation of ML algorithms. Anticipated advancements in mHealth include predictive analytics that can provide personalized intervention recommendations based on users' real-time data (Torous, J., & Keshavan, M., 2018). Additionally, machine learning models are expected to become more adept at understanding and analyzing the complexities of mental health disorders, leading to more effective and personalized treatment strategies.

Suggested Areas for Further Research

Despite the progress made in integrating machine learning into mental health, there remain several gaps in current research that warrant further exploration. One critical area is the evaluation of long-term outcomes associated with machine learning-based interventions. While many studies focus on short-term effectiveness, understanding the long-term impact of these interventions on mental health outcomes is essential for assessing their true value (Philippe, et al 2022). Future research should prioritize longitudinal studies to evaluate the sustainability of therapeutic benefits. Another gap lies in understanding the ethical implications and societal impacts of deploying machine learning in mental health care. As algorithms are increasingly used to inform treatment decisions, it is crucial to investigate issues related to privacy, data security, and algorithmic bias (Obermeyer, Powers, Vogeli, & Mullainathan, 2019). Research should focus on developing frameworks for ethical AI use in mental health, addressing how to safeguard patients' rights while maximizing the benefits of technology. Furthermore, there is a need for research on the effectiveness of various machine learning algorithms across diverse populations. Many existing studies predominantly involve homogeneous samples, limiting the generalizability of findings. Exploring the applicability of machine learning tools across different cultural and demographic groups can inform more inclusive and effective mental health interventions (Sweeney, M. E., & Sullivan, A. H., 2021).

Integration of New Technologies

The integration of new technologies, such as virtual reality (VR), augmented reality (AR), and advanced machine learning techniques, holds substantial promise for the future of mental health care. VR has shown potential in providing immersive therapeutic experiences for patients with conditions such as post-traumatic stress disorder (PTSD) and anxiety disorders. By simulating real-world environments, VR can help individuals confront and manage their fears in a controlled setting, leading to improved therapeutic outcomes (Maples, N. J., & Derlega, V. J., 2021). Similarly, AR can enhance traditional therapeutic practices by overlaying digital information onto the physical environment, providing real-time feedback and support during therapy sessions. For example, AR applications could assist individuals in practicing social skills in real-world scenarios by offering cues and prompts through smart glasses (Lee, J. H., & Choi, M. J., 2020). Incorporating advanced machine learning techniques, such as reinforcement learning and deep learning, can further enhance the effectiveness of digital mental health interventions. These technologies can be utilized to create adaptive systems that evolve with the user's progress, providing personalized interventions that continuously improve over time (Gönül, et al 2021). The future of machine learning in mental health care is ripe with opportunities for innovation. By exploring emerging trends, addressing gaps in research, and integrating new technologies, mental health practitioners can leverage the potential of machine learning to enhance therapeutic outcomes and improve the overall quality of care.

CONCLUSION

The integration of machine learning into mental health care has emerged as a transformative force, offering innovative solutions to address complex challenges in the field. This chapter has highlighted the critical role of machine learning in enhancing digital therapy and personalized support, underscoring the potential for these technologies to improve therapeutic outcomes and accessibility for individuals experiencing mental health issues. Through the application of advanced algorithms, mental health practitioners can analyze vast amounts of data to gain insights into patient behavior and preferences, leading to more tailored interventions that resonate with individual needs. Studies have shown that machine learning models can effectively predict treatment outcomes, enabling practitioners to make data-driven decisions that enhance patient care (Sharma, D., & Kumar, S., 2022). Moreover, the utilization of digital therapy platforms has been revolutionized through machine learning technologies, which allow for real-time monitoring and adjustments based on user engagement and feedback. This adaptability not only fosters a more dynamic therapeutic environment but also empowers patients by promoting self-management of their mental health (Ryu, A., & Lee, J. H., 2021). Case studies of successful implementations reveal that when machine learning is effectively harnessed, it leads to significant improvements in patient engagement, satisfaction, and overall mental health outcomes (Bickmore, T., & Picard, R. W., 2020). Looking towards the future, it is clear that the landscape of mental health care will continue to evolve as machine learning technologies become more sophisticated and accessible. The potential for integrating emerging technologies such as virtual reality (VR) and augmented reality (AR) further extends the capabilities of mental health support systems. These advancements can offer immersive therapeutic experiences that enhance engagement and provide effective treatment for conditions like anxiety and PTSD (Maples, N. J., & Derlega, V. J., 2021). Additionally, the emphasis on personalized approaches will likely lead to a paradigm shift in how mental health services are delivered, prioritizing the unique needs of individuals. As we envision the future of mental health care, it is essential to remain vigilant about ethical considerations surrounding data privacy, algorithmic bias, and inclusivity. Addressing these challenges will be crucial to building trust in digital mental health solutions and ensuring that they serve all populations effectively. By fostering collaboration between technologists, mental health practitioners, and researchers, we can create an ecosystem that supports innovation while safeguarding the rights and well-being of patients. The significance of machine learning in mental health care cannot be overstated. As we embrace these technologies, the potential to reshape mental health support and improve the quality of life for countless individuals is not only promising but imperative for the future of mental health care.

REFERENCES

Alhuwaydi, A. M. (2024). Exploring the Role of Artificial Intelligence in Mental Healthcare: Current Trends and Future Directions – A Narrative Review for a Comprehensive Insight. *Risk Management and Healthcare Policy*, *17*, 1339–1348. DOI: 10.2147/RMHP.S461562 PMID: 38799612

Banaee, H., Ahmed, M. U., & Loutfi, A. (2013). Data mining for wearable sensors in health monitoring systems: A review of recent trends and challenges. *Sensors (Basel)*, *13*(12), 17472–17500. DOI: 10.3390/s131217472 PMID: 24351646

Beck, J. S. (2020). *Cognitive Behavior Therapy: Basics and Beyond* (3rd ed.). Guilford Press.

Bickmore, T., & Picard, R. W. (2020). Establishing and maintaining long-term human-computer relationships. *ACM Transactions on Computer-Human Interaction*, *7*(2), 95–132. DOI: 10.1145/1057237

Binns, R. (2018). Fairness in Machine Learning: Lessons from Political Philosophy. Proceedings of the 1st Conference on Fairness, Accountability and Transparency, in Proceedings of Machine Learning Research 81:149-159 Available from https://proceedings.mlr.press/v81/binns18a.html.

Bishop, C. M. (2006). *Pattern Recognition and Machine Learning*. Springer.

Bockting, C. L., Hollon, S. D., Jarrett, R. B., Kuyken, W., & Dobson, K. (2015). A lifetime approach to major depressive disorder: The contributions of psychological interventions in preventing relapse and recurrence. *Clinical Psychology Review*, *41*, 16–26. DOI: 10.1016/j.cpr.2015.02.003 PMID: 25754289

Braun, V., & Clarke, V. (2006). Using thematic analysis in psychology. *Qualitative Research in Psychology*, *3*(2), 77–101. DOI: 10.1191/1478088706qp063oa

Coppersmith, G., Dredze, M., & Harman, C. (2015). Quantifying mental health signals in Twitter. *Proceedings of the Fifth International Conference on Social Media Technologies, Communication, and Informatics*, 51-60

Creswell, J. W., & Poth, C. N. (2018). *Qualitative Inquiry and Research Design Choosing among Five Approaches* (4th ed.). SAGE Publications, Inc.

Davenport, T., & Kalakota, R. (2019). The potential for artificial intelligence in healthcare. *Future Healthcare Journal*, *6*(2), 94–98. DOI: 10.7861/futurehosp.6-2-94 PMID: 31363513

Gomes, N., Pato, M., Lourenço, A. R., & Datia, N. (2023). A Survey on Wearable Sensors for Mental Health Monitoring. *Sensors (Basel)*, *23*(3), 1330. DOI: 10.3390/s23031330 PMID: 36772370

Gönül, S., Namlı, T., Coşar, A., & Toroslu, İ. H. (2021). A reinforcement learning based algorithm for personalization of digital, just-in-time, adaptive interventions. *Artificial Intelligence in Medicine*, *115*, 102062. DOI: 10.1016/j.artmed.2021.102062 PMID: 34001322

Jordan, M. I., & Mitchell, T. M. (2015). Machine learning: Trends, perspectives, and prospects. *Science*, *349*(6245), 255–260. DOI: 10.1126/science.aaa8415 PMID: 26185243

Kessler, R. C., Berglund, P., Demler, O., Jin, R., Merikangas, K. R., & Walters, E. E. (2005). Lifetime Prevalence and Age of Onset Distributions of DSM-IV Disorders in the National Comorbidity Survey Replication. *Archives of General Psychiatry*, *62*(6), 593–602. DOI: 10.1001/archpsyc.62.6.593 PMID: 15939837

Kolaas, K., Berman, A. H., Hedman-Lagerlöf, E., Lindsäter, E., Hybelius, J., & Axelsson, E. (2024). Internet-delivered transdiagnostic psychological treatments for individuals with depression, anxiety or both: A systematic review with meta-analysis of randomised controlled trials. *BMJ Open*, *14*(4), e075796. DOI: 10.1136/bmjopen-2023-075796 PMID: 38569713

Lee, E. E., Torous, J., De Choudhury, M., Depp, C. A., Graham, S. A., Kim, H. C., Paulus, M. P., Krystal, J. H., & Jeste, D. V. (2021). Artificial Intelligence for Mental Health Care: Clinical Applications, Barriers, Facilitators, and Artificial Wisdom. *Biological Psychiatry: Cognitive Neuroscience and Neuroimaging*, *6*(9), 856–864. DOI: 10.1016/j.bpsc.2021.02.001 PMID: 33571718

Lee, J. H., & Choi, M. J. (2020). Augmented reality for social skills training in adolescents with autism spectrum disorder: A pilot study. *Journal of Autism and Developmental Disorders*, *50*(6), 2211–2220. DOI: 10.1007/s10803-019-04094-3 PMID: 30864059

Lipton, Z. C. (2016). The Mythos of Model Interpretability. *Proceedings of the 2016 ICML Workshop on Human Interpretability in Machine Learning*, 96-100. https://arxiv.org/abs/1606.03490

Liu, S., Zhang, Y., & Huang, J. (2021). Personalized mental health intervention using machine learning: A systematic review. *Health Informatics Journal*, *27*(3), 1460458221999187. DOI: 10.1177/1460458221999187

Luo, M., Guo, L., Yu, Y., & Zhang, L. (2020). The psychological and mental impact of coronavirus disease 2019 (COVID-19) on medical staff and general public: A systematic review and meta-analysis. *Psychological Medicine*, •••, 1–10. DOI: 10.1017/S0033291720001200 PMID: 32563745

Mahmood, M. A., Tazeen, M. F., & Abid, A. (2021). The effectiveness of health belief model-based interventions in changing health behavior: A systematic review. *Journal of Public Health*, *43*(1), 97–108. DOI: 10.1093/pubmed/fdz043

Makhnevich, A., Marefat, A., & Darwish, A. (2021). Evaluating the impact of the AI-driven digital therapy platform Youper on users' emotional health: A pilot study. *BMC Psychiatry*, *21*(1), 305. DOI: 10.1186/s12888-021-03437-0 PMID: 34118905

Maples, N. J., & Derlega, V. J. (2021). Virtual reality exposure therapy for post-traumatic stress disorder: A systematic review. *Computers in Human Behavior*, *115*, 106610. DOI: 10.1016/j.chb.2020.106610

Markert, C., Sasangohar, F., Mortazavi, B., & Fields, S. (2021). The Use of Telehealth Technology to Support Health Coaching for Older Adults: Literature Review. *JMIR Human Factors*, *8*(1), e23796. DOI: 10.2196/23796 PMID: 33512322

Meyer, B., Von Hoebel, J., & Dyb, G. (2020). Digital therapy for mental health: A systematic review. *Journal of Medical Internet Research*, *22*(3), e18371. DOI: 10.2196/18371

Miralles, I., Granell, C., Díaz-Sanahuja, L., Van Woensel, W., Bretón-López, J., Mira, A., Castilla, D., & Casteleyn, S. (2020). Smartphone Apps for the Treatment of Mental Disorders: Systematic Review. *JMIR mHealth and uHealth, 8*(4), e14897. DOI: 10.2196/14897 PMID: 32238332

Mnih, V., Silver, D., & Graves, A. (2015). Human-level control through deep reinforcement learning. *Nature, 518*(7540), 529–533. DOI: 10.1038/nature14236 PMID: 25719670

Naylor, C., Bell, A., & Parsonage, M. (2021). The role of technology in mental health services: A systematic review. *The Lancet. Psychiatry, 8*(2), 98–106. DOI: 10.1016/S2215-0366(20)30466-7

Obermeyer, Z., Powers, B., Vogeli, C., & Mullainathan, S. (2019). Dissecting racial bias in an algorithm used to manage the health of populations. *Science, 366*(6464), 447–453. DOI: 10.1126/science.aax2342 PMID: 31649194

Olawade, D. B., Wada, O. Z., Odetayo, A., David-Olawade, A. C., Asaolu, F., & Eberhardt, J. (2024). Enhancing mental health with Artificial Intelligence: Current trends and future prospects. *Journal of Medicine, Surgery, and Public Health, 3*, 100099. Advance online publication. DOI: 10.1016/j.glmedi.2024.100099

Park, S. Y., Nicksic Sigmon, C., & Boeldt, D. (2022). A Framework for the Implementation of Digital Mental Health Interventions: The Importance of Feasibility and Acceptability Research. *Cureus, 14*(9), e29329. DOI: 10.7759/cureus.29329 PMID: 36277565

Pérez, J. S., Boulton, M., & Van Melle, E. (2021). Natural language processing in mental health: A systematic review. *Journal of Medical Internet Research, 23*(6), e25492. DOI: 10.2196/25492 PMID: 33944788

Philippe, T. J., Sikder, N., Jackson, A., Koblanski, M. E., Liow, E., Pilarinos, A., & Vasarhelyi, K. (2022). Digital Health Interventions for Delivery of Mental Health Care: Systematic and Comprehensive Meta-Review. *JMIR Mental Health, 9*(5), e35159. DOI: 10.2196/35159 PMID: 35551058

Rosenstock, I. M. (1974). Historical origins of the Health Belief Model. *Health Education Monographs, 2*(4), 328–335. DOI: 10.1177/109019817400200403

Ryu, A., & Lee, J. H. (2021). The role of digital mental health interventions in the COVID-19 pandemic: A review of effectiveness. *Asian Journal of Psychiatry, 54*, 102458. DOI: 10.1016/j.ajp.2020.102458

Sadeh-Sharvit, S., Camp, T. D., Horton, S. E., Hefner, J. D., Berry, J. M., Grossman, E., & Hollon, S. D. (2023). Effects of an Artificial Intelligence Platform for Behavioral Interventions on Depression and Anxiety Symptoms: Randomized Clinical Trial. *Journal of Medical Internet Research, 25*, e46781. DOI: 10.2196/46781 PMID: 37428547

Sculley, D., Holtz, S., Guyon, I., & Eick, S. (2015). Machine Learning: The High-Interest Credit Card of Technical Debt. Proceedings of the 28th International Conference on Machine Learning, 151-158. https://www.aaai.org/ocs/index.php/ICML/ICML15/paper/view/10088

Sharma, D., & Kumar, S. (2022). A machine learning framework for personalized mental health interventions: A systematic review. *Journal of Biomedical Informatics, 126*, 103961. DOI: 10.1016/j.jbi.2022.103961

Sheikh, A., Jacob, J., Vostanis, P., Ruby, F., Spuerck, I., Stankovic, M., Morgan, N., Mota, C. P., Ferreira, R., Eruyar, Ş., Yılmaz, E. A., Fatima, S. Z., & Edbrooke-Childs, J. (2024). What Should Personalised Mental Health Support Involve? Views of Young People with Lived Experience and Professionals from Eight Countries. *Administration and Policy in Mental Health, 51*(5), 753–768. DOI: 10.1007/s10488-024-01382-2 PMID: 38907740

Sullivan, A. H., Sweeney, M. E., & Jackson, J. (2021). Designing Accessible Digital Health Solutions for Mental Health: A User-Centric Approach. *Journal of Technology in Behavioral Science, 6*(2), 92–105. DOI: 10.1007/s41347-020-00230-y

Sweeney, M. E., & Sullivan, A. H. (2021). Evaluating the inclusivity of machine learning tools in mental health interventions: A call for diversity in research. *Journal of Health Psychology, 26*(7), 1069–1076. DOI: 10.1177/1359105320958850

Torous, J., & Keshavan, M. (2018). A digital future for mental health care: A brief overview of the current state of technology in mental health. *Psychiatric Services (Washington, D.C.), 69*(3), 250–255. DOI: 10.1176/appi.ps.201700456 PMID: 29334879

Wang, K., Varma, D. S., & Prosperi, M. (2018). A systematic review of the effectiveness of mobile apps for monitoring and management of mental health symptoms or disorders. *Journal of Psychiatric Research, 107*, 73–78. DOI: 10.1016/j.jpsychires.2018.10.006 PMID: 30347316

Wheeler, L., Rosenthal, M., & Lund, L. (2022). Training Mental Health Professionals in Machine Learning: A Path Forward for Innovative Care. *Frontiers in Psychology, 13*, 765712. DOI: 10.3389/fpsyg.2022.765712

World Health Organization. (2022). Mental health: strengthening our response. https://www.who.int/news-room/fact-sheets/detail/mental-health-strengthening-our-response

Xia, Y., Li, J., & Wang, Q. (2020). Unsupervised machine learning for mental health analysis: An overview. *Journal of Biomedical Informatics, 109*, 103529. DOI: 10.1016/j.jbi.2020.103529

KEY TERMS AND DEFINITIONS

Machine Learning: A subset of artificial intelligence that focuses on the development of algorithms that enable computers to learn from and make predictions or decisions based on data. In the context of mental health, machine learning is utilized to analyze user data, enhance therapeutic interventions, and optimize treatment outcomes.

Digital Therapy: Evidence-based therapeutic interventions delivered through digital platforms, which can include mobile applications, websites, and virtual therapy sessions. Digital therapy aims to provide accessible mental health support and is often used to supplement traditional therapeutic approaches.

Personalized Support: Tailored mental health care strategies that cater to the individual needs, preferences, and characteristics of a patient. Personalization is achieved through the analysis of user data, allowing for customized interventions that enhance engagement and efficacy.

Natural Language Processing (NLP): A branch of artificial intelligence that enables machines to understand, interpret, and respond to human language. In mental health applications, NLP is used to analyze text-based communications, such as chat messages, to identify emotional states and provide appropriate support.

Predictive Modeling: A statistical technique used to create models that forecast future outcomes based on historical data. In mental health, predictive modeling can identify patterns that help anticipate patient needs, optimize treatment plans, and improve intervention effectiveness.

Cognitive Behavioral Therapy (CBT): A widely used therapeutic approach that focuses on changing unhelpful cognitive distortions and behaviors. Digital CBT applications leverage technology to deliver therapeutic exercises, enabling patients to work through their challenges in an accessible format.

Health Belief Model: A psychological model that seeks to explain and predict health behaviors by focusing on individuals' beliefs about health conditions. It is often applied in the context of digital interventions to understand how patients perceive the benefits and risks of using digital mental health solutions.

Virtual Reality (VR): An immersive technology that simulates real or imagined environments, allowing users to interact with 3D worlds. In mental health, VR is used for exposure therapy and to create engaging therapeutic experiences that can aid in the treatment of anxiety, phobias, and PTSD.

Augmented Reality (AR): A technology that overlays digital information onto the real world, enhancing the user's perception of their environment. AR has potential applications in mental health treatment by providing interactive experiences that can support therapeutic interventions.

Thematic Analysis: A qualitative research method used to identify and analyze patterns or themes within qualitative data. In the context of mental health research, thematic analysis helps in understanding user experiences, feedback, and the effectiveness of digital therapies.

Chapter 9
Sentiment and Emotion Analysis Using Machine Learning Techniques

Bhavanam Amarnath Reddy
https://orcid.org/0009-0002-9540-0586
Vignana Jyothi Institute of Management, India

Ullal Akshatha Nayak
Nitte Meenakshi Institute of Technology, India

S. Srinivasan
https://orcid.org/0000-0002-9480-2009
AMET University, India

E. Angela Jean Mary
St. Claret College, India

Amit Chauhan
Christ University, India

R. Rajagopal
Alliance University, India

ABSTRACT

Sentiment analysis and text emotion identification have grown in prominence due to their wide range of applications in fields such as psychology, artificial intelligence, human-computer interaction, and so on. There are numerous Machine Learning approaches available for emotion recognition and sentiment analysis. The chapter also delves into the key procedures of data collecting, preprocessing, and emphasising the necessity of good data in training effective models. Real-world applications from a variety of disciplines, including business, healthcare, and entertainment, are investigated to demonstrate the practical utility of these strategies. The chapter also covers the issues of ambiguity, context-awareness, and cross-linguistic disparities, as well as providing suggestions for future study paths. This chapter provides a detailed exploration of machine learning approaches to sentiment and emotion analysis, making it a valuable resource for researchers, practitioners, and students interested in using machine learning to understand and interpret emotional content in textual data.

DOI: 10.4018/979-8-3693-6910-4.ch009

INTRODUCTION

Though they signify distinct things, sentiment analysis and emotion analysis are conceptually comparable. The goal of sentiment analysis is to determine someone's stance on a particular problem or the big picture by collecting and analyzing subjective data derived from human sentiments. Another important field of research along these lines is emotion analysis, which aims to identify the right emotion by categorizing them based on their surroundings. Almost every part of our lives can be enhanced by analyzing text for emotions and sentiments. There is an almost infinite number of possible uses for emotional intelligence. Some examples include: enhancing human-computer interactions; keeping tabs on mental health; adjusting or improving company strategies based on customer sentiment; measuring public sentiment towards any national, international, or political event; identifying possible criminals or terrorists by analyzing social media behavior after a terrorist attack; and many more. As an example, to better understand how to sell a product, it is possible to do product review analysis to find out what buyers think about the product. You can tell a lot about people's general or particular moods by reading their reactions to various public and private events through social media postings and tweets (Sailunaz & Alhajj, 2019). If someone's communications show indications of melancholy, rage, or despair, it could be beneficial to identify them as early warning symptoms of mental health difficulties and send them to specialists before their problems worsen, leading to suicide or other extreme measures. Law enforcement could benefit from using emotional analysis on social media posts mentioning terrorists or terrorism in general to identify possible future terrorists (K & F, 2016). A more precise character analysis can be obtained by tracking a person's emotional state throughout time. Many organizations in business, society, and politics can use emotion analysis to better or adjust their programs and methods; it is not limited to health groups (Patwardhan, 2017). Machine learning's dependable and scalable methods for handling enormous datasets have revolutionized sentiment and emotion research. Emotion and sentiment analysis make extensive use of a wide variety of machine learning approaches, and this chapter will examine only a few of them, including deep learning, hybrid models, and classic algorithms. It will also cover the most current achievements and fundamental issues in the field.

UNDERSTANDING SENTIMENT AND EMOTION ANALYSIS

Sentiment Analysis

Sentiment analysis has been more popular in recent years, and not only among academics. Additionally, it has been extensively utilized by several sectors, including corporations and government entities. Consequently, sentiment analysis has been very popular among academics as of late. Opinion analysis and sentiment mining are two of the many names for sentiment analysis. The sentiment analysis field is growing rapidly (Wankhade et al., 2022). In light of the exponential growth of social media, several new fields of study have emerged, all with a same goal: to decipher the information that these platforms collect and retain. Sentiment analysis examines a document for underlying emotions by analyzing words and phrases. There must be a plethora of seminal articles dealing with sentiment analysis, a branch of natural language processing, given the well-documented role of public opinion in decision-making. Sentiment analysis, on the other hand, will be useful well into the new century. Many up-to-date surveys forego sentiment analysis in favor of lexicon-based methods, machine learning, or transformer learning.

Unlike previous research, this one doesn't try to cover every possible work; instead, it zeroes in on the most common methods. Various surveys delve into sentiment analysis from the perspective of specific tasks, unique obstacles, or worries, like product evaluations. This study covers all the bases in sentiment analysis research, including topics such as challenges, applications, tools, and methodologies. It also provides a thorough evaluation of the area from different angles. Researchers and newcomers alike will benefit from the work's organization of a plethora of subject material into a single, logical book.

Figure 1. Various levels of sentiment analysis

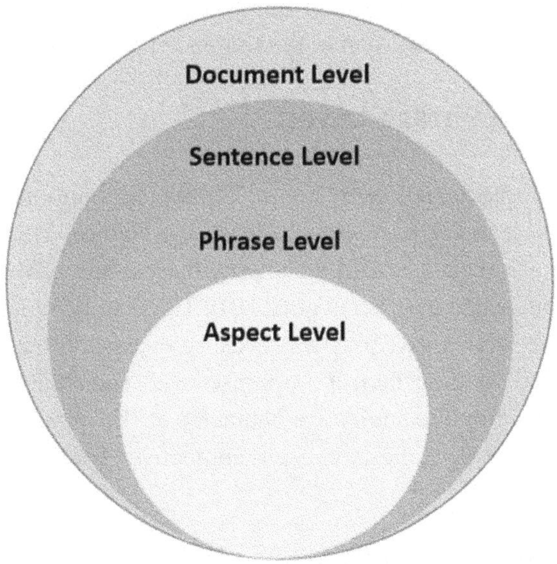

Various levels of sentiment analysis have been studied, including document, sentence, phrase, and aspect levels. At each level, as illustrated in Figure 1, sentiment analysis is performed. These levels include document, sentence, and phrase levels.

Sentiment Analysis at the Document Level:

An entire document is assigned a single polarity for sentiment analysis at the document level. These sentiment analysis methods are hardly used (Bhatia et al., 2007). This feature allows users to place positive, negative, or neutral marks on specific chapters or pages within books. It is possible to classify documents at this level using either supervised or unsupervised learning methods. Cross-domain sentiment analysis and cross-language sentiment analysis are two big obstacles to document-level sentiment analysis (Yang & Cardie, 2014). Preliminary results show that domain-specific sentiment analysis is very accurate and domain-sensitive. The feature vectors for these jobs are required as a collection of limited, domain-specific words.

Sentiment Analysis at the Sentence Level:

Examining each sentence and determining its polarity is part of this level of analysis. Things get better when a document has a wide range of feelings associated to it. An associated level of classification is subjective classification (Behdenna et al., 2018). As with the document level, we can determine the polarity of each sentence independently by making better use of processing resources and training data. Easily know the document's mood just by adding up the polarity of the sentences, or it can use it on its own. Although document-level sentiment analysis is useful, it is not always sufficient (Rao et al., 2018). Most of the earlier research on sentence-level analysis was devoted to finding subjective sentences. Nevertheless, dealing with conditional statements and ambiguous affirmations are among the more difficult tasks. In many cases, sentiment analysis down to the sentence level is crucial.

Sentiment Analysis at the Phrase Level:

It is also possible to do sentiment analysis by mining phrases for opinion terms and then categorizing them. It is possible for every sentence to possess at least one feature. This is an example of a phrase expressing a single aspect; such reviews could be lengthy but useful ones. This is something that academics have lately become interested in (Thet et al., 2010). The document level focuses on labelling the entire document as subjective, which isn't practical because every text has both positive and negative comments. Instead, it's better to examine them at the phrase level. The polarity of a word—the most basic unit of language—is determined by the subjective character of the phrase or text in which it appears. Adjectives practically always show that the sentence is subjective. That is why word is so important for text sentiment analysis.

Sentiment Analysis at the Aspect Level:

Aspect-level sentiment analysis is executed. Sentiment analysis at the aspect level is necessary since every sentence may include more than one aspect (Lu et al., 2011). The sentence's overall sentiment is determined by paying close attention to each and every facet and then giving each one a polarity.

Emotion Analysis

Feelings are an essential part of being human. Humans are able to make better decisions and express themselves more effectively when certain emotions are present (Nandwani & Verma, 2021). "Emotion detection" or "emotion recognition" refers to the act of identifying and categorizing human emotions. Researchers have come a long way in the past few years in their quest to automate emotion recognition. Although it is difficult to infer emotions from text alone, a person's heart rate, sweating, trembling hands, and vocal tone can all be indicators of their emotional state. (Kratzwald et al., 2018). The already challenging task of emotion identification from text is further hampered by the daily appearance of new terminology or terminologies and the proliferation of ambiguity surrounding existing terms. In addition to the three primary emotional states—sadness, joy, and anger—emotion detection can go as high as a 6-or 8-scale, depending on the emotion model.

LITERATURE SURVEY

The detection of emotional duality and mixed emotional experience was accomplished by building a feature level fusion vector and then using a Support Vector Machine (SVM) classifier (Zadeh et al., 2017). Comparable research on feature-level fusion using a self-attention scheme.They supplied a literature review spanning several affect dimensions (e.g., sentiment analysis, emotion analysis, etc.) in advance of the multi-modal study. It provides a multi-modal fusion-based approach to emotion classification (Patwardhan, 2017). The author absolutely depended on high-level audiovisual fusion for this emotion categorization (Hazarika et al., 2018). It introduced the multi-modal dictionary-based approach to better capture the relationship between spoken words and facial expression while conveying emotion (Tzirakis et al., 2016). distinct research also recommended using a Tensor Fusion Network (TFN) to capture the interplay between the distinct multi-modalities. Despite the longevity of uni-modal analysis (Poria, Mazumder, et al., 2017) (Poria, Peng, et al., 2017), numerous assessments have concluded that multi-modal sentiment and emotion analysis are in their early stages. The challenging but necessary process of feature selection (fusion) must be addressed by any multi-modal study (Ghosal et al., 2018). They proposed a multi-kernel learning-based feature selection approach for multi-modal sentiment and emotion detection (Chauhan et al., 2016). Implemented a convolutional deep belief network (CDBN) to acquire important multi-modal characteristics of mild emotional expressions; (Chauhan et al., 2019) utilized a convolutional attention network to acquire representational knowledge of multi-modal features in speech and text data for multi-modal emotion detection (Poria et al., 2016).Their presentation focused on the Urdu SA approach, which utilized KNN, DT, and SVM classifiers. Accuracy levels below 50% indicate that the classifiers are not performing up to par (Morency et al., 2011). Analysis of the results led to adjustments, such as feature extraction, which improved the rating performance to an acceptable level (Ranganathan et al., 2016). When compared to SVM and DT, KNN performed better in respect of accuracy, precision, retrieval, and f-measurement (Al-qablan, 2023). According to the previous section, another SA study was performed that compared a supervised ML approach with a lexicon-based approach (Mukhtar et al., 2018). The results showed that the lexicon approach was more efficient and effective than the ML approach in terms of time saved and effort used, as well as precision, accuracy, and recall (Mukhtar & Khan, 2018). Sentiment distribution prediction using a semi-supervised model based on recursive neural networks (RecursveNN) (Ghosh & Sanyal, 2018). This method used the recursive nature of sentences to learn vector space representations of phrases with more than one word. While looking into unsupervised SA approaches, the writers took a lot of factors into account. For an aspect-based SA problem, they constructed word clusters as features (Socher, R., Pennington, 2011). To add to that, they utilized HPS, an unsupervised stemming method, to assist with the intricate morphology of Polish (Azzouza et al., 2017). We also utilized an unsupervised ML strategy to identify important phrases related to the main topic so that we could investigate thoughts and detect the polarity of tweets. Using the Apache Storm tool, the authors developed a system to track the sentiment on Twitter in real-time.

MACHINE LEARNING TECHNIQUES FOR SENTIMENT AND EMOTION ANALYSIS

Achieving self-learning models is the ultimate goal of machine learning. In the end, these methods revolve around the same basic premise: using a limited amount of data to train a learning model that can independently generate suggestions, predictions, or judgments. The three main methods in machine learning are supervised, unsupervised, and semi-supervised. Unsupervised ML relies on unlabeled data for learning, whereas supervised ML relies on properly labelled data, which is the most notable difference between the three ML methodologies. As a stopgap measure, semi-supervised ML makes advantage of partially tagged data to train on the unlabeled half. Machine learning (ML) techniques have been successful in numerous domains, including pattern identification, finance, learning, emotion detection, and countless more.Promising results have shown that ML systems can extract user emotions from a variety of inputs, such as text, video, and audio (Alslaity & Orji, 2022). In addition, particular ML methods have demonstrated rapid processing and learning speeds, which is perfect for interactive and adaptive systems that require real-time emotion identification. Consequently, since the turn of the century, there has been a flurry of activity in the area of emotion recognition utilizing ML techniques. Many further methods were then suggested. Thanks to the growing interest in it, it is crucial to study this field and give a thorough analysis of its present trends.

Traditional Machine Learning Methods

Classification of emotions and moods typically makes use of more conventional machine learning techniques (Neethu & Rajasree, 2013). To extract features, these algorithms often use classifiers such as Term Frequency-Inverse Document Frequency (TF-IDF) and basket-of-words (BoW).

Nave Bayes Classifier

Sincethe Naive Bayes Classifier treats each feature in the feature vector as an independent variable, it makes use of all of them. Let us consider two categories: c1 = male and c2 = female.

We have an individual whose gender is unknown, referred to as "drew" or "d." Classifying Drew as male or female is tantamount to enquiring whether it is more probable that Drew is male or female, specifically, which is greater: p (male I Drew) or p (female I Drew).

According to one interpretation of Naive Bayes, the conditional probability is:

$$L(K|g_z) = \prod_{s=1}^{t} L(k_s|g_z) \quad (1)$$

The feature vector 'K' is defined as$K = (k_1, k_2, \ldots, k_t)$, while class labels are represented by g_z. Using a plethora of independent data, including emotional keywords, emoticons, the quantity of positive and negative hash tags, and a number of other metrics, the Naive Bayes classifier performed admirably in our investigation. The interdependencies between features are ignored by Naive Bayes. This prevents it from making use of the connections between negation, part-of-speech tags, and emotive keywords.

SVM

Using a massive margin, SVM Classifier performs its classifications. The tweets are divided using a hyper plane. Support vector machines (SVMs) employ a discriminative function, which is just

$$y(K) = c^T \phi(K) + u \tag{2}$$

There are three vectors in this context: the feature vector, the weights vector (K), and the bias vector (u). Input space to high-dimensional feature space is translated non-linearly by$\phi(K)$. Automatic learning of c and u occurs on the training set. In this case, a linear kernel was employed for classification. A large chasm remains between the two groups. For Eaxmple, now examine another instance in which we graph the data and attempt to categorise it, seeing that several hyperplanes can achieve this classification. Which option is superior? The above picture depicts numerous linear classifiers (hyperplanes) that delineate the data. Nonetheless, only one of these attains optimal separation. The necessity arises from the potential for a hyperplane to be disproportionately closer to one dataset than others, which is undesirable; so, the concept of a maximum margin classifier or hyperplane emerges as a viable option.

A Classifier for Maximum Entropy

No specific feature-relationship is assumed by Maximum Entropy Classifier. This classifier aims to maximize the system entropy by estimating the conditional distribution of the class label. The conditional distribution is defined as follows:

$$L_\lambda(g|K) = 1/J(K)\exp\left\{ \sum_s \lambda_s v_s(K, g) \right) \tag{3}$$

Both the feature vector K and the class label g are letters in the English language. In this context, λ_s refers to the weight coefficient, while $J(K)$ is the normalization factor. The feature function, $v_s(K, g)$, is indicated as

$$v_s(K, g) = \begin{cases} 1 & K = k_s \ and \ g = g_s \\ 0 & otherwise \end{cases} \tag{4}$$

Our feature vector makes good use of the relationships between the part-of-speech tag, the emotive keyword, and negation for classification purposes.

LR

Logistic regression is employed to predict the probability of a result possessing a binary value. As an example, values that can be stated as yes/no or true/false do not function with linear regression (OSMANOĞLU et al., 2020). Furthermore, predicting the various applications of a trained classifier might be challenging. For instance, the identical classifier may be utilised by various physicians who opt to conduct distinct medical examinations on their patients, and this knowledge remains unknown during the learning phase. Nonetheless, even a minor absence of data can significantly impair the efficacy of well-trained models, resulting in predictions that markedly diverge from those generated with complete data.

Figure 2. LR model graph

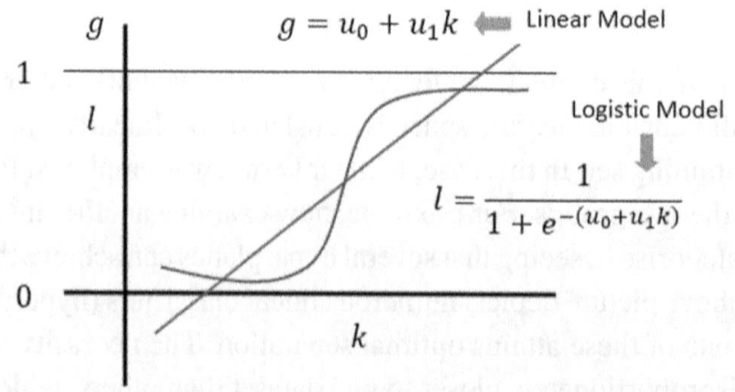

The logistic curve can only take on values between 0 and 1, as that is the range of numbers that logistic regression works with. Instead of using the trajectory probability, as is done in linear and logistic regression, logistic regression uses the natural logarithm of the target variable's probabilities to draw inferences. Think about this if you want to know more about the formula for linear regression:

$$g = u_0 + u_1 K \tag{5}$$

$$logit(l) = log(l(1 - l)) \tag{6}$$

In logistic regression, u_1 determines the curve's slope and u_0 shifts it to the right or left.
Figure 2 shows that the logistic regression equation can be expressed using the probability ratio,*logit*(*l*).

Deep Learning Methods

The accuracy of mood and sentiment identification has been substantially enhanced by the development of deep learning techniques, which have enabled models to learn more complicated textual elements and representations.

RNN

Dedicated memory and a bearing structure define RNNs, a subset of feed-forward neural networks. The network is able to continuously cycle data since it takes in samples from both the past and the present (Peng et al., 2022). A recurrent neural network (RNN) is illustrated in Figure 3 with the following layers: x for input, O for output, H for hidden, and u, V, and w for weights of the layers above each other. It is defined as the output of the buried layer:

Figure 3. RNN framework

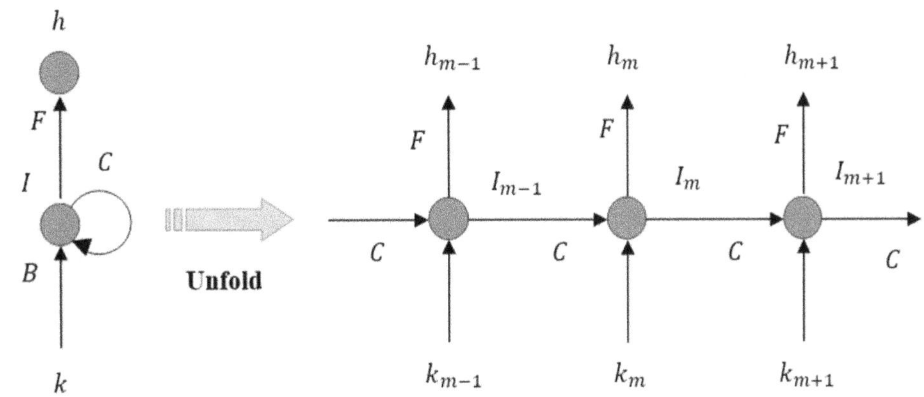

$$o_m = \sigma(c_{ko}k_m + c_{oo}o_{m-1} + u_o) \tag{7}$$

Here is the description of the out-layer's output:

$$h_{m+1} = \sigma(c_{og}o_m + u_g) \tag{8}$$

$$g_m = softmax(h_m) \tag{9}$$

in which g_m stands for the time-varying output, $m, c_{ko}, c_{oo} and c_{og}$ as the weight matrix, u_o and u_g as the bias terms, o_m as the hidden state in m, and k_m as the input data.

CNN

Five layers make up an ANNN, which is a type of feed-forward neural network. These layers are: input, pooling, convolution, and thorough connection. You may view the diagram in Figure 4. Using a vectorized sentence matrix as input, merging it with numerous convolution kernels in the convolution layer, and then extracting different feature representations from the original input are the major processes when applying a CNN to the domain of natural language processing.

In order to create more generalized features, the pooling layer samples from the collected attributes. The last step of matching is for the whole connection layer to apply the matching classification function to the results.

Figure 4. CNN framework

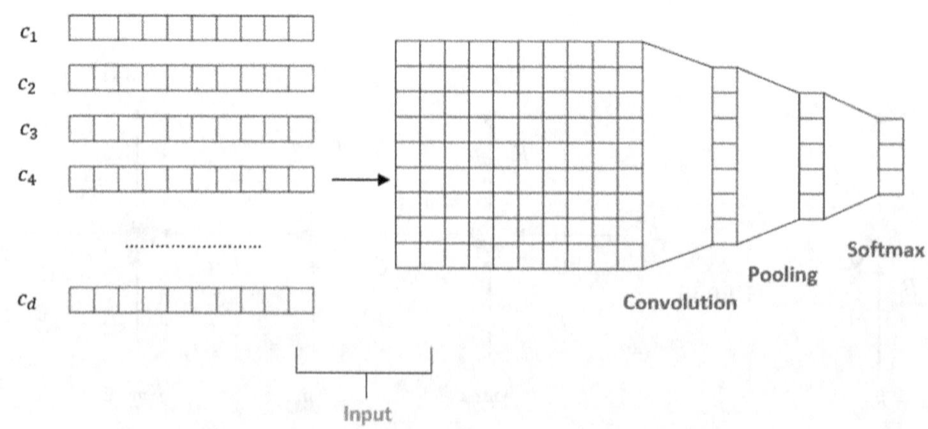

Input Layer:

Vectorizing the input data is its principal use. Take the following d-word statement as an illustration of an input layer matrix:

$$R \in \mathbb{R}^{dxt} \tag{10}$$

in which the dimension of a word vector is represented by t.

Convolution Layer:

A number of convolution kernels are used to obtain the feature maps of the convolution kernel after performing the convolution operation on the input matrix and extracting local features. Here is the description of the specific representation:

$$K_z^p = v\left(\sum_{s=1}^{T} K_s^{p-1} * X_{sz}^p + u_z^p\right) \tag{11}$$

Where K_z^prepresents the output of the convolution operation between the input and the z-th convolution kernel, K_s^{p-1}represents the output of the previous layer (i.e., the s-th local receiving domain of the current convolution layer p), X_{sz}^prepresents the j-th convolution kernel of p, u_z^p represents the offset term of the current convolution kernel, and M represents all the local received domains that the convolution kernel needs to traverse.

Pooling Layer:

Once the convolution process generates a matrix, pertinent features can be extracted using the corresponding sampling function. The goal is to reduce the dimensionality of the matrix. Also, it simplifies calculations without removing any useful features. Two common approaches to data pooling are average pooling and maximal pooling.

Full Connection Layer:

The input categorization, classification result retrieval, and output layer output are all responsibilities of this layer.

DAE

Using an unsupervised learning strategy, a DAE modifies autoencoders. It allows for the formation of DL networks with several stacked layers. The encoder, hidden layer, and decoder are the three parts that make up a denoising autoencoder. By utilizing encoders, the dimensionality of inputs with high dimensions can be reduced v. The input version k, which is input into$v(k)$, is obtained by adding noise to input k. After that, the activation function and linear transformation provide the implicit coding result g. A reconstructed vector k is obtained by using decoder $y(g)$. Here is how the calculations for g and j are detailed:

$$g = v(k) = I_v\left(Ck + u_g\right) \tag{12}$$

$$j = v(g) = I_y\left(C^T g + u_j\right) \tag{13}$$

$$I_v = sigmoid(k) = 1(1 + e^{-k}) \tag{14}$$

$$I_y = sigmoid(g) = 1(1 + e^{-g}) \tag{15}$$

The weight matrix is denoted by C, the offset term is represented by u, the nonlinear transformation activation function is I_v, the decoder activation function is I_y.

Hybrid Approach

The lexicon-based SA method does not need testing but depends on words to represent emotions clearly, whereas the machine learning method needs training and learning. The two methods are not mutually exclusive. Text mining from Twitter and other social media platforms also includes frequently used words and acronyms that are constantly changing. Consequently, new hybrid methods have evolved, which aim to improve accuracy even further by merging lexicon-based methods with machine learning. A hybrid approach was developed by Appel, Chiclana, and Carter to evaluate the strength and polarity of semantic orientation at the sentence level. It combines NLP methods with sentiment lexicons and fuzzy sets (Zucco et al., 2020). The outcome classifier outperforms the competition, including Naive Bayes and maximal entropy, according to the results. To find out how political a single tweet is, a combination of methods is employed. When the preprocessing is complete, we select subjective tweets from the multi-perspective question-answering Lexicon, which includes expressions that include subjectivity information. Once again, the MPQA lexicon is employed for word-level polarity recognition, and the outcomes are subsequently combined for tweets. A support vector machine (SVM) classifier is subsequently trained using these annotated tweets.

CHALLENGES IN SENTIMENT AND EMOTION ANALYSIS

Machine learning models are limited to learning about the domain that was taught to them, hence they cannot address most modern sentiment analysis problems. Consider the challenge of training a model to assess microblog posts rather than product reviews as an example of a domain-unresolved issue. Irony and ambiguity pose significant challenges to sentiment analysis. The use of sarcasm to flatter objects in an effort to create a negative impression is one example where conventional sentiment analysis methods fail. Multiple attempts have been made to discern sarcasm in speech. Since computers have a hard time understanding cultural references due to the peculiarity of comedy, the problem remains unsolved. We argue that multimodal sentiment analysis can be enhanced to better detect sarcastic comments by using voice and facial expressions (Soleymani et al., 2017). To undertake sentiment analysis on human-machine interactions, datasets that are comparable to those used for emotion recognition are necessary. Therefore, it is subject to the same limitations as previously, namely, a lack of robust ground truth and a limited sample size. Just proved that cameras can record a wide range of emotions and behaviours. Although these technologies allow for a level of scale that is impossible in a lab, the audiovisual recording quality is compromised. The fact that only authorized people should have access to lab data adds to the worry that labelling will be a time-consuming and painstaking process. Consequently, we have a hard time classifying huge datasets and can't collect data in the lab. An automated system that can detect emotionally neutral subjective expressions could be a solution. Although multimodal sentiment analysis has shown some promise, text analysis remains the most formidable obstacle in the field of sentiment analysis. Extracting entities and identifying opinion holders are both made possible by natural language processing (Mohammad, 2017). We hope that multimodal sentiment analysis can enhance, and augment existing sentiment analysis approaches rather than completely replace them. Having precise and unambiguous instructions is essential for obtaining high-quality annotations. Sentiment annotation, which comprises labelling instances as neutral, positive, or negative, is one example of an apparently simple annotation job that falls into this category. We will refer to these annotation techniques as basic emotion questionnaires. On the other hand, if there is little data, the annotator can be unsure of how to label statements that Favor one side over another, statements that are critical of others, or retweets. One alternative to sentiment annotation is to just state one's opinion about the target of one's view (Zadeh et al., 2017).. These methods of annotation will be called semantic-role based sentiment surveys from now on. Although this method of sentiment annotation is more complex and comprehensive than the basic sentiment questionnaire method, it still fails in many circumstances. Keep in mind that this method does not take the speaker's emotional condition into account. The authors claim that adequate machine translation quality allows for sentiment analysis to be performed on machine translated texts (Ghosal et al., 2018). Presently, there is simply an overwhelming amount of research in this area to offer a thorough summary. It should be noted, nonetheless, that the intended use case might frequently dictate certain design choices made by the sentiment analysis system.

APPLICATION OF SENTIMENT AND EMOTION ANALYSIS

Businesses are adopting linguistic sentiment analysis to streamline client feedback and evaluations. We can collect this kind of feedback quickly and cheaply, and we can also aggregate ideas on a large scale. In the past, sentiment analysis was a costly and time-consuming alternative to focus groups or polls.

Thanks to the abundance of audiovisual perspectives shared on social media, such passionate reviews on YouTube, sentiment analysis has the potential to be a more collaborative and economical endeavour. All of our connections and interactions are shaped by our feelings toward people and things. Thus, sentiment analysis can be employed to ameliorate interactions between humans and machines. Agents should be able to understand user preferences just by listening to their speech. The user's choices were given context through the use of Langlet and Clavel annotated statement pairs, which include an ECA statement followed by a human response. Among their most recent accomplishments is the capability to correctly identify entities based on topic words extracted from each user's speech. Their strategy revolves around a voice recognition system (ASR), which is crucial for language analysis. Their research was thus comparable to text-based sentiment analysis in terms of both breadth and approach. Autonomous visual content description, however, made extensive use of ANPs as qualities and abstract ideas. Multimedia analytics is one emerging field where sentiment analysis is beginning to make a splash. As an example, automated program analysis and summarization can be achieved by using multimodal sentiment analysis to broadcast video news. Finding politically compelling content can also be accomplished through multimodal sentiment analysis. Using these innovations, we may sift through the innumerable online channels and broadcast TV channels on the Internet to find the thoughts that people have voiced.

FUTURE DIRECTIONS

A more complete picture of the sentiment conveyed across many media types can be achieved using multimodal sentiment analysis, which considers not just text but also pictures, videos, and audio. To examine the tenor of content created in more than one language is the goal of Cross-Lingual Sentiment Analysis. That can only be achieved if models can comprehend and interpret emotional expressions across a wide range of linguistic and cultural settings. Emotion recognition seeks to comprehend human emotional states by recognizing and classifying emotions expressed in written and spoken language (Al-qablan, 2023). Although numerous SA strategies have advanced and grown, there are still certain fundamental issues that need to be resolved. While we do outline some of the major challenges in the field, a fresh viewpoint will be necessary for future SA work to tackle all of them. Although numerous SA strategies have advanced and grown, there are still certain fundamental issues that need to be resolved. While we do outline some of the major challenges in the field, a fresh viewpoint will be necessary for future SA work to tackle all of them.

Fake Reviews and Spam Detection

These days, a lot of people check internet reviews before buying anything. Spammers will use this to their advantage in order to either improve or harm a company's reputation. Knowing the difference between genuine reviews and spam is a challenging undertaking that requires grammar and language skills. The main problem with using a classification strategy to differentiate between spam reviews is that there is no tagged dataset to work with. In order to detect opinion spam, one study sought to establish a collection of characteristics that might be used as trustworthy input with various classifiers. The results showed that when it came to Persian, the best classifiers for identifying spam perspectives were DT and AdaBoost. Review sentiment and comments, content-based factor, and rating variation were the topic of another study. Review spam was detected using this method using data from comments. After applying the ML model to the labelled data, the authors were able to classify the unlabelled data as well.

Fake Reviews and Spam Detection:

The use of sarcasm is considered problematic in South Africa. If what you write or say doesn't fully reflect how you really feel, you're using sarcasm. Using well-chosen words to express negative feelings, or vice versa, is known as sarcasm. It can fool SA models easily if they aren't trained to recognize sarcasm. Even computers have trouble understanding sarcasm sometimes. Many approaches have been proposed to identify sarcasm. Scofflaws in English were attempted to be catalogued in single research. Interjection and unigram features performed better than others when it came to sarcastic tweet classification.

Negations

Reversing the polarity of words (a negation) has a significant impact on identifying the right emotional tone. The refusal set includes the imperative verbs "no," "not," and "can't." Finding negations in a statement entails identifying the words that are impacted by them. "The camera is not perfect, but the phone is nice" is a negative that can only target the word immediately following it since "but" splits the sentence in half. Not only that, but the negation in "The phone has not worked properly for a long time." encompasses the whole statement (Shah et al., 2024). Three separate negations that change the valence of words, phrases, or complete structures have been provided by the writers. After suggesting a personalized reinforcement learning strategy, the authors compared and contrasted both options using a two-pronged test. In the beginning, they compared the predicted performance using a dataset that had been manually categorized. We surpass competing approaches in the literature by achieving a balanced classification accuracy of 70% or higher with reinforcement learning. Secondly, we looked at ways to improve financial news SA by finding negation scopes; this brought a 10% improvement in the association between story mood and stock market performance. This demonstrates the importance of identifying negation scopes for sentiment decision assistance. Introducing a novel approach, we can now calculate the negation travel distance of a statement... The authors suggest a combined method that could get important data for determining if a token is in scope.

Independent domains:

Everyone knows that sentiment classification depends on the domain. Since different domains often use different sentiment expressions, it is quite unlikely that a sentiment classifier trained in one domain will perform badly in another. Also, the same term can evoke diverse emotions depending on the setting. The term "easy" is often used to describe technological devices; for instance, "This tablet is easy to use." However, the security area has a reputation for using the phrase negatively. "This app can be easily hacked." Because of this, we think we require a sentiment classifier that is both more effective and more generalizable. One way to describe this is as SA that spans domains.

CONCLUSION

Sentiment analysis is a large subfield of NLP that aims to automatically identify the sentiments (positive, negative, or neutral) in a given text. Emotion analysis revolves around the process of discovering and extracting pre-programmed emotions from data. Anger, fear, contempt, melancholy, surprise, joy, and trust are some examples of typical different emotions. Sentiment and emotion are sometimes used interchangeably because of their comparable connotations, which is caused by their subjective nature. This is mostly due to the fact that our experiences are formed by the interaction of our emotional, mental, and social selves. Our sentiments and feelings are a result of these interactions (Kumar et al., 2019). Sentiment analysis is a large subfield of NLP that aims to automatically identify the sentiments (positive, negative, or neutral) in a given text. Emotion analysis revolves around the process of discovering and extracting pre-programmed emotions from data. Anger, fear, contempt, melancholy, surprise, joy, and trust are some examples of typical different emotions. Sentiment and emotion are sometimes used interchangeably because of their comparable connotations, which is caused by their subjective nature. This is mostly due to the fact that our experiences are formed by the interaction of our emotional, mental, and social selves. Our sentiments and feelings are a result of these interactions.

REFERENCES

Al-qablan, T. A. (2023). A survey on sentiment analysis and its applications. In *Neural Computing and Applications* (Issue August). Springer London. DOI: 10.1007/s00521-023-08941-y

Alslaity, A., & Orji, R. (2022). Machine learning techniques for emotion detection and sentiment analysis : Current state, challenges, and future directions. *Behaviour & Information Technology*, (December), 1–26. DOI: 10.1080/0144929X.2022.2156387

Azzouza, N., Akli-Astouati, K., Oussalah, A., & Ait Bachir, S. (2017). A Real-Time Twitter Sentiment Analysis using an unsupervised Method. *ACM International Conference Proceeding Series, Part F1294*. DOI: 10.1145/3102254.3102282

Behdenna, S., Barigou, F., & Belalem, G. (2018). Document Level Sentiment Analysis: A survey. *EAI Endorsed Transactions on Context-Aware Systems and Applications, 4* (13), 154339. DOI: 10.4108/eai.14-3-2018.154339

Bhatia, P., Ji, Y., & Eisenstein, J. (2007). *Better Document-level Sentiment Analysis from RST Discourse Parsing.*

Chauhan, D. S., Akhtar, M. S., Ekbal, A., & Bhattacharyya, P. (2019). Context-aware interactive attention for multi-modal sentiment and emotion analysis. *EMNLP-IJCNLP 2019 - 2019 Conference on Empirical Methods in Natural Language Processing and 9th International Joint Conference on Natural Language Processing,Proceedings of the Conference*, 5647–5657. DOI: 10.18653/v1/D19-1566

Chauhan, D. S., Ghosal, D., Poria, S., Ekbal, A., & Bhattacharyya, P. (2016). Multi-task Learning for Multi-modal Emotion Recognition and Sentiment Analysis. *ArXiv:1905.05812v1.*

Ghosal, D., Akhtar, M. S., Chauhan, D., Poria, S., Ekbal, A., & Bhattacharyya, P. (2018). Contextual inter-modal attention for multi-modal sentiment analysis. *Proceedings of the 2018 Conference on Empirical Methods in Natural Language Processing, EMNLP 2018*, 3454–3466. DOI: 10.18653/v1/D18-1382

Ghosh, M., & Sanyal, G. (2018). An ensemble approach to stabilize the features for multi-domain sentiment analysis using supervised machine learning. *Journal of Big Data, 5* (1), 44. Advance online publication. DOI: 10.1186/s40537-018-0152-5

Hazarika, D., Gorantla, S., Poria, S., & Zimmermann, R. (2018). Self-Attentive Feature-level Fusion for Multimodal Emotion Detection. *2018 IEEE Conference on Multimedia Information Processing and Retrieval (MIPR)*, 196–201. DOI: 10.1109/MIPR.2018.00043

Kratzwald, B., Ilic, S., Kraus, M., Feuerriegel, S., & Prendinger, H. (2018). *Deep learning for affective computing: text-based emotion recognition in decision support.* DOI: 10.1016/j.dss.2018.09.002

Kumar, A., Ekbal, A., Kawahra, D., & Kurohashi, S. (2019). Emotion helps Sentiment: A Multi-task Model for Sentiment and Emotion Analysis. *Proceedings of the International Joint Conference on Neural Networks, 2019-July*. DOI: 10.1109/IJCNN.2019.8852352

Lu, B., Ott, M., Cardie, C., & Tsou, B. K. (2011). Multi-aspect sentiment analysis with topic models. *Proceedings - IEEE International Conference on Data Mining, ICDM*, 81–88. DOI: 10.1109/ICDMW.2011.125

Mohammad, S. M. (2017). *Challenges in Sentiment Analysis.*, DOI: 10.1007/978-3-319-55394-8_4

Morency, L. P., Mihalcea, R., & Doshi, P. (2011). Towards multimodal sentiment analysis: Harvesting opinions from the web. *ICMI'11 - Proceedings of the 2011 ACM International Conference on Multimodal Interaction*, 169–176. DOI: 10.1145/2070481.2070509

Mukhtar, N., & Khan, M. A. (2018). Urdu sentiment analysis using supervised machine learning approach. *International Journal of Pattern Recognition and Artificial Intelligence, 32* (2), 1–15. DOI: 10.1142/S0218001418510011

Mukhtar, N., Khan, M. A., & Chiragh, N. (2018). Lexicon-based approach outperforms supervised machine learning approach for Urdu Sentiment Analysis in multiple domains. *Telematics and Informatics, 35* (8), 2173–2183. Advance online publication. DOI: 10.1016/j.tele.2018.08.003

Nandwani, P., & Verma, R. (2021). A review on sentiment analysis and emotion detection from text. *Social Network Analysis and Mining, 11* (1), 1–19. DOI: 10.1007/s13278-021-00776-6 PMID: 34484462

Neethu, M. S., & Rajasree, R. (2013). Sentiment analysis in twitter using machine learning techniques. *2013 4th International Conference on Computing, Communications and Networking Technologies, ICCCNT 2013.* DOI: 10.1109/ICCCNT.2013.6726818

Osmanoğlu, U. Ö., Atak, O. N., Çağlar, K., Kayhan, H., & Can, T. (2020). Sentiment Analysis for Distance Education Course Materials: A Machine Learning Approach. *Journal of Educational Technology and Online Learning, 3* (1), 31–48. DOI: 10.31681/jetol.663733

Patwardhan, A. S. (2017). Multimodal mixed emotion detection. *Proceedings of the 2nd International Conference on Communication and Electronics Systems, ICCES 2017, 2018-Janua* (Icces), 139–143. DOI: 10.1109/CESYS.2017.8321250

Peng, S., Cao, L., Zhou, Y., Ouyang, Z., Yang, A., Li, X., Jia, W., & Yu, S. (2022). A survey on deep learning for textual emotion analysis in social networks. *Digital Communications and Networks, 8* (5), 745–762. DOI: 10.1016/j.dcan.2021.10.003

Poria, S., Chaturvedi, I., Cambria, E., & Hussain, A. (2016). Convolutional MKL based multimodal emotion recognition and sentiment analysis. *Proceedings - IEEE International Conference on Data Mining, ICDM*, 439–448. DOI: 10.1109/ICDM.2016.0055

Poria, S., Mazumder, N., Cambria, E., Hazarika, D., Morency, L. P., & Zadeh, A. (2017). Context-dependent sentiment analysis in user-generated videos. *ACL 2017 - 55th Annual Meeting of the Association for Computational Linguistics, Proceedings of the Conference (Long Papers), 1*, 873–883. DOI: 10.18653/v1/P17-1081

Poria, S., Peng, H., Hussain, A., Howard, N., & Cambria, E. (2017). Ensemble application of convolutional neural networks and multiple kernel learning for multimodal sentiment analysis. *Neurocomputing, 261*, 217–230. DOI: 10.1016/j.neucom.2016.09.117

Ranganathan, H., Chakraborty, S., & Panchanathan, S. (2016). Multimodal emotion recognition using deep learning architectures. *2016 IEEE Winter Conference on Applications of Computer Vision, WACV 2016*. DOI: 10.1109/WACV.2016.7477679

Rao, G., Huang, W., Feng, Z., & Cong, Q. (2018). LSTM with sentence representations for document-level sentiment classification. *Neurocomputing*, *308*, 49–57. DOI: 10.1016/j.neucom.2018.04.045

Sailunaz, K., & Alhajj, R. (2019). Emotion and sentiment analysis from Twitter text. *Journal of Computational Science*, *36*, 101003. DOI: 10.1016/j.jocs.2019.05.009

Schouten, K., & Frasincar, F. (2015). Survey on aspect-level sentiment analysis. *IEEE Transactions on Knowledge and Data Engineering*, *28* (3), 813–830.

Shah, S., Akshata Bhat, M., Sumitra Singh, M., Arya Chavan, M., & Aryan Singh, M. (2024). Sentiment Analysis. *International Journal of Progressive Research in Engineering Management and Science (Ijprems)*, *4* (4), 1542–1546. https://www.doi.org/10.58257/IJPREMS33384 PMID: 38834214

Socher, R., & Pennington, J. (2011). Semi-Supervised Recursive Autoencoders for Predicting Sentiment Distributions. *InProceedings of the 2011 Conference on Empirical Methods in Natural Language Processing*, 151–161. DOI: 10.3109/03630269.2012.746943

Soleymani, M., Garcia, D., Jou, B., Schuller, B., Chang, S. F., & Pantic, M. (2017). A survey of multimodal sentiment analysis. *Image and Vision Computing*, *65*, 3–14. DOI: 10.1016/j.imavis.2017.08.003

Thet, T. T., Na, J. C., & Khoo, C. S. G. (2010). Aspect-based sentiment analysis of movie reviews on discussion boards. *Journal of Information Science*, *36* (6), 823–848. DOI: 10.1177/0165551510388123

Tzirakis, P., Trigeorgis, G., Nicolaou, M. A., Schuller, B., & Zafeiriou, S. (2016). End-to-End Multimodal Emotion Recognition using Deep. *Neural Networks*, *14* (8), 1–9. ArXiv1704.08619v1

Wankhade, M., Rao, A. C. S., & Kulkarni, C. (2022). A survey on sentiment analysis methods, applications, and challenges. In *Artificial Intelligence Review* (Vol. 55, Issue 7). Springer Netherlands. DOI: 10.1007/s10462-022-10144-1

Yang, B., & Cardie, C. (2014). Context-aware learning for sentence-level sentiment analysis with posterior regularization. *52nd Annual Meeting of the Association for Computational Linguistics, ACL 2014 - Proceedings of the Conference, 1* (2008), 325–335. DOI: 10.3115/v1/P14-1031

Zadeh, A., Chen, M., & Cambria, E. (2017). Tensor Fusion Network for Multimodal Sentiment Analysis. *ArXiv:1707.07250v1*. DOI: 10.18653/v1/D17-1115

Zucco, C., Calabrese, B., Agapito, G., Guzzi, P. H., & Cannataro, M. (2020). Sentiment analysis for mining texts and social networks data: Methods and tools. In *Wiley Interdisciplinary Reviews: Data Mining and Knowledge Discovery* (Vol. 10, Issue 1). DOI: 10.1002/widm.1333

Chapter 10
Designing Teaching Strategies Using Artificial Intelligence for Neurodivergent Students in Higher Education

Paola Elvira Velazquez-Solis
https://orcid.org/0009-0005-7301-2103
Universidad Politécnica de Baja California, Mexico

Martha Elena González Correa
https://orcid.org/0009-0000-6073-9585
Universidad Politecnica de Baja California, Mexico

Martha Angelica Martinez
Universidad Politecnica de Baja California, Mexico

Jesus Guillermo Arroyo
Universidad Politecnica de Baja California, Mexico

Maria Ysabel Marquez
Universidad Politecnica de Baja California, Mexico

ABSTRACT

Artificial intelligence (AI) can enhance teaching strategies for neurodivergent students in higher education. Neurodiversity, such as autism, ADHD, and dyslexia, needs tailored educational approaches often missing in traditional systems. As neurodivergent student enrollment rises, inclusive learning environments are essential. AI technologies offer personalized learning through adaptive platforms that adjust content based on real-time feedback, helping students tackle challenges and utilize their strengths. AI tools, such as emotion recognition, support mental health by detecting stress, frustration, or anxiety for timely interventions. A case study highlights integration of an AI-powered chatbot in a Color Theory course for Engineering students. The chatbot provided guidance and used visual aids, improving communication, independence, and confidence. Despite AI benefits, issues like over-reliance and digital divide must be

DOI: 10.4018/979-8-3693-6910-4.ch010

considered. Developing empathetic AI solutions is crucial for supporting neurodivergent students, and collaboration among educators and researchers is essential.

INTRODUCTION

Neurodiversity refers to the natural variations in brain function and cognition and includes conditions such as autism, ADHD, dyslexia, and more (Clouder et al., 2020; Hamilton & Petty, 2023). These conditions, often treated as deficits in traditional education systems, are increasingly recognized as part of the broader spectrum of human diversity. Diagnosis of these conditions is complex, and involves various medical, psychological, and behavioral assessments (Stein et al., 2024). Despite the increasing number of neurodiverse students in higher education, they face systemic barriers that result in lower graduation rates compared to their neurotypical peers, primarily due to educational environments that do not accommodate their unique learning needs. In Latin America, a study by UNESCO, (2021) found that only 4.8% of students with disabilities, including neurodiverse students, have access to higher education, and those who do have access face significant challenges in terms of accessibility, teaching methods, and institutional support. In Mexico, it is estimated that only 2.4% of students in higher education students report having some form of disability, and even fewer are diagnosed as neurodivergent, suggesting that a significant number of students may be undiagnosed or unsupported (INEGI, 2020).

The increasing number of neurodiverse students at universities represents both a challenge and an opportunity. To meet the needs of this growing population, there is an urgent need for educational reforms that move away from rigid, one-size-fits-all teaching methods to more flexible, personalized approaches (Clouder et al., 2020). Neurodiverse students often struggle with conventional assessment methods, independent mastery of learning tasks, social integration, and mental health issues related to their condition. These barriers can severely impact their academic success and contribute to a lower likelihood of graduating, pursuing postgraduate studies or finding a suitable job (Aslan et al., 2024; Hamilton & Petty, 2023). In Mexico, for example, the employment rate of people with disabilities remains significantly lower than that of the general population, and those who do not complete higher education have fewer opportunities (CONAPRED, 2021).

In this context, new technologies, especially artificial intelligence (AI), offer promising solutions. AI can be a powerful tool for designing personalized learning experiences that address the specific needs of neurodiverse students. Adaptive learning platforms powered by AI can customize content delivery in real time based on individual learning profiles and provide tailored support to students with conditions such as autism, ADHD and dyslexia (Chrysochoou et al., 2022; Stein et al., 2024). AI technologies are not limited to personalized instruction. They can also monitor student behavior, detect signs of disengagement or frustration, and intervene in a timely manner (Black et al., 2024). This ability to adapt in real time makes AI an ideal tool for supporting neurodiverse students who may require more differentiated approaches to learning. Although AI has great potential, its implementation in education needs to be aligned with human interaction. AI cannot replace the empathy, understanding and social support that educators provide. For neurodiverse students, human interaction is often critical to emotional and psychological well-being. Therefore, educators and technologists must work together to develop AI-driven tools that complement, rather than replace, human teaching.

The chapter begins with the increasing prevalence of neurodiverse students in higher education and the limitations of traditional teaching methods in accommodating their diverse needs. It highlights the importance of developing innovative AI-integrated teaching strategies that move beyond deficit-oriented models and instead focus on promoting neurodiverse students' strengths, such as their attention to detail, creativity, and unique problem-solving abilities (Le Cunff et al., 2024). In addition, the chapter addresses the ethical considerations of integrating AI into education, particularly in relation to privacy, the risk of over-reliance on the technology and the potential to exacerbate existing educational inequalities. The digital divide remains a major issue, as not all students have equal access to the necessary technology to benefit from AI-driven learning tools. These challenges need to be addressed to ensure that AI-based solutions do not unintentionally widen the gap between neurodivergent and neurotypical students. Finally, this chapter emphasizes the need for ongoing, meaningful collaboration between educators, researchers, and AI developers to create effective, inclusive, and empathetic learning environments. By leveraging the strengths of AI and human educators, higher education institutions can create a supportive and personalized learning experience that helps neurodivergent students succeed.

Background

Neurodiversity is a dimension that has been relatively under-researched in the higher education pedagogical literature, despite the increasing number of neurodiverse students in universities worldwide (Hamilton & Petty, 2023). Many faculty and fellow students may not have a comprehensive understanding of neurodiversity or the special needs of neurodiverse students. This lack of awareness can lead to misunderstanding, stigmatization and inadequate support. Students may feel isolated or misunderstood, which can exacerbate their anxiety and limit their academic engagement (Clouder et al., 2020). Neurodiversity encompasses a wide range of neurodevelopmental differences, such as attention-deficit/hyperactivity disorder (ADHD), autism spectrum disorder (ASD), dyslexia, and other conditions that affect learning and cognitive processing in unique ways. Each of these disorders has unique characteristics and challenges, so educators must use customized approaches to effectively support these students. For example, while some neurodivergent students may have problems with concentration or motor coordination, others may excel at tasks that require analytical thinking, creativity or pattern recognition. These different profiles illustrate the need for an adaptive and inclusive learning environment. To better understand these different conditions, Table 1 provides an overview of the main neurodiverse conditions commonly found in educational settings, along with their defining characteristics. This table provides a clearer picture of the types of challenges students may face and the potential strengths that can be utilized in an appropriate learning framework.

Table 1. Overview of neurodivergent conditions and their key characteristics

Authors	*Neurodiversity*	*Characteristics*
Aslan et al., 2024; Black et al., 2024; Chrysochoou et al., 2022; Le Cunff et al., 2024	Attention Deficit Hyperactivity Disorder (ADHD)	A neurodevelopmental disorder characterized by patterns of inattention, hyperactivity, and impulsivity that can interfere with functioning or development. Individuals with ADHD may struggle with organization, time management, and maintaining focus, but they can also display creativity and high energy levels.
Chrysochoou et al., 2022; Clouder et al., 2020	Autism Spectrum Disorder (ASD)	A developmental disorder that affects communication, behavior, and social interaction. Individuals with ASD may exhibit a range of strengths and challenges, including difficulties in social situations and a preference for routine, but they may also possess exceptional skills in areas such as mathematics, memory, or pattern recognition.
Aslan et al., 2024; Black et al., 2024	Dyslexia	A specific learning disability that affects reading and language processing. Individuals with dyslexia may have difficulty with phonological processing, spelling, and decoding words, which can hinder their reading fluency and comprehension.
Black et al., 2024; Chrysochoou et al., 2022; Le Cunff et al., 2024	Dyscalculia	This is a specific learning disability that affects an individual's ability to understand and work with numbers. It can manifest as difficulties in performing basic arithmetic, understanding number concepts, and recognizing patterns in numerical data.
Black et al., 2024; Clouder et al., 2020	Developmental Coordination Disorder (Dyspraxia)	A neurodevelopmental disorder affecting motor skill development and coordination. It impacts fine motor skills (e.g., writing) and gross motor skills (e.g., running), making tasks involving precise or large movements challenging. It can affect daily activities and self-esteem.
Aslan et al., 2024; Stein et al., 2024	Tourette syndrome	A neurological disorder characterized by repetitive, involuntary movements (motor tics) and vocalizations (vocal tics). Motor tics can include actions like blinking or head jerking, while vocal tics may involve sounds such as grunting or throat clearing. Symptoms typically begin before age 18 and may be accompanied by ADHD or OCD.
Chrysochoou et al., 2022; Clouder et al., 2020	Sensory Processing Disorder (SPD)	Is a condition where the brain has difficulty receiving and responding to sensory information. Individuals may be overly sensitive (hypersensitive) or under-responsive (hyposensitive) to sensory stimuli, such as sounds, lights, textures, or tastes.
Black et al., 2024; Le Cunff et al., 2024	Nonverbal Learning Disorder (NVLD)	A neurodevelopmental disorder marked by difficulty with spatial and nonverbal cues, including trouble understanding body language, visual-spatial tasks, and organizing information. Individuals with NVLD may excel in verbal skills but struggle with motor coordination, social interaction, and interpreting visual-spatial relationships.
Chrysochoou et al., 2022; Stein et al., 2024	Generalized Anxiety Disorder (GAD)	Is characterized by excessive and uncontrollable worry about various aspects of life, including academic performance, social interactions, and future events. Individuals with GAD may experience physical symptoms such as restlessness, fatigue, and difficulty concentrating.
Aslan et al., 2024; Stein et al., 2024	Obsessive-Compulsive Disorder (OCD)	A mental health disorder characterized by unwanted, intrusive thoughts (obsessions) and repetitive behaviors or mental acts (compulsions) performed to alleviate anxiety. Individuals with OCD may feel compelled to perform certain rituals to prevent perceived negative outcomes.

Source: Own elaboration

Given the wide range of cognitive profiles in neurodiverse students, traditional teaching methods that are tailored to a specific size often fail to meet their needs, leading to academic frustration and lower engagement. For example, students with ADHD may benefit from more flexible learning environments that allow for movement and short, structured tasks, while students with ASD require clear, explicit communication and predictability in their schedules (Clouder et al., 2020). Studies confirm that these students need differentiated approaches that are tailored to their strengths and challenges. ADHD students, for example, thrive in an environment that minimizes distractions, promotes engagement through multisensory activities, and allows for frequent breaks (Sarver et al., 2021). Similarly, students with ASD benefit from instructional methods that emphasize structure, routine, and clarity in communication while utilizing their pattern recognition and attention to detail skills (Koegel et al., 2020).

Furthermore, neurodiverse students often excel in certain areas that do not align with traditional learning metrics. For example, students with ASD may demonstrate exceptional ability in tasks that require pattern recognition, memory, or problem solving in structured environments (Chrysochoou et al., 2022). Similarly, students with dyslexia may excel in areas such as creativity, problem solving, and verbal reasoning when learning strategies emphasize their strengths rather than focusing on their reading problems (Serrano & Defior, 2018). These diverse profiles challenge educators to create inclusive learning environments that capitalize on students' unique abilities while providing the necessary scaffolding to address their challenges.

Research suggests that shifting the focus from reducing cognitive differences to optimizing learning opportunities opens the door to more innovative instructional designs. Rather than attempting to "normalize" neurodiverse students, education should accommodate neurodiverse students by designing adaptive learning environments that cater to diverse cognitive profiles (Lawson et al., 2021). The use of adaptive learning technologies, including artificial intelligence (AI), holds promise for creating personalized learning experiences that meet the needs of neurodiverse students. AI-driven tools can help identify students' learning styles, track progress, and adapt instructional content in real time to optimize engagement and understanding (Aslan et al., 2024).

AI is particularly useful when it comes to catering to individual variations by offering flexible learning paths tailored to each student's cognitive strengths and weaknesses. These systems, when developed with neurodiversity in mind, can minimize barriers to learning and improve student engagement through dynamic, data-driven adjustments to content (Zawacki-Richter et al., 2019). AI technologies can also help reduce the administrative burden on educators, allowing them to focus more on relationship building and personalized instruction. Paired with human judgment and creativity, AI can foster more inclusive learning environments that increase equity and improve outcomes for neurodiverse students (Holmes et al., 2021). This evolving understanding of neurodiversity and its integration into education emphasizes the importance of flexible, adaptive teaching methods that go beyond the one-size-fits-all paradigm. By focusing on creating supportive and personalized learning environments, educators can ensure that neurodiverse students are not only included, but also empowered to reach their full potential.

CHALLENGES AND OPPORTUNITIES IN AI-DRIVEN EDUCATION

Concerns About Over-Reliance on AI

The long-term impact of AI integration into education for neurodivergent students is still under exploration, but early signs point to both potential benefits and risks. AI tools, such as Task Reminders and Assistive Technologies, are praised for their ability to provide personalized support, helping students manage tasks and develop organizational skills (Hall et al., 2024). These tools can foster independence, allowing students to utilize their strengths, but the risk of over-reliance remains significant. A situation mentioned by Luckin, (2021) is that an overdependence on AI may result in a diminished capacity for students to develop crucial life skills such as time management, resilience, and problem-solving. Overreliance on AI could also lead to a reduction in teacher-student interaction, which is critical in fostering emotional and social learning, particularly for neurodivergent students who benefit from personalized, human engagement.

A critical balance must be maintained between leveraging AI for its strengths in adapting to individual learning styles and ensuring that the emotional and social dimensions of learning are not overlooked. Cotton et al., (2022) emphasize that human educators provide empathy, trust, and a nuanced understanding of student needs qualities that AI, despite its sophistication, cannot replicate. This underscores the importance of using AI as an aid to human instruction, rather than a substitute. In practice, educators must be conscious of when to lean on AI and when to provide direct, personalized attention.

Challenges in Educator Acceptance of AI

Resistance to AI adoption from educators presents a critical challenge from multiple factors. Many educators worry that AI might diminish their role or, worse, make their jobs redundant (Luckin, 2021). This fear is exacerbated by the growing presence of generative AI, which can create lesson plans, quizzes, and even grade assignments autonomously. In the short term, educators may view these technologies as helpful, but long-term concerns persist about AI's ability to replace critical thinking, adaptability, and the nurturing roles teachers play.

Resistance to AI also originates from a lack of understanding and insufficient professional development opportunities. Verhoeven et al., (2023) argue that comprehensive training is necessary to help educators realize how AI can complement their teaching rather than replace it. Successful implementation requires the development of AI literacy among educators, helping them grasp the nuances of how AI can enhance, but not overtake, the human elements of teaching. Furthermore, Xu & Brown, (2023) show that when educators are properly trained in using AI tools, they are more likely to employ them effectively in ways that enrich the learning experience for neurodivergent students. This includes using AI for administrative tasks, freeing up more time for educators to focus on personalized, one-on-one engagement with students.

The authors Cota-Rivera et al., (2024) highlight how AI can adapt teaching methods to meet the specific needs of individuals, including those who face discrimination in traditional approaches due to neurological differences. The implementation of artificial intelligence resources can be utilized in two key areas: simplifying administrative tasks performed by teachers—such as assignment management, knowledge evaluation, and assessment review—and engaging with content interactively through augmented reality (AR), virtual reality (VR), and immersive reality (RX) activities. Regarding the first area, generative AI, such as ChatGPT 3, has been successfully used to develop study guides, questionnaires, topic sheets, syllabi, and exams.

Fostering Inclusivity and Collective Creativity

Incorporating AI into neurodivergent learning environments presents unique opportunities for fostering inclusivity and collective creativity. Numerous studies have demonstrated that diverse teams, including neurodiverse individuals, are often more creative and capable of solving complex problems (Aslan et al., 2024; Hamilton & Petty, 2023). In classrooms, neurodivergent students bring unique perspectives that, when nurtured in a supportive environment, can significantly enhance group problem-solving.

For instance, incorporating team-based assessments and collaborative projects can shift the focus from individual achievement to collective innovation. This approach aligns with the concept of "collective intelligence," which argues that groups with diverse cognitive perspectives are better at solving problems than homogenous teams (Jones & Müller, 2023).

However, traditional support models for neurodivergent students often focus on diagnosing deficits and offering accommodations rather than nurturing strengths. Sharma, (2020) criticizes this "disability services" model for being overly bureaucratic and slow, which can result in students not receiving the support they need in a timely manner. Instead, educators should adopt flexible teaching strategies that promote collective creativity and value each student's unique contributions. When supported by AI, these strategies can ensure that neurodivergent students are fully integrated into the learning process, rather than being sidelined or stigmatized due to their differences.

Digital Divide in AI Education

Despite the potential benefits of AI-driven educational tools, not all students have equal access to the necessary technology. According to a UNESCO Report (2023), the digital divide is a growing concern, particularly for students from low-income backgrounds and those with disabilities. These students often lack access to high-quality internet, up-to-date devices, and digital literacy programs, creating a significant barrier to benefiting from technological advancements in education. Neurodivergent students, who already face unique learning challenges, are especially vulnerable, as they may encounter additional obstacles in accessing the resources required to benefit from AI-powered tools. This inequality not only impacts their academic performance but also their ability to engage with peers and educators in a meaningful way.

The digital divide not only exacerbates existing inequalities but also limits the effectiveness of AI as a universal educational solution. (Meyer et al., 2022) found that students with limited access to digital resources are often left behind, unable to take full advantage of AI tools designed to enhance learning. These students miss out on adaptive learning systems, personalized feedback, and other features that could significantly improve their educational experience. For neurodivergent learners, this gap is even more pronounced, as they are more likely to require specialized tools and support mechanisms that address their individual needs. In the absence of these resources, they may struggle to keep pace with their peers.

This gap highlights the importance of implementing policies that ensure equitable access to technology, particularly for neurodivergent learners. Schools and policymakers must prioritize digital inclusion by providing the necessary infrastructure, devices, and training to ensure that all students, regardless of their socioeconomic status or neurodivergence, have the opportunity to benefit from AI-enhanced education. Moreover, Kline et al., (2023) argue that AI tools should be flexible enough to accommodate a wide range of learning preferences and styles, ensuring that they serve, rather than hinder, neurodivergent students. These tools must be designed with adaptability in mind, allowing for adjustments in both interface and content delivery based on individual needs. For example, some students may benefit from visual aids, while others may require text-to-speech features or additional time to process information.

Without adequate customization and access, AI tools may fail to address the unique needs of neurodivergent students, perpetuating the very challenges they are meant to resolve. The risk is that, rather than leveling the playing field, AI-driven educational solutions could further entrench disparities, leaving the most vulnerable students even further behind. Therefore, the development and deployment of AI in education must be accompanied by robust strategies to close the digital divide, ensuring that all students, particularly those who are neurodivergent, can access and benefit from these transformative tools.

AI IMPLEMENTATION STRATEGIES

The traditional education system is becoming less effective in meeting the diverse needs of learners, particularly neurodivergent students, as it often adheres to one-size-fits-all teaching approaches. As such, there is a growing recognition that emerging technologies, especially artificial intelligence (AI) and natural language processing (NLP), hold transformative potential for reshaping education. These technologies offer personalized learning environments tailored to the unique ways neurodiverse individuals think, process, and learn, providing a more inclusive educational model (Luckin, 2021). For instance, AI can adapt textual information, evaluate individual learners' strengths and weaknesses more precisely, and develop customized intervention programs that enhance students' cognitive and emotional skills (Gogoulou et al., 2022). One significant advantage of AI in education is its ability to analyze large volumes of data to provide personalized feedback. AI-based systems can continuously evaluate how neurodivergent students are engaging with content, offering real-time adaptations to support their learning process. According to Xu and Brown, (2023), AI tools like adaptive learning platforms can modify educational materials based on student performance, ensuring that content remains challenging but achievable. This not only enhances student engagement but also promotes skill development at an individual pace, a crucial aspect of supporting neurodivergent learners.

Additionally, multidisciplinary collaboration between educators, AI developers, and cognitive scientists presents new research opportunities. This collaboration is key to ensuring that AI solutions are developed in ways that prioritize inclusivity and accessibility for neurodivergent students (Aslan et al., 2024). AI's potential to revolutionize neurodiverse education requires the input of experts across fields to ensure that the technology does not reinforce existing educational biases but instead enhances equity. Several AI-based tools have been designed to support inclusivity and address the unique needs of neurodiverse individuals, especially regarding career opportunities. AI-enhanced screening and matching platforms, for example, are designed to measure cognitive traits and performance metrics, which are often overlooked in traditional hiring and educational assessment methods (Walkowiak, 2021). These platforms use AI and neuroscience applications to match neurodiverse individuals with appropriate career paths, focusing on their cognitive strengths and personality traits, rather than on areas of weakness. This not only promotes inclusivity in the workplace but also helps educational institutions create tailored educational paths for students that align with their natural abilities. AI-powered platforms can also mitigate biases in traditional educational assessments. For neurodivergent students, standard testing methods may not accurately reflect their true potential, as these tests are often structured around neurotypical thinking patterns. Walkowiak, (2021) highlights how AI can assess a broader range of cognitive abilities, offering a more accurate picture of neurodiverse students' potential, which is essential for career development and academic success.

Beyond cognitive assessments, AI and related technologies, such as Virtual Reality (VR), are being increasingly used for neurodiversity training and support. VR offers immersive environments that can help neurodivergent students practice and prepare for stress-inducing situations or tasks, such as public speaking or navigating social interactions (Walkowiak, 2023). VR environments are particularly beneficial for individuals with autism or ADHD, as they allow for safe, controlled, and repetitive practice of real-world scenarios without the risk of failure or overwhelming sensory input (Ma & Lai, 2021). This type of intervention can help reduce anxiety and improve self-confidence in neurodivergent learners. Moreover, AI-enhanced emotion recognition technologies are being developed to support neurodiverse students' emotional well-being. These technologies monitor stress, anxiety, and other emotional states through facial expressions, voice analysis, and physiological signals. When emotional distress is detected, the technology can intervene by providing coping strategies, like relaxation exercises or notifications for a break. Walkowiak, (2023) argues that these tools offer a valuable resource for improving mental health support in educational settings, where neurodivergent students may struggle with managing emotions in high-pressure situations.

Integration of AI Tools in the Color Theory Class: Case Example

As educational environments continue to evolve, the integration of AI tools has gained traction as a method to address diverse learning needs. AI-powered tools like chatbots and intelligent tutoring systems are playing a critical role in enhancing the learning experience, particularly for students facing instructional challenges. These systems provide personalized, on-demand support, which is especially useful in courses with complex topics like Color Theory. This case example explores the use of an AI-powered chatbot in the Color Theory course within the Animation and Visual Effects Engineering program at the Universidad Politécnica de Baja California. During the 2024-2 semester, it was identified that students in the second semester of the program faced significant difficulties following instructions for completing exercises. Given the diverse range of learning barriers identified in the group, an educational intervention was introduced: a chatbot capable of responding to frequently asked questions that arose when completing exercises for the class.

According to Cota-Rivera et al. (2024), chatbots are AI-driven tools that provide a personalized and adaptive learning environment. These tools are computer-based systems grounded in artificial intelligence that can support students by adapting to their individual learning needs. For this case, the free POE chatbot tool was implemented. The chatbot was integrated into the course via the Google Classroom platform, where students could ask questions and receive guidance outside of regular class hours. The chatbot was fed with the course bibliography, instructions for exercises, and relevant supplementary material, which enabled it to provide detailed answers and instructions to students as they worked through their assignments. The use of the POE chatbot provided several notable benefits. As highlighted by Cota-Rivera et al. (2024), some students with learning difficulties require more detailed instructions than what can often be provided during class time. The chatbot addressed this need by offering step-by-step guidance, expanded explanations of terms, and additional clarifications on processes related to Color Theory exercises. This reduced the number of email inquiries from students and allowed them to work more independently, boosting their confidence in their ability to complete the assignments.

Beyond simple text-based responses, the chatbot was capable of answering questions with visual aids. For example, when students asked about the process of completing a specific exercise, the chatbot could not only provide instructions but also display relevant images and diagrams pulled from the sources it was trained on (see Figure 1). This multimodal support proved essential for students who learn better through visual aids and who needed to see the practical steps alongside textual instructions.

Figure 1. Example of a teacher's assistant on the CLASSROOM platform

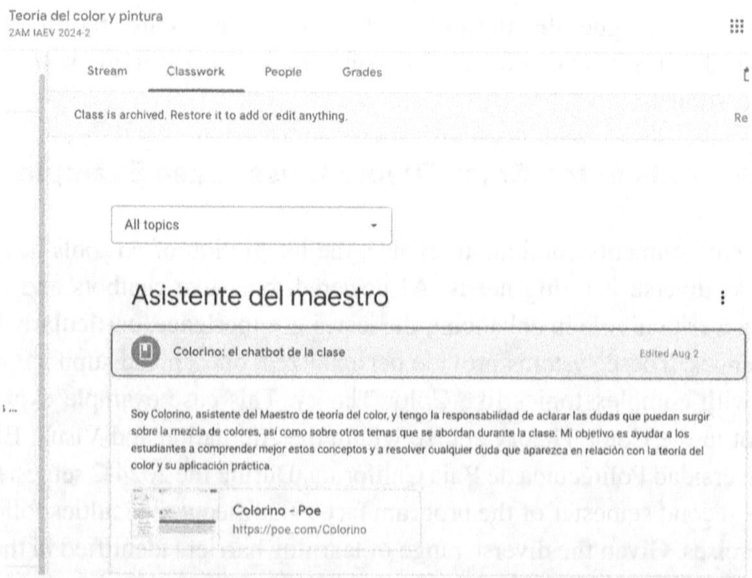

As explored in Cota-Rivera et al.'s study (2024), AI tools like these are reshaping how education is delivered in higher education institutions. In this context, chatbots represent an adaptive learning system that can help address gaps in traditional teaching methods. The use of AI in the Color Theory course allowed for the creation of a learning environment that was both inclusive and responsive to individual student needs. AI's ability to provide on-demand, personalized support helps ensure that students who may otherwise struggle to keep up with instructions can still succeed in their studies. The chatbot also aligns with broader trends in AI in education, which emphasize the reduction of teacher workloads through the automation of repetitive tasks. As mentioned by Cota-Rivera et al. (2024), AI tools such as these allow teachers to focus more on mentoring and developing meaningful interactions with students, as the AI takes over routine queries and support tasks. Additionally, this form of integration demonstrates AI's potential for fostering greater educational inclusivity by accommodating students with different learning preferences, as highlighted in the broader context of AI's impact on education.

Poe allows the creation of chatbots and the feeding of relevant texts and references for the class, so that it adapts to the types of frequently asked questions that students may have when completing their exercises.

Some students with learning difficulties require detailed instructions, which many times the teacher does not provide at the level of detail these students need. Therefore, the chatbot can deliver detailed instructions and expand explanations of terms and processes. In the Color Theory class, the use of the chatbot produced positive results: email inquiries were reduced, and students were able to complete their work independently with the support of this tool (Figure 2). The assistant Colorino is capable of answering questions with images, for example, by asking about the process of creating a particular exercise, providing instructions for its elaboration, and showing text references from the sources the assistant was trained on.

Figure 2. Colorino teacher's assistant on the classroom platform

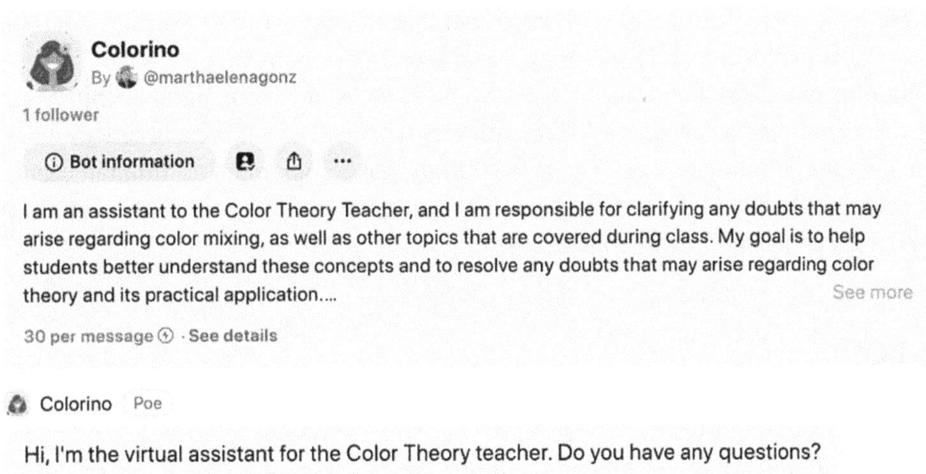

(Own Elaboration)

FUTURE RESEARCH DIRECTIONS

Future research in AI for the education of neurodiverse students offers many promising avenues that can reshape and improve educational outcomes. One important area to explore is the potential of AI to provide personalized learning environments that dynamically adapt to students' evolving needs. Neurodiverse students, such as students with ADHD, autism spectrum disorder (ASD) or dyslexia, have different cognitive profiles and learning styles that often cannot be accommodated by traditional educational models. By using AI, researchers could explore how machine learning algorithms can be fine-tuned to provide adaptive teaching strategies in real time, ensuring that lesson content and methods are always aligned with a student's cognitive and emotional state.

In addition, future research could investigate the role of AI in improving assessment methods for neurodiverse students. Traditional tests often fail to capture the true abilities of these students because they rely on neurotypical measures. AI-driven testing could introduce more inclusive assessment tools that take into account each student's strengths and allow for alternative methods of demonstrating knowledge and skills. For example, students with ASD who excel in pattern recognition or problem solving could benefit from performance-based assessments that focus on their unique abilities rather than being limited by rigid, standardized tests.

Another promising research direction lies in the integration of AI with immersive technologies, such as virtual reality (VR) and augmented reality (AR). These tools, in combination with AI, can create customized, low-stress learning environments in which neurodivergent students practice real-world scenarios such as social interactions or complex problem solving in controlled environments. For students with ASD, this could significantly reduce the anxiety associated with unpredictable environments and improve their confidence and overall learning experience.

Ethical considerations must also be taken into account in future research. When developing AI technologies, the privacy and autonomy of neurodiverse students must be taken into account. It must be ensured that data is used in an ethical manner and that excessive use of AI does not replace human interaction, which remains crucial for emotional and social development.

Finally, interdisciplinary collaboration between educators, AI developers and cognitive scientists will be critical to drive innovation that aligns AI technologies with educational best practices. Future research can focus on how this collaboration can produce AI tools that complement traditional teaching methods and ensure that neurodivergent students are not only accommodated, but also thrive in an environment tailored to their strengths and learning needs.

CONCLUSION

The chapter provides an in-depth analysis of the challenges and opportunities faced by neurodiverse students in traditional education systems and highlights the shortcomings of conventional teaching methods. These systems often fail to address the diverse needs of neurodiverse students, such as students with ADHD or autism spectrum disorders (ASD), who require more flexible, structured environments that accommodate their unique cognitive profiles. Rather than focusing exclusively on their deficits, the text argues for a pedagogical shift toward recognizing and utilizing the specific strengths of neurodivergent students, such as attention to detail and pattern recognition.

The potential of adaptive learning technologies, especially artificial intelligence (AI), is highlighted as a promising solution for creating inclusive and personalized educational experiences. AI offers the ability to tailor lessons to individual learning profiles and make real-time adjustments to support different cognitive needs. However, it is important not to over-rely on AI as this risks losing the essential human interaction required to deal with the emotional and social needs of neurodiverse students. A balanced approach is crucial, combining the efficiency of AI with the empathy and understanding of educators.

An illustrative example of the integration of AI into education is the color theory course at the Universidad Politécnica de Baja California. In this class, an AI-driven chatbot was implemented to support students who have difficulties with the instructions for completing exercises. Students in the Animation and Visual Effects Engineering program faced challenges similar to those of neurodiverse learners, such as difficulty following complex, sequential tasks. The chatbot, based on the POE tool, provided personalized on-demand support via Google Classroom. By answering frequently asked questions and providing visual aids, the chatbot played a crucial role in promoting students' understanding and independence. This intervention demonstrates how AI can complement teaching by providing tailored support that adapts to students' individual needs and enables them to engage with the material more confidently. This case illustrates how adaptive technologies can be a crucial aid in courses that require visual and conceptual understanding, particularly for students who may struggle in traditional contexts.

In addition to personalization, AI and new technologies such as natural language processing (NLP) and virtual reality (VR) offer new possibilities for education. These tools can be used to create immersive learning environments and career-focused platforms that recognize and build on students' cognitive strengths, providing a more effective path to academic success and career fulfillment. AI's ability to analyze and adapt large amounts of data in real time can improve engagement and skill development and foster inclusive learning environments that address individual needs.

The chapter also addresses broader systemic issues, such as resistance to AI adoption among educators, which often stems from concerns about job security and ignorance of the potential benefits of AI. Current support structures for neurodiverse students, built within a "disability support" model, can be slow, bureaucratic and stigmatizing, focusing more on weaknesses than strengths. The text calls for a collaborative, team-based approach to learning that values collective intelligence and in which neurodivergent and neurotypical students can thrive together in a dynamic and supportive environment.

Furthermore, the chapter underlines the urgent need to address the growing digital divide and inequalities in access to technology. While AI-driven tools hold great promise, their effectiveness will be significantly limited if students do not have access to the necessary technologies, resources and support systems. Without sufficient customization options, AI tools may fail to meet diverse individual learning styles and needs, potentially exacerbating the very problems they are designed to solve. Neurodiverse students continue to face significant barriers to academic success. These include lower graduation rates and fewer opportunities for further study or stable employment in their chosen fields.

In response to these challenges, the chapter offers several forward-looking recommendations for the integration of AI and other new technologies into education. These include AI-powered screening platforms for career guidance, virtual reality (VR) tools to prepare students for stressful tasks, and assistive technologies such as text-to-speech software for students with dyslexia or ADHD. Multidisciplinary collaboration between educators, AI developers and cognitive scientists is essential to ensure that these technologies promote inclusion without reinforcing existing educational biases. Ultimately, the chapter emphasizes the importance of a coordinated and comprehensive effort to effectively integrate these technologies in a way that promotes inclusion, supports cognitive diversity, and improves educational outcomes for all neurodiverse learners, leading to better educational experiences and future success.

REFERENCES

Aslan, S., Durham, L. M., Alyuz, N., Okur, E., Sharma, S., Savur, C., & Nachman, L. (2024). Immersive multi-modal pedagogical conversational artificial intelligence for early childhood education: An exploratory case study in the wild. *Computers and Education: Artificial Intelligence*, *6*(April), 100220. DOI: 10.1016/j.caeai.2024.100220

Black, M. H., Helander, J., Segers, J., Ingard, C., Bervoets, J., de Puget, V. G., & Bölte, S. (2024). Resilience in the face of neurodivergence: A scoping review of resilience and factors promoting positive outcomes. *Clinical Psychology Review*, *113*(May), 102487. Advance online publication. DOI: 10.1016/j.cpr.2024.102487 PMID: 39178757

Chrysochoou, M., Zaghi, A. E., & Syharat, C. M. (2022). Reframing neurodiversity in engineering education. *Frontiers in Education*, *7*(November), 1–12. DOI: 10.3389/feduc.2022.995865

Clouder, L., Karakus, M., Cinotti, A., Ferreyra, M. V., Fierros, G. A., & Rojo, P. (2020). Neurodiversity in higher education: A narrative synthesis. *Higher Education*, *80*(4), 757–778. DOI: 10.1007/s10734-020-00513-6

CONAPRED. (2021). *Situación de las personas con discapacidad en México.*

Cota-Rivera, E. I., & Gonzalez Correa, M. E. Martinez, M., Bernal Marin, L. A., & Marquez Montenegro, M. Y. (2024). Transforming Education With the Power of Artificial Intelligence: Case Studies. In *Enhancing Higher Education and Research With OpenAI Models* (pp. 113–140). IGI Global.

Cotton, M., Smith, L., & Johnson, T. (2022). Empathy in Education: The Role of Human Interaction. *Educational Leadership*, *79*(5), 23–29.

Gogoulou, A., Salta, K., & Efstathiou, N. (2022). AI in Education: Supporting Personalized Learning for Neurodiverse Students. *Computers & Education*, *176*, 104253. https://doi.org/https://doi.org/10.1016/j.compedu.2022.104253

Hall, R., Peterson, A., & Sandoval, J. (2024). Assistive Technology for Neurodivergent Students: Empowering Independence Through AI. *Learning and Instruction*, *55*, 120–133. https://doi.org/https://doi.org/10.1016/j.learninstruc.2024.101421

Hamilton, L. G., & Petty, S. (2023). Compassionate pedagogy for neurodiversity in higher education: A conceptual analysis. *Frontiers in Psychology*, *14*(February), 1–9. DOI: 10.3389/fpsyg.2023.1093290 PMID: 36874864

Holmes, W., Bialik, M., & Fadel, C. (2021). AI in education: Opportunities and risks. *International Journal of Artificial Intelligence in Education*, *31*(1), 5–27. https://doi.org/https://doi.org/10.1007/s40593-021-00235-7

INEGI. (2020). *Censo de Población y Vivienda.* https://www.inegi.org.mx

Jones, F., & Müller, G. (2023). Collective Intelligence in Neurodiverse Learning Environments. *Journal of Creativity and Innovation*, *15*(1), 33–47. https://doi.org/https://doi.org/10.1057/s41117-023-00285-8

Kline, S., Taylor, B., & Huang, Z. (2023). AI and Neurodivergent Learning: The Case for Personalization. *Technology and Education, 22*(4), 501–520.

Koegel, L. K., Ashbaugh, K., & Koegel, R. L. (2020). Improving social communication in ASD through structured interventions. *Journal of Autism and Developmental Disorders, 50*(4), 1254–1267. https://doi.org/https://doi.org/10.1007/s10803-019-04382-x

Lawson, W., Beckett, A. E., & Fernandes, R. (2021). Embracing cognitive diversity: A framework for inclusive pedagogy in higher education. *Educational Research Review, 34*, 100392. https://doi.org/https://doi.org/10.1016/j.edurev.2021.100392

Le Cunff, A. L., Giampietro, V., & Dommett, E. (2024). Neurodiversity and cognitive load in online learning: A systematic review with narrative synthesis. *Educational Research Review, 43*(June 2023), 100604. DOI: 10.1016/j.edurev.2024.100604

Luckin, R. (2021). AI in Education: Opportunities and Challenges for Educators. *AI & Society, 36*(1), 1–11. https://doi.org/https://doi.org/10.1007/s00146-020-01050-0

Ma, W., & Lai, Y. (2021). Using Virtual Reality for Neurodiversity Training: A Case Study in Higher Education. *Virtual Reality & Intelligent Hardware, 3*(2), 192–204. https://doi.org/https://doi.org/10.1016/j.vrih.2021.04.003

Meyer, A., Graham, D., & Sharp, M. (2022). The Digital Divide in Education: Addressing the Barriers to Equitable Access. *Computers & Education*, •••, 176.

Sarver, D. E., McCabe, K. A., & Mashburn, J. A. (2021). Classroom engagement strategies for ADHD learners: A longitudinal study. *Journal of Attention Disorders, 25*(9), 1237–1248. https://doi.org/https://doi.org/10.1177/1087054720974822

Serrano, F., & Defior, S. (2018). Dyslexia beyond reading: Addressing creativity and verbal reasoning strengths. *Annals of Dyslexia, 68*(1), 59–75. https://doi.org/https://doi.org/10.1007/s11881-018-0158-0

Sharma, R. (2020). Rethinking Disability Services in Higher Education. *Disability & Society, 37*(5), 1049–1064.

Stein, B., Hoeft, F., & Richter, C. G. (2024). Stress, resilience, and emotional well-being in children and adolescents with specific learning disabilities. *Current Opinion in Behavioral Sciences, 58*, 101410. DOI: 10.1016/j.cobeha.2024.101410 PMID: 39157263

UNESCO. (2021). *Informe mundial sobre la inclusión educativa: América Latina y el Caribe*. https://www.unesco.org

UNESCO. (2023). *The Impact of AI on Equity in Education: A Global Perspective.*

Verhoeven, T., de Graaf, S., & van Dijk, L. (2023). Teacher Training for AI-Enhanced Education: Addressing the Challenges and Opportunities. *Education and Information Technologies, 28*(5), 3215–3236.

Walkowiak, E. (2021). Neurodiversity of the workforce and digital transformation: The case of inclusion of autistic workers at the workplace. *Technological Forecasting and Social Change, 168*(March), 120739. DOI: 10.1016/j.techfore.2021.120739

Walkowiak, E. (2023). Digitalization and inclusiveness of HRM practices: The example of neurodiversity initiatives. *Human Resource Management Journal, November 2022*, 578–598. DOI: 10.1111/1748-8583.12499

Xu, M., & Brown, S. (2023). AI and the Automation of Administrative Tasks in Education. *Computers in Human Behavior*, •••, 143.

Zawacki-Richter, O., Marin, V. I., Bond, M., & Gouverneur, F. (2019). Systematic review of research on artificial intelligence applications in higher education: Emerging trends and future research directions. *International Journal of Educational Technology in Higher Education, 16*(1), 39. https://doi.org/https://doi.org/10.1186/s41239-019-0171-0. DOI: 10.1186/s41239-019-0171-0

ADDITIONAL READING

Chen, L., Chen, P., & Lin, Z. (2020). Artificial intelligence in education: A review. *Computers & Education, 156*, 103930.

Cota-Rivera, E. I., & Gonzalez Correa, M. E. Martinez, M., Bernal Marin, L. A., & Marquez Montenegro, M. Y. (2024). Transforming Education With the Power of Artificial Intelligence: Case Studies. In *Enhancing Higher Education and Research With OpenAI Models* (pp. 113–140). IGI Global.

Heffernan, N., & Koedinger, K. (2020). Intelligent Tutoring Systems: Harnessing AI to Optimize Student Learning. In *The Cambridge Handbook of Artificial Intelligence in Education* (pp. 211-231). Cambridge University Press.

Khalique, F., Khan, N., Sarna, S., & Saini, K. (2024). Diverse Minds, Infinite Potential: Navigating Neurodiversity in Today's Workplace. In Global Practices on Effective Talent Acquisition and Retention (pp. 96-108). IGI Global.

Spector, J. M., & Ma, W. (2018). Artificial Intelligence in Instructional Systems: The Road Ahead. In *Handbook of Research on Educational Design and Cloud Computing in Modern Classroom Settings* (pp. 112-134). IGI Global.

Tariq, M. U. (2024). Neurodiversity Inclusion and Belonging Strategies in the Workplace. In *Resilience of Multicultural and Multigenerational Leadership and Workplace Experience* (pp. 182–201). IGI Global. DOI: 10.4018/979-8-3693-1802-7.ch009

KEY TERMS AND DEFINITIONS

Adaptive Learning Platforms: AI-based systems that personalize educational content in real-time according to a student's individual progress, learning style, and needs.

Artificial Intelligence (AI): A branch of computer science that enables machines to perform tasks typically requiring human intelligence, such as learning, reasoning, problem-solving, and adaptation.

Chatbot: An AI-driven software application that simulates human conversation, providing personalized responses to user inquiries, often used in education to support learning outside of regular class hours.

Emotion Recognition Technologies: AI tools designed to detect and interpret human emotions based on data such as facial expressions, voice tone, or physiological responses, often used to provide emotional support in learning environments.

Inclusion: The practice of designing educational environments and strategies that accommodate the diverse needs of all learners, particularly those with neurodivergent conditions, to ensure equal access to learning opportunities.

Neurodiversity: A concept recognizing the natural variation in brain function and behavior among individuals, encompassing conditions such as autism, ADHD, and dyslexia, each requiring distinct educational approaches.

Personalized Learning: An educational approach that tailor's instruction to the individual needs, abilities, and interests of each student, often supported by AI to deliver customized content and support.

Real-Time Feedback: Immediate responses provided to students based on their actions or performance, commonly facilitated by AI systems in educational settings to enhance learning and engagement.

Visual Aids: Tools such as images, diagrams, and videos used in educational settings to enhance understanding, especially for students who benefit from visual learning methods.

Chapter 11
Machine Learning for Natural Disaster Prediction and Prevention

V. Vijayagopal
https://orcid.org/0000-0001-8897-3242
Dr. M.G.R. Educational and Research Institute, India

N. Chidambararaj
St. Joseph's College of Engineering, India

M. Kavitha
Sathyabama Institute of Science and Technology, India

Deepak Sundrani
https://orcid.org/0000-0002-7217-8911
NICMAR University, India

Karthik Kumar Vaigandla
Balaji Institute of Technology and Science, India

Bhoopathy Varadharajan
https://orcid.org/0000-0003-2175-6328
Sree Rama Engineering College, India

ABSTRACT

This chapter delves at the ways in which ML enhances the prediction and prevention of natural disasters. Disaster forecasts can be more accurate and timely with the use of machine learning algorithms that combine massive amounts of data from satellite images, seismic activity, and weather patterns. It is possible to foretell the occurrence of natural catastrophes using ML techniques, such as supervised, unsupervised, and deep learning models. As part of its ML implementation, it checks for issues with data quality, model interpretability, and processing limitations in real time. This chapter takes a look at the potential directions that machine learning ethics, real-time decision-making, and multimodal data integration could go in the future. Using case studies and real-world examples, this chapter discusses the possibilities and constraints of machine learning in mitigating natural disasters

INTRODUCTION

When catastrophic events occur as a result of Earth's inherent natural processes, we call it a natural disaster. Natural disasters cause enormous disruption to people's lives by destroying property, killing individuals, and injuring others. Every year, catastrophes strike, causing misery to millions of people

DOI: 10.4018/979-8-3693-6910-4.ch011

and putting regional development goals in jeopardy. Tragic loss of life, displacement of entire communities, and economic devastation are all consequences of natural disasters. Due to data scarcity, experts in the field, including researchers, legislators, and business owners, require methods to evaluate possible consequences and propose long-term fixes. They roll back advancements done throughout time and also impede future development. The World Health Organization reports that more than 90,000 individuals lose their lives annually as a result of natural disasters around the world. The impacts of natural catastrophes on various animals, including humans, are catastrophic. They wreak havoc on ecosystems by removing plants and animals and changing features like water and light availability. The atmospheric, hydrologic, and geological components of natural disasters like landslides, tsunamis, earthquakes, floods, and droughts can cause damage or loss. Devastation to property and revenue as well as intangible costs like physical and physiological impairments are some of the many negative outcomes of natural disasters (Barnes et al., 2019). Every year, natural catastrophes in industrialized nations like the US cause many deaths and damage worth billions of dollars to property and infrastructure. Worldwide, millions of people are impacted by catastrophes, both natural and caused by humans, each year. Tragically, human lives are frequently cut short by these disasters. In addition to the human casualties, disasters cause massive damage to buildings and other property. Reducing human casualties, protecting people and infrastructure, minimizing economic damages, and restoring normalcy are the goals of disaster management efforts. The complexity and criticality of catastrophic operations make robust decision-making, assisted by AI, absolutely necessary. With the help of machine learning, disaster management has become more knowledgeable and efficient, which helps deal with the magnitude and effects of disasters. Disaster management is a long-term strategy for dealing with disasters (Hassini & Belbachir, 2014). Mitigation, readiness, reaction, and recovery are the four stages that have gained the most traction. The goal of catastrophe mitigation is to lessen or eliminate their impact. Communities can better respond to and mitigate the effects of catastrophes through preparedness operations such as emergency planning, prepositioning of supplies, training, and education. The community's socioeconomic structure, as well as its members, their property, and the environment, are the targets of response efforts. Medical treatment, shelter management, supply distribution, emergency planning, and damage assessment are all part of disaster relief and response operations. Research on this area is in high demand at this era since both people and infrastructure need help quickly. At this stage, every second counts, so methods focus on being as accurate and efficient as possible. What we call "recovery" or "reconstruction" are the long-term efforts to get a community back to normal. Providing financial aid and repairing damaged structures and infrastructure are two of the main tasks at this stage. Resilience can be enhanced by incorporating local communities into crisis management efforts.

Predicting, mitigating, and supervising earthquakes, floods, hurricanes, and wildfires through the application of machine learning (ML) approaches is the focus of the "Machine Learning for Natural Disaster Prediction and Prevention" chapter. The chapter takes a systematic look at how machine learning is being used to manage disasters, with a focus on how it can be used for early warning systems, risk assessment, allocating resources, and real-time monitoring.

DISASTER PREVENTION

There is no one solution or field that can solve the problem of disaster reduction; rather, it requires a multifaceted combination of technology and societal initiatives. Do not expect technology to solve all of your disaster problems, say disaster researchers. Lessening susceptibility to natural disasters is the goal of disaster prevention. The goal of these actions is to detect, contain, or anticipate potential calamities so that they can be prevented. The goal of catastrophe risk management, which includes preparation, is to lessen vulnerability while accepting that some consequences are inevitable (Usman, 2017). The term "preparedness" describes the strategies and actions taken in advance of a potential danger in order to guarantee an adequate response. The priority moves to response operations, which include coordinating emergency services, following a disaster. To recover means to put a community's infrastructure, means of subsistence, and living conditions back in order after a tragedy has struck. Providing adequate social services, rebuilding or upgrading physical infrastructure, and ensuring access to food and resources are all part of this. In the aftermath of a disaster, it is crucial to restore and rebuild in order to save lives, provide immediate needs, get life back to normal as soon as possible, and reduce the likelihood of another disaster.

THE ROLE OF MACHINE LEARNING IN DISASTER MANAGEMENT

The problems of catastrophe management need to be able to be addressed by disaster prediction systems. An outline of machine learning approaches to disaster and pandemic management is given in this section. In order to effectively manage and respond to disasters, it is essential to allocate resources. During disasters, decision makers can make smarter, more economical choices with the help of machine learning, which optimizes the allocation of resources. Allocating resources in the event of a disaster can benefit from machine learning in multiple ways, such as: Natural catastrophes such as hurricanes, earthquakes, and wildfires can be predicted with some degree of accuracy using machine learning algorithms. In order to provide early warning systems for allocating resources, these models examine past data and current environmental conditions. Employing data from past disasters, machine learning can predict how much food, water, medical supplies, and emergency personnel will be needed. With this information, authorities can better prepare for potential danger zones by deploying resources ahead of time. Considering factors like population density, road conditions, and traffic, machine learning algorithms determine the best evacuation routes. In order to ensure a smooth evacuation, these models can optimize the distribution of resources. When tragedy strikes, machine learning can assess data from faraway sensors, social media, and Internet of Things devices to give an accurate picture of the situation in real time. Proper allocation of resources can be aided by this data. With the help of machine learning, we can build models for allocating resources that take accessibility, infrastructure, vulnerability, and population density into consideration. Prioritizing essential demands through optimal resource allocation is the goal of these methods. With the use of satellite images and aerial surveys, machine learning can evaluate the damage from disasters. Based on this data, we may direct our resources to the areas where they will have the biggest influence. Using machine learning, the supply chain may be optimized to provide resources to areas hit by disasters. In order to ensure timely deliveries, it takes into account transportation routes, vehicle availability, and demand in real-time. By analyzing both past and present data, machine learning algorithms can estimate the probability and impact of disasters. Unusual data patterns that can indicate a disaster can be found by anomaly detection systems. Predicting the demand for resources like food, medical supplies, and emergency workers is made possible by machine learning approaches. Analyzing past data and current environmental conditions optimizes the distribution of resources to high-risk places. Machine learning has revolutionized numerous industries, including disaster prediction. Machine learning algorithms improve the accuracy of natural disaster incidence and severity predictions by analyzing past data, current environmental factors, and real-time inputs. Various ML algorithms, including supervised, unsupervised, and reinforcement learning, are discussed in this section as they pertain to catastrophe prediction. Additionally, it examines the application of these algorithms to different types of natural catastrophes through the use of case studies and examples.

DATA SOURCES

Use of high-quality and diverse data is crucial for machine learning to be effective in disaster prediction. Catastrophe prediction makes use of a wide variety of data sources, including satellite imagery, meteorological reports, seismic data, social media feeds, and sensor networks, among others. An outline of the data sources is provided in this section.

Satellite Imaginary

Several operations can be aided quantitatively and qualitatively by satellite remote sensing technologies. These include assessing disaster damage, offering operational support, and more (Yamazaki et al., 2016), and reducing risk. Remote sensing photography is particularly useful for assessing post-disaster damage and detecting changes (Pradhan et al., 2016). Disaster management makes use of remote sensing in a variety of ways, such as high resolution, multidimensional, and multi-technique. For the purpose of detecting changes in texture both before and after natural disasters, high-resolution satellite imagery is collected. After a crisis, it is critical to get this data. In disaster-stricken regions, high-resolution satellite photos can show altered structures, bodies of water, and destroyed buildings (Yuei-An et al., 2010). The application of high-resolution remote sensing technology allows for the reliable localization of lost persons and the commencement of rescue operations. Recognizing human settlements, evaluating flood risk, and decreasing landslides are just a few examples of the many ways disaster risk reduction makes use of satellite imaging.

Data Simulation

Predicting natural disasters produced by events involving the weather, land surfaces, and pollutants relies heavily on numerical modelling and forecasting. Damage assessment and damage prediction using 3D modelling has been helpful after disasters. Models for disaster management can be developed, validated, and improved using the massive amounts of observation data produced by natural disasters. Using high-resolution data, agent-based models can examine human behaviour and changes in the environment at a low level. (Goldenberg et al., 2015) developed a backup strategy for rescuing people in the event of a landslide using an agent-based framework. The system employs a three-dimensional geological model to predict landslides, high-resolution remote sensing images for data extraction, and a multi-agent simulation method to model human behaviour in real-time catastrophe scenarios. Efficient evacuation with fewer casualties is achieved through simulations.

Crowd Sourcing

Both social media data and crowdsourcing are contributed to by the general public. Due to contributors' lack of knowledge about the data collection process, social media data is usually passively contributed, in contrast to crowdsourcing data, which is actively contributed. The idea behind crowdsourcing systems is to get people to voluntarily provide needed data. The sheer amount, inherent noise, and lightning-fast streaming rate of crowdsourced disaster data need automated processes and specialized technologies. Based on the location and analysis of the event, crowdsourced data might be prioritized. Online platforms and mobile applications can be used to gather data from the public in the aftermath of disasters. According to (dos Santos Rocha et al., 2017), there are several strategies that disaster managers might employ to boost the efficiency of digital volunteers. Dispersed intelligence, participatory engagement, and self-mobilization are three areas where the authors believe crowd sensing and collaborative mapping could be useful. In order to enhance communication among stakeholders, they suggested implementing a feedback mechanism.

Social Media

When it comes to crisis management, social media platforms like Flickr, Foursquare, Twitter, and YouTube have been invaluable. Through their application programming interfaces (APIs), social media sites make it possible to access geotagged data. By making it easier to communicate information, social media has been essential in disaster management. There has been a recent uptick in the use of social media by government and non-profit disaster management groups to assess public sentiment and response. With the help of online crisis crowdsourcing platforms, information can flow in both directions, improving the efficiency of reaction and recovery efforts. Despite the possibility of crowdsourcing, crisis management is making effective use of social media. According to (Granell & Ostermann, 2016), Twitter is more successful for detection and prediction than recovery and response. Spatial video, UAV, and phone call data can provide valuable proof of post-disaster conditions, including in-situ images.

PREPROCESSING

Data preparation is defined as a collection of methods for correctly initializing data in order for it to be input to a particular ML algorithm throughout this chapter. Instead of using the vague terminology utilized in earlier works to describe all data pretreatment procedures, it favors the data preparation notation for building data preprocessing sections. Although we acknowledge the validity of this terminology, we think it's more useful to differentiate between data reduction and data preparation because the latter has grown in significance over the past few years and has clear implications for other areas of study. Preparing the data is usually a crucial part of the process. It updates old data so machine learning algorithms can make better use of it. As demonstrated in Figure 1, properly preparing data is essential for ML algorithms to operate correctly and prevent errors while running. If the algorithm is effective, it doesn't guarantee that the output will be useful or correct.

Figure 1. Data preparation different forms

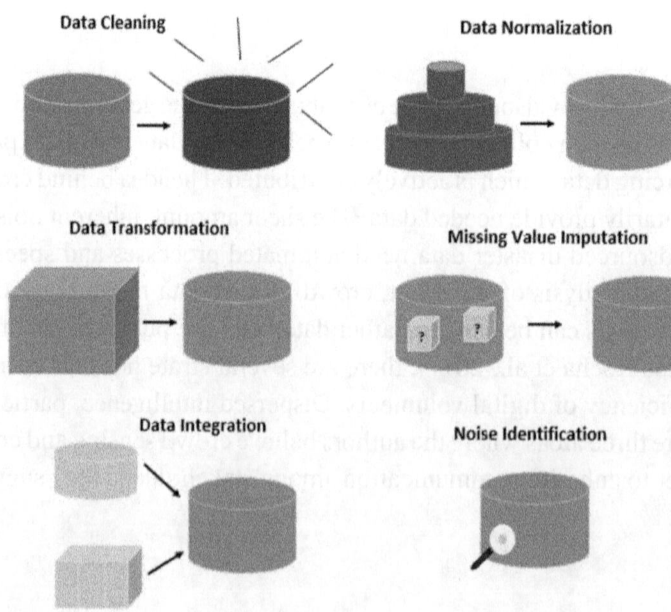

Data Cleaning

Data cleansing involves correcting faulty data, removing incorrect data, and reducing extraneous detail. This idea encompasses or overlaps with several widely used data preparation approaches. Missing and noisy data are treated separately in this book to allow for a more in-depth study of intelligent proposals (García et al., 2015). Other data cleaning responsibilities include detecting discrepancies and filthy data (fragments that do not make sense). Human auditing is typically required for jobs that demand a deeper comprehension of original data.

Data Transformation

Preprocessing involves converting or consolidating data to improve mining efficiency. Data transformation includes subtasks such as smoothing, feature generation, aggregation, normalization, discretization, and generalization. Data transformation procedures, including data cleansing, are typically separated into discrete tasks because they are part of a larger data preprocessing family. Traditional data transformation techniques, such as report generation, attribute aggregation, and categorical attribute generalization, require human supervision and rely heavily on data. Examples include replacing complete dates in databases with year numbers only.

Data Integration

It comprises merging data from multiple data repositories. This approach necessitates meticulous implementation to eradicate duplication and irregularities in the generated data set. Common steps in data integration include discovering and combining domains and variables, studying the link between attributes, creating duplicate tuples, and finding inconsistencies in data values from various sources.

Data Normalization

Data analysis might be impacted by the measuring unit utilized. Consistency in measurement units and scale/range should be maintained across all properties. Because it gives all attributes the same weight, normalizing data is useful for statistical learning methods. When dealing with training and testing datasets, normalization or standardization is necessary to eliminate any learning biases caused by differences in numerical features (Al-Utaibi & El-Alfy, 2018). Common normalizing approaches include minmax and z-score (zero mean).

Min-Max Normalization

The min-max normalization transforms a d-dimensional feature vector $V = (v_1, v_2, \ldots, v_n)$ into a normalized feature vector $V^* = (v_1^*, v_2^*, \ldots, v_n^*)$, where each feature value v_g is translated to the range [0, 1] by the following equation:

$$v_g^* = \frac{v_g - v_g^{min}}{v_g^{max} - v_g^{min}}, g = 1,2,\ldots,n$$

The minimum and maximum values for the g-th feature in v are represented by v_g^{min} and v_g^{max}, respectively.

Z-Score Normalization

The z-score (or z-mean) normalization turns feature vector data into a distribution with an average of 0 and a standard deviation of 1, as indicated in the equation below:

$$v_g^* = \frac{v_g - \rho}{\varepsilon_g}, g = 1,2,\ldots,n$$

Where ε_g and ρ are the mean and standard deviation of the g-th feature in v.

Missing Data Imputation

This is a data cleansing technique that involves adding understandable data to variables with MVs. Providing an estimate of an appropriate data value is often preferable to leaving it blank.

Reduction of Noise

Smoothing is a phase in data cleaning that aims to detect random errors or variances in measured variables. We use the term "noise detection" rather than "noise removal," which is more relevant to the IS task in data reduction. When a noisy example is found, we can use a correction-based procedure that may require an underlying operation.

MACHINE LEARNING TECHNIQUES FOR DIFFERENT NATURAL DISASTER

Earthquake Prediction

Massive earthquakes tend to follow a predictable pattern, with extended periods of aseismic quiet (decades or centuries) between each occurrence. Earthquakes, which only last a few minutes at most, can be recorded by seismographs as the free surface moves. Surface movements in seismically active areas are sluggish, according to geodetic studies; annual horizontal and vertical displacements are less than a few centimeters. It has long been acknowledged that there are substantial social and economic ramifications to earthquake prediction. Earthquakes can cause enormous damage. We can potentially limit the loss of life and property if we can accurately foresee the next major seismic event (Vasanthi & Satish Kumar, 2016). Earthquake prediction is a vital tool for seismologists. Talks about how convolutional neural networks (CNNs) and other deep learning models are being used to analyze seismic data and make earthquake predictions. Additionally, it delves into how early warning systems can be optimized through the use of reinforcement learning.

CNN

CNNs, similar to traditional ANNs, learn to optimize themselves by utilizing neurones. The building blocks of many artificial neural networks are neurons, which can take in data and process it using operations like scalar products and non-linear functions. In order to generate a class score, the network takes in raw image vectors as input and applies a single perceptual scoring function (weight). In the last layer, we'll implement class-specific loss functions while keeping with the tried-and-true methods of artificial neural networks. CNNs are distinct from other types of ANNs since they are designed to identify patterns in images. Network performance for image-centric tasks is improved and model setup parameters are reduced by the incorporation of image-specific features into the design (Saxena, 2022). As shown in Figure 2 CNNs consist of three types of layers. These include convolutional, pooling, and fully-connected layers.

Figure 2. Simple five-layer CNN architecture

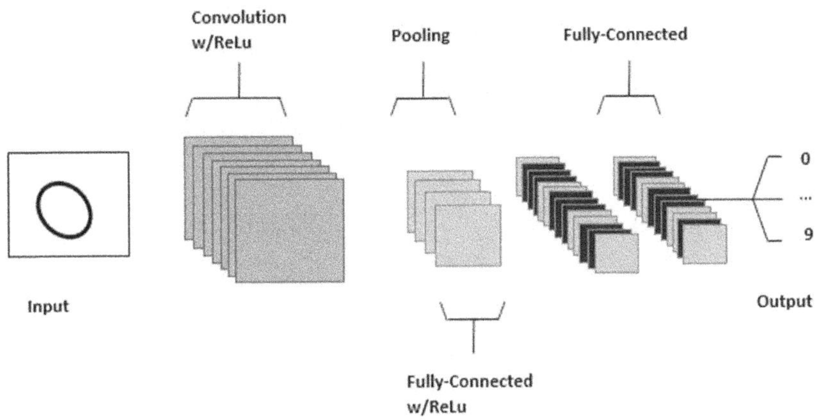

Stacking these layers forms a CNN architecture. Figure 2 illustrates a simplified CNN architecture for MNIST classification. The CNN example has four important areas of functionality.

1. Similar to previous ANN models, the input layer stores the image's pixel values.
2. The convolutional layer calculates the scalar product of neurones' weights and the input volume to determine their output. The rectified linear unit (ReLu) applies a 'elementwise' activation function like sigmoid to the output of the previous layer.
3. The pooling layer will then simply execute downsampling along the spatial dimension of the input, lowering the number of parameters within that activation.
4. Fully-connected layers conduct conventional ANN tasks and generate class scores for categorisation. Using ReLu between these layers may enhance performance.

CNNs use convolutional and downsampling techniques to alter input layers, producing class scores for classification and regression.

Reinforcement Learning

The three main parts of deep reinforcement learning—the agent, the environment, and the reward—make it a goal-oriented learning system. Through iterative interactions with its environment, the agent optimizes its decision-making policy to maximize total rewards. Here we provide the following definitions of the agent, environment, and reward: (1) The agent selects an action direction from six possible ones: west, east, north, south, down, and up. (2) The environment records the agent's position and the travel-time-corrected waveforms in response to the agent's actions. (3) The reward is determined by the three-dimensional distance (Kuang et al., 2024). This results in the closed loop of a deep reinforcement learning system, as seen in Figure 3, where the agent constantly decides to explore potential locations based on environmental observations, and the environment generates new observations in reaction to the agent's actions. Finding the best possible decision-making technique that maximises overall rewards is the end goal for the agent.

Figure 3. Reinforcement learning architecture

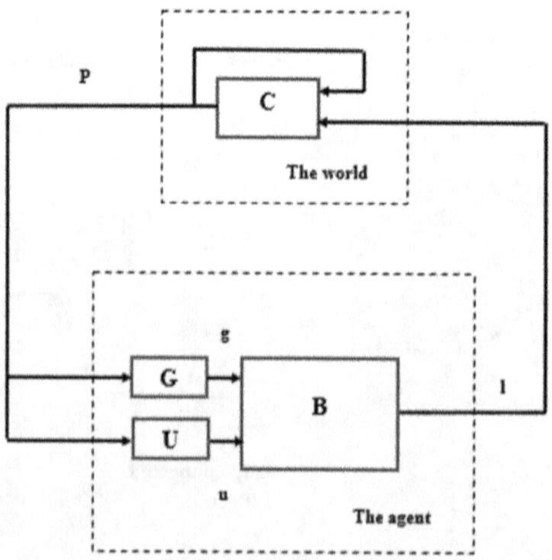

The reinforcement learning framework (Figure 2) includes two main entities: the learning agent and the world. The world is modelled as a dynamic system, with agents' actions causing transitions between states. The agent has three functional parts: behaviour (B), input (I), and reinforcement (R). The input function converts world outputs (world states) into agent inputs (Kokar & Reveliotis, 1993). The reinforcement function assigns a value to each condition of the world. The behaviour function updates the agent's information and generates its actions.

Flood Prediction

It is usual practice to use real-time data from rain gauges or other sensing devices for different return periods, in addition to previous flood records, when creating an ML prediction model. Modern remote-sensing technology, such as satellites, multisensory systems, and radars, as well as terrestrial rain gauges, provided the data used to compile the dataset. When it comes to real-time, high-resolution data acquisition, remote sensing is invaluable. Because of their superior resolution, weather radar observations are not always inaccurate when compared to rain gauges. The general accuracy of rainfall predictions is enhanced when a radar rainfall dataset is used in their construction. One can utilize ground gauges or a dataset based on radar to build a prediction model. For building and evaluation, the historical dataset is separated into discrete sets of hourly, daily, or monthly values. Model is trained using separate data sets, which are then confirmed, verified, and tested. The prediction of floods makes use of a number of machine learning methods, such as artificial neural networks (ANNs), neuro-fuzz, multilayer perceptrons, adaptive neuro-fuzzy inference systems (ANFIS), support vector machines (SVM), and wavelet neural networks (WNN). The following sections give an overview and background on some of the most basic algorithms used in machine learning.

SVM

It is usual practice to use real-time data from rain gauges or other sensing devices for different return periods, in addition to previous flood records, when creating an ML prediction model. Modern remote-sensing technology, such as satellites, multisensor systems, and radars, as well as ground rain gauges, provided the data used to compile the dataset. When it comes to real-time, high-resolution data acquisition, remote sensing is invaluable. Because of their superior resolution, weather radar observations are not always inaccurate when compared to rain gauges. The general accuracy of rainfall predictions is enhanced when a radar rainfall dataset is used in their construction. One can utilize ground gauges or a dataset based on radar to build a prediction model. For building and evaluation, the historical dataset is separated into discrete sets of hourly, daily, or monthly values. Data sets undergo the following steps: training, validation, verification, and testing.model. For flood prediction, a number of machine learning methods are employed, such as artificial neural networks (ANNs), neuro-fuzzy networks, multilayer perceptrons, support vector machines (SVMs), and adaptive neuro-fuzzy inference systems (ANFIS). The following sections give an overview and background on some of the most basic algorithms used in machine learning.

$$\sum_{g \in PX} e_g \beta_g Z(v_g, v) + m$$

The term β_g refers to the dual coefficients, which are upper-bound by W. The independent variable m requires estimation. Z (v is the kernel, g is the input, v) v vector.

MLP

MLP is likely the most prevalent artificial neural network architecture (Rezaeianzadeh et al., 2014). Perceptrons are a network of basic neurones. The perceptron produces a single output from many real-valued inputs by combining linear relationships based on input weights with nonlinear transfer functions (see Figure 4).

Figure 4. Architecture of MLP

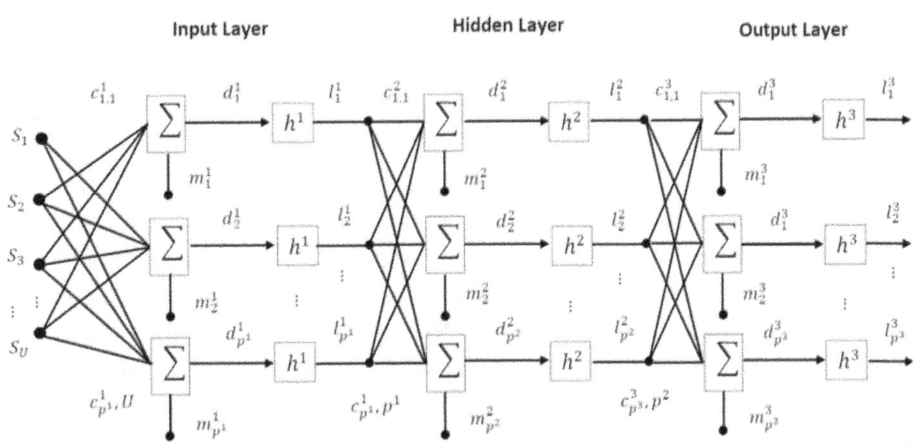

The MLP can be mathematically represented as follows:

$$e = h\left(\sum_{g=1}^{d} c_g s_g + m\right)$$

c_g is the weight vector, s_g is the input vector ($g = 1,\ldots,d$), m is the bias, h is the transfer function, and e is the output. (Zadeh et al., 2010) found that the tangent sigmoid transfer function outperformed the logistic sigmoid function. (Yonaba et al., 2010) discovered that using tangent sigmoid was more effective for streamflow forecasting. (Maier & Dandy, 1998)found that training with the hyperbolic tangent function is faster than training with the logistic sigmoid transfer function. Additionally, hyperbolic tangent networks produce slightly better predictions than logistic sigmoid transfer functions. Based on previous research, we utilised the tangent sigmoid transfer function, defined for every variable p as:

$$h(p) = \frac{2}{(1 + e^{-2p})} - 1$$

MLPs are commonly trained using the back error propagation approach, which involves iteratively changing the network's interconnecting weights to minimise the preset error (root-mean-square error).

Hurricane and Cyclone Prediction

Extreme weather phenomena, such as tropical cyclones, gales, rainstorms, and storm surges, can inflict significant damage in coastal locations around the world. Meteorologists and warning centres have made significant advances in observational technology, intensification physics, atmospheric interactions, atmospheric boundary layer, air-sea interface, ocean responses, and forecasting techniques over the past century (Chen et al., 2020). However, there are still significant issues with predicting TC genesis, intensity, and danger. Most tropical cyclone forecast models have low accuracy due to faulty vortex initialisation, insufficient modelling of complicated processes, and coarse resolution. Investigates the use of recurrent neural networks (RNNs) and long short-term memory (LSTM) networks to forecast the route and severity of hurricanes and cyclones. The part also covers the integration of machine learning algorithms and meteorological simulations.

Machine Learning Algorithms in TC Forecast

Machine learning applications in TC forecasts are categorised into five categories. The ultimate goal of TC genesis forecasts is to create real-time probabilistic and quantitative forecasts for a specific location, allowing for better monitoring of the tropical ocean. Currently, machine learning can only predict whether precursors can mature into TCs and the seasonal frequency of TC genesis in each area. These tasks are classified as classification and regression in machine learning. Researchers typically employ algorithms such as DT, LR, SVM, AdaBoost, and RF to predict TC genesis. Although ensemble algorithms are theoretically superior to solo algorithms, their effectiveness must be evaluated on an individual basis. Deep learning algorithms like MLPs and CNNs improve TC genesis forecasting by better fitting complex functions and processing visual data. Nonlinear mapping is used to anticipate the position of TCs based on their characteristics and associated meteorological and oceanographic data. MLP and RNN, the researchers' first choices, were also found to be more effective than traditional approaches.

Spatio-temporal models, such as ConvLSTM, have shown success in predicting TCs as they vary over time and place. GANs, deep belief networks (DBNs), and clustering algorithms such as FMM can be used to generate predicted TC cloud images and locate the TC centre, in addition to traditional methods. While these methods produced promising experimental results, it needs to be seen if they can outperform existing strategies in operational forecasting. DT, a rule-mining algorithm, can effectively anticipate TC landfall and recurvature, setting the groundwork for future track forecast modelling.

Wildfire Prediction

Wildland fires are caused by a combination of elements such as ignition source, fuel composition, weather, and topography, making them a complex process. Fire activity can be studied at several scales, from ignition and combustion in centimetres over seconds to spread and growth over minutes to days in meters and kilometres. Fire frequency can be measured over years to millennia on regional, continental, and planetary dimensions. Combustion and fire behaviour are physical phenomena that can be described in mechanistic (physics-based) models at small scales. This research uses ML models to predict fire growth based on a dataset of wind speed, humidity, location, temperature, and other variables. The prediction yields the burning area in hectares (Jain et al., 2020). Describes the use of machine learning to anticipate wildfires using data from vegetation indices, weather conditions, and human activities. The part also investigates the use of clustering algorithms to identify high-risk locations.

Machine Learning for Disaster Prevention and Mitigation

Disasters in India have been exacerbated by both artificial and natural factors. Natural disasters have been increasingly vulnerable to climate change adaptation. The country's infrastructure has deteriorated due to climate-related natural disasters including as earthquakes, floods, droughts, and famines, negatively impacting lives and property. Man-made disasters such as gas explosions, oil fires, and train accidents have caused significant devastation in both hilly and plain sections of the country (Usman et al., 2014). The development of Natural Disaster Management (NDM) has been underway for about three decades. However, the approach developed a decade ago has been more proactive in carrying out on-site actions. Machine learning is essential for catastrophe prevention and mitigation, in addition to prediction. This section investigates how ML models are used to create methods to mitigate the effects of natural disasters. The topics discussed include:

Vulnerability Assessment and Disaster Management

Risk assessment is crucial for catastrophe risk reduction, guiding risk management strategies and mitigation methods. Quantitative risk analysis and GIS-based risk assessments can identify high-risk locations and devise targeted mitigation measures (Labis, 2024) . Assessing hazards and zoning areas based on vulnerability allows authorities to develop specific disaster mitigation methods. The complexity and diversity of natural and man-made hazards make effectively assessing and zoning disaster risks problematic. Traditional risk assessment approaches rely on historical data and expert judgement, which may limit their reach and fail to account for the dynamic nature of catastrophic threats. Machine learning (ML) can improve risk assessment by analysing complex datasets to identify high-risk areas and vulnerabilities. Integrating ML algorithms into risk assessment techniques allows for more detailed risk maps and zoning policies, enabling focused mitigation efforts.

Resource Allocation

Fires, explosions, terrorist attacks, criminal acts, and traffic accidents are some examples of social phenomena that can cause urban emergencies. People in the area need to be removed from the dangerous area as soon as possible because they need medical attention. Predicting how many individuals would be impacted and killed in the event of a disaster is the main objective of urban emergency demand forecasting. Shelter placement, evacuation route, and resource allocation are just a few examples of how this data might improve disaster preparedness and response. The casualty count is comparable to the resident population at a makeshift shelter for evacuees (Huang et al., 2021). People who have been forced from their homes due to natural catastrophes might find a place to stay in permanent shelters. Consider important effect factors including event kind, population distribution, geographic location, and land utilization to estimate the number of victims assigned to each shelter in the affected area.

Real Time Monitoring

One of the most important uses of machine learning (ML) in catastrophe prevention and prediction is real-time monitoring. Quick actions can be taken to lessen the effect of natural catastrophes when environmental conditions can be monitored in real-time. Machine learning models have the ability to activate early warning systems when potential dangers are detected through real-time monitoring. Authorities, emergency services, and the public can all benefit from these alerts, which provide crucial time for preparation and response to potential dangers. Time spent finding and responding to potential crises is reduced through the automation of monitoring and alert systems using ML. Having this skill to respond quickly is essential for reducing the impact of natural disasters and potentially saving lives and property.

Damage Prediction

To better prepare for and react to natural disasters, governments, organizations, and communities can use damage prediction, an essential application of machine learning (ML) in disaster management. Many different types of natural disasters, such as hurricanes, floods, and earthquakes of varying intensities, can be modeled using ML algorithms. More accurate predictions of possible harm in different scenarios are made feasible by these simulations, which help to understand the spectrum of possible outcomes. By calculating the potential damage to important infrastructure, homes, farms, and other sectors, ML models may also estimate the monetary effect of a disaster. Policymakers can use this data to better allocate resources and plan for economic recovery. Based on their specific location and the predicted path or severity of a disaster, individuals and companies can obtain tailored risk assessments. Better local preparedness, including the ability to secure buildings, evacuate high-risk areas, and safeguard valuable assets, is made possible by this. In conclusion, a proactive strategy for disaster management is made possible by damage prediction utilizing machine learning. Governments and organizations can benefit from machine learning's disaster impact projection capabilities to better prepare for and respond to potential disasters, protecting lives, property, and economies.

CHALLENGES

Natural disasters, including earthquakes, fires, and flooding, present unique challenges beyond the typical ones.

General Challenge

This section discusses broad issues in disaster management systems, followed by short descriptions of specific challenges within each category.

- Allocating limited resources, such as manpower, equipment, and financing, can be challenging in disaster management, especially during many events.
- Effective communication and coordination among stakeholders is crucial to overcome challenges including complicated networks, language hurdles, and coordination concerns between agencies and organisations.
- During disasters, inadequate data, misinformation, and communication breakdowns can make it difficult to gather, analyse, and disseminate accurate information in a timely manner.
- Promoting community resilience can be challenging due to cultural differences, distrust, lack of information, and limited resources.
- Disaster management systems must adapt to changing disaster features and repercussions (Khan et al., 2023), such as those caused by climate change and urbanisation.
- Overcoming problems in collaborating across disciplines with different terminologies, methodologies, and priorities, such as emergency management, engineering, social sciences, and public health, can be challenging.
- Beyond the immediate response phase, issues include ensuring sustainable and equitable recovery, addressing vulnerabilities, and incorporating disaster risk reduction into development plans.

Flood Prediction Challenge

Flood management systems confront the following challenges:

- Improve flood-related areas such as model-driven, machine learning, and GIS to improve model speed and interpretability.
- Inadequate information on the processing time involved with flood modelling, limiting the evaluation of model applicability in disaster management.
- Limited development of feature selection strategies to improve decision support in flood management.
- Few research have compared the computing efficiency of various flood management simulation platforms.

Challenges in Wildfire

Challenges in wildfire management include

- lack of comprehensive datasets for training models, which may not adequately represent intense wildfire scenarios.
- Improved landscape dynamics data collection is necessary to characterise geographic patterns and changes in wildfire management.
- Addressing model overfitting in wildfire forecast and management.
- Integrating monitoring, learning, and decision-making in wildfire control.
- Creating comprehensive models for effective wildfire management across multiple stages.

Earthquake Prediction Challenges

Earthquake management systems have special obstacles, in addition to general ones.

- Limited research on global earthquake prediction models and their potential for generalisation across areas and datasets.
- Earthquake protection is not fully integrated into multi-functional landscape management planning, despite the importance of ecosystem services.
- Inadequate data analysis for earthquake modelling, especially during active disasters.
- Uncertainty in earthquake predicting models might affect management decisions.

GLOBAL IMPACT

Human and monetary resources are both devastated by catastrophic events. The annual cost is $520 billion, and 26 million people are pushed into poverty as a result. Common disasters cost the world economy 313 billion USD in 2022. Different types of disasters, such as those with a long start, like drought or coastal disintegration, or those with a rapid start, such preemptive evacuations from high-risk areas, can trigger uprooting in different ways. Poverty, food insecurity, and job loss might result from this. (Bhanumurthy et al., 2014). The climate emergency is causing upheaval and hardship for those forced to flee. Climate change is affecting the entire population, but vulnerable individuals in fragile and conflict-affected countries are disproportionately impacted (Cashin et al., 2017).

Between 2000 and 2019, there were 7,348 significant disasters that claimed 1.23 million lives, affected 4.2 billion people, and resulted in approximately US 2.97 trillion in economic losses5. This is a significant increase over the past two decades(Chansa Chanda et al., 2024). The global impact of natural disasters is a growing worry, particularly in terms of job loss and dislocation. Understanding the frequency, severity, and impact of disasters is crucial for better planning and ensuring people's safety and livelihoods.

- **Role of climate change:** Climate change leads to more frequent and severe natural disasters. There are various variables contributing to this.
- **Overview of Global Surface Temperatures:** As global surface temperatures rise, the likelihood of more droughts and stronger storms is expected to increase. As additional water vapour is evaporated into the atmosphere, it serves as fuel for the formation of increasingly intense storms DMMU.
- **Sea Levels:** Higher locations are increasingly vulnerable to the destructive power of rising sea levels, as they are exposed to the erosive forces of waves and currents.

CASE STUDIES

Kochi Floods, India

Because of its inadequate drainage systems and heavy rainfall, the Indian city of Kochi is prone to frequent flooding, particularly during the monsoon season. Encroachment on bodies of water raises the probability of flooding, which in turn disturbs everyday life and causes significant damage to dwellings and infrastructure. Improvements to drainage systems, restoration of bodies of water, and the application of sustainable urban planning concepts are all potential responses to the problem of flooding (Mechler, 2016).

World Bank Support

Over the previous decade, disaster risk reduction initiatives have decreased the number of individuals harmed and killed by catastrophes, according to an analysis of World Bank financing from FY10–20. There are long-term benefits to investing in mitigation strategies. Communities are better protected, more secure, and economically prosperous when disasters are lessened (Khaspuria et al., 2024). Additionally, it encourages resilience, which helps communities get ready for future disasters. Disaster prevention should be a top priority for governments and organizations, and they should allocate resources accordingly (Mpolomoka, 2021).

FUTURE DIRECTIONS

Organizational, political, geographical, occupational, topical, and sociological barriers are just a few of the many that disasters traverse. Multiple levels of access to different kinds of data at different times are essential components of any effective disaster management system. The disaster information infrastructure comprises of three subsystems: knowledge infrastructure, interconnection infrastructure, and integration infrastructure. Data management, visualization, analysis, forecasting, and modeling are all parts of the knowledge infrastructure. All of the means of exchanging information and obtaining data are part of the interconnectivity subsystem. Included in the integration infrastructure are operating systems, standards, protocols, methods for evaluating reliability and quality, and necessary training for staff. Highlighting the importance of using evolving technology to improve results, the "Future Directions" section outlines the next steps in improving the use of machine learning (ML) for catastrophe prediction and prevention (Pachar et al., 2023). Here's a breakdown of the main points.

- **Integration of AI and IoT:** The integration of Artificial Intelligence (AI) with the Internet of Things (IoT) will allow for more accurate and timely disaster prediction. IoT devices can collect massive volumes of data from diverse sources, which AI may then use to find patterns and predict impending calamities with increased accuracy.
- **Development of Interpretable Machine Learning Models:** As ML models become increasingly complicated, it is critical to create interpretable models, which means that the decision-making processes can be easily comprehended. This transparency is critical for increasing trust in

AI-powered predictions and ensuring that stakeholders can make informed decisions in disaster circumstances.

- **Harnessing the Power of Machine Learning with Emerging Technologies:** The future of disaster prediction may also involve combining ML with cutting-edge technologies like quantum computing and blockchain. Quantum computing might significantly improve the speed and efficiency of data processing, whereas blockchain could improve the security and reliability of catastrophe prediction data, resulting in more resilient systems.
- **Federated Learning Strategies for Secure Data Sharing and Analysis:** Federated learning enables several organisations or regions to interact and exchange insights without revealing raw data, hence protecting data privacy. This method can improve the accuracy and reliability of disaster predictions by using a larger dataset while maintaining secrecy.
- **Secure Collaboration: Federated Learning for Data-Driven Insights:** Federated learning enables several organisations or regions to interact and exchange insights without revealing raw data, hence protecting data privacy. This method can improve the accuracy and reliability of disaster predictions by using a larger dataset while maintaining secrecy.

This section discusses the bright future of machine learning in disaster management, which is being driven by technology advancements and an emphasis on enhancing predictive models' interpretability, accuracy, and collaboration potential.

CONCLUSION

Many people have been hurt or killed, and a lot of money and resources have been lost due to recent natural catastrophes like storms, floods, earthquakes, and fires. In the aftermath of a disaster, people's daily lives may be irreparably altered for years to come. In order to lessen the effects of disasters, modern disaster management practices emphasize intervention, rescue, and recovery. By utilizing big data, social networks can provide vital real-time information during and after disasters. A flood of media is swiftly shared on social media during these kinds of occurrences. Reduced disaster losses may be the consequence of improved disaster response strategies made possible by the timely and precise categorization of this media data. Managers in charge of disaster coordination will have access to accurate and timely information if social media data can be recognized and categorized quickly and efficiently. By automating detection, analysis, and summarization, AI-powered evaluation of massive data sets speeds up catastrophe management. Our capacity to identify and lessen the impact of natural catastrophes might be greatly enhanced by machine learning. Early warnings, optimized resource allocation, and solutions to lessen the impact of these catastrophic events can be devised by ML models using complex algorithms and vast amounts of data. While problems remain, there are encouraging avenues for future research and development that could help communities become more resilient and lessen the impact of natural disasters. This chapter has provided a thorough introduction to machine learning's function in catastrophe prediction and prevention, including its potential advantages and upcoming difficulties.

REFERENCES

Al-Utaibi, K. A., & El-Alfy, E. S. M. (2018). Intrusion detection taxonomy and data preprocessing mechanisms. *Journal of Intelligent & Fuzzy Systems, 34*(3), 1369–1383. DOI: 10.3233/JIFS-169432

Barnes, B., Dunn, S., & Wilkinson, S. (2019). Natural hazards, disaster management and simulation: A bibliometric analysis of keyword searches. *Natural Hazards, 97*(2), 813–840. DOI: 10.1007/s11069-019-03677-2

Bhanumurthy, V., Rao, K. V., Rao, S. S., Rao, K. R. M., Chandra, P. S., Vidhyasagar, J., Diwakar, P. G., & Dadhwal, V. K. (2014). Enabling heterogenous multi-scale database for emergency service functions through geoinformation technologies. *International Archives of the Photogrammetry, Remote Sensing and Spatial Information Sciences - ISPRS Archives, XL–8*(1), 7–14. DOI: 10.5194/isprsarchives-XL-8-7-2014

Campolo, M., Andreussi, P., & Soldati, A. (1999). River flood forecasting with a neural network model. *Water Resources Research, 35*(4), 1191–1197. DOI: 10.1029/1998WR900086

Cashin, P., Mohaddes, K., & Raissi, M. (2017). Fair weather or foul? The macroeconomic effects of El Niño. *Journal of International Economics, 106*, 37–54. DOI: 10.1016/j.jinteco.2017.01.010

Chansa Chanda, T., Edwin Phiri, V., Chitondo, L., Chansa Thelma, C., & Vinandi Phiri, E. (2024). Ensuring National Food Security in Southern Africa: Challenges and Solutions. *International Journal of Novel Research in Interdisciplinary Studies, 11*(2), 1–18. DOI: 10.5281/zenodo.10785042

Chen, R., Zhang, W., & Wang, X. (2020). Machine learning in tropical cyclone forecast modeling: A review. *Atmosphere (Basel), 11*(7), 1–29. DOI: 10.3390/atmos11070676

dos Santos Rocha, R., Widera, A., van den Berg, R. P., de Albuquerque, J. P., & Helingrath, B. (2017). Improving the involvement of digital volunteers in disaster management. *IFIP Advances in Information and Communication Technology, 501*, 214–224. DOI: 10.1007/978-3-319-68486-4_17

García, S., Luengo, J., & Herrera, F. (2015). Data Preprocessing in Data Mining. In *Intelligent Systems Reference Library* (Vol. 72). DOI: 10.1007/978-3-319-10247-4

Goldenberg, S. B., Gopalakrishnan, S. G., Tallapragada, V., Quirino, T., Marks, F., Trahan, S., Zhang, X., & Atlas, R. (2015). The 2012 triply nested, high-resolution operational version of the Hurricane Weather Research and Forecasting Model (HWRF): Track and intensity forecast verifications. *Weather and Forecasting, 30*(3), 710–729. DOI: 10.1175/WAF-D-14-00098.1

Granell, C., & Ostermann, F. O. (2016). Beyond data collection: Objectives and methods of research using VGI and geo-social media for disaster management. *Computers, Environment and Urban Systems, 59*, 231–243. DOI: 10.1016/j.compenvurbsys.2016.01.006

Hassini, A., & Belbachir, A. H. (2014). Thermal method of remote sensing for prediction and monitoring earthquake. *2014 1st International Conference on Information and Communication Technologies for Disaster Management, ICT-DM 2014*. DOI: 10.1109/ICT-DM.2014.6917790

Huang, D., Wang, S., & Liu, Z. (2021). A systematic review of prediction methods for emergency management. *International Journal of Disaster Risk Reduction*, *62*, 102412. Advance online publication. DOI: 10.1016/j.ijdrr.2021.102412

Jain, P., Coogan, S. C. P., Subramanian, S. G., Crowley, M., Taylor, S., & Flannigan, M. D. (2020). A review of machine learning applications in wildfire science and management. *Environmental Reviews*, *28*(4), 478–505. DOI: 10.1139/er-2020-0019

Khan, S. M., Shafi, I., Butt, W. H., Diez, I. de la T., Flores, M. A. L., Galán, J. C., & Ashraf, I. (2023). A Systematic Review of Disaster Management Systems: Approaches, Challenges, and Future Directions. *Land (Basel)*, *12*(8), 1–37. DOI: 10.3390/land12081514

Khaspuria, G., Ranjan, A., Sahil, , Soni, P., & Dadhich, K. (2024). Natural Disaster Mitigation Strategies: A Comprehensive Review. *Journal of Scientific Research and Reports*, *30*(8), 20–34. DOI: 10.9734/jsrr/2024/v30i82221

Kokar, M. M., & Reveliotis, S. A. (1993). Reinforcement learning: Architectures and algorithms. *International Journal of Intelligent Systems*, *8*(8), 875–894. DOI: 10.1002/int.4550080805

Kuang, W., Yuan, C., Zou, Z., Zhang, J., & Zhang, W. (2024). Autonomous Earthquake Location via Deep Reinforcement Learning. *Seismological Research Letters*, *95*(1), 367–377. DOI: 10.1785/0220230118

Labis, P. M. (2024). Machine Learning for Disaster Risk Reduction - Review and Research Directions. *Sciences, June*, 0–10.

Maier, H. R., & Dandy, G. C. (1998). The effect of internal parameters and geometry on the performance of back-propagation neural networks: An empirical study. *Environmental Modelling & Software*, *13*(2), 193–209. DOI: 10.1016/S1364-8152(98)00020-6

Mechler, R. (2016). Reviewing estimates of the economic efficiency of disaster risk management: Opportunities and limitations of using risk-based cost–benefit analysis. *Natural Hazards*, *81*(3), 2121–2147. DOI: 10.1007/s11069-016-2170-y

Mpolomoka, D. L. (2021). Using Participatory Approaches To Improve Environmental Sustainabiluity Inselected Parts of Zambia. *Milungushi University Multidisciplinart Journal, 2*(1).

Pachar, S., Dudeja, D., Batra, N., Tomar, V., Bhimavarapu, J. P., & Singh, A. K. (2023). Recent Developments in Machine Learning Predictive Analytics for Disaster Resource Allocation †. *Engineering Proceedings*, *59*(1), 1–10. DOI: 10.3390/engproc2023059019

Pradhan, B., Tehrany, M. S., & Jebur, M. N. (2016). A New Semiautomated Detection Mapping of Flood Extent from TerraSAR-X Satellite Image Using Rule-Based Classification and Taguchi Optimization Techniques. *IEEE Transactions on Geoscience and Remote Sensing*, *54*(7), 4331–4342. DOI: 10.1109/TGRS.2016.2539957

Rezaeianzadeh, M., Tabari, H., Arabi Yazdi, A., Isik, S., & Kalin, L. (2014). Flood flow forecasting using ANN, ANFIS and regression models. *Neural Computing & Applications*, *25*(1), 25–37. DOI: 10.1007/s00521-013-1443-6

Saxena, A. (2022). An Introduction to Convolutional Neural Networks. *International Journal for Research in Applied Science and Engineering Technology, 10*(12), 943–947. DOI: 10.22214/ijraset.2022.47789

Usman, A. (2017). Integrated disaster risk management in Indian environment: Prediction, prevention and preparedness. *GHTC 2017 - IEEE Global Humanitarian Technology Conference, Proceedings, 2017-Janua*, 1–6. DOI: 10.1109/GHTC.2017.8239246

Usman, A., Dutta, R., Usman, A., Azmee, F., & Divakar, B. P. (2014). Fire disaster management in trains using a new technique of water pipelines-first aid mechanism. *Procedia Engineering, 78*(Dm), 112–119. DOI: 10.1016/j.proeng.2014.07.046

Vasanthi, A., & Satish Kumar, K. (2016). Understanding Conspicuous Gravity Low Over the Koyna–Warna Seismogenic Region (Maharashtra, India) and Earthquake Nucleation: A Paradigm Shift. *Pure and Applied Geophysics, 173*(6), 1933–1948. DOI: 10.1007/s00024-015-1237-0

Yamazaki, F., Liu, W., & Yamazaki, F. (2016). Remote Sensing Technologies for Post-Earthquake Damage Assessment: a Case Study on the 2016 Kumamoto Earthquake. *ASIA Conference on Earthquake Engineering, 6ACEE*, 22–24. https://www.researchgate.net/publication/307351403

Yonaba, H., Anctil, F., & Fortin, V. (2010). Comparing Sigmoid Transfer Functions for Neural Network Multistep Ahead Streamflow Forecasting. *Journal of Hydrologic Engineering, 15*(4), 275–283. DOI: 10.1061/(ASCE)HE.1943-5584.0000188

Yuei-An, L., Kar, S. K., & Chang, L. (2010). Use of high-resolution formosat-2 satellite images for post-earthquake disaster assessment: A study following the 12 may 2008 Wenchuan Earthquake. *International Journal of Remote Sensing, 31*(13), 3355–3368. DOI: 10.1080/01431161003727655

Zadeh, M. R., Amin, S., Khalili, D., & Singh, V. P. (2010). Daily Outflow Prediction by Multi Layer Perceptron with Logistic Sigmoid and Tangent Sigmoid Activation Functions. *Water Resources Management, 24*(11), 2673–2688. DOI: 10.1007/s11269-009-9573-4

Chapter 12
Enhancing Industrial Employee Performance by Integrating Machine Learning and Human Psychology

Saurabh Chandra
https://orcid.org/0000-0003-4172-9968
Bennett University, India

R. Muthukumaran
https://orcid.org/0000-0002-3297-784X
O.P. Jindal University, India

B. Uma Maheswari
https://orcid.org/0000-0002-5063-8542
PSG College of Arts and Science, India

S. Logeswaran
KPR Institute of Engineering and Technology, India

Shweta Tiwari
https://orcid.org/0000-0002-3806-8102
IILM University, Gurgaon, India

Sampath Boopathi
https://orcid.org/0000-0002-2065-6539
Muthayammal Engineering College, India

ABSTRACT

Machine learning has the potential to improve employer/employee performance in industries by inducing human psychology in understanding these psychological factors is crucial for optimizing production and workplace satisfaction in the rapidly changing work environment. Machine learning algorithms enable organizations to analyze large data sets to identify patterns in employee performance, work engagement, and well-being, enhancing overall organizational performance and overall well-being. The integration of predictive analytics and targeted interventions in workforce management helps in informed decision-making, reducing issues like turnover and burnout, and promoting a proactive approach to employee needs, ultimately improving the overall performance of the corporation. This chapter provides practical guidance for organizations seeking an efficient, resilient, and psychologically informed working environment, backed by real case studies and empirical evidence. The main message emphasizes the need for integrating technology with human understanding for sustainable industrial growth.

DOI: 10.4018/979-8-3693-6910-4.ch012

INTRODUCTION

This is opening a whole new frontier in optimizing employer performance and work dynamics with industrial implementation of machine learning along with human psychology. With the recent spurt of rapid technological advancement and constantly changing workforce dynamics, organizations need to use their best resources of innovation in relating technology with human insight to stay competitive. This domain of technology has a lot to offer in the study and amendment of human behavior in the working place, particularly in a factory setup. It has therefore been pretty instrumental in providing answers to various intricate work-related problems, such as employee engagement and overall mental well-being, thereby fostering business development in the long term (Rathore & Rathore, 2024).

Indeed, machine learning has revolutionized the processing of information in such a way that vast amounts of information are processed in a matter of real-time; otherwise, invisible patterns start arising. Machine learning applied to human psychology will revolutionize the ability to see workforce behavior and motivations on a scale and depth unprecedented till now. Companies can now focus interventions on the heart of psychological drivers that might be able to have an impact on productivity, collaboration, and morale to improve performance, reduce turnover, and strengthen resilience in the workforce. Psychological data-tailored models of machine learning provide companies with the ability to predict trends and behavior, allowing companies to prepare and act preventively against the problems that can occur-burnout or disengagement (Fan et al., 2023). The integration of psychology with technology leads to a more empathetic approach toward management since it improves the efficiency of operations and also brings business goals in harmony with employee well-being.

Probably, the most important advantages offered by interdisciplinary techniques are the better ability to make more informed and qualified decisions. Interdisciplinary techniques usually let management make decisions regarding the workforce and choices for performance enhancement based on intuition, experience, or limited data that traditional management relies on. Data-driven information from algorithms of machine learning supplants bias and therefore provides an objective method of processing workplace trends. For example, through the analysis of the stress, job satisfaction, and productivity metrics by machine learning, some recommendations could therefore be derived on how to make a workplace supportive (Norrman, 2020). It may point out motivating factors for employees; the better understanding of which helps in pointing out areas of the workplace that require correction-avoiding issues from flaring up into problems and thus avoiding productivity loss and ensuring job satisfaction.

Machine learning is also now making it possible to personalize the employee experience in ways that were once considered impossible. Just like retail or media recommendation algorithms are set to the individual's taste, machine learning can be utilized to personalize experiences and professional development opportunities at work. For example, by taking into account the psychological factors of strengths, weaknesses, and aspirations at a career level, the organization could tailor appropriate training programs, projects aligned to their interests, and constructive feedback accordingly (Rodriguez et al., 2021). Beyond personal performance, personalized development plans boost engagement and loyalty since the special needs and career goals of staff are recognized and valued. It is through personalization that a culture for inclusions building becomes a determinant factor in talent retention in a competitive labor market.

One other critical application of machine learning in industrial psychology is through evaluating sources of stress and possible means to mitigate them. As understanding on mental health at work continues to develop, it's the responsibility of organizations to be proactive about employee's psychological well-being at work. In this sense, it can then be applied in monitoring and tracking manifestations of stress, such as over-time, low productivity, or sudden changes in behavior which may lead to burnout or disengagement (Fukui et al., 2023). Putting this knowledge together with psychological models then creates for businesses the opportunity to craft supportive policies and practices: flexible work arrangements, for example, or access to mental health care. Such an approach to mental health would work to benefit not just employees but give practical benefits for employers in terms of decreased absenteeism and health care costs, while increasing the general level of productivity (Qamar, 2022).

The third area where the value of putting machine learning into psychology is clearly seen, namely workforce management. Companies can predict anything-from expected turnover rates and team performance under various circumstances to eventual outcomes-through predictive analytics, making companies better armed in terms of resource allocations and changing strategies around expected changes. For example, when predictive models reflect a lot of turnover likely to happen in a certain department, management can stage interventions long before the occurrence of turnover-for example, through focused engagement activities or even changing workloads (Hall, 2021). Predictive gives the organization the ability to be responsive and flexible with change, thereby identifying problems before they can affect performance. In that sense, a company makes data-driven changes in ensuring performance, thus building a healthy and stable environment where success is possible in the long run (Qamar, 2022).

Although infusion of machine learning and psychology does raise some benefits there are a few challenges that come from this. For example, psychological data analysis involves management of confidential information related to employees. It turns out to be a messy knot of laws and ethical issues to address when firms collect, process, and store all this information within the boundaries drawn by laws like GDPR. That would diminish trust if there were not transparent data governance and informed consent. Incomes with the advent of machine learning is a need for explainable AI, an advantage to all stakeholders, who are made to understand how the decisions are reached. Explainable AI of prime importance is instilling trust between the employees and employers and the ethical approach in how AI should drive management (Pan, 2021).

The question is how it interprets the insights delivered by a machine that can only sometimes offer trends and patterns but can never fully give the context of the human emotion and social dynamic. Human psychology, in its very nature, is tough, ridden with complexity from so many variables that cannot be quite represented in data alone. For example, sometimes to a large extent, changes in the personal lives of employees or economic conditions affect the behavior of employees and their performance, but exactly, cannot be determined. This is why machine learning in workforce management should be viewed as augmenting human judgment rather than a replacement for it. Basically, machine learning in industrial psychology will thrive if the data scientists collaborate with psychologists and HR personnel to ensure the insights are applied with a sensitivity of human behavior complexities (Liu et al., 2023).

Integration of machine learning with human psychology presents the excellent tool package towards elevation of the performance level of industrial employers. Leaning on machine learning to understand and predict the behavior of employees will enable the organizations to build more individualized, supportive working environments; make better decisions; and take proactive efforts to begin looking at strategies for mental health and well-being (Chujfi & Meinel, 2019). But that would depend on very careful consideration over concerns related to privacy and ethics of practice, and limits in the scope of insights revealed by algorithms. Organizations embracing this interdisciplinary approach will, thus, navigate complexities toward a modern form of workforce management that is innovative, inclusive, and resilient. Where companies utilize machine learning responsibly in industrial psychology, they can achieve larger performance gains in better ways and yet remain positive contributors to the life of employees, opening up a sustainable pathway forward within an increasingly digital environment (Bastani et al., 2021).

Scope

This chapter scopes out an intensive analysis into how machine learning might be applicable in industrial psychology to advance employee experience and organizational outcomes. Included in such coverage are topics such as psychological drivers of work behavior, predictive analytics, turnover, and performance management, alongside the development of tailormade interventions aimed at engagement and well-being. The current chapter also discusses ethical considerations such as data privacy and algorithmic bias while at the same time being all aware of the use of technology together with human insight. It hence investigates these topics and then elaborates on this relationship, which is dynamically evolving between machine learning and industrial psychology, towards the discovery of innovation and growth through organizational change at the work environment (Chandana et al., 2024).

Objectives

This chapter delves into the nexus of machine learning and industrial psychology in efforts at improving employee work performance and well-being. It is useful for both predictive analytics on workforce management and personalized interventions and for ethical considerations in issues of AI. It will also spot future trends and opportunities for innovation in this field of discourse.

BACKGROUND: THE PSYCHOLOGICAL DRIVERS OF INDUSTRIAL PERFORMANCE

Understanding the psychological impetus of performance is therefore critical for an effective and highly motivated workforce in the modern industrial field. Psychological factors influence work behavior to the greatest degree since they impact the way employees perceive and interact with their jobs, colleagues, and organizational goals. In return, better insight into employee engagement, motivation, morale, and emotional intelligence may develop a more productive and more harmonious workplace environment toward improved overall performance (Bormah, 2023). Figure 1 illustrates various Psychological Factors drivers of industrial performance.

Figure 1. Various psychological factors drivers of industrial performance

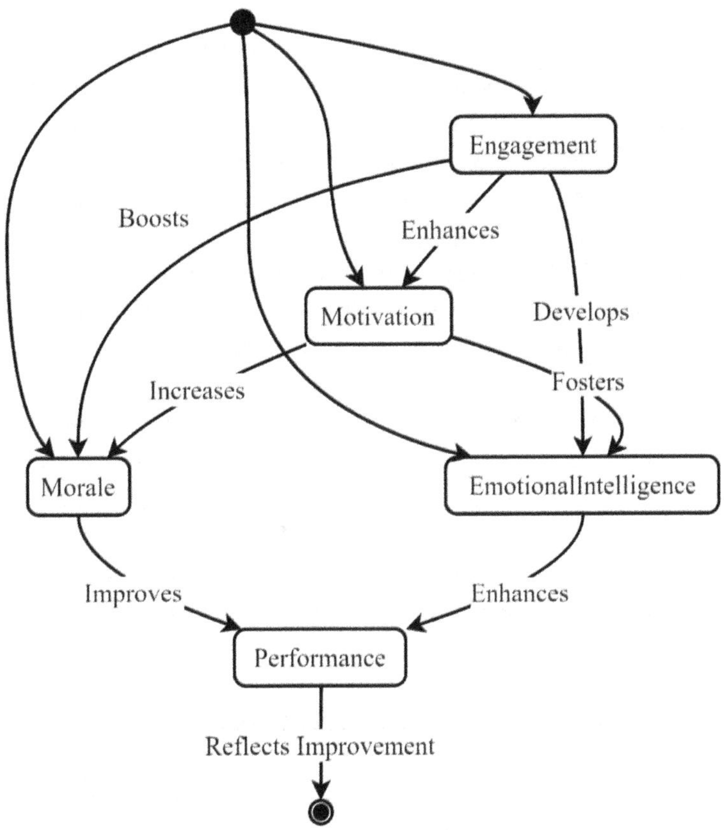

Psychological Factors Involved in Workplace Behavior

There are many psychological factors, including individual perceptions, attitudes, and emotions, that manage the workspace environment. All these elements can influence work outcomes in terms of productivity, creativity, and collaboration. Of all the crucial psychological factors is perception, which works as how the workers perceive and realize their working environment. For example, a good perception regarding the culture of the company will lead to more job satisfaction, but negative perceptions might foster disengagement and frustration (Gon et al., 2019).

The attitudes towards work and the organization play a crucial role also. Workers who are optimistic generally possess commitment and resilience, which links with high performance levels. Negative attitudes hinder teamwork and innovation while creating a poisonous atmosphere that can affect the team dynamics. The emotional well-being factor is also a very significant psychological driver. Workers who believe that they are extremely stressed or pressured may be unable to focus on work and, consequently, perform well enough to match their expectation levels. As such, organizations must be aware of such psychological factors and must make available suitable strategies that help the employees manage their mental health.

Employee Engagement, Motivation, and Morale

Employee engagement is one factor directly related to industrial performance. Engaged employees are those who fully invest in their work and make dynamic contributions to the organization's success. A study showed that when this engagement level is at its height, productivity will be much higher and absenteeism reduced along with improvements in customer satisfaction. It is characterized by an environment in which employees feel valued and recognized for their efforts. This situation can include regular feedback opportunities to develop and grow along with clear communication around company goals and values (Gon et al., 2019).

Another very influential variable on performance is motivation. There are two kinds of it: intrinsic and extrinsic motivation. Intrinsic motivation is within the individual, many times from his personal needs-satisfaction, curiosity, or interest of the activities themselves. Extrinsic is dictated by outside influences: salary, benefits, promotions, and recognition. It's often those organizations that are able to keep a balance between the two where employees report higher fulfillment and loyalty.

In addition, morale represents a general state of emotional and psychological condition of staffs present in an organization. The workplace is more effective when morale is high that translates into employees who are more energetic at work, enjoy their job and productive interaction with others. Factors affecting morale comprise the culture of the working place, style of leadership, and existence of good relation among team members. Organizations promoting initiatives such as teaming or wellness programs can significantly enhance overall industrial performance.

Role of Emotional Intelligence in the Industrial Activity

The definition of emotional intelligence makes up the ability to understand and recognize emotions in oneself and others and to manage those emotions. This significant application of emotional intelligence at the workplace leads the way to moderation in interactions between coworkers, communication, conflict resolution, and teamwork. Adaptation is vital in industrial settings, and this requires cooperation. EI can determine the performance of the individual or team (Manzini et al., 2024).

The most inspiring and motivated leaders are those with strong emotional intelligence. They can easily form a culture of trust and respect, lead with emotional intelligence, which helps them notice signs of disengagement or stress among their workers and take immediate action so that the morale and overall performance do not decline. Emotionally intelligent individuals will more often accept feedback and adapt themselves as required to ensure an organization's continued improvement.

As reported by several studies, the better the organization is placed for focusing on emotional intelligence in hiring and training, the better the employees are, along with performance, plus employee satisfaction. For instance, organizations that provide training about emotional intelligence skills will record achievements in terms of effective leadership and integration within the teams. Emotionally, emotional intelligence is increasingly being regarded as an integral part of organizational success, thereby calling for a 360° approach to employee development (Mohanty et al., 2023; Senthil et al., 2023).

Therefore, understanding the psychological drivers of industrial performance is the crucial mandate of organizations if they want to optimize engagement, motivation, and morale. Psychological factors can influence workplace behavior, and the recognition of such influences in the workplace can create targeted strategies for enhanced performance outcomes. Furthermore, creating an emotionally intelligent workforce leads not only to improved interpersonal relationships but also to a collaborative and adaptive culture. Therefore, it will become very important for such organizations navigating the complexities of modern industrial landscapes to focus on such psychological dimensions in order to obtain sustainable growth and success.

MACHINE LEARNING TECHNIQUES FOR ANALYZING WORKFORCE DATA

These more advanced analytical methods assist firms to discover insights about improving decisions and turning overall organizational performance. Two primary areas of study are predictive analytics, as well as behavior forecasting along with the respective data gathering, processing, and ethics. This chapter explores these areas that have shown critical in how to effectively tap into workforce potential by way of using workforce data (Ullal et al., 2022). Figure 2 depicts various components in machine learning techniques used for analyzing workforce data.

Figure 2. Various components in machine learning techniques for analyzing workforce data

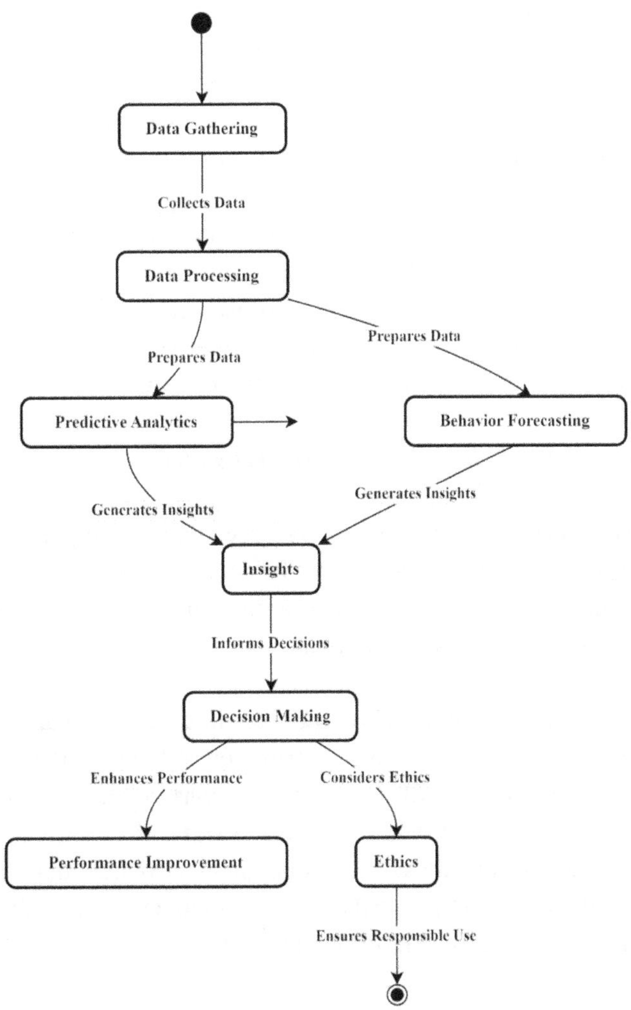

Predictive Analytics and Behavior Forecasting

Predictive analytics is an application of statistical algorithms and machine learning models used to forecast the probability or likelihood of certain events happening in the future based on past data. In workforce management, predictive analytics may be crucial in forecasting a wide range of employee-based metrics, such as turnover, productivity, and performance. It may help the organizations check those potential issues and improve their human resource strategy in advance by making use of trends established in employees' data (Jaisawal, 2022).

For example, by determining an employee's propensity to leave through predictive models, organizations can recognize at-risk employees by blending historical turnover data with scores for employee engagement, demographic data, and performance reviews. Arming such insight allows HR teams to work on retention efforts in the precise areas: development plans aligned to the individual or increased engagement initiatives, therefore optimizing a lower rate of turnover and also reduced recruitment and training expenses.

Behaviour forecasting changes the play in predictive analytics from what employees would have done given the conditions to what they would do next based on their behavior. This is where machine learning algorithms are applied on the behavioral data: work habits, communication patterns, and collaboration styles. For instance, an organization may employ the clustering technique to categorize their workers in terms of behavior; one may find high performing teams and those that might need some further support. This in turn enables organizations to recognize and identify these behavioral patterns, which may then help them devise interventions and then couple those with resources to enhance a more cohesive, productive work environment.

Apart from behavior prediction, the organization can use forecasting to predict employee performance results. Using historical performance, organizations can find KPIs and formulate a model in forecasting future performance related to many factors including completion of training, involvement in projects, or peer evaluations. It could be highly valuable information when making promotion decisions, allowing people onto projects, or other performance review recommendations.

Data Collection, Processing, Ethical Considerations

Effective applications of machine learning to workforce analytics require sound methodologies for data gathering and processing. Relevant data needs to be sourced from employee surveys, performance evaluations, attendance records, and even social media, to name a few. This broad aggregate can bring about comprehensive analysis on the dynamics of an organization's workforce (Kannimuthu & Premalatha, 2024).

However, at the same time, data should also be collected in an ethical manner and with utmost transparency. Therefore, employees must know what information is being gathered, how it is going to be applied, and ways in which their privacy will be protected. In particular, ethical considerations of utmost importance relate to sensitive information, such as mental health assessments or personal demographics. Organizational data governance policies must be transparent; in particular, that involve employee consent as well as the protection of their information so that the collected data is also observant of the regulation practices such organization uses, be it the GDPR or the CCPA.

Once data has been collected, it is then processed and prepared for analysis. In processing it, errors within the data are cleaned, meaning all records being the same in form so easily read and analyzed, and no missing values exist. Thus, the data should be preprocessed with techniques like normalization and feature engineering for data quality improvement. Feature engineering is actually the process of manufacturing new variables from the existing data to make prediction capabilities better. Hence, an organization may extract its features based on completion rates of employee engagement scores with respect to project outcomes or training. In the final stage, any workforce data can be passed through a number of machine learning algorithms for analysis. Common techniques include regression analysis, decision trees, random forests, and neural networks. Each has benefits, so an organization can select the one suited to the objectives that the analyst wishes to achieve with their analysis. For example, regression models are very useful in cases when continuous outcome is predicted, and decision trees are very helpful to visualize complex processes of decision.

Machine learning provides an exciting potential space for workforce analysis, but organizations will need to stay awake to the ethical issues attached to their models. An algorithm with bias can lead to a kind of unfair treatment of employees if data used in modeling has been slightly biased towards historical discrimination. For example, if hiring data history is against certain demographic groups, such biases may perpetuated through future hiring decisions through machine learning predictive models. To mitigate this, organizations should be conducting regular audits of their algorithms and datasets to determine where potential biases may exist and correct them to make determinations of outcomes more fair.

Workforce data in their approach is changing through machine learning techniques: the predictive analytics and behavior forecasting. It allows organizations through these advanced methodologies to unlock excellent insights into employee behavior and performance, which in turn improves decision-making processes and operational efficiencies. Efforts toward these needs, however, will highly depend on ethical data collection and processing practices. It is only making transparency, consent, and fairness an everyday practice that enables organizations to create a data-driven culture that respects employee privacy yet harnesses analytics power. As businesses continue growing in this ever-increasing complexity of the landscape, the effort in integrating machine learning into workforce analytics will catalyze sustainable growth and performance.

PERSONALIZED INTERVENTIONS: ENHANCING EMPLOYEE EXPERIENCES

Enhancing employee experiences remains paramount in a highly competitive and dynamic business environment. Of late, the quest for engagement, satisfaction, and performance has led organizations to seek more innovative ways, such as personal interventions driven by machine learning, to create a culture that is truly empowering and allows for collective success through developing work environments, employee development programs, and performance optimization strategies. This chapter introduces how personalization of work environments and optimum employee development leads to a better deal in general employee experience (Sharma & Dhingra, 2024). Figure 1 depicts personalized interventions aimed at improving employee experiences.

Figure 3. Personalized interventions for enhancing employee experiences

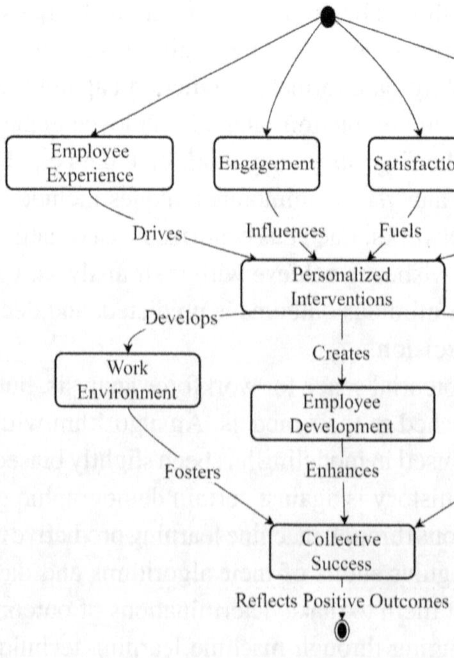

Machine Learning for Customizing Work Environments

It is now experiencing rapid change where employees expect to have a personal experience answering to their distinct needs and preferences. Applications of machine learning in tailoring work environments will be very considerable as they can indicate the behavior of employees, their preferred practices, and the productivity patterns during or outside work hours. Organizations can then curate workspaces and practices tailored to meet various employee needs through vast data analysis (Kannimuthu & Premalatha, 2024; Sharma & Dhingra, 2024).

Tailored Workspace Design: By analyzing data generated by employee surveys, usage of the workspace, and productivity metrics, machine learning algorithms can identify the optimal configurations for a workspace. For instance, it can determine the kind of workspace—open offices, collaborative spaces, or quiet zones—that yields efficiency combined with satisfaction. By clustering, companies can identify employees with preferences such as collaboration, focus, or flexibility to facilitate development workspaces that enhance comfort while supporting productivity.

Integration of Smart Technology: Introducing smart technologies at the workplace can bring in vast improvements in employee experiences. Further features of machine learning might be able to push the developments of smart office solutions toward adaptation according to employees' preferences. Examples include smart lighting and climate control systems that adjust themselves according to employee preferences or the occupancy level of employees. Analyzing the data about employees' habits and preferences, the organization can thus craft environments that improve comfort and well-being and consequently bring increased focus and productivity.

Flexible Work Arrangements: The character of work has become so dynamic that flexibility at the workplace has become the need of the hour. Machine learning can inspire flexible work arrangements that might cater to the personal preferences and circumstances of employees. Organizations could thereby build models in light of the analysis of employee performance, work-life balance, and levels of engagement that may be used to optimize the work schedules and remote work policies. For example, predictive analytics might be used to predict which employees prefer working from home or with others, and consequently, an organization may tailor its approach to workforce management more specifically.

Employee Development and Performance Optimization

Personalized interventions extend beyond the physical workplace and also include employee development and performance optimization (Li et al., 2023; Pereira, 2024). As organizational learning opportunities can be enhanced, as well as development programs focusing specifically on the individual's needs, using machine learning methods:

Learning Paths: It could automatically be used to transform the old traditional program of training into a dynamic, personalized experience for learning. With employee performance data analysis and skills assessment, besides learning preferences, organizations can generate customized learning paths in line with individual career goals and organizational needs. For instance, if an employee has a great aptitude for data analysis, the learning platform should recommend advanced courses in data science, but another one may benefit from leadership training. This focused strategy not only develops the skills of the employees but also creates an ownership for skill development by the employee.

Real-Time Feedback and Coaching: Feedback is very important for the growth of employees and results in the optimization of performance. Machine learning algorithms can analyze performance data in real-time, enabling organizations to provide immediate feedback and coaching. For instance, it could be performance management systems powered by machine learning, which track employee progression against KPIs and deliver personal insights and recommendations on employee improvement. This allows the employee to make changes and improve on work, leading to continuous development of a culture.

Performance Predictive Analytics: Predictive analytics enables organisations to determine their high potentials and discover disengagement and underperformance propensity. The machine learning model does this based on historical data related to performance and scores on engagement, among others, which enables it to make more accurate predictions on potential performance results in the future. With such knowledge, the organisations intervene appropriately to encourage support through mentoring programs or training that positively impacts the overall performance. An organization provides a more helpful and effective working environment by identifying problems before they become even more difficult to address.

Employee Welfare Initiatives: The core experience enhancer in an organization is wellness. Based on the needs of the employees, machine learning can influence wellness programs across organizations. Organizations can develop customized wellness initiatives for the employees by analyzing the data of health metrics regarding an employee's health, engagement levels, and stress indicators. For example, if the data shows that certain teams are under greater stress during peak project times, those organizations can provide stress management workshops or mindfulness programs for those employees.

Personalized interventions, propelled by machine learning, are changing the way organizations better employee experiences. As organizations make work environments fit the individual's preferences, so productivity and satisfaction culture will be cultivated. At the same time, tailoring the employee's development program and performance optimization strategy ensures that their employees feel supported in their growth and aspirations. In an era where employee experience is the new bellwether, the only means by which organizations would truly unlock the real potential of machine learning to craft individualized interventions for employees would not be a choice but a necessity for organizations that are desirous of sustainable success in addition to competitive advantage in the business world. As businesses continue to evolve, the needs of their workforce will be the ultimate priority to ensure keeping the best talent while driving the overall performance of the organization.

Figure 4. Application of machine learning in addressing workplace mental health

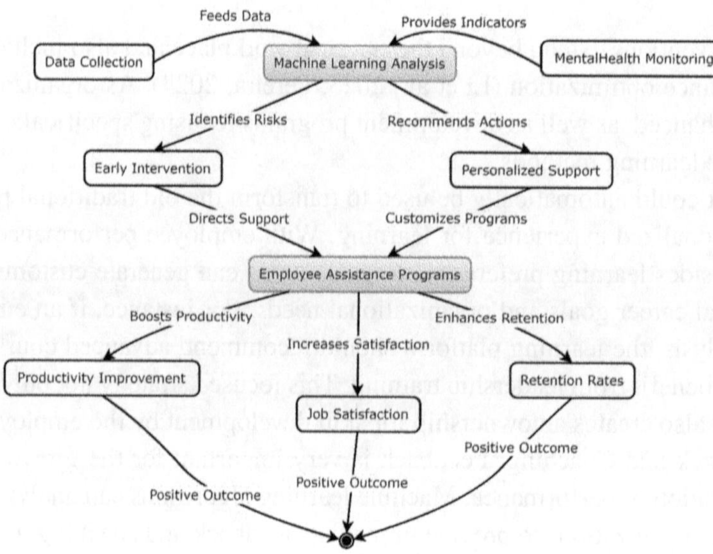

ADDRESSING WORKPLACE MENTAL HEALTH WITH MACHINE LEARNING

Mental health is increasingly becoming paramount in the workplace as research shows that mental health issues significantly impede productivity, job satisfaction, and retention among employees. Organizations are now understanding that these problems must be tackled proactively, and machine learning has emerged as one of the most potent tools for enhancing mental health initiatives at the workplace (Fan et al., 2023; Norrman, 2020; Rodriguez et al., 2021). Organizations can create a healthier workplace putting great importance on employee wellbeing through the detection and prediction of mental health risks by the implementation of proactive measurements in reducing burnout and stress and through the integration of support and resources into the mental health of employees. The application of machine learning in addressing workplace mental health is depicted in Figure 4.

Detection and Prediction of Mental Health Risks

One of the most ubiquitous applications of machine learning within the realm of mental health is in relation to employee mental health risks detection and prediction. Using various sources, for example, about data, organizations can track patterns and signals indicating mental health challenges.

Employee Data Analysis: Algorithms of machine learning enable organizations to process all that massive employee information, from absenteeism rates and productivity metrics to engagement scores. Suddenly or rapidly rising rates of absenteeism or rapidly falling performance could thus indicate early mental health issues. Predictive analytics can enable an organization to build models that flag employees at risk based on their behavior, thus allowing HR departments to diagnose problems before a crisis develops.

Sentiment Analysis: NLP is the subset of machine learning. It can be used to send employees' communications for sentiment analysis. Analyzing electronic mail, chat messages, or responses in surveys will let an organization know the general attitude of its workforce. Negative sentiment trends or language indicating distress may serve as an early indicator of potential mental health issues. Generally, performing sentiment analysis regularly lets the organizations stay attuned to the employees' well-being and identify emerging mental health issues.

Wearable Technology and Mobile Apps: Wearables and mental health applications have become very trendy lately, which has led to an overflow of huge amounts of data that machine learning algorithms can process. Wearables would monitor physiological signals that are highly correlated with mental health- heart rate variability, sleep patterns, and physical activity. Organizations would be able to identify at-risk employees for stress or burnout by analyzing this data. It would also allow the frequency of self-assessments for reporting employees' mental health status. Machine algorithms can aggregate this data to offer insights into the general mental health landscape in the organization.

Proactive Measures in Prevention of Burnouts and Stress

To address mental health at work, proactive measures are necessary in reducing the risks associated with the occurrence of burnouts and stress. Through machine learning, it's possible to inform targeted interventions aimed at enhancing employee well-being (Kannimuthu & Premalatha, 2024; Pereira, 2024).

Tailor-made Wellness Programs: Organizations can use machine learning to build well-being programs for particular employee needs. Mining responses from employees, performance metrics, and engagement scores will help an algorithm determine which kind of stress certain employee segments experience. On the biggest workplace pressure type, an employee might need to have some workshop on the management of stress, but another needs flexible working arrangements not to lose the balance in personal life. Customized well-being will ensure employees receive support in thriving at work.

Workload Balancing: Predictive analytics also allows organizations not to burn out employees. Based on a project timeline, the performance of a team, and individual workloads, machine learning algorithms provide suggestions concerning changes in project assignments and resource allocation. This preview process not only controls employee stress levels but also prevents teams from being flooded with a workload.

Training and Development: Machine learning can enhance the training programs designed especially for mental health awareness and stress management. Using machine learning, one can analyze employee engagement with the training materials as well as the effectiveness of different programs in assessing the best strategy for training. For instance, data can show that workshops stimulate better engagement compared to lectures. Employee training programs can be so designed to enable employees to acquire stress management skills.

Provision of Mental Health Support and Resources

A healthy workplace will comprise not only the risk assessment and mitigation but also the access to all mental health resources for all employees.

AI-Powered Support Systems: Organizations can introduce AI-powered chatbots and virtual assistants, which will provide employees with immediate access to mental health support. The resources that are provided can range from comprehensive counseling on how to cope, connecting the employees with mental health professionals, and making referrals to other places. The integration of such support systems in communication channels can help organizations foster a culture of openness regarding mental health and make employees feel comfortable seeking help at any point in time.

Anonymity for Employees: Organizations may also allow employees to seek mental health support from behind the anonymity of computers, as awareness of mental health matters is growing. Machine learning may also assist in creating more effective and holistic mental health resource libraries for workers. An organization will be better placed to understand what topics and materials are in the highest demand among employees if it gathers information about employees' comments and views on mental health resources. Such information will then inform the design of training modules, workshops, and information products that will resonate with employees. It will promote employees' enhanced capacity to address mental health challenges if they have access to the right resources.

Continuous monitoring and feedback: The incorporation of support for mental health requires an ongoing feedback loop in order to constantly monitor its impact across the organization. Therefore, machine learning can be applied to analyze data from feedback surveys by employees, usage rates of mental health support, and global employee sentiment in order to measure the impact of mental health initiatives. This allows organizations to perceive their "readiness" and make necessary adjustments in strategies to stay responsive to the needs of their employees.

Addressing workplace mental health is not only a moral imperative but also a strategic necessity for organizations aiming to enhance employee well-being and productivity. With the application of machine learning in detecting and predicting mental health risks, designing proactive interventions against burnout and stress, and infusing robust mental health support systems, organizations can create a healthier work environment. As the complexities of modern work life continue to evolve, the future resilience, engagement, and overall satisfaction of employees are closely tied to the overall mental health balance achieved within an organization. Investment in mental health initiatives demonstrates an organization's commitment to employees' well-being and better positions it for success in the longer term, as it competes in this increasingly competitive playing field.

PREDICTIVE ANALYTICS FOR WORKFORCE MANAGEMENT

Predictive analytics serves as a necessary function in smart workforce management, and organizations use the information to make the right decisions that enhance employee performances, decrease turnover rates, and optimize resources. Organizations can use techniques such as machine learning to uncover insights about employee behavior and performance trends and work towards more strategic workforce planning. This chapter discusses three strategic areas of predictive analytics in workforce management: using machine learning to predict turnover, enhancement of performance of the team through data-driven decisions, and optimal resource distribution for strategic planning (Manzini et al., 2024; Najafi-Zangeneh et al., 2021). Figure 5 showcases the efficacy of predictive analytics, a combination of data collection, machine learning analysis, and workforce optimization, in enhancing employee performance, reducing turnover, and optimizing resources.

Figure 5. Components - predictive analytics in workforce management

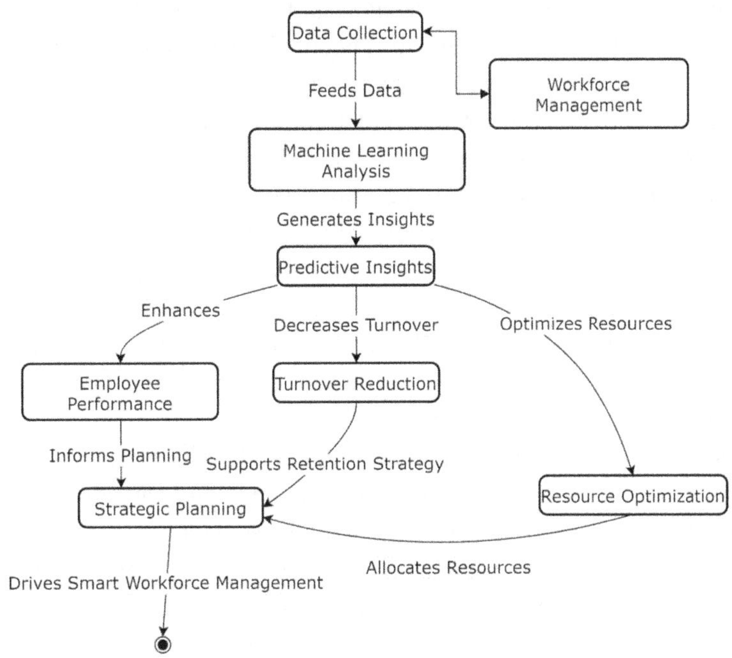

Using Machine Learning for Turnover Prediction

This turnover may come as a significant challenge for the organizations involved, bringing extra recruitment costs and loss of valuable knowledge embedded in the organization and low morale among the other employees remaining at work. Predictive analytics, especially through applying machine learning algorithms, can be of great help to the organizations in predicting and preventing such risks of turnover (Hall, 2021; Qamar, 2022).

Identification of Risk Factor: Machine learning algorithms can analyze the historical employee data across demographics, job performance, engagement levels, and tenure to check the presence of any key factors associated with turnover. Organizations can build a model using classification techniques like logistic regression or a decision tree to predict an individual's likelihood of turnover-for instance, where lack of career advancement opportunity, scoring low on engagement, or having frequent absences are definitely telling predictors of the potential for turnover.

Predictive Models and Insights: With predictive models, the organization will be able to determine which employees are most at risk for leaving. Models can give a score or a probability of chances that signals or indicates how likely they would leave. Thus, the team in the HR will concentrate their intervention efforts on such people and tailor retention strategies on individuals. For instance, high-risk employees could be approached by personalized career development plans or engagement initiatives so as to minimize the rate of change.

Continuous Monitoring and Enhancement: Predictive analytics is not an exercise performed one-time. The organization will need to refresh the predictive models from time to time, using new data, to keep abreast of changes in workforce dynamics and external factors that may impact staff satisfaction. Reinforcement learning, a type of machine learning, assists in enhancing models over time to be even more predictive and realistic. It can predict turnover, but predictive analytics can make a huge difference in team performance in the most fundamental way by making data-driven decisions.

Performance Metrics and Analysis: Machine learning enables organizations to discover various performance metrics across teams. From KPIs such as productivity levels, project completion rates, collaboration frequency, organizations can pinpoint high-performing teams and understand which factors contribute to such success. For example, using cluster analysis techniques, organizations can group their different teams according to established performance characteristics and then replicate successful best practices across other teams.

Targeted Training and Development: Furthermore, predictive analytics can be used to provide inputs for appropriate target training and development initiatives. Data from team performance may allow the determination of skill gaps and what areas need training. For instance, if a team regularly misses project deadlines, possibly analysis may reveal that the lack of skills in time management or prioritization has resulted in such a situation. The organization can then develop a customized training program for particular needs, thus improving team performance and job satisfaction.

Collaboration Work Space: This could create collaboration at work through predictive analytics. It focuses on communication patterns, how often a team collaborates, and the conflicts that happen within a team. An organization can pinpoint areas requiring intervention based on this analysis. For example, when communication is low among a certain group of team members, predictive models can alert the manager, thus enabling team-building activities or training communication. Such an approach will be able to enhance teamwork and cooperation in practice.

Resource Assignment and Strategic Planning

Optimal performance in an organization can only be achieved by the effective allocation of resources. It enables organizations to strategically assign resources because of predictive analytics by applying information-driven insights (Rathore & Rathore, 2024; Sharma & Dhingra, 2024).

Prediction of required workforces: Organizations will be able to predict and plan ahead in terms of when and where they may need extra staff or when they might have a surplus of employees, thus enabling HR departments to anticipate recruitment and training needs ahead of time and employ the right personnel at the right time through time-series analysis.

Predictive analytics can also be applied in the decision-making process about different levels of the allocation of organizational resources, including individual roles and an entire team. Data analysis about the workload, productivity level, and demands of projects will help organizations make the most optimal use of their resources. For example, if predictive models are indicating that a certain project requires more people to achieve a deadline, then HR can transfer these workers or hire contract employees so the project will be completed. This models of allocation are focused on dynamics and help organizations remain agile and responsive to constantly changing business conditions.

Strategic Planning and Decision-Making: Predictive analytics helps in strategic planning through data-driven insights. Workforce trends, performance metrics, and turnover risk can be obtained from it to determine organizational structure, talent management, and long-term growth strategies. For example, predictive models may indicate that one department is likely to face turnover; so management acts to rectify or solve problems at such an early stage that the department is stable and productive.

Predictive analytics changes how the workforce is managed since it presents organizations with valuable insights to enhance performance of employees, reduce turnover rates, and even appropriate resources. Applying the machine learning algorithms in turnover prediction will protect corporations from potential issues threatening a stable workforce by having an early warning or taking chances to mitigate risks related to such matters. This aspect of data-driven decisions improves the performance of teams through various targeted training initiatives and collaboration. Predictive analytics helps organizations strategize and plan to prepare and provision required resources so that it acts promptly in adapting the new business scenarios. As business strategies rely more and more on data-driven processes, inclusion of predictive analytics in workforce management will play a crucial role in establishing a responsive and involved workforce at the workplace for the new business scenario.

Figure 6. Challenges and ethical considerations in using machine learning for workforce analysis

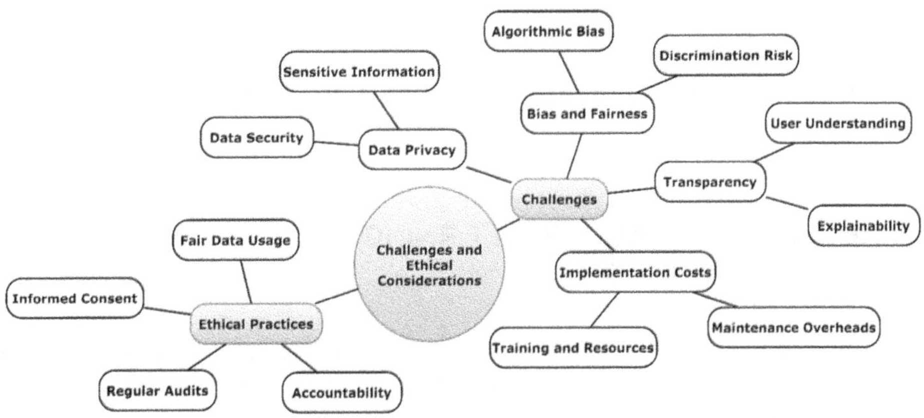

CHALLENGES AND ETHICAL CONSIDERATIONS

As the organizations increasingly embrace machine learning techniques in workforce analysis, they must face a few challenges and considerations through responsible and ethical practices. The key concerns in this list include respect of privacy and data protection, ethical AI practices, and that the proper balance between insights from machine learning and human judgement is achieved. Figure 6 highlights the ethical considerations and challenges linked to the use of machine learning for workforce analysis (Chandana et al., 2024; Kannimuthu & Premalatha, 2024).

Privacy and Data Protection Workforce Analysis: Respecting Privacy and Protecting Data is one very important challenge when undertaking workforce analysis about employee information. Organisations are also bound by regulations such as the GDPR and CCPA among others. The provisions of these laws demand strict measures concerning data handling and consent. Sensitive employee information subjected to collection and analysis risks privacy breaches when it is mishandled. Therefore, the employer must have very sound data governance frameworks that will ensure anonymization and proper use of data in protecting rights from employees.

Ethical AI and Explainability: While using AI in decision-making processes raises questions on bias, fairness, and transparency, machine learning algorithms will, by design, perpetuate biases in data training unless controlled and corrected from happening. Using such algorithms will lead to discriminatory hiring, promotions, or resource allocations. In this regard, some organizations must put emphasis on ethical AI use by conducting frequent audits in algorithms on bias and on fairness checks. Explainability is another important aspect. Stakeholders need to know how AI-derived decisions are made, so there will be trust and accountability. The transparency of decision-making gives workers the feeling that they are valued or treated as well as they deserve to be.

Human Judgment vs. Machine Learning: Machine learning will shed light on new information, but it cannot substitute entirely for human judgment. Decisions about the careers and well-being of employees should be made under human scrutiny, taking into consideration context, emotional intelligence, and ethical considerations that are often out of the reach of algorithms. Good balance is needed between exploiting the functionality of machine learning versus human intuition so as not to compromise the creation of a fair workplace environment. Organizations can thus use machine learning while adhering to ethics and building trust with employees through the benefits offered by combined technology and human expertise.

FUTURE DIRECTIONS IN MACHINE LEARNING AND INDUSTRIAL PSYCHOLOGY

Machine learning and industrial psychology are about to undergo an unparalleled revolution based on AI emerging trends and human-centered design, where an increasingly growing recognition knows that workplace dynamics and employee well-being need to improve, and thus will find both disciplines informing each other to create novel solutions to modern challenges (Manzini et al., 2024; Pereira, 2024).

Emerging Trends in AI and Human-Centered Design: Since human-centered design focuses on empathy and user experience, it can have a huge bearing on AI application development within industry. The future trends may therefore be strictly based on the needs and preferences of employees, leading to intuitive interfaces and thus personalized interaction. In this case, integrating user feedback in the design will be important as organizations continue incorporating AI-driven tools. This makes it easier to use but also builds a culture of inclusiveness and engagement, while the technology becomes an extension of human capabilities.

Role of Machine Learning in Workforce Dynamics Evolution: The role of machine learning in workforce dynamics will continue to evolve with deeper insights into employee behavior and performance. As these algorithms become more and more advanced, they will not only predict the ability and willingness of employees to leave (turnover) or become engaged but also the impact of organizational changes on employee morale. This development shall allow for proactive management strategies targeting potential issues before problems become unmanageable and create a healthier work environment. Another shift in this approach could be the integration of NLP capabilities for better communication between employees and management to improve feedback mechanisms and overall collaboration.

Multiple opportunities for innovation and growth would be presented by integrating machine learning in industrial psychology. Organizations can indeed use data analytics for identifying the emerging requirements and needs of employees in terms of professional development, mental health support, and workplace culture and can consequently come up with related programs that meet their requirements. With an emphasis on the well-being and productivity of employees, synergies between machine learning and industrial psychology will become channels to mold future workplaces - channels that will spur sustainable growth in a competitive advantage in the changing labor market.

CONCLUSION

The combination of machine learning and industrial psychology has enormous promise for restaging workplace dynamics and employee well-being. Applying big-data analytics and human-centered design principles, organizations can create environments that balance performance with mental health. The prediction of turnovers, personalization of interventions, and optimization of resource allocation help employers take on decisions based on explicit data to leverage increased overall productivity and job satisfaction. With time, the push for ethics in data privacy and AI fairness will see organizations infuse more openness and human decision-making skills into their pursuit of AI and ML. Here, the future of machine learning and industrial psychology will shape today's quest to address a modern workplace issue, innovation, and a hardy workforce. This mutual collaboration will make better, but more importantly prepare, organizations for long-term growth in an ever more competitive landscape.

REFERENCES

Bastani, H., Bastani, O., & Sinchaisri, W. P. (2021). Learning best practices: Can machine learning improve human decision-making. *Proceedings - Academy of Management, 1* (1), 14006. DOI: 10.5465/AMBPP.2021.14006abstract

Bormah, A. M. (2023). *Productivity Forecasting of Employees Performance Using Machine Learning and Adaptive Neuro-Fuzzy Inference System* [PhD Thesis]. KING ABDULAZIZ UNIVERSITY.

Chandana, C., Sarkar, A., Deshmukh, K., Kulkarni, P., Acharjee, P. B., & Lourens, M. (2024). Analysing Employee Management Using Machine Learning Techniques and Solutions in Human Resource Management. *2024 4th International Conference on Innovative Practices in Technology and Management (ICIPTM)*, 1–6.

Chujfi, S., & Meinel, C. (2019). Machine learning and human cognition combined to enhance knowledge discovery fidelity. *2019 IEEE First International Conference on Cognitive Machine Intelligence (CogMI)*, 1–7. DOI: 10.1109/CogMI48466.2019.00010

Fan, X., Zhao, S., Zhang, X., & Meng, L. (2023). The impact of improving employee psychological empowerment and job performance based on deep learning and artificial intelligence. [JOEUC]. *Journal of Organizational and End User Computing, 35* (3), 1–14. DOI: 10.4018/JOEUC.321639

Fukui, S., Wu, W., Greenfield, J., Salyers, M. P., Morse, G., Garabrant, J., Bass, E., Kyere, E., & Dell, N. (2023). Applying Machine Learning to Human Resources Data: Predicting Job Turnover among Community Mental Health Center Employees. *The Journal of Mental Health Policy and Economics, 26* (2), 63. PMID: 37357871

Gon, S., Ma, H., Wan, Y., & Xu, A. (2019). Machine learning in human-computer nonverbal communication. In *NeuroManagement and Intelligent Computing Method on Multimodal Interaction* (pp. 1–7). DOI: 10.1145/3357160.3357670

Hall, O. P. (2021). Managing employee turnover: Machine learning to the rescue. *International Journal of Data Science, 6* (1), 57–82. DOI: 10.1504/IJDS.2021.117472

Jaisawal, R. (2022). Machine learning to evaluate important human capital (HC) determinants impacting IT compensation. *Ramanujan International Journal of Business and Research, 7* (2), 16–25. DOI: 10.51245/rijbr.v7i2.2022.797

Kannimuthu, S., & Premalatha, K. (2024). Human Resource Analytics: Performance Analysis of Machine Learning Algorithms for Employee Attrition Analysis. *Proceedings of the 2024 Asia Pacific Conference on Computing Technologies, Communications and Networking*, 105–113. DOI: 10.1145/3685767.3685785

Li, C., Zhang, Y., Niu, X., Chen, F., & Zhou, H. (2023). Does artificial intelligence promote or inhibit on-the-job learning? Human reactions to AI at work. *Systems, 11* (3), 114. DOI: 10.3390/systems11030114

Liu, Q., Wan, H., & Yu, H. (2023). The Application of Deep Learning in Human Resource Management: A New Perspective on Employee Recruitment and Performance Evaluation. *Academic Journal of Management and Social Sciences, 3* (1), 101–104. DOI: 10.54097/ajmss.v3i1.9573

Manzini, T., Perali, P., & Murphy, R. R. (2024). Three Challenges in Utilizing Machine Learning to Predict Human Behavior from Observational Data. *Companion of the 2024 ACM/IEEE International Conference on Human-Robot Interaction*, 737–739.

Mohanty, A., Venkateswaran, N., Ranjit, P., Tripathi, M. A., & Boopathi, S. (2023). Innovative Strategy for Profitable Automobile Industries: Working Capital Management. In *Handbook of Research on Designing Sustainable Supply Chains to Achieve a Circular Economy* (pp. 412–428). IGI Global.

Najafi-Zangeneh, S., Shams-Gharneh, N., Arjomandi-Nezhad, A., & Hashemkhani Zolfani, S. (2021). An improved machine learning-based employees attrition prediction framework with emphasis on feature selection. *Mathematics*, *9* (11), 1226. DOI: 10.3390/math9111226

Norrman, F. (2020). *Predicting employee attrition with machine learning on an individual level, and the effects it could have on an organization.*

Pan, T. (2021). Performance Evaluation Method of Enterprise Human Resource Management Based on Machine Learning. *2021 IEEE International Conference on Industrial Application of Artificial Intelligence (IAAI)*, 100–105. DOI: 10.1109/IAAI54625.2021.9699954

Pereira, L. A. S. (2024). *Using machine learning to predict employee turnover: A case study of the Willis Towers Watson Lisbon Hub* [PhD Thesis].

Qamar, N. (2022). THE IMPACT OF MACHINE LEARNING ON HUMAN RESOURCE MANAGEMENT PRACTICE. *NeuroQuantology: An Interdisciplinary Journal of Neuroscience and Quantum Physics*, *20* (13), 2424.

Rathore, R., & Rathore, S. P. S. (2024). Machine Learning Applications in Human Resource Management: Predicting Employee Turnover and Performance. *International Journal for Global Academic & Scientific Research*, *3* (2), 48–59. DOI: 10.55938/ijgasr.v3i2.77

Rodriguez, W., Angle, P., & Snyder, M. (2021). Can Machine Learning Enhance Human Learning in Times of Disruption? *Ubiquitous Learning. International Journal (Toronto, Ont.)*, *14* (1).

Senthil, T., Puviyarasan, M., Babu, S. R., Surakasi, R., Sampath, B., & Associates. (2023). Industrial Robot-Integrated Fused Deposition Modelling for the 3D Printing Process. In *Development, Properties, and Industrial Applications of 3D Printed Polymer Composites* (pp. 188–210). IGI Global.

Sharma, R., & Dhingra, L. (2024). Advancing Human Resource Strategies with Deep Learning: Predictive Analytics for Improving Employee Retention Rates. *2024 2nd World Conference on Communication & Computing (WCONF)*, 1–4.

Ullal, M. S., Nayak, P. M., Dais, R. T., Spulbar, C., & Birau, R. (2022). Investigating the nexus between Artificial Intelligence and machine learning technologies in the case of Indian services industry. *Business: Theory and Practice*, *23* (2), 323–333. DOI: 10.3846/btp.2022.15366

KEY TERMS AND DEFINITIONS

GDPR: is an abbreviation for General Data Protection Regulation, a general data protection law in EU - how personal data is collected, stored, and processed.

AI: is an abbreviation for Artificial Intelligence. It refers to the simulation of human intelligence in machines. AI equips machines with the ability to perform tasks that typically require human minds, such as learning and problem-solving.

HR: is an abbreviation for Human Resources and defines the department that manages the employee's relationship, recruitment, training, and organizational development.

EI: Emotional Intelligence - They are able to recognize, understand, and control their own emotions and those of others. This is very much significant to communicate effectively and work as a team.

KPI (Key Performance Indicator): It is a measurement benchmark for assessing the success of an organization in achieving its goals and objectives.

CCPA: California Consumer Privacy Act is data privacy legislation in California, which enhances the rights to privacy and consumer protection concerning the privacy of residents.

NLP: stands for Natural Language Processing It is a subset of AI dealing with computer interaction through human language. It lets machines understand, interpret, and even generate text and speech.

ML: stands for Machine Learning It is a subset of AI that trains algorithms based on observations or experience collected from data in order to make the predictions that matter.

Chapter 13
Advancing Geographic Information Systems With Machine Learning:
Innovations in Spatial Analysis and Environmental Management

E. Ivette Cota-Rivera
https://orcid.org/0009-0002-6240-9269
Universidad Politécnica de Baja California, Mexico

ABSTRACT

A Geographic Information System (GIS) is a technological tool that allows for the capture, storage, analysis, and visualization of geographically referenced data. These systems integrate various forms of spatial and non-spatial data, facilitating the analysis of geographic phenomena and patterns.The integration of Machine Learning (ML) into Geographic Information Systems (GIS) has revolutionized the way geospatial data is analyzed and used. Machine Learning, with its ability to learn from large volumes of data and make accurate predictions, complements the analytical capabilities of GIS, allowing for the extraction of complex patterns and the performance of advanced predictions that were not previously possible. The purpose of this chapter is to explore the applications of Geographic Information Systems (GIS) empowered by the use of machine learning, highlighting their impact on spatial analysis and environmental management.

INTRODUCTION

A Geographic Information System (GIS) is a technological tool that allows for the capture, storage, analysis, and visualization of geographically referenced data. These systems integrate various forms of spatial and non-spatial data, facilitating the analysis of geographic phenomena and patterns. GIS is used in a wide variety of fields, including urban planning, natural resource management, agriculture, public health, transportation, and environmental research, among others. Their ability to handle large

DOI: 10.4018/979-8-3693-6910-4.ch013

volumes of spatial data and perform complex analyses makes them an essential tool for location-based decision-making (Chang, 2016).

The history of GIS dates back to the 1960s, with the development of the first computerized mapping system in Canada. The pioneer in this field was Roger Tomlinson, known as the "father of GIS," who developed the Canada Geographic Information System (CGIS) in 1962. CGIS was designed to store, analyze, and manipulate geographic data and was primarily used for land census and natural resource management (Bryan & Gorge, 2003).

Throughout the 1970s and 1980s, advances in computer technology and the availability of digital data drove the development of more advanced GIS. During this period, several major GIS companies were founded, such as Esri (Environmental Systems Research Institute), which released its ARC/INFO software in 1981, becoming an industry standard (Bryan & Gorge, 2003).

In the 1990s, the proliferation of personal computers and the increasing accessibility of the Internet facilitated the use and dissemination of GIS. This allowed a wider range of users to access these tools for various purposes. Today, GIS is integrated into web and mobile applications, allowing users to access and share geographic data in real-time (Maliene et al., 2011).

The evolution of GIS continues with the development of advanced technologies such as artificial intelligence and machine learning, which promise to further improve spatial analysis capabilities and location-based decision-making (Maliene et al., 2011).

Machine Learning (ML) is a subdiscipline of artificial intelligence that focuses on the development of algorithms and techniques that allow computers to learn from data and make predictions or decisions without being explicitly programmed for each task. This approach is based on creating mathematical models from training data, which are then used to make predictions or decisions on new data (Sharma & Chaudhary, 2023).

The history of Machine Learning dates back to the mid-20th century. In 1950, Alan Turing proposed the idea of a "machine that can learn" in his influential work "Computing Machinery and Intelligence." In the 1950s, Arthur Samuel developed one of the first machine learning programs: a chess program that improved its performance by playing against itself (Sharma & Chaudhary, 2023).

During the 1960s and 1970s, the main focus was on the development of basic algorithms such as the perceptron, introduced by Frank Rosenblatt in 1958, and other supervised and unsupervised learning algorithms. However, the field experienced a pause in significant progress during the 1970s and 1980s due to computational limitations and the lack of large datasets (Sharma & Chaudhary, 2023).

The resurgence of Machine Learning began in the 1990s with the availability of greater computational capabilities and large volumes of data. Algorithms such as Support Vector Machines (SVMs) and Random Forests gained popularity. The 2000s saw exponential growth in research and applications of Machine Learning, with technology companies like Google, Amazon, and Microsoft heavily investing in this area.

In the last decade, the development of Deep Learning has revolutionized the field, enabling significant advances in tasks such as image recognition, natural language processing, and autonomous driving. Researchers such as Geoffrey Hinton, Yann LeCun, and Yoshua Bengio have pioneered the use of deep neural networks, pushing Machine Learning to new frontiers. Today, Machine Learning is an omnipresent, see Figure 1, technology that drives a wide range of applications in various industries, from healthcare and finance to consumer technology and engineering. Its ability to transform data into practical knowledge remains a key driver of innovation in the 21st century (Sharma & Chaudhary, 2023).

Figure 1. Relations among AI, machine learning, and deep learning

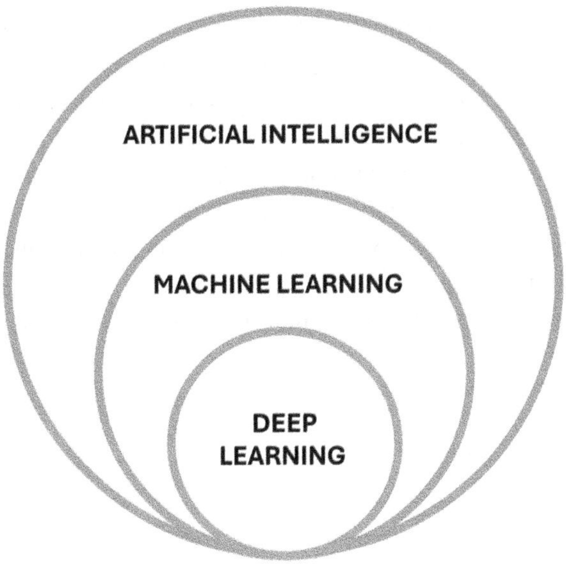

(Hu et al., 2019)

The integration of Machine Learning (ML) into Geographic Information Systems (GIS) has revolutionized the way geospatial data is analyzed and used. Machine Learning, with its ability to learn from large volumes of data and make accurate predictions, complements the analytical capabilities of GIS, allowing for the extraction of complex patterns and the performance of advanced predictions that were not previously possible (Choi, 2023).

One of the most important applications of this integration is in the field of environmental management. ML algorithms can analyze geospatial data to predict species distribution, identify areas at risk of natural disasters such as floods or wildfires, and monitor changes in land use. This allows environmental managers to make informed and proactive decisions for the conservation and sustainable management of natural resources (Choi, 2023).

In urban areas, the combination of ML and GIS is used for traffic analysis, urban planning, and infrastructure management. For example, ML models can predict traffic patterns in real-time and suggest alternative routes to reduce congestion. They can also analyze data from urban sensors to optimize the management of public services such as water and electricity supply. Another critical area is precision agriculture, where GIS, powered by ML, helps farmers make data-driven decisions about planting, irrigation, and fertilization. Algorithms can process satellite images and climate data to predict crop yields and detect plant diseases before they spread, thereby improving agricultural efficiency and sustainability (Metz, 2023).

In the field of public health, the integration of ML and GIS enables the tracking and prediction of disease spread, which is crucial for emergency response planning and the implementation of preventive measures (Metz, 2023).

In summary, the integration of Machine Learning into Geographic Information Systems has greatly amplified the analysis and prediction capabilities of GIS. This not only improves the accuracy and efficiency of existing applications but also opens up new possibilities for addressing complex problems in various industries. The synergy between ML and GIS is essential for advancing innovative solutions that benefit both society and the environment (Metz, 2023).

The purpose of this chapter is to explore the applications of Geographic Information Systems (GIS) empowered by the use of machine learning, highlighting their impact on spatial analysis and environmental management. In a world where environmental challenges are becoming increasingly complex, the integration of these technologies offers new insights and innovative solutions. This chapter will examine how GIS and machine learning are combined to enhance the accuracy and efficiency in the collection and analysis of geospatial data. From early detection of deforestation and monitoring air quality to predicting natural disasters, these technologies transform our ability to understand and manage the environment. Additionally, case studies and practical examples will be presented illustrating the application of these systems in various areas, underscoring their potential to contribute to a more sustainable and resilient future. Figure 2.

Figure 2. Integration of machine learning in gis

(Own work, 2024)

METHODS FOR INTEGRATING ML ALGORITHMS INTO GIS PLATFORMS

The fusion of machine learning algorithms with geospatial data, science and technology to solve spatial problems is known as GeoAI (Geospatial Artificial Intelligence) (Openshaw., 1992). This combination allows users of these systems to solve spatial problems with techniques for pattern detection, prediction and spatio-temporal forecasting in a more efficient and accurate way. There are different ML methods and algorithms for Geographic Information System platforms, and we will analyze some of them below.

Native Integration

It refers to the integration of tools within the geographic information systems platform that allow the user to solve classification, regression, data pattern detection and prediction problems without the need for add-ons or additional software. Esri's ArcGIS is an example of this as it has integrated a wide range of geoprocessing and exploratory analysis tools, as well as a series of pre-trained deep learning (DL) models that simplify some aspects of the training process that require more resources and time (GeoAI, n.d.).

Integration through APIs and libraries According to Amazon Web Services (AWS, n.d.), an application programming interface (API) is a mechanism that allows two software components to communicate with each other through a set of definitions and protocols to exchange data, features, and functionality. APIs are very useful tools for developers because they allow developers to integrate data, services, and capabilities from other applications without having to start from scratch. In addition, users can write scripts tailored to their requirements to combine GIS capabilities with ML tools, developing highly specialized work processes. As mentioned above, this approach can provide many advanced ML tools that users can take advantage of, however, to use it, a certain degree of knowledge in programming and basic ML principles is required, which can sometimes be a limitation for some users (Google Maps Platform Documentation, n.d.).

Cloud Integration

In their work Qian et al (2009) mention that cloud computing is the ability to access IT resources and services over the Internet without the need for physical storage. There are mainly three cloud computing service models: Infrastructure as a Service (IaaS), where the client has no control over the underlying cloud, but does have control over the operating systems, storage and deployed applications, e.g. Amazon Web Services (AWS), Microsoft Azure and Google Compute Engine (GCE), Platform as a Service (PaaS), where the client does not manage or control the underlying cloud including network, operating systems or storage, but does have control and configuration over the applications offered e.g. Google App Engine, Software as a Service (SaaS), where the client only has limited access to configuration settings for specific applications, e.g. Google Apps, Salesforce, Dropbox storage solutions. Cloud computing has facilitated the integration of ML algorithms into GILs, facilitating access to large volumes of information without the need for a robust infrastructure.

Integration Through Hybrid Environments

Hybrid environments have revolutionized the way information technology resources are managed, offering solutions that integrate local and cloud systems ensuring responsiveness and scalability as well as better cost management and data security (Breiter & Naik, 2013).

The combination of local and cloud processing allows data preprocessing in the local environment using traditional geographic information system tools and more robust analysis is performed with ML tools in the cloud. The advantage of hybrid environments is their adaptability; however, they tend to increase their degree of complexity due to the different workflows (Change Americas, 2024).

SPATIAL ANALYSIS TECHNIQUES ENHANCED BY ML

Geospatial Clustering

Clustering is a machine learning technique whose main objective is to associate data into groups of similar objects (clusters) and thus discover hidden structures or patterns in the data. Objects are commonly described as vectors or attributes that can be numerical or categorical. Several common algorithms can be named for this technique such as K-means, BDSCAN (density-based clustering), Agglomerative Hierarchical (Folini, 2020). Associating this technique with spatial analysis provides a reference structure to join and analyze large-scale data in different geographic contexts contributing to efficient decision making (Mouraz, 2022).

Classification and pattern recognition

According to Lee and Choi (2004), AI algorithms are used for processing satellite images, which have been very useful in pattern recognition analysis. This is because they allow the identification of natural groups based on spatial distribution and attribute similarities. In addition, these algorithms analyze large amounts of information or satellite images, detecting complex patterns for the analysis of historical data, predicting future data and detecting water bodies, vegetation types, and urban areas, which is important for both urban and environmental planning.

Sensor Data Analysis

According to Krishna et al. (2024) in recent years various investigations have highlighted the possible applications and benefits of integrating wireless sensor networks enabled by IoT, geospatial technology, and machine learning in the management of urban infrastructure systems, the health sector, air quality, water resources and a wide range of fields. These technologies have the ability to analyze data in real time to be monitored and managed.

Machine learning (ML) algorithms have increased their importance in the health field, as the wireless sensor network environment based on IoT and geospatial technology is capable of detecting hidden patterns, establishing correlations and providing useful and timely information. For example, smart watches with a physical activity tracker can monitor vital signs, sleep patterns and activity levels providing relevant information about the health status of patients which can be observed in real time.

APPLICATIONS IN THE INTEGRATION OF MACHINE LEARNING IN GEOGRAPHIC INFORMATION SYSTEMS

The large-scale advancement of technology in geographic information systems, from automatic image classification, 3D models and terrain analysis, to advanced real-time interpolation functions and Big Data support, has caused Artificial Intelligence to gradually displace traditional tools and methods, as well as manual processes. In addition, the integration of machine learning has been a fundamental pillar in the development of increasingly robust technologies and tools for the storage, processing and analysis of the growing amount of data that is constantly generated (Siabato, 2028), benefiting various areas.

Below, we describe how AI has revolutionized GIS in various areas, such as environmental analysis, urban planning, and disaster or emergency response.

Prediction and modeling.

Machine learning solves complex problems in the generation of predictive models that are useful for planning and strategic decision-making in response to different geographic phenomena such as soil erosion, species conservation, and the expansion of urban areas, among others. By using Convolutional neural networks (CNNs), complex patterns of climatic and meteorological images can be analyzed on a large scale in real time, offering solutions to problems such as climate change, environmental pollution, and water pollution (Abdalla & Esmail, 2018).

Route Optimization and Logistics

Machine learning (ML) is of great use in the optimization of transportation routes based on historical patterns and real-time conditions. Yfang et al. (2019) conducted a novel study based on GCN (Global Convolutional Neural Networks). The study used real GPS trajectory data provided by a transportation service with the aim of detecting intersections from trajectories to facilitate the construction of high-quality road networks. This was achieved by using GCN at multiple scales to generate various levels of graphs by applying logistic regression to discriminate between intersections and non-intersections. This graphical representation can have multiple applications, not only in the detection of intersections but also in different geospatial modeling applications.

Change Detection

Machine learning (ML) has proven to be a key technique in detecting variations over time in geospatial data, which, among many other applications, can provide more accurate maps, detect changes in urban expansion, invasions of agricultural land, deforestation, or the retreat of glaciers. Through these techniques, it is possible to detect the time periods in which the growth of a determined area was faster and periods in which the expansion turns out to be slower. In their study, Atef et al. (2019) used maximum likelihood (MLH) and machine learning algorithms (random forest (RF) algorithm and support vector machine (SVM)) for image classification to monitor changes in land use and coverage in El-Fayoum. Faster and more efficient processing of vast data sets was achieved, as well as a system with greater classification accuracy. This study is a valuable contribution that can contribute to more informed decision-making regarding the well-being of society and the environment in general.

Disaster Management

Artificial intelligence has enhanced the capacity of Geographic Information Systems in the area of analysis and response to catastrophes. Machine learning algorithms can identify patterns by analyzing historical disaster data in order to predict future calamities and the impact they cause. Additionally, artificial intelligence can be supported by real-time monitoring systems to detect anomalies in a timely manner or identify possible disasters and provide timely alerts, significantly improving response times and emergency preparedness. It is also possible to analyze large data sets in real time to identify critical areas, assess damage, and account for population density, supporting the distribution of resources and effective decision making (Emami & Marzban, 2023).

AI TOOLS AND TECHNOLOGIES THAT ARE INTEGRATED INTO GIS

The capacity for the analysis and management of geospatial data has evolved at a rapid pace, largely due to the fact that AI tools in GIS have been developing by leaps and bounds in recent years. In this section, the most promising artificial intelligence tools that are integrated into geographic information systems will be reviewed.

ArcGis Pro: Is a desktop GIS application that allows the creation and exploitation of 2D maps and 3D scenarios, along with data visualization, and can also be combined with ARCGIS Online (Corbin, 2015). ArcGis Pro contains a tool for extracting image features through Artificial Intelligence models. These models can be pre-trained using external ArcGis databases or with deep learning model packages. The tool allows users to choose between several detection or classification models in order to extract features and detect pixels in the images (Lau, 2019; Esri, n.d).

ArcGIS API for Python: It is a library that allows the user to perform spatial data science. It is a simple and lightweight library that can access already cleaned geographic data and can also improve efficiency by automating the administration of your web GIS such as creating or managing groups, monitoring server usage, and transferring item ownership between applications (Toms, 2022; Esri, n.d.).

QGIS Automatic Digitizing Plugin: In the world of GIS, one of the most tedious tasks is the digitization of maps and cartographic documents by converting raster images to vectors. A practical solution to this problem is the plugin called Bunting Labs AI Vectorizer. This plugin allows QGIS users to digitize shapes such as lines and polygons, as well as more complex structures such as contour lines, among others. In addition, through AI tools, this plugin can identify and trace shapes with great precision, significantly reducing the time and effort required for the digitization process (Acolita, n.d.).

Google Earth Engine: This application has a built-in capability that allows the use of ML to perform various tasks, such as classifying satellite image pixels into two or more categories, which is highly useful in applications like land cover and land use mapping. With this application, different ML techniques, such as supervised and unsupervised classifications, and regression, can be applied. Although Earth Engine does not support more advanced techniques like deep learning or convolutional neural networks, it is possible to access them externally through other platforms for example TensorFlow or PyTorch (Google, n.d; Velastegui-Montoya, 2023).

Erdas Imagine

This application incorporates convolutional neural networks (CNNs), which are extensively used in image classification tasks. Erdas Imagine proves to be a highly effective tool in remote sensing applications as it includes image interpretation features such as multispectral analysis, advanced algorithms, and seamless integration with other software (Nijhawan, 2017).

Envi+IDL

This is a robust tool that includes a deep learning module capable of solving problems in urban development, disaster response, and transportation, among others. One of the main attractions is the ML module's intuitive tools and workflows which do not require programming knowledge, making it accessible to a wide range of end users (NV5, n.d).

Orfeo Toolbox

It operates within the QGIS environment, and it can be used in the implementation of segmentation and classification algorithms. One of its key features is its ability to process large images, even on computers with limited resources or laptops (Gidahatari, n.d.; De Luca et al., 2019).

ETHICAL CONSIDERATIONS IN THE USE OF MACHINE LEARNING (ML) IN GEOGRAPHIC INFORMATION SYSTEMS (GIS)

It is not new that the use of artificial intelligence raises ethical questions, and the use of AI in geographic information systems is no exception. As systems become more autonomous, questions arise about who is responsible for decision-making. If an algorithm produces an incorrect or biased result, who is responsible? The programmer, the end user, or the organization that implemented the algorithm. In this section, we will address some of the main questions about the use of machine learning (ML) in geographic information systems, particularly concerning privacy and bias in algorithmic decision-making.

Privacy Concerns

The use of Machine Learning in Geographic Information Systems involves the collection and processing of massive amounts of data. This includes the handling of sensitive information about the security of individuals, such as their exact home address, work location, frequent routes, movement patterns, and daily activities, among others. The exposure of all this information brings with it several risks to people's privacy. The uncontrolled use of large volumes of information can lead to security risks. Currently, the use of data for criminal activities is one of the main problems of society, which results in the loss of privacy and personal autonomy caused by the misuse of GIS for activities such as the systematic and non-consensual monitoring of people. To mitigate these risks, it is essential to establish relevant legal regulations that guarantee the privacy and security of individuals, in addition to implementing information control and protection practices. Boyd & Crawford (2012), in their study address the technological and academic implications of Big Data, stating that accessibility does not equate to ethics, so the use of data that is public without considering ethical issues is not justifiable.

Concerns related to bias in algorithmic decision-making

One of the main concerns that is becoming increasingly important when implementing machine learning algorithms in geographic information systems is the transparency of data when making decisions based on AI algorithms. There are various negative factors that can influence decision-making when bias exists, such as training with data that reflects social inequalities or historical prejudices, which can affect the results derived from training biased algorithms, perpetuating and propagating the biases of the training data. When working with deep learning neural networks, the activation function can become very complex and uninterpretable, making it difficult to correct errors in the detected biases (Burrell, 2016).

What strategies can be implemented to avoid biases in ML algorithms? One of the suggested practices is algorithmic transparency, which refers to the practice of making algorithms more accessible and understandable to those in charge of the strategic decision-making process, allowing for oversight and insight into the algorithms that influence decision-making in increasingly automated environments, such as Machine Learning. Another strategy would be to include interdisciplinary work teams that support bias mitigation (Diakopoulus, 2016; Doshi-Velez & Kim, 2017).

EQUITABLE ACCESS TO GEOGRAPHIC INFORMATION SYSTEMS AND MACHINE LEARNING TECHNOLOGIES FOR ENVIRONMENTAL JUSTICE AND SUSTAINABLE DEVELOPMENT

Equal access to GIS and ML technologies

With the rise of Big Data, the belief that analyzing large data sets provides completely truthful and reliable information has become more prevalent, due to the idea that "the data speaks for itself." Unfortunately, this is not entirely true. Both data and data sets are human creations, meaning they are given significance through the interpretations of the analyst in charge. However, from the collection and analysis stages, this data can harbor biases that pose significant threats to the reliability of the analysis to be carried out. The main challenges to be faced in equitable access to GIS and ML technologies are the digital divide and education training.

Digital divide

In different parts of the world, the most disadvantaged communities are the least equipped with technological tools that can be decisive in facing climate problems or natural disasters. Although there are studies that demonstrate the importance of GIS to analyze the behavior of patterns in geographic contexts (Mennies & Manson, 2011), a study showed that in 2012, during Hurricane Sandy, approximately 20 million tweets were generated in less than a week. Twitter data combined with information from the Foursquare platform revealed very interesting insights, such as grocery shopping peaking the day before the storm, or unexpected information such as that nightlife resumed the next day. However, this data does not represent the overall picture of the disaster. Most of the tweets were generated in the Manhattan area, giving the impression that this area was the epicenter of the hurricane. However, the reality was that there were areas that were much more affected and due to constant blackouts and the gradual depletion of cell phone batteries, could not counteract the massive sending of information that was generated from more privileged areas. This is without considering some much more disadvantaged communities that had little or no signal to transmit information that would have been relevant (Crawford, 2023).

Training and education

Observations on unequal access to data and technology clearly indicate that conventional GIS applications are not fully accessible to the broader community. In their study, Kyem & Kyem (2021) discuss the need to introduce geographic information systems to communities that would otherwise be excluded from research or policy discussions involving GIS analysis. Access to technological tools without adequate knowledge and training to use them may also contribute to perpetuating inequality in participation and inclusion. Investing in the training and education of vulnerable communities in GIS and ML technologies is critical, as it will enable inclusion to be achieved through more efficient monitoring of the environment to deal with catastrophes and environmental disasters (OECD, 2023).

GIS and ML for environmental justice

Environmental justice implies that every person, regardless of their origin, has access to comprehensive health conditions, a healthy environment and protection from climatic conditions and environmental disasters. This concept emphasizes the progressive deterioration of the natural environment and the lack of equity in terms of policies and laws to address it. It highlights that the least favored social groups continue to be those with the least opportunities for the development of democratic and inclusive principles in decision-making processes (López, 2014). This raises different questions on the subject: How can GIS and ML technologies help achieve objectives related to environmental justice?

There are different strategies in which GIS and machine learning can be involved in achieving environmental and equitable justice goals for disadvantaged communities. The most important ones are mentioned below.

Pollution Monitoring

Geographic information systems and remote sensing are fundamental tools in providing a complete picture of large areas that can be analyzed. Machine learning algorithms offer promising resources for implementing data monitoring technologies at more affordable prices, providing the opportunity to examine how human activity is affecting the environment (Jain et al., 2022). In their study, Goswami and Sarma (2020) provide a detailed analysis of how GIS and machine learning are crucial tools in environmental research, such as environmental pollution monitoring, vegetation studies and land cover use that contribute to solving problems like reducing hazardous environments, detecting levels of polluting particles in a given perimeter, and validating satellite data. More recent literature highlights additional ML models that contribute to addressing environmental issues, such as random forests (RF) (the most widely used), decision trees (DT), support vector machines (SVM), artificial neural networks (ANN), convolutional neural networks (CNN), and principal component analysis (PCA) (Binetti et al., 2024).

Inclusive urban planning

Machine learning models play a critical role in sustainable urban development as they contribute to better planning decisions by predicting urban development patterns which improve the quality of life in growing cities. The predictors of these models include land use, population density, socioeconomic characteristics, among others. In their study, Nagappan & Daud (2021) propose that to avoid uncontrolled urbanization and contribute to sustainable development, it is essential to carefully choose the predictors of machine learning models. Furthermore, by identifying areas with fewer resources, cities can invest in largely underserved areas, thereby improving the quality of life of their inhabitants. More efficient analysis of big data results in higher classification accuracy, helping planners make more informed decisions about city planning and development, and leads to better predictions of environmental impacts, which is crucial to minimizing damage to the environment and achieving improved long-term sustainability (Choi, 2023).

Climate Change Impact Assessment

The integration of GIS and ML is important to achieve better analysis in understanding climate change. These technologies can help identify areas that are especially vulnerable to climate change, such as coastal communities prone to flooding or, conversely, areas prone to drought, in addition to generating and providing crucial information for mitigation plans (Fekadu & Dullo, 2020).

EXPLORING THE POTENTIAL OF ML IN GIS: APPLICATIONS ACROSS VARIOUS SECTORS

In this section we will present a series of practical case studies where geographic information systems (GIS) powered by machine learning (ML) and artificial intelligence (A) were used.

Case 1: Wildfire susceptibility mapping

In the article titled "Machine learning-based forest fire susceptibility mapping using remote sensing and GIS fire data: a case study of Adana and Mersin provinces, Turkey" Iban & Seek Ertekin (2022) conducted a study in two provinces highly susceptible to forest fires, with the main objective of generating using data from the MODIS satellite and GIS. The authors evaluated 7 different algorithms to predict fire risk considering 13 factors influencing fires such as topography, vegetation, meteorological factors, anthropogenic factors and more. These factors were evaluated in terms of information gain and multicollinearity.

The tested algorithms include logistic regression (LR), support vector machine (SVM), linear discriminant analysis (LDA) and ensemble algorithms such as Random Forest (RF), Gradient Boosting (GB), eXtreme Gradient Boosting (XGB) and AdaBoost (AB). As a result, Random Forest was the model that showed the highest accuracy (7.2%) identifying the highest risk fires zones. This study highlights the importance of generating these maps, which are valuable tools for people responsible for strategic decision-making in the management of forest fires, which have become increasingly frequent in recent decades.

Case 2: ML and GIS in Healthcare

The article titled "Projecting the criticality of COVID-19 transmission in India using GIS and machine learning methods" makes a significant contribution to understanding the role of technologies such as machine learning and geographic information systems addressing public health challenges. In their study conducted in India, Khan et al. (2021) provided a framework for real-time monitoring, risk assessment, and strategic planning in times of public health crises like COVID-19. The authors employed three ML models: decision trees (DT), support vector machines (SVM), and Gaussian process regression (GPR) to predict the transmission and severity of COVID-19 in different regions of India, which showed a high predictive accuracy, especially GPR.

An index based on population density, daily positive cases, and healthcare system availability was developed to assess the transmission rate in different areas, classifying regions into high, moderate, and low risk zones. This analysis proved invaluable in identifying when the case numbers began to decline, allowing for the formulation of strategies for the gradual lifting of confinement measures, based on projected trends.

Case 3: Groundwater Potential Assessment

Arabameri et al., (2021) conducted a study on the application of machine learning techniques such as multilayer perceptual models (MLP), classification and regression tree (CART) and naive bayesian tree (NBTree) and geographic information systems for groundwater potential (GWP) assessment. The study analyzed more than 200 spring locations using geospatial data on various groundwater determinants such as elevation, slope, drainage density among others. These models demonstrated a high predictive accuracy, leading to the conclusion that these models are valuable tools for sustainable management of hydric resources. Furthermore, they provide critical information in decision-making related to water distribution and conservation efforts.

CONCLUSION

The integration of machine learning with geographic information systems, has transformed the way geospatial data is analyzed. ML has the capacity to process vast amounts of data and make highly accurate predictions. It can identify complex patterns and generate forecasts that were previously unattainable. ML is particularly valuable in various fields, such as environmental management and large-scale data analysis. For instance, it can estimate locations of certain species or identify risk zones for disasters like floods or wildfires. Moreover, in urban environments these techniques are crucial for supporting traffic analysis, urban planning and infrastructure management.

However, several challenges still need to be addressed. One of the primary concerns is privacy, because the data a GIS collects can include sensitive information about individuals, such as their home address, work location, usual routes, and daily activity which could compromise personal privacy. Additionally, a controversial challenge relies on how GIS and ML technologies can contribute to achieving goals related to environmental justice, which remains a topic of ongoing debate

REFERENCES

Abdalla, R., & Esmail, M. (2018). *WebGIS for disaster management and emergency response*. Springer.

Acolita. (n.d.). *Digitalización automática en QGIS usando inteligencia artificial*. Acolita. https://acolita.com/digitalizacion-automatica-en-qgis-usando-inteligencia-artificial/

Amazon Web Services. (n.d.). *What is an API?* Amazon. https://aws.amazon.com/es/what-is/api/

Arabameri, A., Chandra Pal, S., Rezaie, F., Asadi Nalivan, O., Chowdhuri, I., Saha, A., Lee, S., & Moayedi, H. (2021). Modeling Groundwater Potential Using Novel GIS-Based Machine-Learning Ensemble Techniques. *Journal of Hydrology. Regional Studies*, *36*, 100848. Advance online publication. DOI: 10.1016/j.ejrh.2021.100848

Atef, I., Ahmed, W., & Abdel-Maguid, R. H. (2023). Modelling of land use land cover changes using machine learning and GIS techniques: A case study in El-Fayoum Governorate, Egypt. *Environmental Monitoring and Assessment*, *195*(6), 637. DOI: 10.1007/s10661-023-11224-7 PMID: 37133528

Binetti, M. S., Massarelli, C., & Uricchio, V. F. (2024). Machine Learning in Geosciences: A Review of Complex Environmental Monitoring Applications. *Machine Learning and Knowledge Extraction*, *6*(2), 1263–1280. DOI: 10.3390/make6020059

Boyd, D., & Crawford, K. (2012). Critical questions for big data: Provocations for a cultural, technological, and scholarly phenomenon. *Information Communication and Society*, *15*(5), 662–679. DOI: 10.1080/1369118X.2012.678878

Breiter, G., & Naik, V. K. (2013, March). *A framework for controlling and managing hybrid cloud service integration. In 2013 IEEE international conference on Cloud engineering (ic2e)* (pp. 217-224). IEEE.

Bryan, K. N., & George, R. (2003). *Geographic information systems. The methods and materials of demography*, 733.

Burrell, J. (2016). How the machine 'thinks': Understanding opacity in machine learning algorithms. *Big Data & Society*, *3*(1), 2053951715622512. DOI: 10.1177/2053951715622512

Chang, K. T. (2016). *Geographic information system. International encyclopedia of geography: people, the earth, environment and technology*, 1-10.

Choi, Y. (2023). GeoAI: Integration of artificial intelligence, machine learning, and deep learning with GIS. *Applied Sciences (Basel, Switzerland)*, *13*(6), 3895. DOI: 10.3390/app13063895

Crawford, K. (2013, April 1). *The Hidden Biases in Big Data*. Harvard Business Review, https://hbr.org/2013/04/the-hidden-biases-in-big-data

De Luca, G. N., Silva, J. M., Cerasoli, S., Araújo, J., Campos, J., Di Fazio, S., & Modica, G. (2019). Object-based land cover classification of cork oak woodlands using UAV imagery and Orfeo ToolBox. *Remote Sensing*, *11*(10), 1238. DOI: 10.3390/rs11101238

Diakopoulos, N. (2016). Accountability in algorithmic decision making. *Communications of the ACM*, *59*(2), 56–62. DOI: 10.1145/2844110

Doshi-Velez, F., & Kim, B. (2017). *Towards a rigorous science of interpretable machine learning*. arXiv preprint arXiv:1702.08608

Emami, P., & Marzban, A. (2023). The synergy of artificial intelligence (ai) and geographic information systems (gis) for enhanced disaster management: Opportunities and challenges. *Disaster Medicine and Public Health Preparedness*, *17*, 1–3. DOI: 10.1017/dmp.2023.174 PMID: 37694385

Esri. (n.d.). *ArcGIS API for Python*. Esri. https://www.esri.com/es-es/arcgis/products/arcgis-python-libraries/libraries/arcgis-api-python

Esri. (n.d.). *Get started with ArcGIS Pro*. ArcGIS Pro Documentation. https://pro.arcgis.com/es/pro-app/latest/get-started/get-started.htm

Fekadu, A., Soromessa, T., & Dullo, B. W. (2020). GIS-based assessment of climate change impacts on forest habitable Aframomum corrorima (Braun) in Southwest Ethiopia coffee forest. *Journal of Mountain Science*, *17*(10), 2432–2446. DOI: 10.1007/s11629-019-5722-2

Folini, A. (2020). *Integrating machine learning techniques into GIS software: development of a comprehensive and versatile QGIS plugin for cluster analysis on geospatial data*. GeoAI. (n.d). *ArcGIS-Pro*https://pro.arcgis.com/es/pro-app/latest/help/analysis/ai/geoai.htm

Gidahatari. (n.d.). *Conozca Orfeo Toolbox (OTB) para procesamiento de imágenes satelitales*. Gidahatari. https://gidahatari.com/ih-es/conozca-orfeo-toolbox-otb-para-procesamiento-de-imagenes-satelitales

Google. (n.d.). *Machine learning guide*. Google Earth Engine. https://developers.google.com/earth-engine/guides/machine-learning

Google Maps API Developer Guide. [Online]. Available: https://developers.google.com/maps/documentation [Accessed: Sep. 3, 2024].

Goswami, T., & Sarma, H. (2020*). Intelligent computing for air pollution monitoring using gis, remote sensing and machine learning*. In Emerging Trends in Electrical, Communications, and Information Technologies: Proceedings of ICECIT-2018 (pp. 125-133). Springer Singapore."

Hu, Y., Li, W., Wright, D., Aydin, O., Wilson, D., Maher, O., & Raad, M. (2019). *Artificial intelligence approaches.arXiv preprint arXiv:1908.10345.*

Iban, M. C., & Sekertekin, A. (2022). Machine learning based wildfire susceptibility mapping using remotely sensed fire data and GIS: A case study of Adana and Mersin provinces, Turkey. *Ecological Informatics*, *69*, 101647. DOI: 10.1016/j.ecoinf.2022.101647

Jain, S., Kaur, N., Verma, S., Kavita, , Hosen, A. S. M. S., & Sehgal, S. S. (2022). Use of machine learning in air pollution research: A bibliographic perspective. *Electronics (Basel)*, *11*(21), 3621. DOI: 10.3390/electronics11213621

Khan, F. M., Kumar, A., Puppala, H., Kumar, G., & Gupta, R. (2021). Projecting the Criticality of COVID-19 Transmission in India Using GIS and Machine Learning Methods. *Journal of Safety Science and Resilience = An Quan Ke Xue Yu Ren Xing (Ying Wen)*, *2*(2), 50–62. DOI: 10.1016/j.jnlssr.2021.05.001

Krishna, E. P., Praveena, N., Manju, L., Malathi, N., Giri, R. K., & Preetha, M.E. S. Phalguna Krishna. (2024). IoT-Enabled Wireless Sensor Networks and Geospatial Technology for Urban Infrastructure Management. *Journal of Electrical Systems*, *20*(4s), 2248–2256. DOI: 10.52783/jes.2395

Kyem, P. A. K., & Kyem, P. A. K. (2021). *The Development of Community-Based GIS Applications. Managing Natural Resource Conflicts with Participatory Mapping and PGIS Applications*, 101-133.

Lee, S., & Choi, J. (2004). Landslide susceptibility mapping using GIS and the weight-of-evidence model. *International Journal of Geographical Information Science*, *18*(8), 789–814. DOI: 10.1080/13658810410001702003

LinkedIn. (n.d.). *Entornos híbridos: Estrategias para integrar infraestructuras*. LinkedIn. https://es.linkedin.com/pulse/entornos-h%C3%ADbridos-estrategias-para-integrar-infraestructuras-kvbmc

López, I. (2014). *Justicia ambiental*. EUNOMÍA. Revista en Cultura de la Legalidad, (6), 261-268.

Maliene, V., Grigonis, V., Palevičius, V., & Griffiths, S. (2011). Geographic information system: Old principles with new capabilities. *URBAN DESIGN International*, *16*(1), 1–6. DOI: 10.1057/udi.2010.25

Mennis, J., & Mason, M. J. (2011). People, places, and adolescent substance use: Integrating activity space and social network data for analyzing health behavior. *Annals of the Association of American Geographers*, *101*(2), 272–291. DOI: 10.1080/00045608.2010.534712

Metz, R. K. (2023, junio 29). *The power of machine learning in geospatial analysis: Enhancing decision-making and insights*. Medium. https://ryankmetz.medium.com/the-power-of-machine-learning-in-geospatial-analysis-enhancing-decision-making-and-insights-287cf1bebc51

Mouraz, C. P., Almeida, R. M., & Silva, J. M. (2022). Combining cluster analysis and GIS maps to characterise building stock: Case study in the historical city centre of Viseu, Portugal. *Journal of Building Engineering*, *58*, 104949. DOI: 10.1016/j.jobe.2022.104949

NV5 Geospatial. (n.d.). *ENVI deep learning*. NV5 Geospatial. https://www.nv5geospatialsoftware.com/Products/ENVI-Deep-Learning

Nagappan, S. D., & Daud, S. M. (2021). Machine learning predictors for sustainable urban planning. *International Journal of Advanced Computer Science and Applications*, *12*(7). Advance online publication. DOI: 10.14569/IJACSA.2021.0120787

Nijhawan, R., Sharma, H., Sahni, H., & Batra, A. (2017, December). *A deep learning hybrid CNN framework approach for vegetation cover mapping using deep features*. In 2017 13th international conference on signal-image technology & internet-based systems (SITIS) (pp. 192-196). IEEE. DOI: 10.1109/SITIS.2017.41

OECD. (2023). *The potential impact of Artificial Intelligence on equity and inclusion in education*. OECD iLibrary. https://www.oecd-ilibrary.org/education/the-potential-impact-of-artificial-intelligence-on-equity-and-inclusion-in-education_181source

Openshaw, S. (1992). Some suggestions concerning the development of artificial intelligence tools for spatial modelling and analysis in GIS. *The Annals of Regional Science*, *26*(1), 35–51. DOI: 10.1007/BF01581479

Qian, L., Luo, Z., Du, Y., & Guo, L. (2009). *Cloud computing: An overview. In Cloud Computing*: First International Conference, CloudCom 2009, Beijing, China, December 1-4, 2009. [Springer Berlin Heidelberg.]. *Proceedings, 1*, 626–631.

Sharma, S., & Chaudhary, P. (2023). *Machine learning and deep learning*. Quantum Comput. Artif. Intell. Train. Mach. Deep Learn. Algorithms Quantum Comput, 71-84.

Siabato, W. (2018). *Sobre la evolución de la información geográfica: Las bodas de oro de los SIG*. Cuadernos de Geografía. *Cuadernos de Geografia, 27*(1), 1–9. DOI: 10.15446/rcdg.v27n1.69500

Toms, S., Parker, B., Tucker, C., & Rubalcava, R. (2022). *Python for ArcGIS Pro: Automate cartography and data analysis using ArcPy, ArcGIS API for Python, Notebooks, and pandas*. Packt Publishing Ltd.

Tripp Corbin, G. I. S. P. (2015). *Learning ArcGIS Pro*. Packt Publishing Ltd.

Velastegui-Montoya, A., Montalván-Burbano, N., Carrión-Mero, P., Rivera-Torres, H., Sadeck, L., & Adami, M. (2023). Google Earth Engine: A global analysis and future trends. *Remote Sensing (Basel), 15*(14), 3675. DOI: 10.3390/rs15143675

Yin, Y., Sunderrajan, A., Huang, X., Varadarajan, J., Wang, G., Sahrawat, D., & Ng, S. K. (2019, November). *Multi-scale graph convolutional network for intersection detection from gps trajectories*. In *Proceedings of the 3rd ACM SIGSPATIAL International Workshop on AI for Geographic Knowledge Discovery* (pp. 36-39). DOI: 10.1145/3356471.3365234

ADDITIONAL READING

Hu, Y., Gao, S., Lunga, D., Li, W., Newsam, S., & Bhaduri, B. (2019a). GeoAI at ACM SIGSPATIAL: Progress, Challenges, and Future Directions. *SIGSPATIAL Special, 11*(2), 5–15. DOI: 10.1145/3377000.3377002

Huang, B., Zhao, B., & Song, Y. (2018). Urban Land-Use Mapping Using a Deep Convolutional Neural Network with High Spatial Resolution Multispectral Remote Sensing Imagery. *Remote Sensing of Environment, 214*, 73–86. DOI: 10.1016/j.rse.2018.04.050

Janowicz, K., Gao, S., McKenzie, G., Hu, Y., & Bhaduri, B. (2020). GeoAI: Spatially Explicit Artificial Intelligence Techniques for Geographic Knowledge Discovery and Beyond. *International Journal of Geographical Information Science, 34*(4), 625–636. DOI: 10.1080/13658816.2019.1684500

Li, T., Shen, H., Yuan, Q., Zhang, X., & Zhang, L. (2017). Estimating Ground-Level PM2.5 by Fusing Satellite and Station Observations: A Geo-Intelligent Deep Learning Approach. *Geophysical Research Letters, 44*(23), 11–985. DOI: 10.1002/2017GL075710

Openshaw, S. (1992). Some Suggestions concerning the Development of Artificial Intelligence Tools for Spatial Modelling and Analysis in GIS. *The Annals of Regional Science, 26*(1), 35–51. DOI: 10.1007/BF01581479

Yuan, Q., Shen, H., Li, T., Li, Z., Li, S., Jiang, Y., Xu, H., Tan, W., Yang, Q., Wang, J., Gao, J., & Zhang, L. (2020). Deep Learning in Environmental Remote Sensing: Achievements and Challenges. *Remote Sensing of Environment, 241*, 111716. DOI: 10.1016/j.rse.2020.111716

KEY TERMS AND DEFINITIONS

Convolutional Neural Networks (CNN): A type of neural network especially useful for image analysis. In GIS, CNNs are used to interpret satellite image data, such as detecting areas affected by natural disasters.

Deep Learning: A subset of machine learning based on deep artificial neural networks. It is particularly useful for recognizing complex patterns and classifying images in GIS, such as detecting changes in land use or identifying natural hazards.

Geographic Information Systems (GIS): A set of tools and technologies that enable the capture, storage, manipulation, analysis, and visualization of geospatial data. GIS is essential for managing and analyzing location-related information.

Geospatial Analysis: The process of interpreting and analyzing data that has a geographic reference. Integrating ML into this process helps find complex patterns and improve decision-making in areas such as land management and disaster control.

Machine Learning (ML): A branch of artificial intelligence that allows machines to learn from data and make predictions or decisions without being explicitly programmed for each task. In GIS, it is used to analyze large geospatial datasets and improve prediction accuracy.

Random Forest: A machine learning algorithm based on the creation of multiple decision trees, which are combined to improve prediction accuracy. It is widely used in GIS for the classification and prediction of geospatial data.

Remote Sensing: A technique for collecting information about objects or areas from a distance, typically using satellites or airplanes. Data obtained through remote sensing is a key source for geospatial analysis in GIS, and ML helps interpret this data with greater accuracy.

Spatial Data: Data that has a geographic reference and is used in GIS to represent the location and characteristics of objects on Earth. Analyzing spatial data with ML allows for more accurate predictions of geographic phenomena.

Supervised Learning: A type of machine learning where a model is trained with labeled data, meaning it is provided with both input and correct output examples. In GIS, it is used for tasks like satellite image classification.

Unsupervised Learning: A machine learning method where the model is not given labeled output data and must find patterns in the data on its own. In GIS, it is useful for clustering or segmenting geographic areas without prior classification.

Chapter 14
Cultivating Sustainable Growth and Well-Being in Digital Financial Education

Shrabani Mukherjee
https://orcid.org/0000-0002-4803-0896
Shiv Nadar University, India

Debdulal Thakur
https://orcid.org/0000-0003-2854-9495
Vinayaka Missions University, India

Payal Kanwar Chandel
Central University of Haryana, India

Saurabh Chandra
https://orcid.org/0000-0003-4172-9968
Bennett University, India

Anishkumar Karia
https://orcid.org/0009-0009-4897-801X
PCET's S.B. Patil Institute of Management, India

ABSTRACT

This chapter explores about strategies of creating sustainable growth and wellbeing through digital financial education. It stresses how access to knowledge about finance by individuals on these same platforms can be democratized through digital means-the best tools in making informed financial decisions. Focusing on equity, this chapter looks at how digital education can fill the gaps of financial literacy of those disenfranchised groups as a way to resilient economic growth. The integration of principles of well-being to financial learning develops sound economic stability along with personal well-being. The chapter seeks personalized adaptive technology learning and interactive content to engage and retain these learners. This chapter has taken forward AI, gamification, and data analytics by such digital financial education to promote behavioral change and financial responsibility.

DOI: 10.4018/979-8-3693-6910-4.ch014

INTRODUCTION

Quite literally, today's fast-paced digital world has transformed financial education dramatically-from tradition to more innovative and technology-based strategies. Of course, through this point, digital financial education has been an emerging power tool to improve the financial literacy of people at large levels across the globe by providing available resources that otherwise would have been withheld from the majority. Going by the increasingly digital learning platforms, knowledge of finance will be democratized, where citizens from all walks of life can acquire a critical financial skill leading to sustainable economic growth and further improved personal well-being. This chapter deals with the key ingredients of digital financial education and how these tools may be used to foster sustainable growth and well-being at the individual and collective level (Bayuk & Altobello, 2019).

In essence, financial literacy empowers individuals to make decisions that serve their goals and values rather than handling the concept of money and budgeting or investments. Financial education has generally not been of high quality for the masses but only for those who have the means and resources to seek such education. Unfortunately, these are marginalized and underserved communities left behind. This inequality is addressed by digital platforms that make financial education accessible and affordable at scale. Mobiles and apps, online courses, and interactive content with possibilities for increased reach across geographical, economic, and social divides now create a new habitat to close the financial literacy gap. Digital financial education, as it promises knowledge, gives tools that can personalize to meet the needs of diverse learners and empowers the learners to take care of their financial futures (Sorgente & Lanz, 2019).

Sustainable growth is one of the prominent advantages digital financial education offers. Digital education helps in building stronger and more resilient economies by equipping the population with skills and knowledge on how to make reasonable financial decisions. Thus, for example, information regarding savings, investing, and debt management enables the individual to be better prepared in times of economic downturn and financial challenge; hence, contributing to greater stability, both individually and socially. Even for small business owners and entrepreneurs, digital tools can prove to be necessary by providing much-needed finance education to grow sustainable businesses (Lyons & Kass-Hanna, 2021). The tools aid entrepreneurs in managing their resources more productively and hence reduce the risk of financial failure and pave a way for long-term growth. These digital finance education tools have been used to improve personal well-being besides ensuring economic resilience.

Financial stress is, in particular, a major contributor to issues related to mental and emotional health, as financially illiterate people end up succumbing to debt, fails at budgeting, and do not properly plan their long-term finance. Eventually, through digital platforms, well-being principles can be used in learning financial education toward getting the individual to feel financially secure, thereby reducing stress levels. Examples include interactive learning modules on budgeting, debt management, and retirement planning. Not only do such modules provide knowledge on the critical part of finance, like budgeting and debt management, but they also encourage users to build good habits toward healthy finances, which help users maintain steady financial stability and have a peace of mind in the long run. For access to educational resources on-demand, people can understand things at their own pace and reduce pressure and anxiety that most times tag along with financial education (Sabri et al., 2024).

Another important determinant of effective digital financial education is adaptive learning technologies. It is quite different from traditional classroom settings, where a one-size-fits-all approach leaves the different and varied needs of the learners unfulfilled, as digital platforms can provide various learning experiences that satisfy individual skill levels, learning styles, and financial goals. Such platforms will use AI and data analytics to determine the rate of a learner and adjust the delivery in real-time so that every learner receives the support they individually require for success. Gamification-Quizzes, Simulations, and Reward systems further increase the engagement of such learners with the subject matter, thus the learning process becomes more fun and exciting. These new approaches not only enhance the retention of knowledge but also spur behavioral change; people are likely to apply what they have learned into their everyday financial decisions (Burr et al., 2020).

While digital financial education has very many advantages, it's not problem-free. Its significant problem is ensuring equity in high-quality access across different populations. Such new digital channels, despite bringing the outreach to such untapped communities have so far failed to crack other barriers in terms of internet access, technological savvy, and language barrier. The answer to all these lies in public-private partnerships, government initiatives, and policy frameworks. What is needed would be a collaboration among governments, educative institutions, and private sector entities to ensure that those regardless of their socio-economic background or where they live also receive a digital- financial education. This entails the building of infrastructure, access to technology at affordable costs, and content that is culturally appropriate and in the language of the target group (Lusardi, 2019).

The quality and credibility of digital financial education will also pose another challenge. With an increasing number of users accessing online financial education resources, it calls for regulation and oversight of such sources to ensure that information being imparted is reliable, current, and not biased. Educators, policymakers, and technology developers should align to set standards for digital financial education, such that the content comes not only informative but also trusted and aligned to best practices on financial literacy. As digital platforms increase their use of AI and data analytics in delivering personalized learning experience, ethics, for example, relating to privacy, security of data, and algorithmic bias, will be tested in maintaining trust and protecting users (Panos & Wilson, 2020).

Innovation in improving effectiveness is on its way for the future of digital financial education. New technologies that are increasingly being integrated with financial education platforms include virtual reality, augmented reality, and blockchain that will introduce new ways to learn how to interact and apply acquired knowledge in practice. A variety of examples include VR simulating investment decisions in a virtual financial environment, but AR tools overlap financial information from physical environments for real-time insights in everyday transactions. Blockchain technology, with its availability in providing secure and transparent financial transactions, can make equal impacts on financial education by enabling learners to grasp complex financial systems, creating trust in digital financial tools (Philippas & Avdoulas, 2021).

In general, digital financial education is a transformative force for sustainable growth and personal well-being. Technology can be used to open up access to a wealth of financial knowledge. This free access to knowledge will enable individuals to work towards controlling their own financial futures, thereby increasing economic resilience and decreasing financial stress. However, digital financial education will unlock its full potential only if some key challenges are addressed: equal access, quality of content, and ethical concerns. Continued development of digital financial education will thus provide future generations with innovative, personalized learning experiences through collaboration between governments, private sector organizations, and educators to support financial literacy, sustainable growth, and well-being for all (Abad-Segura & González-Zamar, 2019).

Scope

The scope of this chapter covers exploring the potential impact of financial education on individuals and society at large. Some of the topics covered include digital platforms as facilitators for improving financial literacy, facilitating sustainable economic growth, and improving personal well-being. The issues around emerging technologies, such as artificial intelligence, adaptive learning, and gamification, applied to create adaptive learning experiences, are discussed in this chapter. Apart from presenting the specific issues related to access, credibility of the content as well as ethical considerations as concerns for data privacy, the chapter discussed the future prospect of digital financial education and its implications at a societal level, grounded on the triangular aspects of technology, education, and finance.

Objectives

This chapter focuses on the role of digital financial education in sustainable economic growth and personal well-being. It does shine light on the potential of digital platforms to democratize access for people to more financial knowledge, especially with regard to underserved populations. The chapter is also an attempt to investigate how the best adaptive learning technologies, AI, and gamification can be used to stimulate increased engagement, behavioral change, and improved financial decision-making. The solution addresses other issues such as accessibility and credibility in the content of quality solutions for digital financial education, considering collaboration in schemes by policymakers, educators, and the private sector.

THE ROLE OF DIGITAL PLATFORMS IN FINANCIAL EDUCATION

Access to needed financial information across digital platforms has changed the financial education landscape, broadened literacy gaps, and reached the majority of underserved populations. The world is getting more digitized, and financial literacy is a critical skill set needed to navigate complex financial systems, make good decisions, and achieve economic well-being. Traditionally, financial education has been earmarked for formal or professional stations or is available only to those who pay for the services of professional financial advisory services. However, through digital platforms, the nature of accessing and interacting with financial knowledge has drastically changed (Friedline et al., 2021). Digital platforms offer scalable, personalized, and accessible solutions that could meet the needs of a wide variety of learners.

Democratizing Access to Financial Knowledge

It is thoroughly democratized. Access to financial education today is much more open and democratic in nature as compared to the access that existed in the past. Most recently, financial literacy called for attending actual seminars, formal courses, or costly financial consultancy, which was impossible for many people. Nowadays, easy-to-access, affordable, and even free digital platforms offer the same resources to many diverse groups of people with varying levels of financial knowledge.

With the increasing trend of mobile applications, online courses, and web-based resources, millions of people worldwide can now have financial education. These online resources offer an extensive variety of content-from video lessons, articles, quizzes, simulations, to engaging tools-that would aid people in important ideas on personal finance, including budgeting, saving, investing, debt management, and retirement planning. For example, apps like Mint and YNAB (You Need A Budget) can provide budgeting tools combined with educational resources to guide people in their personal management. Social media applications, and other content-sharing platforms such as YouTube and Instagram host democratized financial information. Influencers, educators, and experts talk about bite-sized financial advice and tutorials in these avenues to reach people more greatly. It is free and alluring content, mostly custom-made according to the daily financial problems of an individual. Hence, digital media have made it possible for financial literacy to cross the traditional constraints of cost, time, and distance.

The other factor that has contributed to greater access through personalization is the nature of digital financial education. The new digital platforms use a user's needs, financial goals, and learning speed for the adaptation of the content. The application of AI-driven algorithms in the analysis of the user data may suggest modules for learning that are tailored to their specific needs more than what would be met in one-size-fits-all settings applied through traditional education. This personalization ensures that learning users interact with content appropriate and meaningful enough for their specific financial situation, enhances learning outcomes, and boosts financial competencies.

Bridging Financial Literacy Gaps

Digital financial education plays a tremendous role in filling and bridging gaps related to the financial literacy of populations that have heretofore been excluded from access to general financial systems. Global financial literacy statistics highlight the existence of a core deficit in the foundational knowledge required to operate effectively upon personal finances. Even though this deficit is typical of the financial landscape, it disproportionately afflicts the low-income, rural, and frequently marginalized sectors of society, not only because of lack of adequate access to financial services but also education opportunities as well.

Digital provides scalable answers in addressing the gaps. It ensures a wide array of content- from the very basics to more advanced in finance that equips enough knowledge in using money on their personal finance. Fintechs, for instance, increasingly integrate financial literacy modules into their services to make it so that while users can use the platforms for a transaction, they can at the same time learn about financial issues. These platforms can further change the content to fit the financial facts of reality for the demographic. This may include a young professional, retirees, or small business people. This way, digital platforms help to balance the financial literacy gap among different socio-economic groups through the right information. Also, the digital platforms are an engaging and participative mode of learning that most diversified population, especially the younger one that is believed to be native, will celebrate. Utilizing gamification, such as rewards, badges, or challenges, improves user engagement and retention of knowledge. Platforms such as the financial literacy called Duolingo, or the gamified investment apps simulating real-world markets investing, push users to learn more by making the learning process enjoyable and interactive. Additionally, they teach not only the financial skills but also encourage users to seek to apply them in real-world practices, which will reinforce behavioral change.

Online platforms offer the opportunity to learn multiple and dynamic methods; thus, it allows every individual to fill those gaps regarding financial knowledge irrespective of their level of experience or education. This gives way to a more financially literate society where individuals are best prepared for making well-informed choices, living financially responsibly, and, subsequently, leads to economic stability.

Reaching Underserved Groups

Perhaps the most significant positive impact of digital financial education is that it is accessible to previously underserved populations. Traditionally, financial education was concentrated much in towns or in formal institutions, and nearly everywhere else remained deprived of this very much-needed resource. The platform stretches geographical and financial barriers to make it reach people anywhere who have internet access or access to a mobile device.

It is true that mobile technology has outdone its counter interloper, and access to computerization and conventional banking systems has been undermined by reaching populations unable to access computers and conventional banking systems. For example, in the developing world, mobile phones are actually the biggest sources of internet. For example, fintech solutions that target mobile platforms have enabled individuals in remote places to enjoy the services and education offered by financial service providers. Apps that provide financial education locally, along with suitably low literacy levels, are especially empowering the locals to acquire essential financial skills. Public-private partnerships and government schemes have ensured the use of ICT in delivery of digital financial education to the underprivileged. Different governments and banks have been entering into agreements with technology companies to reach out to the underprivileged to deliver financial literacy programs through digital modes in the far-flung areas and villages. For example, the World Bank in conjunction with multiple NGOs, partnered with a number of fintech firms to roll out mobile-based financial literacy products involving resources on savings, loans, and credit management in developing countries.

Additionally, it is evident how channels are getting modernized including features that are particularly needed for women, migrants, and older people. Initiatives directed at achieving financial inclusion for these individuals usually provide for digital education so that they can learn how to be independent and self-sufficient as far as the management of their future finances is concerned. An example is the kind of financial education platforms targeting women, offering resources on overcoming barriers to female financial independence, such as pay disparities or access to financial services.

Hence, digital platforms have transformed financial education in order to provide democratized access and close gaps of literacy to the previously underserved populations. Digital platforms make financial knowledge more accessible, personalized, and engaging; thereby, individuals from every background acquire the necessary skills to reach financial stability and well-being. Digital platforms, using technology, will enable every global society to become financially literate and resilient.

Figure 1. Promoting sustainable growth through digital financial education

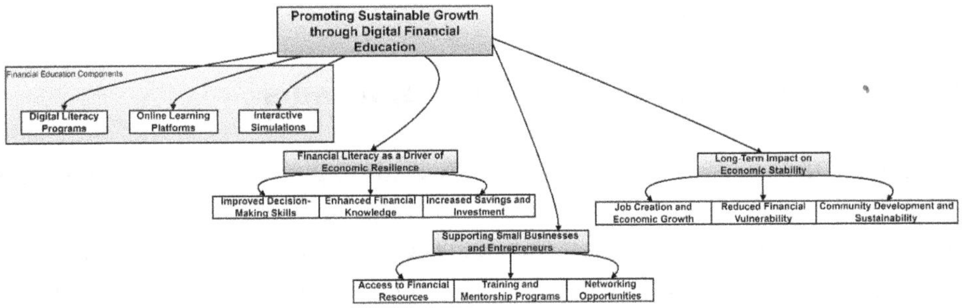

PROMOTING SUSTAINABLE GROWTH THROUGH DIGITAL FINANCIAL EDUCATION

Such education, digital financial education, has become a critical enabler of more sustainable economic development. Individual economic resilience and the growth of small businesses or entrepreneurs have broader implications, but the achievement of long-term economic stability is only possible with a solid base. Evidently, considering the growing complexity and digitalization of the financial sector, commensurate financial literacy becomes more and more necessary too. Digital channels are an efficient and scalable enabler of making accessible, adaptive, and effective financial education. This chapter explores how the process of financial education in the digital age is supposed to drive sustainable growth with its ability to empower the individual and the business to make informed financial decisions to foster resilience as well as long-term stability (Voukelatou et al., 2021). Figure 1 illustrates the concept of fostering sustainable growth through digital financial education, focusing on financial literacy, business support, and long-term economic stability.

Financial Literacy as an Accelerator of Economic Resilience

Individual's financial literacy is at the heart of sustainable economic growth since it represents an accelerator of economic resilience. Financially literate persons can readily and effectively manage personal finance, chart or navigate possible financial dangers, and respond to shifts in economic patterns. Understanding a need to save, budget, invest, and manage debt influence decisions that promote long-term financial stability, avoid debt, and even improve one's ability to respond to unexpected economic shocks.

In such an environment, digital financial education platforms play a very vital role because they provide scalable and accessible tools to help people build up their capacity in financial literacy. Interacting learning modules, quizzes, videos, and gamified experiences facilitate this process on digital platforms and enable users to come out well-equipped to handle all aspects of personal finance. Such flexibility, therefore, becomes crucial for those who didn't have access to formal financial education in the past. Besides that, financial literacy would help in building economic resilience through positive financial behaviors, among them regular savings and prudence in investments. Some online investment platforms and apps offer users personal recommendations on investments to guide them in prudent decisions on where and how to invest their savings. It could help individuals build their wealth overtime and prepare for future financial needs to contribute to broader economic growth by helping them raise their financial security and participation in the financial system.

Contributions to Small Businesses and Entrepreneurs

Besides promoting the financial capabilities of individuals, digital financial education advances the development of small businesses and entrepreneurs—the main agents of sustainable economic growth. Small businesses are core job creation, innovation, and economic diversification agents, although some entrepreneurs are grounded with such massive financial challenges related to a lack of financial literacy. While without a proper understanding of business finance, cash flow management, and financial planning, many small businesses struggle for survival and scale in competitive markets.

Digital financial education platforms are bridging this gap by offering firms-specific resources and equipment tailored to the small business owner/entrepreneur. Where an online business educational platform equips an aspiring entrepreneur with knowledge of management of business finance, access to funding, tax understanding, and how to steer around the regulatory framework in which they operate, applications such as QuickBooks, Wave, and Xero combine book-keeping and other accounting software with ancillary learning materials to help entrepreneurs handle their financial books and make and prepare for growth. Third, digital channels offer scaleable solutions to entrepreneurs with finance advice and mentorship. There are online business incubators and accelerators with a network of experts and investors to facilitate the better orientation of entrepreneurs on financial management and a business strategy. Such systems help entrepreneurs avoid pitfalls, access capital, and properly scale up their businesses as part of broader economic development.

The education coming through digitized financial services brings support to entrepreneurs in low-access markets or low-economic regions without access to traditional financial institutions. Mobile banking and fintech platforms open up for small business owners the source of finance such as loan and payment services that offer educational resources to improve their own financial literacy. The support to small business becomes critical because small businesses are going to drive innovation, which will help lead long-term sustainable economic growth in an environment where they thrive.

Long-term effects on the economy

Thus, the impact of financial education through digital channels on an individual and company basis can, therefore, be said to be more profound, for it nourishes overall economic stability. A financially well-educated population is most likely to conduct themselves responsibly in a financial aspect, thereby greatly reducing the risks that result in economic crisis due to poorly made financial decisions at the household or business level. Examples can even be shown such as prevention from overborrowing, lowering loan default rates, and lowering the vulnerability of individuals and businesses to economic downturns by virtue of mass financial literacy (Nurunnabi et al., 2020).

Such education also leads to more participation in the formal financial system, much needed to be settled in an economy. People and small entities better understand financial products and services, yet they are more likely to engage with banks, credit unions, and investment platforms. Thus, deepened financial markets, enhanced liquidity, and support for sustainable economic growth through a strengthened financial sector are beneficiary results. The role of financial literacy is increasingly recognized by governments and policy makers in promoting long-term economic stability, and, as such, they are supportive of digital financial education initiatives. It has also been identified that public-private partnerships are an effective approach toward financial education, where governments collaborate with fintech companies and educational institutions to make financial education accessible to all. Financial capacity building among citizens and small businesses makes them more resilient in the overall economy.

These would be radical changes in developing countries with low levels of financial literacy. Mobile money platforms should develop low-cost accessible financial education for the regions for probable improvements in financial inclusion, upliftment from poverty, and resultant stability and growth in the global economy. Such programs as the financial literacy efforts by the World Bank, now undertaken through digital means in Africa and Asia, have shown to have a positive long-term effect on both individual welfare and macroeconomic stability. Financial education also promotes a culture that embraces long-term financial planning and investments. Such a person who is informed about retirement savings, long-term investment forms, and risk management is already putting in place measures to safe guard his or her financial futures. Preparation in this manner takes some pressure off the government-sponsored safety nets and social security systems, making their economic policies sustainable.

Thus, digital financial education would be a powerful tool to encourage sustainable economic growth, stimulate the economy's adaptability in terms of financial resilience, provide support to small enterprises, and lead to long-term economic stability. Digital resources make this financial literacy reachable to all, enabling persons and businesses to make the most appropriate decisions toward an economy stable and resilient. With the ever-changing financial systems, the digital education of financial factors will eventually become even more crucial in helping to foster a sustainable pattern of growth and stability, hence better prepared to face prospective challenges.

Figure 2. Financial education enhances well-being through various interconnected components

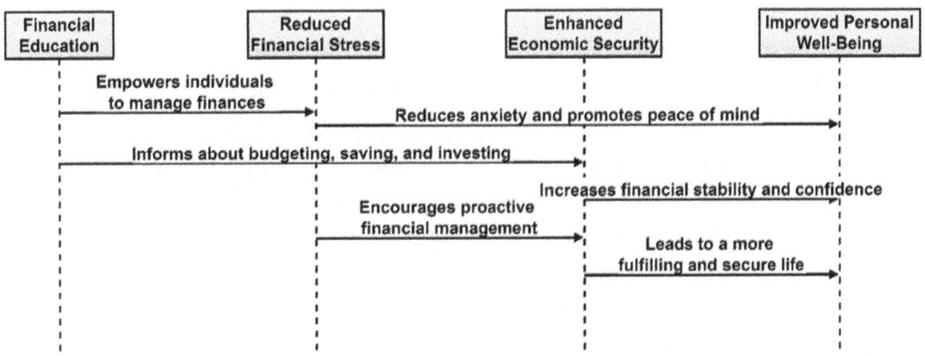

ENHANCING WELL-BEING THROUGH FINANCIAL EDUCATION

Financial education has more meaning than a budget, invest, or save. Rather, it would mean being an empowered individual to enhance one's overall well-being by helping alleviate financial stress, enhancing economic security, and linking the personal well-being into financial choices. As online platforms or digital ways of financial learning grow then individuals can access more than just having increased financial literacy but also resulting in emotional or psychological well-being. This chapter shall discuss how significant it is for financial education to play an important role in reducing financial stress and anxiety, promoting long-term financial security, and engraving well-being as a core theme in the process of financial learning. Figure 2 illustrates the positive impact of financial education on financial stress reduction, economic security, and overall personal well-being through interconnected components (Nurunnabi et al., 2020).

Reducing Financial Stress and Anxiety

The very first and perhaps one of the most pressing impacts of financial education could be that it can reduce financial stress and anxiety. Financial stress is among the most prevalent reasons behind mental health issues-both for individuals and families. People are usually stressful when they do not know the basic money management skills. Even simple bills payments, savings for the future, or managing debt can be stressful. Due to this pressure, people may suffer sleepless nights, anxiety disorder, and even depression. They contribute significantly to a severe dislodge in the quality of life.

Digital finance education could be able to curb these problems by providing accessible comprehensive tools of financial literacy. Thus, the persons should feel confident to manage their finances. These types of platforms typically offer modules on debt management, financial planning, and expense tracking: step-by-step user guides to overcoming many of the common difficulties found in finance. Budgeting app YNAB (You Need a Budget) or Mint will allow the user to set financial goals, keep track of expenses, and make well-informed financial decisions. Such practical advice can help lighten the often-heavy load of stress that comes with money management by breaking down complex finance into doable things. Financial literacy also helps people learn not to make certain common errors that typically include running up large amounts of credit card debt or other poor investment choices. The same thing which is possessed by these people after learning such information and right guidance helps them feel more confident regarding the management of their finances, a decrease in anxiety that also results in a greater feeling of financial freedom. With more surety in managing one's finances, people are not as likely to suffer from chronic financial stress, thereby also promising better overall mental health.

Financial Security and Peace of Mind Financial security is an important personal issue while education in financial matters helps achieve this. Financial security can be defined as a state of having ample resources to satisfy all basic needs, emergencies, and planning in the long term. Many people face financial insecurity due to bad savings practices and poor debt management and lack of proper planning over the long run. Such factors may make one constantly worry about the stability of their finances, breeding an uncomfortable feeling of being helpless. Digital financial education is the way to build financial security because it teaches a person how to save wisely, invest intelligently, and prepare for short-term or long-term requirements. Many current digital financial services can be accessed through tools that help to create individual savings plans, retirement plans, or emergency funds. An individual can learn how to allocate their income toward necessary expenses by using structured learning paths and saving some for other needs.

For example, services like Acorns and Betterment enable one to invest little amounts of money in diversified portfolios. Investment solutions, in this way, can become as easy as possible to build wealth over time without requiring an in-depth understanding of financial markets. With such easy learning tools, people will be more likely to create a financial cushion that eventually implies long-term financial security and peace of mind.

Besides that, financial literacy equips a person with the basic idea on how to shield herself from financial dangers such as unemployment, medical emergencies, and falls in the economy. One will only be further equipped with a sense of security for the unknown associated with financial matters by simply knowing the different insurance avenues, diversifying investments, and having an emergency savings fund. Financial security, in the sense that one no longer lives with constant anxiety about the future of how they will handle their finances, means peace of mind.

Inclusion of Well-Being within Financial Education

With all these knowledge findings related to the conjunction of financial well-being and general well-being, there should be no omission of the integration of well-being within financial education as a primary element. Financial literacy programs need to be more than enabling individuals to run their households and finances; they need to equip them with the knowledge of how these decisions made about money will impact their mental, emotional, and social well-being. Since the subjective concept of well-being is included in financial education, individuals are motivated to think about finances from a more holistic perspective, considering not only financial goals but also how this financial decision will change the quality of their life (Chambers & Sandford, 2019).

Many digital financial education platforms have begun including wellness in their models by providing resources that deal with the emotional part of money management. Some offer mindfulness techniques to help people better deal with the stress of making financial choices, and others made available coaching services where a user gets to work with a financial advisor to coach them through a difficult financial situation while maintaining focus on emotional well-being. Third, well-being is linked with financial education in the process of instructing individuals on how personal values and goals need to be aligned with financial decision-making. Financial well-being is not about wealth alone but is more about satisfaction and fulfillment from decisions regarding finance. An example would be spending money on experiences that create happiness, like travel or hobbies, instead of saving every penny. These value-based spending programs especially tend to educate people on how to manage finances that are in balance with achieving their set financial goals and taking good care of themselves.

These programs of increased financial wellness, provided to employees, indicate an understanding of the link between literacy in finances and well-being. Such programs attempt to reduce the level of financial stress with respect to an employee by engaging him or her in workshops, one-on-one financial counseling, and access to financial education platforms. Such initiatives not only enhance the level of the employees' financial knowledge but also enhance overall well-being, productivity, reduced absenteeism, and job satisfaction. Financial literacy enhances well-being to the point of decreasing financial stress and anxiety, and increasing one's security about their finances together with the integration of well-being into financial decisions. Scalability, accessibility, and customization in digital financial education platforms are what will help to achieve these ends. Through a focus on financial literacy and emotional well-being, these methods show greater promise for the development of holistic financial management approaches, leading to higher satisfaction and peace of mind overall. As the sphere of financial education evolves in the near term, such roles will be highly central in helping individuals live with peace of mind and balance in a productive and financially secure manner.

LEVERAGING TECHNOLOGY FOR EFFECTIVE FINANCIAL EDUCATION

Financial education is no longer restricted to the pace at which the changes occur in terms of the way technology has its impact on access, delivery, and personalization. The scope of the implementation of adaptive learning technologies, AI, data analytics, and gamification will have a sense to make financial education more effective to access diverse audiences. It may enrich the learning experience, increase user engagement, and build a level of financial literacy by tailoring educational content according to learners' needs. In this chapter, we shed light on how adaptive learning technologies, AI, data analytics, and gamification are revolutionizing financial education and how they should be applied for even more influential learning experiences (Ingale & Paluri, 2022). Figure 3 illustrates how technology enhances financial education through access, delivery, and personalization, illustrating key transitions and states in the integration process.

Figure 3. Technology leverages effective financial education through access, delivery, and personalization

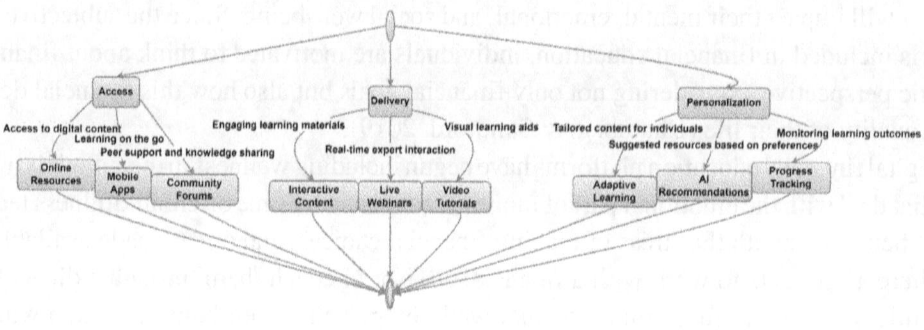

Adaptive Learning Technologies

Adaptive learning technologies provide a tailored method for financial education so that the person learns content with regard to specific learning needs and pace. With algorithms and machine learning, adaptive learning technologies can adjust curriculum according to learner's progression and strengths and weaknesses. The adaptation can be pretty useful in the scope of financial education because persons coming from different backgrounds will require different levels of financers.

For instance, Coursera or Khan Academy adaptive learning technologies first diagnose the current knowledge of the users and then produce a tailored learning trajectory. This means a learner knowing very little about budgeting gets an educational track of foundational education, whereas more advanced learners might look into investment strategies or retirement planning. This customized approach prevents storing information that is either too advanced or at a level that is lower than what learners need to retain and become engaged (Gabor & Brooks, 2020).This technology will not only offer adaptability learning and personalization but also instantaneous feedback, such that the mistakes of the learners become obvious and can be corrected in the shortest time possible. In financial education, this is pivotal because in real life, financial mistakes have huge implications. Adaptive platforms simulate financial decision-making and provide instant feedback; thus, learners are able to make better decisions regarding money handling in a safe and controlled environment.

Role of AI and Data Analytics

There is a revolution of AI and data analytics in the modern world of financial education in that they give insights to learner behavior, performance, and needs. Using an AI-driven platform, researchers can scale large amounts of data to tell whether people learn in patterns in similar ways or in patterns that are unique to them, where they make mistakes, and what content appeals most to them. It therefore gives educators power to make decisions in designing and improving financial education programs based on data (Burr et al., 2020).

For example, AI could predict which topics a learner may have difficulties coping with based on his/ her interaction history and thus be able to tailor what it presents. For example, within the financial education domain, AI can provide tailored suggestions in savings, investment, or debt management based on the individual's financial situation, goals, and behaviour. A user who continually accesses content on credit management can receive more materials on improving credit scores to avoid falling into traps of debt. Other AI-driven online personal finance learning platforms, such as Zogo and Everfi, applies NLP to enable users to engage in a conversation and answer any questions while providing the user with personalized advice about their finance. It mimics the process that would be involved between a financial advisor and provides users with customized suggestions on financial planning, retirement strategy, and investment. With every interaction, the AI continues learning and enhances the relevance and appropriateness of its suggestions to the fullest extent.

Data analytics is also very crucial for measuring the effect of financial education programs. For instance, based on user data, an instructor would track learner activity and hence pinpoint the areas that need more content modification, which teaching methodology to adopt, and much more. As an example, if it is a fact that users drop out of a specific financial education course once learners read certain modules. The educator should dig further to determine whether the module is too tough to understand, unclear, or less interesting. This leads to continuous improvement in course designs, which, in turn, will ultimately reflect in better learning outcome.

Gamification and User Engagement

Gamification is one of the most important elements for enriching user engagement in financial education. The introduction of game-like elements such as rewards, leaderboards, and challenges supports the making of financial education platforms interesting for users. As gamification encourages learning further, it keeps all the necessary financial concepts running in users' minds in an entertaining and memorable way (Ingale & Paluri, 2022). For example, there are habit-forming and financial literacy apps - similar to Duolingo or Habitica, games of developing the productive learning habits. The reward - for example, points, badges, or in-game currency - could be designed to motivate people to do this by setting an exercise plan, completing lessons, taking quizzes, or mastering challenges. The same would apply in financial education: motivate users to track spending, save money, or complete certain educational modules on the topics of budgeting or investing. So, Obtaining rewards for completing tasks creates a feeling of accomplishment and encourages the student to continue along the curriculum. Gamifying has the ability to incorporate competitive components like a leaderboard. Here, the goal is to challenge a person to improve their best score by bettering somebody else or be better than the next highest score on the leaderboard. That creates a sense of community and friendly competition, which should be enough to motivate on its own. For example, a site about financial literacy may conduct a competition of users in simulated investment challenges, where virtual stock trades are made and the highest returns attained. This presents opportunities for learners to apply theoretical knowledge in a practical yet risk-free environment while engaging with the idea.

Gamification enhances building financial habits itself. For example, virtual platforms like Squirrel or Bankaroo help people, especially young audiences, know how to put up budgets and plans in saving through the reward of meeting savings goals or staying within the budget. It becomes more relatable and less intimidating when learning about finances with this method, however, might intimidate a few traditional ways of learning about finances. Hence, technology plays a transformative role in reshaping financial education. Adaptive learning technologies, AI, data analytics, and gamification are all powerful tools to personalize and enhance financial literacy programs to the highest level. These innovations will increase access to financial education, improve individual engagement, and enhance the impact of the acquired skills and knowledge toward financial independence and long-term well-being. At the same time, as new digital technologies are developed, the level of integration of these technologies will grow, and learning will become even more engaging and effective with a greater reach in democratizing access to financial knowledge and guidance toward making responsible financial decisions.

CHALLENGES IN DIGITAL FINANCIAL EDUCATION

Digital financial education holds the promise to change the whole process of gaining financial knowledge and skills. However, there are few challenges in the present scenario that hinder its development, which are fairness in accessing the same, quality and credibility of the contents, ethical issues in collecting and storing data and privacy and security (Lyons et al., 2020). Figure 4 presents challenges in digital financial education, highlighting key areas of concern that may hinder its effective delivery and uptake.

Figure 4. Challenges in digital financial education

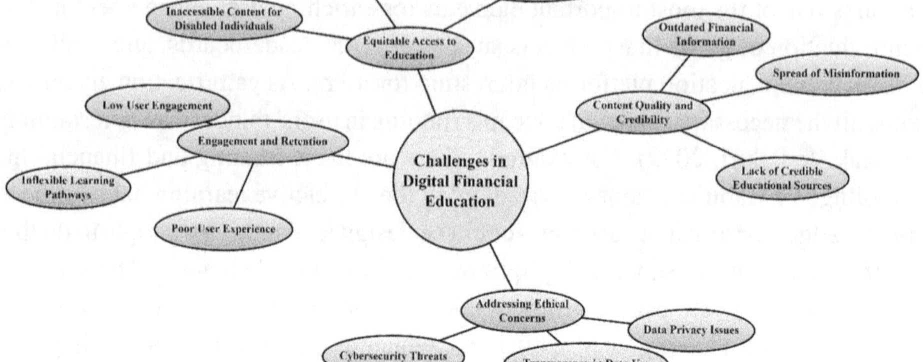

Equitable Access to Education

The most significant challenge of the digital finance is education while furnishing equal access to the resources for everybody. While online and digital devises remain the dominant face of this world, significant disparities exist and persist, including low-income communities, rural populations, and marginalized groups who may not have reliable internet access or digital devises for online financial education. Thus, the digital divide reinforces and extends existing inequalities in financial literacy in a way that it is very hard to reach underserved populations more effectively (Gabor & Brooks, 2020; Ingale & Paluri, 2022).

Moreover, even when access is enabled, lack of digital literacy can also present a barrier to the effective usage of any financial education resources. Most people without the relevant skills to engage with digital means tend to feel frustrated and out of touch. To this end, stakeholders should invest more in up-grading digital infrastructure and delivering literacy programs in these disadvantaged areas so that everyone gets an opportunity to unleash digital financial education.

Quality and Credibility of Content

A third challenge is that, with this sheer volume of information online, determining or even guaranteeing that the financial education content is high quality and believable becomes increasingly complicated. Many provide advice on money management, and it is challenging for learners to differentiate between reliable and bad information or incorrect advice. Misinformation leads to wrong financial decisions with devastating consequences to the financial health of more or less people (Abad-Segura & González-Zamar, 2019; Goyal & Kumar, 2021; Tay et al., 2022). Digital financial education should, therefore, have clearly outlined guidelines regarding its content development and evaluation. Institutions of learning, financial houses, as well as supervisory bodies, can pool their efforts to create and distribute quality material. The platforms should also have verification mechanisms to ensure that any information served is relevant and up to date to increase the chances of engaging with the content.

Addressing Ethical Issues: Data Privacy and Security

Because digital financial education relies on capturing sensitive information, data privacy and security issues arise right at the top. The consumers may not want to use the online channels due to fear of their financial data being exploited or compromised. Data breaches and breach of confidential data destroy trust in digital financial education and hence have a lower participation and student engagement. These risks can be mitigated by giving prime importance to data protection measures, as well as establishing transparent privacy policies. Strong encryption methods along with regular security audits and clear user consent to get access to their personal information may protect this information from hackers and thieves. Moreover, educating the users on safe data handling and security practices will enable them to make better decisions over their engagement with digital financial education.

While there is enormous promise in digital financial education, big challenges exist that must be addressed in order to increase its impact. Focusing on equitable access, the quality of content, and issues of ethics will ensure that everyone can benefit from valuable resources for financial education.

POLICY AND COLLABORATIVE FRAMEWORKS

A successful policy framework for digital finance requires collaboration, public-private partnerships, digital literacy-focused policy efforts, and substantial infrastructure and technology investments. These elements can create an inclusive ecosystem where financial literacy is fostered, enabling individuals to make informed financial decisions, promoting financial literacy and inclusivity (Dienlin & Johannes, 2020; Ingale & Paluri, 2022; Lyons et al., 2020).

Figure 5. Policy and collaborative frameworks for digital finance

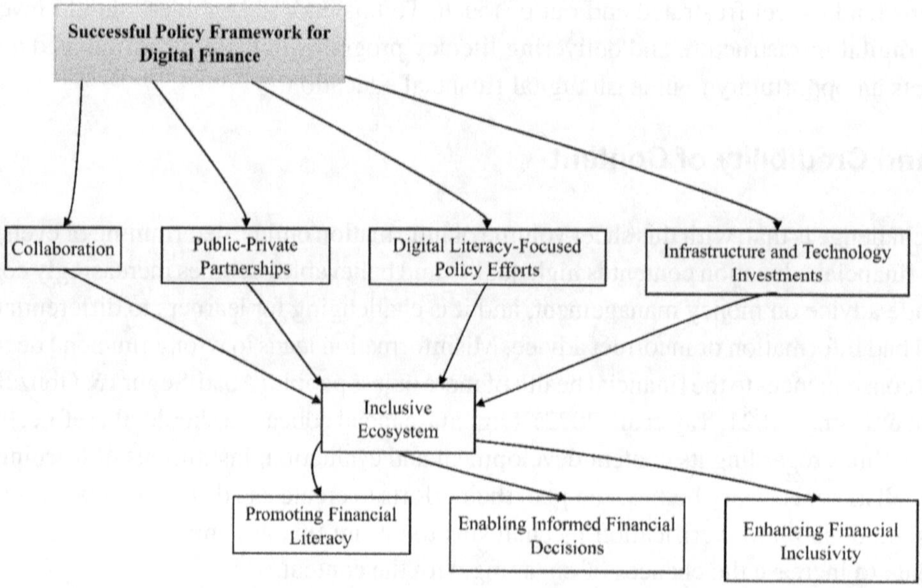

Public-Private Partnerships in Financial Education

The improvement of digital financial education relies on public-private partnerships. With both sectors pooling their resources, skills, and influence, each sector benefits accordingly. In this respect, through the partnerships, various players such as the government agencies, institutions of learning, non-profit organizations, and private companies join forces to develop comprehensive curricula for the financial education systems of specific communities.

For instance, a public-private partnership can create an online resource center with free financial education material targeted to financial institutions' needs. A public-private partnership would guarantee accuracy, relevance, and accessibility to all the stakeholders. It could also pool resources for outreach activities meant to educate the public about the availability of financial education tools and the underserved populations. Public-private partnerships could be designed for innovative solutions, such as mobile applications or gamified learning experiences, to meet the different styles and preferences of learners. Such collaborations can go further to opportunities in research and evaluation to be able to assess the impact of financial education programs so that the policymakers can continuously improve and adapt financial education to the changing needs of the learner.

Policy Initiatives for Digital Literacy

Effective policy initiatives are mandatory to be able to promote digital literacy, which forms the basis for high-quality digital financial education. Governments must target the development of comprehensive digital literacy programs that could equip people with the confidence to access the online resources for financial education. The programs should be sensitive to vulnerable populations so that everybody has a fair chance at increasing their levels of digital competencies.

Polcies may entail including digital literacy in school curriculums, adult training programs, and resources for community-based organizations to offer digital literacy training. Partnerships with the technology firms will also enhance the impact of these initiatives through their very specific knowledge and training in the use of that technology. They will be able to engage with, use, and benefit from the digital financial education resources.

Infrastructure and Technology Investments

Such infrastructure and technology investment is vital for creating an enabling environment for digital financial education. There is a need to bolster internet connectivity infrastructure, especially in rural underserved areas, to make online financial education resources available to all. Subsidized internet services for the poor also constitute one of the efforts to increase the availability of broadband. Simultaneously, investment in the technology solutions for delivering enhanced financial education will be required. These consist of accessible front ends, mobile applications, and appealing interactive media appropriate to various learning styles. Through such tools, technology can assure a smooth, fun, and holistic user experience which may make significant contributions toward the reach and effectiveness of the financial education initiative.

There is also a need for funding in terms of research and development regarding digital financial education. Conducting explorations into novel approaches and technologies will help the stakeholders continually be able to improve quality resources for enhancing financial education, appropriately responding to the changing needs of learners.

Hence, the promotion of digital financial education would require full policy and collaborative undertakings. Such can be created by the development of public-private partnerships in implementing targeted policy initiatives on digital literacy, investment, and infrastructure and technology. In this regard, through the development of an environment that fosters personal improvements in financial knowledge and skill, stakeholders will be able to improve financial literacy, economic resilience, and overall well-being in society.

FUTURE DIRECTIONS AND INNOVATIONS

Future developments of digital financial education promise exciting opportunities through emerging technologies such as augmented reality (AR), virtual reality (VR), and blockchain to positively enhance the experience of learning and access.These advancements can potentially transform the relationship between a person and financial concepts, ushering in even more immersive, effective educational approaches. The figure 6 depicts the future directions and innovations in digital financial education, focusing on emerging technologies (Trivedi et al., 2024).

Figure 6. Future directions and innovations in digital financial education, focusing on emerging technologies

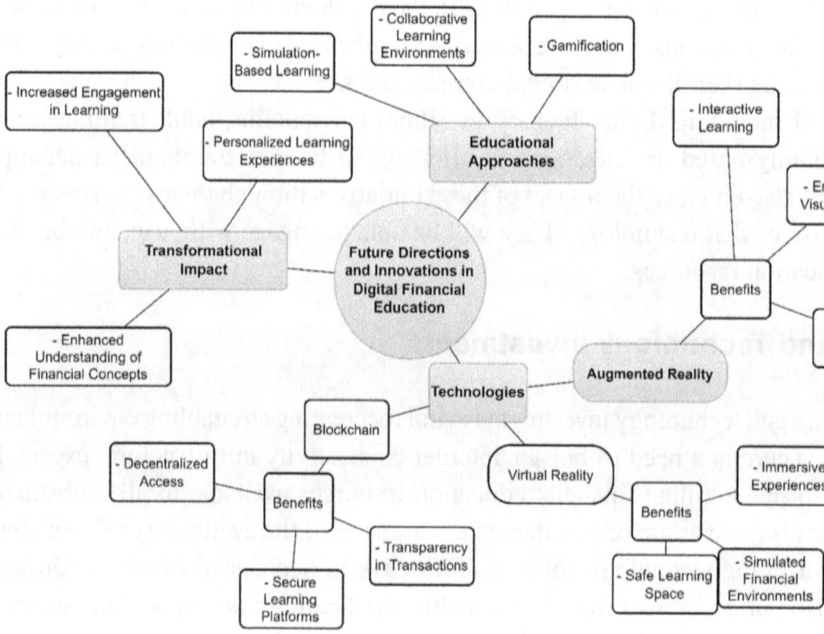

Emerging Technologies: AR, VR, and Blockchain

The prospect of AR and virtual reality in transforming digital financial education lies in the fact that these may offer interactional immersive experiences which could teach learners in an entirely distinctive manner. For instance, AR technology allows information and scenarios about finance to be superimposed over the real world, such that complex financial concepts can be visualized within tangible contexts. It can also depict real-world financial settings, within which learners can make financial choices in a risk-free environment. The technologies can help unfurl the mysteries surrounding financial concepts and make learning more fun, thus increasing learner engagement and retention (Sekhar et al., 2024). Blockchain technology further enables innovative options in digital financial education. It can provide a safe and transparent method for keeping track of transactions. This makes it possible for decentralized education hubs to be created. Such hubs will have verifiable credentials for courses that are completed, thus allowing learners to recognize the achievements they have made. There is also improved data privacy; using blockchain technology, it is possible to maintain control over personal information when accessing financial education resources.

Real-World Applications and Simulations

Real-world applications and simulations can heighten digital financial education considerably. For example, interactive tools simulating investment scenarios or budgeting exercises will allow the learner to apply the theory in the practical scenario. This will help learners sharpen their critical thinking and problem-solving skills, making them competent to make reasoned decisions in their everyday lives (Priyanka et al., 2024; Trivedi et al., 2024; Vijaya Lakshmi et al., 2024).

These financial education programs can further partner with finance companies to provide virtual accounts and investment platforms to the learners. Through this mechanism, users can start interacting with real-time market data and know very well what the implications of their financial decisions would be without really taking a risk on such investments in the real world. Such real-world applications make theory and practice finally click into each other's arms, making financial education more relevant in the real world.

Breaking Down National Barriers Using Technology

Technology forms a very essential component in expanding global access to digital financial education, owing to the fact that the internet heavily connects people on different continents of the earth. Financial education resources will thus be accessible for most persons irrespective of their geographical location, while online platforms can make provision for multilingual content.

Another area in which mobile technology can impact accessibility in developing regions, where most people use smartphones to access the internet, and educational content especially tailored for mobile users will encourage better learning on-the-go. Technology such as AR, VR, and blockchain are bright signs of emerging innovation in digital finance. Integration of real-world application with expanded global reach through technology would be applied in making financial education more effective for empowering people to boldly navigate their financial futures (Dass james & Boopathi, 2016; Haribalaji et al., 2014).

CONCLUSION

In conclusion, digital financial education plays a good role today in an increasingly fast-changing financial world, where sustainable growth and well-being are born. By exploiting the capabilities of technology and innovative approaches, we can democratize access to financial knowledge by bridging literacy gaps and empowering the underserved populations. Financial education becomes more enjoyable and interesting if integrating emerging technologies, such as AR, VR, and blockchain, into it. This makes it possible to make complex financial concepts understandable and readable through life stories.

Plus, public-private partnerships would be important to elaborate fully coherent and enabling frameworks that would support the cause of digital literacy, address the barriers of equitable access, and content credibility. Investment in infrastructures and technologies would mean that financial education shall be provided with free accessibility to all socio-economic classes.

In recent years, responding to emerging needs has involved learning activities in which one is sensitive and responsive to the needs, interests, and aptitudes of learners. Real-life applications or simulations could be useful in supporting practice-based learning where the learner applies what he or she learns to actual contexts where he or she makes long-term financial decisions. It is only when the digital financial education enables people to become more capable of making well-informed financial decisions that we will have gotten the potential of digital financial education into its actual prime. This commitment goes a long way in making people individually more financially resilient and also economists more resilient, because through these people, the economies and the communities get to be strong and pave the way for a fully financially literate society.

REFERENCES

Abad-Segura, E., & González-Zamar, M.-D. (2019). Effects of financial education and financial literacy on creative entrepreneurship: A worldwide research. *Education Sciences*, *9* (3), 238. DOI: 10.3390/educsci9030238

Bayuk, J., & Altobello, S. A. (2019). Can gamification improve financial behavior? The moderating role of app expertise. *International Journal of Bank Marketing*, *37* (4), 951–975. DOI: 10.1108/IJBM-04-2018-0086

Burr, C., Taddeo, M., & Floridi, L. (2020). The ethics of digital well-being: A thematic review. *Science and Engineering Ethics*, *26* (4), 2313–2343. DOI: 10.1007/s11948-020-00175-8 PMID: 31933119

Chambers, F., & Sandford, R. (2019). Learning to be human in a digital world: A model of values fluency education for physical education. *Sport Education and Society*, *24* (9), 925–938. DOI: 10.1080/13573322.2018.1515071

Dass, A., & Boopathi, S. (2016). Experimental Study of Eco-friendly Wire-Cut Electrical Discharge Machining Processes. *International Journal of Innovative Research in Science, Engineering and Technology*, *5* (1), 666–675.

Dienlin, T., & Johannes, N. (2020). The impact of digital technology use on adolescent well-being. *Dialogues in Clinical Neuroscience*, *22* (2), 135–142. DOI: 10.31887/DCNS.2020.22.2/tdienlin PMID: 32699513

Friedline, T., Chen, Z., & Morrow, S. (2021). Families' financial stress & well-being: The importance of the economy and economic environments. *Journal of Family and Economic Issues*, *42* (S1), 34–51. DOI: 10.1007/s10834-020-09694-9 PMID: 32837140

Gabor, D., & Brooks, S. (2020). The digital revolution in financial inclusion: International development in the fintech era. In *Material Cultures of Financialisation* (pp. 69–82). Routledge. DOI: 10.4324/9781351002103-6

Goyal, K., & Kumar, S. (2021). Financial literacy: A systematic review and bibliometric analysis. *International Journal of Consumer Studies*, *45* (1), 80–105. DOI: 10.1111/ijcs.12605

Haribalaji, V., Boopathi, S., & Balamurugan, S. (2014). Effect of Welding Processes on Mechanical and Metallurgical Properties of High Strength Low Alloy (HSLA) Steel Joints. *International Journal of Innovation and Scientific Research*, *12* (1), 170–179.

Ingale, K. K., & Paluri, R. A. (2022). Financial literacy and financial behaviour: A bibliometric analysis. *Review of Behavioral Finance*, *14* (1), 130–154. DOI: 10.1108/RBF-06-2020-0141

Lusardi, A. (2019). Financial literacy and the need for financial education: Evidence and implications. *Swiss Journal of Economics and Statistics*, *155* (1), 1–8. DOI: 10.1186/s41937-019-0027-5

Lyons, A. C., & Kass-Hanna, J. (2021). A methodological overview to defining and measuring "digital" financial literacy. *Financial Planning Review*, *4* (2), e1113. DOI: 10.1002/cfp2.1113

Lyons, A. C., Kass-Hanna, J., Liu, F., Greenlee, A. J., & Zeng, L. (2020). *Building financial resilience through financial and digital literacy in South Asia and Sub-Saharan Africa*. ADBI Working Paper Series.

Nurunnabi, M., Almusharraf, N., & Aldeghaither, D. (2020). Mental health and well-being during the COVID-19 pandemic in higher education: Evidence from G20 countries. *Journal of Public Health Research, 9* (1_suppl), jphr-2020.

Panos, G. A., & Wilson, J. O. (2020). Financial literacy and responsible finance in the FinTech era: Capabilities and challenges. []). Taylor & Francis.]. *European Journal of Finance*, *26* (4–5), 297–301. DOI: 10.1080/1351847X.2020.1717569

Philippas, N. D., & Avdoulas, C. (2021). Financial literacy and financial well-being among generation-Z university students: Evidence from Greece. In *Financial Literacy and Responsible Finance in the FinTech Era* (pp. 64–85). Routledge. DOI: 10.4324/9781003169192-5

Priyanka, N., Sethi, S., Sahai, A., Srivastava, A., Sambathkumar, M., & Boopathi, S. (2024). Reality for Human Experience in AI in the Digital Economy. In *Multidisciplinary Applications of Extended Reality for Human Experience* (pp. 374–400). IGI Global., DOI: 10.4018/979-8-3693-2432-5.ch016

Sabri, M. F., Anthony, M., Law, S. H., Rahim, H. A., Burhan, N. A. S., & Ithnin, M. (2024). Impact of financial behaviour on financial well-being: Evidence among young adults in Malaysia. *Journal of Financial Services Marketing*, *29* (3), 788–807. DOI: 10.1057/s41264-023-00234-8

Sekhar, K. Ch., Ingle, R. B., Banu, E. A., Rinawa, M. L., Prasad, M. M., & Boopathi, S. (2024). Integrating VR and AR for Enhanced Production Systems: Immersive Technologies in Smart Manufacturing. In *Advances in Computational Intelligence and Robotics* (pp. 90–112). IGI Global. DOI: 10.4018/979-8-3693-6806-0.ch005

Sorgente, A., & Lanz, M. (2019). The multidimensional subjective financial well-being scale for emerging adults: Development and validation studies. *International Journal of Behavioral Development*, *43* (5), 466–478. DOI: 10.1177/0165025419851859

Tay, L.-Y., Tai, H.-T., & Tan, G.-S. (2022). Digital financial inclusion: A gateway to sustainable development. *Heliyon*, *8* (6), e09766. DOI: 10.1016/j.heliyon.2022.e09766 PMID: 35785228

Trivedi, P., Ravisankar, A., Kumar, G. P., Krishnamoorthy, V., Shukla, S., & Boopathi, S. (2024). Finance and Technology-Integrated Digital Economy: Circular Age of a Sustainable Digital Revolution. In *Navigating the Circular Age of a Sustainable Digital Revolution* (pp. 91–120). IGI Global. DOI: 10.4018/979-8-3693-2827-9.ch004

Vijaya Lakshmi, V., Mishra, M., Kushwah, J. S., Shajahan, U. S., Mohanasundari, M., & Boopathi, S. (2024). Circular Economy Digital Practices for Ethical Dimensions and Policies for Digital Waste Management. In *Harnessing High-Performance Computing and AI for Environmental Sustainability* (pp. 166–193). IGI Global., DOI: 10.4018/979-8-3693-1794-5.ch008

Voukelatou, V., Gabrielli, L., Miliou, I., Cresci, S., Sharma, R., Tesconi, M., & Pappalardo, L. (2021). Measuring objective and subjective well-being: Dimensions and data sources. *International Journal of Data Science and Analytics*, *11* (4), 279–309. DOI: 10.1007/s41060-020-00224-2

KEY TERMS AND DEFINITIONS

AR (Augmented Reality): Technology that overlays digital content onto the real world, enhancing experiences by blending physical and virtual environments.

VR (Virtual Reality): Immersive technology that creates a fully digital environment, simulating real-world or imaginary scenarios for users.

Blockchain: A decentralized digital ledger that securely records transactions, ensuring transparency and preventing tampering.

Chapter 15
K-Nearest Neighbors Regression and Applications

Dayanna Ortiz-Villaseñor
https://orcid.org/0009-0004-9652-7073
Universidad Autónoma de Baja California, Mexico

Moises J. Castro-Toscano
https://orcid.org/0000-0003-2240-5204
Universidad Autónoma de Baja California, Mexico

Gabriel Trujillo-Hernández
https://orcid.org/0000-0003-1556-387X
Universidad Politécnica de Baja California, Mexico

Leonardo Daniel Medina-Madrazo
https://orcid.org/0009-0000-7319-3630
Universidad Autónoma de Baja California, Mexico

Oscar Real-Moreno
https://orcid.org/0000-0002-6064-5084
Universidad Estatal de Sonora, Mexico

D. Barrera-Román
https://orcid.org/0009-0005-6228-4357
Universidad Politécnica de Baja California, Mexico

ABSTRACT

k-Nearest Neighbors Regression (kNN regression) is a non-parametric, instance-based learning algorithm widely employed for predicting continuous target variables. In contrast to conventional regression methods assuming specific functional forms, kNN regression operates by averaging the target values of the k nearest data points, determined by a chosen distance metric such as Euclidean or Manhattan distance. This method exhibits high adaptability to diverse data distributions, rendering it a versatile tool for modeling in finance, healthcare, and environmental contexts. The selection of k, representing the number of neighbors considered, significantly influences the model's performance by striking a balance between bias and variance. The simplicity and effectiveness of kNN regression are counterbalanced by challenges such as computational complexity with large datasets and susceptibility to noise. This chapter delves into the mathematical foundations and fundamentals of kNN regression, offering a comprehensive guide for leveraging this algorithm in predictive modeling endeavors.

DOI: 10.4018/979-8-3693-6910-4.ch015

INTRODUCTION

The k-Nearest Neighbors algorithm (kNN) performs as an effective method for addressing both classification and regression challenges. The kNN algorithm determines a data point's classification based on its nearest neighbors' classifications. It operates assuming that similar entities are likely nearby, which is essential for the algorithm's practical application and overall effectiveness. The kNN algorithm was developed in 1951 by Evelyn Fix and Joseph Hodges, specifically for discriminant analysis, at a time when determining probabilistic densities through parametric estimation presented considerable challenges (Bansal, Goyal, & Choudhary, 2022). They authored a technical analysis report for the United States Air Force, which introduced discriminant analysis as a non-parametric classification method. Unfortunately, this important paper was never formally published, likely due to confidentiality concerns stemming from the aftermath of World War II. This work marks the first known introduction of non-parametric classification, which means that it doesn't make any assumptions about the underlying data distribution and doesn't build a model until the time of prediction. Instead, it memorizes the entire training dataset. In consequence, this characteristic would later contribute to the development of the well-known kNN algorithm. In 1967, Thomas Cover and Peter Hart demonstrated an upper bound on the error rate for multiclass kNN classifications. Since then, the field has continued to advance, with numerous improvements and innovative approaches emerging.

The kNN regression stands as a foundational and extensively utilized algorithm in machine learning and data science, specifically for predicting continuous numerical outcomes. Unlike traditional parametric models, kNN regression functions as a non-parametric, instance-based learning approach. This characteristic indicates that kNN regression does not depend on a fixed mathematical formula or equation; instead, it refrains from making assumptions regarding the relationship between input data and output. Furthermore, the term “instance-based” highlights that kNN regression does not create a predictive model in advance. Rather, it generates predictions by directly comparing a new data point to existing examples, utilizing the most similar instances to derive the output. In consequence, this relies on the similarity between data points rather than presupposing a predefined relationship between input features and the target variable. This adaptable approach enables kNN regression to effectively model intricate, non-linear data relationships, thus rendering it a pivotal tool across various applications.

While kNN regression presents an intuitive and straightforward implementation, it is accompanied by its challenges. The algorithm's performance is notably sensitive to the selection of hyperparameters and may incur significant computational costs, particularly in handling large datasets. Furthermore, the curse of dimensionality, wherein the meaningfulness of the distance between data points diminishes as the number of features increases, can markedly deteriorate the model's accuracy.

This chapter presents a comprehensive examination of kNN regression, starting with a detailed description of its foundational concepts and principles. Subsequently, an in-depth analysis of the critical determinants impacting its efficacy is undertaken, encompassing distance metrics and methodologies for alleviating the challenges associated with high-dimensional data, commonly called the curse of dimensionality.

BACKGROUND

kNN regression functions on the premise that comparable data points yield consistent outcomes. Upon introducing a new data point, the algorithm identifies the nearest k data points in the feature space, which are neighbors. It computes the predicted value as the average or weighted average of these neighbor points' target values. The selection of k, the choice of distance metric used for measuring similarity, and the dimensionality of the data play a significant role in influencing the model's performance.

Recent research has delved into various facets of kNN regression, emphasizing its strengths and limitations. Truong et al. (2020) underscore kNN's resilience in addressing non-linear data structures, specifically noting its prevalent application in real estate price prediction. In this area, traditional linear models often prove inadequate. Similarly, Sangeetha and Alfia (2024) expound on kNN's efficacy in financial forecasting, particularly in capturing transient trends in volatile markets. However, the choice of k is critical and requires careful tuning. A small k value can lead to high variance, as the model becomes sensitive to noise in the data. Conversely, a large k value can introduce bias, resulting in oversmoothing and underfitting. Du et al. (2024) propose using cross-validation to systematically select the optimal k value, ensuring a balance between bias and variance.

Another key aspect of kNN regression is the distance metric used to determine the similarity between data points. While Euclidean distance is the most common choice, alternative metrics like Manhattan or Minkowski distance can be more appropriate depending on the data's nature. A study by El Morr et al. (2022) highlights that selecting an appropriate distance metric is crucial, particularly in applications involving heterogeneous data types or skewed distributions.

When comparing the kNN regression algorithm to Decision Trees (DT) and Support Vector Machines (SVM), it offers a unique balance of interpretability and flexibility, especially for small to medium-scale datasets (Jadhav & Channe, 2016). For example, DT learning utilizes a decision tree as a predictive model that maps observations about an item to determine the object's mark value. The DT technique is comprised of two primary phases: tree building and tree pruning. During the tree building phase, a top-down approach is utilized. This phase involves the recursive partitioning of the tree until all data items are categorized under the same class label. It is important to note that this process is computationally intensive, as the training dataset is examined multiple times. The subsequent phase, tree pruning, is conducted in a bottom-up manner. The purpose of this phase is to enhance the prediction and classification accuracy of the algorithm by addressing the issue of overfitting. For this reason, overfitting in decision trees may result in misclassification errors, which can significantly impact the effectiveness of the model. Furthermore, Lambora, A., Gupta, K., & Chopra, K. (2019) present the precedents of SVM. Alexey ya. Chervo nenkis and Vladimir N. Vapnik created SVM in 1963. Since the discovery of SVM, this technique has been widely adopted for image, hypertext, and text segregation and classification problems. These algorithms are quite advanced and can be utilized for handwritten text and protein sorting in biological laboratories. They are used in many other domains, like self-driving cars, chatbots, face recognition, etc. Among one of the highly prevalent supervised type learning algorithms, the Support Vector Machine algorithm is intended for regression and classification problems. The SVM algorithm aims to form the finest suitable decision limit or boundary, known as a hyperplane. It separates n-dimensional space into various classes, making placing a different point in the apt category easy. The SVM algorithm chooses extreme vector points called Support Vectors to help create an appropriate hyperplane.

Despite its apparent simplicity, kNN regression encounters challenges related to computational efficiency, the curse of dimensionality, and sensitivity to hyperparameter choices (Timbers, T., Campbell, T. & Lee, M., 2023). Recent literature underscores the ongoing efforts to tackle these challenges, including utilizing dimensionality reduction techniques, optimal distance metrics, and hybrid models to fortify the algorithm's resilience and versatility. As research advances, kNN regression is anticipated to persist as a pertinent and progressive method in predictive modeling.

INTRODUCTION TO K-NEAREST NEIGHBOHRS ALGORITHM

The kNN algorithm represents a non-parametric, supervised learning classifier that leverages proximity to make classifications or predictions regarding the grouping of individual data points (Ratnam Dodda D et al., 2023). This method stands as one of the most widely used and straightforward classification and regression classifiers in contemporary machine learning. The kNN algorithm utilizes feature similarity to make predictions for the values of new data points. Essentially, the algorithm assigns a value to a new point based on how closely it resembles the points in the training set.

The initial step involves calculating the distance between the new point and each training point. There are several methods for computing this distance, the most widely recognized being the Euclidean, Manhattan (for continuous data), and Hamming distance (for categorical data).

The kNN algorithm has its origins in the 1950s and was developed as a method for pattern classification and regression. The pioneering research conducted by Evelyn Fix and Joseph Hodges in 1951 introduced kNN as a non-parametric approach to classification, which relies on the proximity of data points within a feature space (Silverman & Jones, 1989). Following this, Thomas Cover formalized the kNN algorithm in 1967 and conducted an analysis of its statistical properties, thereby establishing a robust foundation for its application in the field of machine learning. The kNN algorithm is frequently employed in search applications where the objective is to identify similar items. This process, known as kNN search, entails locating items that are comparable or virtually identical to a specified reference item. A significant aspect of the kNN algorithm is its sensitivity to the local structure of the data, which can influence search results.

In the field of medicine, the role of prediction is paramount. Healthcare professionals frequently engage in predicting the likelihood of subsequent heart attacks in patients who have been admitted for an initial heart attack, as well as evaluating the risk of cardiac arrest in individuals with elevated cholesterol levels. Additionally, predictions regarding cancer risk are made based on an array of factors, including geographic region, dietary habits, hereditary conditions, height, weight, and other clinical metrics specific to each patient (Taunk et al., 2019). Over the past decade, heart disease has remained one of the leading causes of mortality worldwide (Li et al., 2019). Healthcare providers accumulate vast quantities of patient data, which can be leveraged for statistical analysis and data mining techniques to improve the diagnosis and management of heart disease.

On the other hand, the task of predicting stock prices has become increasingly complex in recent years due to many factors influencing market behavior. These factors include the socioeconomic conditions of various nations, the prevailing political climate, natural disasters, and additional elements beyond the financial performance of individual companies. In 2024, Sangeetha and Alfia conducted an investigation into the effectiveness of linear regression models for predicting stock market trends. The primary objective of their study was to evaluate the viability of linear regression as a forecasting tool for stock prices through the analysis of historical market data. The findings indicate that linear regression models can successfully capture the underlying trends in stock price movements, thereby providing a foundational baseline for the development of more complex predictive models. Nonetheless, the authors acknowledge that, while linear regression offers advantages in terms of simplicity and interpretability, its predictive accuracy may be constrained in environments characterized by high volatility or non-linear market conditions. Further research suggests that the integration of linear regression with machine learning techniques, such as kNN regression, can enhance the identification of historical instances that reflect similarities to current stock price movements or market conditions. By averaging the outcomes (e.g., future prices) of the most analogous instances (neighbors), kNN regression can provide predictions for the subsequent price or trend.

THE CONCEPT OF REGRESSION

Regression is a statistical method used in machine learning, data science, and other fields to model and analyze the relationship between a dependent variable (often called the target or response variable) and one or more independent variables (also called features or predictors). Regression aims to predict the value of the dependent variable based on the values of the independent variables.

In the domain of predictive analysis, regression, like classification, denotes a problem-solving approach that utilizes historical data to anticipate future observations. However, in regression scenarios, the primary objective involves predicting numerical values as opposed to categorical ones. The variable aimed for prediction is commonly referred to as the response variable.

FUNDAMENTALS OF K-NEAREST NEIGHBORS REGRESSION

Key Parameters

Distance metrics

The Euclidean distance in Eq. (1) is computed as the square root of the sum of the squared variances between a new point (**x**) and an existing point (**y**) (Shabani, et al., 2020).

$$d(x,y) = \sqrt{(x_2 - x_1)^2 + (y_2 - y_1)^2 + \ldots + (x_n, y_n)^2} \quad (1)$$

where

- $d(x, y)$ is the Euclidean distance.
- (x_1, y_1) is the coordinate of the first point.
- (x_2, y_2) is the coordinate of the second point.
- (x_n, y_n) is the coordinate of the n point.

Example 1

In a 2-dimensional space, let us proceed with the calculation of the Euclidean distance between two distinct points denoted as **A** and **B**:

Point A:$(2,3)$

Point B:$(5,7)$

Step 1: Identify the coordinates of the points:

- $A_{x,y}:(2,3)$
- $B_{x,y}:(5,7)$

Step 2: Apply the Euclidean distance equation in (1):

$$d\left(A_{x,y}, B_{x,y}\right) = \sqrt{(5-2)^2 + (7-3)^2} \tag{1.1}$$

$$d\left(A_{x,y}, B_{x,y}\right) = \sqrt{(3)^2 + (4)^2} \tag{1.2}$$

$$d\left(A_{x,y}, B_{x,y}\right) = \sqrt{9+16} \tag{1.3}$$

$$d\left(A_{x,y}, B_{x,y}\right) = \sqrt{25} \tag{1.4}$$

$$d\left(A_{x,y}, B_{x,y}\right) = 5 \tag{1.5}$$

The Euclidean distance between points **A** and **B** is 5. In Figure 1, a visualization of calculating the Euclidean distance between the points **A** and **B** is shown.

Figure 1. Euclidean distance between pointsA and **B**

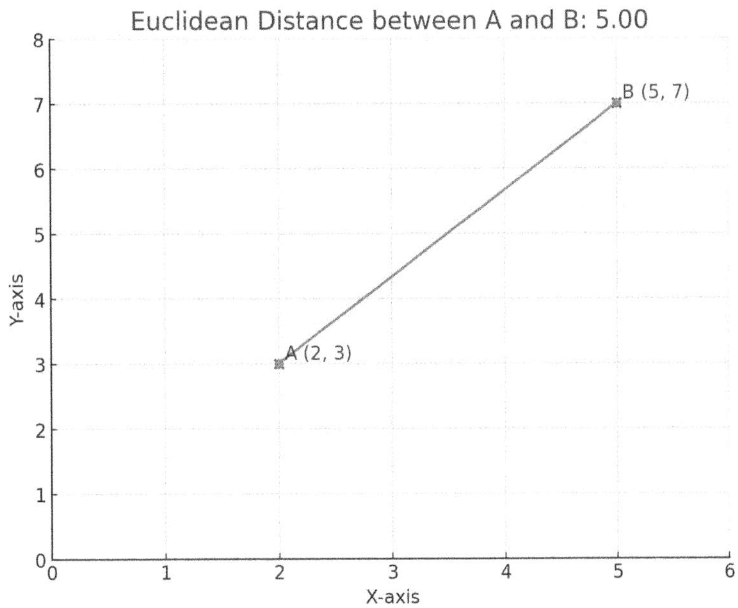

The line illustrates the straight-line distance between the two points, and the calculated distance is 5.

The Manhattan distance, also known as taxicab distance or city block distance, is a commonly used distance metric that measures the absolute value between two points. It is often visualized with a grid to illustrate how one might navigate from one address to another via city streets. The mathematical expression for the Manhattan distance between two points $x = (x_1, x_2, \ldots x_n)$ and $y = (y_1, y_2, \ldots y_n)$ in an n-dimensional space is in Eq. (2):

$$d(x, y) = \sum_{i=1}^{n} |x_i - y_i| \tag{2}$$

where

- $d(x, y)$ is the Euclidean distance between points x and y.
- **x** and **y** are the two points between which the distance is calculated, each point has **n** coordinates in an **n**-dimensional space.
- x_i and y_i represent the i-th coordinate of points$x$ and y, respectively.

Example 2

In a 3-dimensional space, let **Point A**:$(1, 2, 3)$ and **Point B**:$(4, 6, 8)$.

Step 1: Identify the coordinates of the points:

- $A_{x,y,z}:(1,2,3)$
- $B_{x,y,z}:(4,6,8)$

Step 2: Apply the Manhattan distance equation in (2) modified for three dimensions in Eq (3):

$$d(x,y) = |a_1 - b_1| + |a_2 - b_2| + |a_3 - b_3| \quad (3)$$

Step 3: Calculate the absolute differences and sum the total:

$$d(x,y) = |1 - 4| + |2 - 6| + |3 - 8| \quad (4)$$

$$d(x,y) = |-3| + |-4| + |-5| \quad (5)$$

$$d(x,y) = 3 + 4 + 5 \quad (6)$$

$$d(x,y) = 12 \quad (7)$$

The Manhattan distance between points A and B is 12. In Figure 2, a visualization of calculating the Manhattan distance between the points A and B is shown in a 3D space. The dashed lines represent the paths taken from point A to point*B* along the x, y, and z axes. The resulting calculated Manhattan distance is 12.

Figure 2. Manhattan distance between points A and B

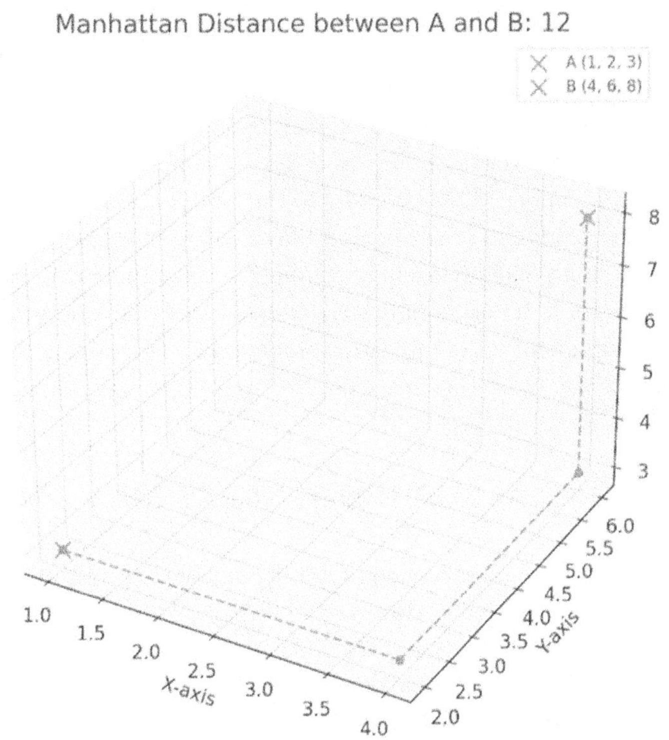

The Hamming distance is frequently utilized with Boolean or string vectors to pinpoint the discrepancies between the vectors. As a consequence, it has also been denoted as the overlap metric. This concept can be mathematically expressed using the Eq. (8):

$$d(s,t) = \sum_{i=1}^{n} I(s_i \neq t_i) \tag{8}$$

where

- $d(s,t)$ is the Hamming distance between the strings s and t.
- s, t are the strings of equal length being compared.
- s_i, t_i represent the i-th character (or element) of the strings s and t respectively.
- $s_i \neq t_i$ is an indicator function that equals 1 if s_i is not equal to t_i, and 0 otherwise.

Defining k factor

The next step is to define a k value. In the context of kNN algorithms, k value denotes the quantity of data points from the training set that are nearest to the new data point and are utilized for prediction. These closest neighbors are identified based on a chosen distance metric, usually the Euclidean distance.

For instance, when k=1, the instance will be assigned to the same class as its nearest neighbor. Determining the value of k is a critical consideration, as different values can result in overfitting or underfitting choices (Timbers, T., Campbell, T. & Lee, M., 2023). Lower values of k can yield high variance but low bias, whereas larger values can result in high bias and lower variance. The selection of k will predominantly hinge on the input data; datasets containing more outliers or noise may perform more effectively with higher values of k. It is generally advisable to employ an odd value for k to circumvent ties in classification, and adoption of cross-validation techniques can facilitate the identification of the optimal k for a given dataset.

K-NEAREST NEIGHBOHRS ALGORITHM REGRESSION

In the context of machine learning, regression problems involve the utilization of a concept similar to that employed in classification problems. However, in the case of regression, the prediction about a continuous value is based on the average of the kNN. This method utilizes the proximity of neighboring data points to make predictions. It is important to note that while classification is used for discrete values, regression is specifically designed to handle continuous variables.

In the same way, the process of constructing a regression model, it is imperative first to partition the data into training and test sets. This division allows for assessing the model's performance based on previously unseen data. Furthermore, cross-validation implementation facilitates the evaluation of different model parameters, such as k, in a kNN model (Zhao, M., & Chen, J. 2015). Importantly, this method involves predicting numerical variables instead of categorical ones.

DATA PREPROCESSING

Data Collection

Collection of data in kNN regression involves the meticulous gathering and preparation of the dataset used for training and evaluating the kNN model. Given that kNN is an instance-based learning algorithm, the caliber and quantity of the gathered data hold significant sway over the model's performance. Below is an outline of the data collection process in kNN regression:

1. **Identify the Problem Discipline:**
   a. Defining the objective. Clearly outlining the regression problem, such as predicting house prices, estimating temperature, or forecasting sales.
   b. Determining Relevant Features. Identifying the variables (features) likely to influence the target variable.
2. Data Sources

a. Primary Data Collection. Involves gathering data directly from experiments, surveys, or sensors that are pertinent to the issue at hand.
b. Secondary Data Collection. Involves making use of existing datasets from databases, online repositories, APIs, or publicly available datasets.

3. Data Gathering Techniques
   a. Manual Data Entry. Data is collected manually through observations, measurements, or interviews.
   b. Automated Data Collection: Sensors, web scraping, or data logging systems are used to automatically collect large volumes of data.
4. Data Quality Considerations
   a. Accuracy. Ensure that the data gathered is precise and devoid of errors.
   b. Completeness. Compile a comprehensive dataset encompassing all pertinent features and observations.
   c. Consistency. Verify that the data maintains uniformity across diverse sources and formats.
   d. Relevance. Prioritize the collection of data directly pertinent to the predictive task.
5. Data Preprocessing
   a. Handling Missed Data. Address missing values through imputation, deletion, or substitution to effectively handle missing data.
   b. Feature Scaling. Normalize the data is necessary to ensure that features with different scales do not disproportionately influence distance calculations (Zhao, M. & Chen, J. 2015).
   c. Data Cleaning. Enhance data quality by removing outliers, correcting inconsistencies, and filtering noise during the data cleaning process.
6. Sampling Techniques
   a. Random Sampling. Obtain a randomized sample from the population to ensure that the dataset is statistically representative (Zhao, M. & Chen, J. 2015).
   b. Stratified Sampling. Ensure that the dataset encompasses proportional representations of various subgroups within the population.
7. Data Splitting
   a. Training Set. Allocate a portion of the collected data for the purpose of training the kNN model.
   b. Test Set. Reserve a distinct portion of the data to assess the model's performance.
   c. Validation Set. Optionally, employ a validation set to refine hyperparameters, such as the value of k.

Dimensionality Reduction Techniques

Dimensionality reduction is an essential preprocessing step in kNN regression, particularly when confronted with high-dimensional datasets. High-dimensional data presents challenges, notably the "curse of dimensionality," wherein the distance metrics used by kNN lose informativeness as the number of features increases (R. Aafrein & Chinnaiyan, 2023). This section delves into crucial dimensionality reduction techniques utilized to alleviate these challenges and enhance the performance of kNN regression, citing recent studies and applications for reference.

1. **Principal Component Analysis (PCA).** This technique is a widely used linear dimensionality reduction technique that transforms the original features into a new set of uncorrelated variables called principal components. These components are ordered by the amount of variance they cap-

ture in the data, allowing the first few components to retain most of the information. In the context of kNN regression, by reducing the dimensionality of the data, PCA helps kNN regression focus on the most informative features, thereby improving accuracy and reducing computational costs. Recent studies have demonstrated the effectiveness of PCA in various domains. For example, Ma et al. (2019) applied PCA to preprocess data in a remote sensing image prediction model using kNN, leading to improved prediction accuracy and reduced computational time.

2. **Linear Discriminant Analysis (LDA)**. Traditionally used for classification, can also be adapted for regression problems. LDA reduces dimensionality by projecting data onto a lower-dimensional space that maximizes class separability (or range separability in regression tasks). LDA is beneficial in scenarios where different ranges of the target variable are influenced by different feature combinations. By focusing on these combinations, LDA can enhance the performance of kNN regression. For example, in a study by Shariatnia et al. (2022), the LDA was used to preprocess angiography data, showing superior performance to other methods such as Quadratic Discriminant Analysis (QDA) and Logistic Regression (LR) for the prediction of coronary artery disease.
3. **Feature Selection.** Involves selecting a subset of the most relevant features from the original dataset, rather than transforming the data into a new space as with PCA or LDA. Feature selection can be particularly effective in reducing dimensionality while maintaining model interpretability. Feature selection reduces the noise introduced by irrelevant or redundant features, thereby enhancing the performance of kNN regression. Moreover, the emphasis should be different for different types of extreme datasets. For Cai et al. (2018), the time complexity of an algorithm is nearly as important as its accuracy when performing with high-dimensional datasets or large samples. Meanwhile, feature selection methods should focus on the accuracy of their recognition capacities for minority classes when working with imbalanced datasets.

Data Splitting: Test, Training and Validation Datasets

The test dataset comprises a set of data points utilized to assess the performance of the regression model. Essentially, it consists of data points that were not employed during the model's training phase, and encompasses input features (Dini et al., 2023). The test dataset is an essential tool for evaluating the model's capacity to generalize to new, unobserved data and for assessing the accuracy of the regression model. The test dataset has the following essential characteristics:

1. **Unrecognized Data.** The test set comprises data points not employed during the training phase to ensure an impartial evaluation of the model's performance.
2. **Input Features and Target Values.** It encompasses the identical features (independent variables) as the training set along with the corresponding target values (dependent variable) that the model is meant to predict.
3. **Purpose.** The primary objective of the test set is to assess the model's generalization ability by predicting the target values for the test set and contrasting them with the actual values.
4. **Evaluation Metrics**. The performance of the kNN regression model on the test set is commonly appraised using regression metrics such as Mean Squared Error (MSE), Mean Absolute Error (MAE), Root Mean Squared Error (RMSE), or R-squared (R^2) to offer insights into the accuracy of the model's predictions.

On the other hand, the training dataset is utilized to "train" the kNN model. In contrast to conventional machine learning models, which undergo a training phase to learn parameters, kNN operates as an instance-based learning algorithm (Dini et al., 2023). As a result, the training dataset is directly stored and utilized for predictive purposes, obviating the need for an explicit learning process. The training dataset

1. **Labeled Data.** The training dataset comprises input features (independent variables) and their corresponding target values (dependent variable). These known target values are employed by the model to make predictions on new, unseen data.
2. **Data Storage.** In kNN, the training dataset is stored as-is. The algorithm utilizes this stored data to establish the nearest neighbors of any new data point during the prediction phase.
3. **Size and Quality**. The size of the training dataset can significantly impact the model's performance. A larger, more diverse dataset has the potential to enhance the model's generalization ability, whereas a smaller or biased dataset could result in inaccurate predictions.

Finally, the validation dataset represents a subset of the original dataset. Its purpose is to refine the model's hyperparameters and evaluate its performance before its application to the test dataset (Dini et al., 2023). While the training dataset allows the model to internalize data points, and the test dataset measures the final model performance, the validation dataset serves to optimize and choose the best model configuration during the model development phase. In kNN regression, the validation dataset is used for the following purposes:

1. **Data Segmentation.** Following data collection and preprocessing, it is customary to segment the data into three parts: training, validation, and test datasets. A typical segmentation involves allocating 60% for training, 20% for validation, and 20% for testing.
2. **Hyperparameter Optimization**. The kNN model is trained using the training dataset, after which various hyperparameter configurations are tested using the validation dataset. The objective is to identify the configuration that yields the best performance, such as the lowest Mean Squared Error, on the validation data.
3. **Cross-Validation (Optional).** In certain scenarios, cross-validation, such as k-fold cross-validation, is used in place of a single validation set (Li et al., 2019). This approach involves dividing the data into multiple folds and using each as a validation set while training on the remaining folds. This provides a more robust estimate of model performance.
4. **Iteration and Refinement.** The model may undergo multiple iterations, during which different hyperparameters settings are evaluated using the validation set. The best model based on the validation data is then selected for final testing.
5. **Prevention of Data Leakage.** It is imperative to ensure that the validation data is not utilized in the training process to prevent data leakage, which could result in overly optimistic performance estimations.

EVALUATING MODEL PERFORMANCE

Mathematical Foundations

Performance metrics are crucial for evaluating the effectiveness of a kNN regression model. These mathematical foundations quantify the accurateness and rate of the predictions made by the model, providing insights into how well the model generalizes to unseen data. The most common performance metrics for kNN regression include:

1. **Mean Squared Error (MSE).** Consist in the average of the squares of the differences between the predicted values and the actual target values. MSE penalizes larger errors more heavily due to the squaring of the residuals, making it sensitive to outliers (Singh, D. K. & Rawat, N, 2023). The mathematical expression for MSE is shown in Ec. (9):

$$MSE = \frac{1}{n}\sum_{i=1}^{n}(y_i - \hat{y}_i)^2 \tag{9}$$

where y_i is the actual value, $\hat{y}_i$ is the predicted value, and n is the number of observations.

2. **Mean Absolute Error (MAE).** Calculates the average of the absolute differences between the predicted values and the actual target values, shown in Ec. (10). Unlike MSE, MAE does not square the errors, making it less sensitive to outliers (Singh, D. K. & Rawat, N, 2023).

$$MAE = \frac{1}{n}\sum_{i=1}^{n}|y_i - \hat{y}_i| \tag{10}$$

3. **Root Mean Squared Error (RMSE).** Obtains the square root of the MSE in Ec. (11). RMSE has the same units as the target variable, making it more interpretable than MSE (Singh, D. K. & Rawat, N, 2023).

$$RMSE = \sqrt{\frac{1}{n}\sum_{i=1}^{n}(y_i - \hat{y}_i)^2} \tag{11}$$

4. **Coefficient of Determination (R^2).** Measures the proportion of the variance in the dependent variable that is predictable from the independent variables. It provides a measure of how well the observed outcomes are replicated by the model in Ec. (12).

$$R^2 = 1 - \frac{\sum_{i=1}^{n}(y_i - \hat{y}_i)^2}{\sum_{i=1}^{n}(y_i - \bar{y})^2} \tag{12}$$

where $\bar{y}$ is the mean of the observed data.

Applications of kNN Regression

kNN Regression is a versatile machine learning algorithm widely used in various fields due to its simplicity and effectiveness. Below are some of the prominent applications of kNN Regression.

1. **Predicting Housing Prices.** The algorithm's ability to make predictions based on the closest similar properties makes it well-suited for this task. The k-nearest neighbor algorithm has found extensive application in housing price prediction due to its simplicity and effectiveness in dealing with non-linear relationships between variables (Truong et al., 2020). One of the core advantages of kNN in this context is its ability to predict housing prices based on the proximity of similar properties. By considering factors such as location, size, number of rooms, and historical prices, kNN identifies the most comparable properties (the "neighbors") and bases its prediction on the average prices of these neighbors. This method is particularly useful in markets where housing prices can vary significantly within small geographic areas, providing a localized estimate that adapts to specific neighborhood characteristics. The study explores the enhancement of traditional kNN models by integrating them with other machine learning techniques, leading to improved accuracy and robustness in predictions. The paper demonstrates that by refining the feature selection process and optimizing the distance metrics used in kNN, it is possible to reduce prediction errors significantly. Additionally, the combination of kNN with techniques like regression models or ensemble learning methods, such as random forests or boosting algorithms, further enhances the prediction performance, making it a valuable tool for real estate analytics. Moreover, the versatility of kNN allows it to be easily adapted for various types of housing data, whether it's structured data like numerical and categorical variables or unstructured data such as text descriptions and images. This adaptability, combined with its intuitive approach, makes kNN a preferred choice for housing price prediction models, particularly in scenarios where interpretability and transparency of the model are crucial for stakeholders in the real estate industry.
2. **Energy Consumption Forecasting.** The application of machine learning techniques for predicting energy consumption in smart buildings has gained significant traction, particularly in the context of sustainable energy management. A case study conducted in Malaysia highlights the effectiveness of various machine learning models, including kNN, SVM, and Artificial Neural Networks (ANN), in forecasting energy usage within commercial buildings (Shapi et al., 2021). The study utilized Microsoft Azure Machine Learning Studio as the platform for model development, taking advantage of its cloud-based environment to handle large datasets and complex computations efficiently. The research focused on two major tenants within a commercial building, where IoT meters provided real-time data on electrical power usage, including variables such as power factor, voltage, and current. These inputs were essential in training the predictive models to estimate energy consumption accurately. The findings demonstrated that each machine learning model had varying degrees of success, with SVM showing the best performance for certain tenants in terms of Root Mean Square Error (RMSE) and Mean Absolute Percentage Error (MAPE). However, kNN also performed robustly, particularly in scenarios where the data exhibited complex, non-linear patterns. One of the most important insights from the study is the importance of handling missing data and the impact of data preprocessing on the overall accuracy of the predictive models. The use of Probabilistic Principal Component Analysis (PPCA) within Azure ML for imputing missing data proved crucial in enhancing model performance. Furthermore, the study underscores the adaptability of cloud-based platforms like Azure ML in deploying machine learning solutions for real-world applications,

offering scalability, ease of use, and integration with various programming environments. This case study not only demonstrates the practical application of machine learning in energy management but also provides a roadmap for future research in optimizing energy consumption prediction models, which is vital for advancing the smart building concept in Malaysia and beyond.

3. **Adaptive Thermal Comfort.** The development of adaptive thermal comfort models using machine learning algorithms, such as kNN, represents a significant advancement in creating personalized indoor environments (Xiong & Yao, 2021). This study focuses on the application of the kNN algorithm to develop a dynamic thermal comfort model that adjusts to individual preferences based on real-time environmental conditions. Unlike traditional models like the Predicted Mean Vote (PMV), which assumes a static relationship between environmental factors and thermal comfort, the kNN-based model adapts to the specific thermal responses of individuals, thereby providing a more personalized comfort experience. The kNN algorithm was selected for its simplicity and effectiveness in classification tasks. By analyzing data collected from 34 volunteers, the study demonstrated that the kNN-based model could accurately predict thermal comfort levels with an accuracy of up to 88.31% after training on 1,000 datasets. The model's accuracy is enhanced through continuous learning from new data, allowing it to refine the thermal comfort boundaries specific to each individual. This adaptability makes the kNN algorithm particularly suitable for developing personalized air-conditioning systems that cater to the varying preferences of different occupants in a shared environment. Moreover, the study incorporated an artificial intelligence (AI) environmental controller, the "Wukong" robot, to implement and test the kNN-based thermal comfort model. This robot collected real-time environmental data, such as temperature, humidity, and air quality, and adjusted the indoor conditions accordingly. The integration of kNN with AI-enabled controllers presents a promising approach to reducing energy consumption in buildings while maintaining optimal comfort levels, highlighting the potential of machine learning in advancing smart building technologies. This research underscores the importance of adaptive models in the field of thermal comfort and energy efficiency, providing a foundation for future studies to explore the integration of more complex algorithms and broader datasets in the pursuit of sustainable building solutions.
4. **Aquilaria Malaccensis Oil Qualities Classification.** The kNN algorithm has been effectively employed in the classification of Aquilaria Malaccensis oil qualities, offering a more accurate, efficient, and scalable alternative to traditional grading methods (Amidon et al., 2024). Historically, the grading of agarwood oil, a valuable product used in perfumes, medicine, and religious ceremonies, relied heavily on human sensory evaluation, which involved assessing color and scent. However, this approach presented several challenges, including subjective assessments, long processing times, and health risks for graders. In response to these limitations, the study introduced a kNN-based grading model that leverages the chemical composition of agarwood oil to classify its quality into four distinct categories: high, medium-high, medium-low, and low quality. The research utilized a dataset of 660 samples of agarwood oil, each containing 11 significant chemical compounds identified through Principal Component Analysis (PCA). The Euclidean distance metric was selected as the most appropriate method for determining the proximity between samples in the kNN model, which was set to consider five nearest neighbors. The dataset was divided into training and testing sets using an 80:20 split, ensuring robust model development and validation. The model demonstrated outstanding performance, achieving 100% accuracy in classifying the quality of the agarwood oil samples, as confirmed by a 4x4 confusion matrix. This success underscores the suitability of the kNN algorithm for handling complex, nonlinear data in the field of agarwood oil quality classification. By replacing

the subjective and labor-intensive traditional methods with a machine learning-based approach, the study not only enhanced the precision of the grading process but also significantly reduced the time and cost associated with it. The findings of this study pave the way for future research in optimizing grading models for other valuable natural products, potentially integrating kNN with other advanced machine learning techniques to further improve classification accuracy and efficiency.

5. **Phishing Prevention.** The increasing prevalence of phishing attacks has prompted the need for more sophisticated detection systems that can accurately distinguish between legitimate and fraudulent websites (Ariffin et al., 2024). This study investigates the efficacy of various machine learning classification algorithms, including kNN, decision trees, and logistic regression, in developing an intelligent phishing website detection system. The research utilized a dataset comprising 11,054 websites sourced from Kaggle, with an even split between phishing and legitimate sites. The dataset underwent rigorous preprocessing, including feature selection through a correlation matrix, to enhance model accuracy by eliminating noisy data. The kNN algorithm, known for its simplicity and effectiveness in classification tasks, was evaluated alongside decision trees and logistic regression to determine the most accurate method for phishing detection. The dataset was split into 80% for training and 20% for testing, with the primary focus on tuning key parameters for each algorithm to optimize performance. The results indicated that while kNN achieved a high accuracy rate of 92.99%, decision trees outperformed the other models with an accuracy of 94.89%, making it the most reliable method for detecting phishing websites in this study. Logistic regression, although commonly used in binary classification tasks, showed the lowest accuracy at 91.04%. The decision tree model's superior performance can be attributed to its ability to handle complex decision-making processes and its robustness in dealing with various features, such as URL length, domain registration length, and the presence of HTTPS tokens. The study's findings highlight the decision tree as the most effective classifier for phishing website detection, providing a foundation for developing more advanced detection systems. Future work could explore the integration of deep learning techniques to further enhance the detection accuracy and develop a fully operational web-based system for real-time phishing prevention.
6. **kNN on GPUS.** The utilization of GPUs for billion-scale similarity search has revolutionized the performance and scalability of kNN algorithms, particularly in handling large datasets in high-dimensional spaces (Johnson, et al., 2019). This paper presents a novel approach to implementing kNN search on GPUs, significantly outperforming previous methods by optimizing key operations such as k-selection and leveraging the massive parallelism inherent in GPUs. The researchers developed a k-selection algorithm that operates at up to 55% of the theoretical peak performance of the GPU, enabling a kNN implementation that is 8.5 times faster than the prior state-of-the-art GPU methods. The core of this achievement lies in the efficient design of both exact and approximate kNN search on GPU architectures. The study focused on the challenges associated with similarity search in large collections of high-dimensional data, where traditional CPU-based methods struggle due to the curse of dimensionality and memory constraints. By introducing a GPU-friendly k-selection mechanism and optimizing memory usage through techniques like product quantization, the researchers were able to process datasets of unprecedented scale. For instance, their implementation constructed a high-accuracy kNN graph on a dataset of 95 million images in just 35 minutes using four GPUs, and managed to scale up to a billion vectors in under 12 hours. The implications of this work are profound, as it opens the door to real-time, large-scale similarity search applications in various domains, including image retrieval, recommendation systems, and bioinformatics. The

open-sourcing of their implementation ensures that the broader research community can build on these advancements, pushing the boundaries of what is possible with large-scale data processing on GPUs. This study not only highlights the potential of GPUs in overcoming the limitations of traditional methods but also sets a new benchmark for performance in the field of similarity search.

7. **Heart Disease Identification.** The identification of heart disease using machine learning techniques has become increasingly crucial in e-healthcare systems (Li et al., 2020). This study explores the application of several classification algorithms, including kNN, SVM, and ANN, to create an efficient heart disease diagnosis system. The proposed method leverages various feature selection algorithms, such as Relief, Minimal Redundancy Maximal Relevance (mRMR), and the Least Absolute Shrinkage and Selection Operator (LASSO), to enhance the predictive accuracy by eliminating irrelevant and redundant features. Additionally, the study introduces a novel Fast Conditional Mutual Information Maximization (FCMIM) algorithm for feature selection, aiming to further improve the model's performance. The research is grounded in the Cleveland Heart Disease dataset, which includes 303 instances with 75 attributes, although the experiments focused on a reduced subset of 14 features. The study employed a leave-one-subject-out cross-validation (LOSO) technique to assess the classifiers' performance, ensuring that the models were rigorously tested for generalization. Among the classifiers, SVM demonstrated the highest accuracy when combined with the proposed FCMIM feature selection algorithm, achieving an impressive 92.37% accuracy. This performance surpassed traditional methods, highlighting the effectiveness of the FCMIM-SVM combination in accurately diagnosing heart disease. Furthermore, the study compared the proposed system with existing machine learning-based diagnostic methods, revealing significant improvements in both accuracy and computational efficiency. The findings suggest that the integration of advanced feature selection techniques with robust classifiers like SVM can lead to more reliable and efficient heart disease detection systems in e-healthcare, ultimately contributing to better patient outcomes through early and accurate diagnosis. The research paves the way for further exploration into hybrid models that combine deep learning techniques with traditional machine learning to enhance diagnostic accuracy.

CONCLUSION

The kNN regression is a robust non-parametric algorithm that thrives in scenarios where data relationships defy easy definition by a specific model. Its strength lies in its simplicity and capacity to generate predictions based on the proximity of data points, rendering it particularly advantageous in various applications, such as real estate pricing, energy consumption forecasting, environmental monitoring, medical data analysis, and financial market prediction. However, kNN regression encounters limitations, notably its sensitivity to the choice of k value and the curse of dimensionality, wherein performance may deteriorate with increasing features. As the number of features within a dataset increases, the significance of the distance between points within the feature space diminishes, potentially resulting in decreased performance for the kNN algorithm in high-dimensional spaces lacking appropriate dimensionality reduction techniques. The selection of a small k value can render the model sensitive to noise, potentially leading to overfitting, while a large k value may overly smooth the data, causing underfitting. Identifying the optimal k value is of paramount importance, albeit often necessitating extensive cross-validation and experimentation, which can be a time-intensive endeavor. Additionally, it may be computationally burdensome for large datasets, as it necessitates the storage of the entire training set and the computation of distances between the test point and all training points.

Nevertheless, kNN regression remains a valuable asset in the machine learning toolkit through judicious preprocessing techniques such as dimensionality reduction, normalization, and optimized algorithms. Its proficiency in handling non-linear data without presuming any underlying distribution underscores its versatility for numerous regression tasks. In summary, while kNN regression may not invariably be the optimal choice for all regression problems, its ease of implementation and effectiveness in specific applications ensures its continued relevance in both research and practical utilization.

REFERENCES

Aafrein, R., & Chinnaiyan, D. R. (2023). Detection of pneumonia using chest X-ray images with deep learning techniques-review. *International Journal of Scientific Research in Science and Technology*, •••, 793–800. DOI: 10.32628/IJSRST523103147

Amidon, A. F. M., Yusoff, Z. M., Ismail, N., Abd Almisreb, A., & Taib, M. N. (2024). kNN Euclidean Distance Model Performance on Aquilaria Malaccensis Oil Qualities. *Journal of Advanced Research in Applied Sciences and Engineering Technology*, *48*(2), 16–28. DOI: 10.37934/araset.48.2.1628

Ariffin, N. H. M., Iqbal, M. I. M., Yusoff, M., & Zulkefli, N. A. M. (2024). A Study on the Best Classification Method for an Intelligent Phishing Website Detection System. *Journal of Advanced Research in Applied Sciences and Engineering Technology*, *48*(2), 197–210. DOI: 10.37934/araset.48.2.197210

Bansal, M., Goyal, A., & Choudhary, A. (2022). A comparative analysis of K-Nearest Neighbor, Genetic, Support Vector Machine, Decision Tree, and Long Short Term Memory algorithms in machine learning. *Decision Analytics Journal*, *3*(100071), 100071. DOI: 10.1016/j.dajour.2022.100071

Cai, J., Luo, J., Wang, S., & Yang, S. (2018). Feature selection in machine learning: A new perspective. *Neurocomputing*, *300*, 70–79. DOI: 10.1016/j.neucom.2017.11.077

Dini, P., Elhanashi, A., Begni, A., Saponara, S., Zheng, Q., & Gasmi, K. (2023). Overview on intrusion detection systems design exploiting machine learning for networking cybersecurity. *Applied Sciences (Basel, Switzerland)*, *13*(13), 7507. DOI: 10.3390/app13137507

Dodda, R.. (2023). Precision in classification: A comparative study of logistic regression, Naive Bayes, LSTM, and CNN for spam email detection. *International Journal on Recent and Innovation Trends in Computing and Communication*, *11*(9), 2276–2280. DOI: 10.17762/ijritcc.v11i9.9233

Du, Z., Sun, X., Zheng, S., Wang, S., Wu, L., An, Y., & Luo, Y. (2024). Optimal biochar selection for cadmium pollution remediation in Chinese agricultural soils via optimized machine learning. *Journal of Hazardous Materials*, *476*(135065), 135065. DOI: 10.1016/j.jhazmat.2024.135065 PMID: 38943890

El Morr, C., Jammal, M., Ali-Hassan, H., & El-Hallak, W. (2022). K-Nearest Neighbors. In *International Series in Operations Research & Management Science* (pp. 301–318). Springer International Publishing.

Jadhav, S. D., & Channe, H. (2016). *Comparative study of K-NN, Naive Bayes and decision tree classification techniques*. DOI: 10.21275/v5i1.NOV153131

Johnson, J., Douze, M., & Jégou, H. (2019). Billion-scale similarity search with GPUs. *IEEE Transactions on Big Data*, *7*(3), 535–547. DOI: 10.1109/TBDATA.2019.2921572

Lambora, A., Gupta, K., & Chopra, K. (2019). Genetic algorithm- A literature review. *2019 International Conference on Machine Learning, Big Data, Cloud and Parallel Computing (COMITCon)*, 380–384.

Li, J. P., Haq, A. U., Din, S. U., Khan, J., Khan, A., & Saboor, A. (2020). Heart disease identification method using machine learning classification in e-healthcare. *IEEE Access : Practical Innovations, Open Solutions*, *8*, 107562–107582. DOI: 10.1109/ACCESS.2020.3001149

Li, Z., Yang, Z., Shen, C., Xu, J., Zhang, Y., & Xu, H. (2019). Integrating shortest dependency path and sentence sequence into a deep learning framework for relation extraction in clinical text. *BMC Medical Informatics and Decision Making*, *19*(S1), 22. Advance online publication. DOI: 10.1186/s12911-019-0736-9 PMID: 30700301

Ma, L., Liu, Y., Zhang, X., Ye, Y., Yin, G., & Johnson, B. A. (2019). Deep learning in remote sensing applications: A meta-analysis and review. [ISPRS]. *ISPRS Journal of Photogrammetry and Remote Sensing*, *152*, 166–177. DOI: 10.1016/j.isprsjprs.2019.04.015

Sangeetha, J. M., & Alfia, K. J. (2024). Financial stock market forecast using evaluated linear regression based machine learning technique. *Measurement. Sensors*, *31*(100950), 100950. DOI: 10.1016/j.measen.2023.100950

Shabani, S., Samadianfard, S., Sattari, M. T., Mosavi, A., Shamshirband, S., Kmet, T., & Várkonyi-Kóczy, A. R. (2020). Modeling pan evaporation using gaussian process regression k-nearest neighbors random forest and support vector machines; Comparative analysis. *Atmosphere (Basel)*, *11*(1), 66. DOI: 10.3390/atmos11010066

Shapi, M. K. M., Ramli, N. A., & Awalin, L. J. (2021). Energy consumption prediction by using machine learning for smart building: Case study in Malaysia. *Developments in the Built Environment*, *5*(100037), 100037. DOI: 10.1016/j.dibe.2020.100037

Shariatnia, S., Ziaratban, M., Rajabi, A., Salehi, A., Abdi Zarrini, K., & Vakili, M. (2022). Modeling the diagnosis of coronary artery disease by discriminant analysis and logistic regression: A cross-sectional study. *BMC Medical Informatics and Decision Making*, *22*(1), 85. Advance online publication. DOI: 10.1186/s12911-022-01823-8 PMID: 35351098

Silverman, B. W., & Jones, M. C. (1989). E. Fix and J.L. Hodges (1951): An Important Contribution to Nonparametric Discriminant Analysis and Density Estimation: Commentary on Fix and Hodges (1951). *International Statistical Review / Revue Internationale de Statistique*, *57*(3), 233–238. DOI: 10.2307/1403796

Singh, D. K., & Rawat, N. (2023). *Decoding optimization algorithms for convolutional neural networks in time series regression tasks*. Mecs-press.org. https://www.mecs-press.org/ijitcs/ijitcs-v15-n6/v15n6-4.html

Taunk, K., De, S., Verma, S., & Swetapadma, A. (2019). A brief review of nearest neighbor algorithm for learning and classification. *2019 International Conference on Intelligent Computing and Control Systems (ICCS)*, 1255–1260. DOI: 10.1109/ICCS45141.2019.9065747

Timbers, T., Campbell, T., & Lee, M. (2023, December 23). *Chapter 7 regression I: K-nearest neighbors*. Datasciencebook.Ca. https://datasciencebook.ca/regression1.html

Truong, Q., Nguyen, M., Dang, H., & Mei, B. (2020). Housing price prediction via improved machine learning techniques. *Procedia Computer Science*, *174*, 433–442. DOI: 10.1016/j.procs.2020.06.111

Xiong, L., & Yao, Y. (2021). Study on an adaptive thermal comfort model with K-nearest-neighbors (kNN) algorithm. *Building and Environment*, *202*, 108026. DOI: 10.1016/j.buildenv.2021.108026

Zhao, M., & Chen, J. (2015). Improvement and comparison of weighted k nearest neighbors classifiers for model selection. *Journal of Software Engineering*, *10*(1), 109–118. DOI: 10.3923/jse.2016.109.118

Chapter 16
Machine Learning in Latin America:
Transforming Industrial Processes and Value Chains

Jorge Ignacio Anguiano Lizaola
https://orcid.org/0000-0003-3602-2650
Universidad Politécnica de Baja California, Mexico

Miriam Arlin Tong Delgado
Universidad Politecnica de Baja California, Mexico

Araceli Garate Garcia
Universidad Politecnica de Baja California, Mexico

Pedro Parra Rubio
Universidad Politecnica de Baja California, Mexico

Angel Daniel Castro Franco
Universidad Politecnica de Baja California, Mexico

ABSTRACT

The manufacturing industry has motivated the development of the region in which it is located. In the last few years manufacturing has adopted innovations and advanced digital process technology, among which can be mentioned Machine Learning (ML), a subset of artificial intelligence (AI) to make industrial processes smarter, more efficient, and highly optimized operations. In this review, research studies related with the application of ML algorithms to problems and situations of manufacturing value chain are explored. After presenting the applications, benefits, and challenges, this chapter recalls the future trends of machine learning in this context, like in predictive maintenance, quality control, process optimization, and supply chain management, accomplished in Latin America. In that way it explores industrial process optimization.

DOI: 10.4018/979-8-3693-6910-4.ch016

INTRODUCTION

Industrial progress, historically, has implied economic growth and development of the region where it happened, because industry brings with it the creation of new formal jobs, development of regional infrastructure, and innovation. In 2023, high income countries, and therefore more developed, appeared mostly in Europe, North America, Australia, Japan, Republic of Korea, Singapore and Israel, none of them is in Latin America and the Caribbean, regions classified as "Developing and Emerging" (IMF, 2023). Since 2020, the world has faced various impacts, COVID-19 pandemic that interrupted global economic production in 2020 and 2021, the conflict between Russia and Ukraine and global warming. 2023 was the hottest year since records were kept. These polycrisis have changed the global value chain. Global industrial production was rapidly regaining its pre-pandemic levels but in countries with low income, like some in Latin America and the Caribbean, this level is still below (UNIDO, 2024). However, smart manufacturing, particularly through the adoption of machine learning (ML), offers new hope for industrialization and economic growth in the region. For instance, Ferreira de Rezende et al. (2023) demonstrated the use of digital twins and deep learning models for predictive maintenance in offshore riser systems, showcasing how ML can optimize industrial processes in Latin America. Similarly, Gómez-Ramírez and Soler Anguiano (2023) applied ML to optimize drilling equipment logistics in the oil and gas industry, a critical sector for many Latin American economies. Furthermore, Komori-Zevallos et al. (2023) highlighted the integration of Lean and Industry 4.0 principles in small and medium-sized enterprises (SMEs) in Peru, illustrating how ML can drive efficiency and productivity in the region's manufacturing sector.

The traditional concept of Manufacturing refers to the making of products from raw materials using various processes, equipment, operations and manpower according to a detailed plan (Scallan, 2003). However, this industry is changing in the era of The Fourth Industrial Revolution, also known as Industry 4.0 (i4.0) which appeared for the first time in 2011 in Germany, of smart manufacturing in North America, as a recent concept there are a broad range of definitions, however several authors agree with nine pillars (Suleiman et al, 2022): Big data, autonomous robots, simulation, additive manufacturing, horizontal and vertical integration, Internet of Things (IoT), cloud computing, cybersecurity, and augmented reality (AR). The integration of these transformed the traditional manufacturing systems to improve them (Issa et al., 2018 in Suleiman et al., 2022).

Manufacturing plants nowadays use several potent data acquisition tools to gather and electronically transfer data from the processes of the factory. This great amount of information constitutes a form of big data. According to Dogan and Birant (2021), the collected data describe characteristics and capabilities of machinery through sensors attached to the machines, machine failures / maintenances; in addition to features and properties of products, production line, the human resources, raw materials that are used in the process, the environment, product quality and other significant manufacturing factors.

Along with big data, artificial intelligence (AI), specifically its subset, machine learning (ML) and data mining (DM) integrated into manufacturing can be beneficial for increasing efficiency and optimization of processes without great changes in cost and resources. Some of the tasks in which ML has an impact are predictive maintenance in machinery and facilities, quality improvement in products and processes by fault detection, intelligent and continuous inspection reducing human participation, supply chain management and task scheduling (Rai, Tiwari, Ivanov & Dolgui, 2021).

Early studies on ML applied to manufacturing, focused mostly on planning and intelligent scheduling of production (Shaw, Park & Raman, 1992; Horvath and Rudas, 1993; Aytug, Bhattacharyya, Koehler & Snowdon, 1994); some others to support manufacturing decision (Canz & Jagdale, 1995); defect detection in robotic assembly (Seabra Lopez and Camarinha-Matos, 1995), proactive maintenance of machinery and equipment (Lee, 1995). Some of them focus on only one field of manufacturing such as the semiconductor manufacturing (Irani, Cheng, Fayyad and Qian, 1990; Perez, Lilkendey & Koh, 1994), aerospace industry (Riddle, Segal and Etzioni, 1994), electronics industry (Suarez, Cusumano & Fine, 1995), sheet metal industry (Cser, Geiger, Gresca & Hoffmann, 1991).

However, during 2023 and 2024, additive manufacturing (AM) is addressed in several studies (Liu, Ye, Silva Izquierdo et al, 2023; Parsazadeh, Sharma & Dahotre, 2023). Including topics of defect detection (Herzog, Brandt, Trinchi, et al, 2023), prediction of life time in materials and products (Hornas, Behal, Homola, Senck, et al., 2023), performance optimization of products (Cheepu, 2023; Duquesnoy, Liu, Zapata Dominguez Kumar, et al, 2023), development of smart composite materials (Babu, Mourad, Harib, & Vijayavenkataraman, 2022), improvement of process parameters in metal mechanic industry (Ko, Lu, Yang, Ndiaye et al., 2023; Zhu, Jiang, Guo, Wang, et al., 2023), optimization of settings in fabrication of thermoplastics in plastics industry (Ranjan, Kumar, Kumar, Kaur, et al., 2023). Aside from AM, for the semiconductor industry, there are studies about fault diagnosis in processes (Nahu, Zeeshan, Safei & Shahzad, 2023), reduction of ambiental impact by optimization of cooling/lubricant fluids in machining industries (Cica, Sredanovic, Tesic & Kramar, 2024), predictive maintenance for manufacturing equipment (Ghasemkhani, Aktas & Birant, 2023), detection and assessment of ergonomic risks, (Ciccarelli, Corradini, Germani, Menchi, et al., 2023), just to mention a few applications.

These advanced digital process technologies present great opportunities for organizations worldwide, the degree of adoption of these innovations depends on companies own characteristics, size, sector, organizational structure, human and financial resources, relationships with the customers and competitive pressure, and others (Grossman et al., 2021). However, the incorporation of these technologies implies investments in the companies, so in addition to being opportunities represents challenges. Hence, there is a great amount of research and review studies about the relationship between manufacturing and machine learning, more than 26000 just for 2023 and the first half of 2024 in a quick look up in Google Scholar search engine. Specifically, in 2021, Vergara Villegas et al. published a review titled "Artificial Intelligence for Industry 4.0 in Iberoamerica", to address the adoption of the technologies of the eight pillars of AI, in Latin America and the Caribbean, including Spain, and many fields not only industry and manufacturing.

As can be seen based on the previous paragraphs, the use and applications of ML in manufacturing have extended over time, from addressing a few types of situations to tackling a wide variety of problems. Similarly, with the advent of additive manufacturing and automation, the utilization of ML solutions is spreading around the world. In Latin America, ML is increasingly being adopted to optimize industrial processes and address unique regional challenges. For example, Ferreira de Rezende et al. (2023) demonstrated the use of digital twins and deep learning models for predictive maintenance in offshore riser systems, showcasing how ML can optimize industrial processes in the region's oil and gas sector. Similarly, Gómez-Ramírez and Soler Anguiano (2023) applied ML to optimize drilling equipment logistics, highlighting the potential of ML to enhance efficiency in the energy industry. Furthermore, Komori-Zevallos et al. (2023) integrated Lean and Industry 4.0 principles in small and medium-sized enterprises (SMEs) in Peru, illustrating how ML can drive productivity and innovation in Latin America's manufacturing sector. These examples underscore the growing relevance of ML in addressing the region's industrial challenges and fostering economic growth. Therefore, this chapter addresses particularly the adoption and presence of machine learning in the manufacturing value chain around Latin America and the Caribbean. In accordance with that, the chapter aims to answer the following research questions:

RQ1. In Latin America, which industrial sector has adopted the ML more extensively?
RQ2. What kind of manufacturing value chain processes problems are solved using ML techniques in Latin America?
RQ3. Which machine learning methods are commonly used to handle manufacturing value chain tasks in Latin America?
RQ4. Which manufacturing value chain processes datasets are popular and used as a benchmark in Latin America?
RQ5. Which performance metrics are generally used to evaluate ML models constructed in manufacturing value chain studies?
RQ6. In comparison with North America, Europe and Asia, how advanced is the adoption of ML techniques in the manufacturing value chain in Latin America?
RQ7. What are the main challenges of ML that are faced in manufacturing in Latin America?
RQ8. What are the possible research directions for ML studies in manufacturing shortly?

MANUFACTURING VALUE CHAIN

In 1985 Michael Porter introduced and defined the concept of value chain in which a manufacturing (or service) organization is considered as a system with inputs, transformation processes, and outputs, that involves resources (IfM, 2024). The activities and resources in the value chain determine the profits. The process of converting inputs and outputs consists in hundreds or thousands of activities, primary and secondary or of support. The primary activities are inbound logistics are all the relationship with suppliers, reception, storing and transport of inputs for transformation process; operations are the activities of transformation inputs in outputs; outbound logistics are the same of inbound logistics but for outputs; marketing and sales presents the products or services to the buyers, induce them to buy and ease the purchase; and, service are the activities of support the products or services after is sold and delivered. The secondary activities include procurement of acquisition of resources for the organization, obtaining raw materials from vendors; human resource management is all the activities of the organization with personnel like recruiting, hiring, training, developing, compensating and dismissing; technological development includes all the equipment, software, procedures and knowledge used in the organization; and, infrastructure are all the support functions in the company, like finance, accounting, legal, planning, public affairs, government, quality assurance and general management (Porter, 1985).

In Luther (2024), the elements of Manufacturing value chain are suppliers and vendors, raw materials, production, storage and warehousing, distribution and transportation, retailers, maintenance and repair, and recycling. Suppliers and vendors provide the resources to create products, and research by comparing multiple vendors allows organizations to identify areas for improvement. ML is being applied to optimize supplier selection and vendor management in Latin America. For example, Reyes et al. (2023) developed a methodology for supplier selection using a combined grouping and multi-criteria approach, which leverages ML to improve decision-making in the procurement process. This study demonstrates how ML can enhance the efficiency and reliability of supplier selection, which is critical for maintaining a robust manufacturing value chain in the region. Raw materials imply that the products are made with materials of quality, the purpose is to create a balance between costs and quality by market research. Latin America is using ML to optimize raw material selection and utilization. To optimize the use of basic materials like water and fertilizer, Cruz et al. (2022) used ML in smart strawberry farming. The agricultural and manufacturing sectors in the area depend heavily on the efficiency and sustainability of raw material utilization, which this study demonstrates how ML may enhance. Production includes all the procedures to convert raw material into products, the variety of these activities require a separate description. However, in order to prevent waste and quality issues, these procedures must be completed quickly and efficiently in the workflow. As an illustration of how ML can increase productivity and decrease waste in manufacturing, Saldaña et al. (2023) employed ML to optimize the grinding and milling processes in the Chilean copper mining industry. Similarly, Rojas-Rodríguez et al. (2023) demonstrated how ML may improve quality control and lower production errors by creating an ML model to forecast injection molding process failures. The potential of machine learning to enhance regional production efficiency and product quality is demonstrated by this research.

Luther (2024) continued with the elements of the value chain, storage and warehousing involving the storage of finished products before they are sent to buyers. Some organizations choose to adopt a Just-in-Time (JIT) approach, where products are sold as soon as they are produced, while others maintain an inventory of finished products to meet future demand. In Latin America, Paredes Rodríguez et al. (2023) used a fuzzy AHP TOPSIS methodology for multicriteria ABC inventory classification, which leverages ML to improve inventory management. This study demonstrates how ML can enhance the efficiency of storage and warehousing processes, reducing costs and avoiding overstocking or shortages in the region's manufacturing sector. Distribution and transportation involve the movement of products within the organization or to clients. Modern businesses have tools that allow real-time tracking of merchandise and inventory levels. ML is being applied to optimize distribution and transportation in LATAM. For example, Escobar-Naranjo et al. (2023) used deep Q-networks (DQN) for autonomous robot navigation, demonstrating how ML can improve the efficiency of transportation within manufacturing facilities. Similarly, Maldonado-Romo and Aldape-Pérez (2021) applied ML to sustainable path planning, showcasing how ML can optimize transportation routes and reduce environmental impact. These studies highlight the potential of ML to enhance distribution and transportation efficiency in the region. Retailers are intermediaries who make wholesale purchases from manufacturing organizations with the purpose of reselling products to customers. Good communication with retailers provides valuable information for demand forecasting, inventory management, customer satisfaction, and marketing. In LATAM for instance, Komori-Zevallos et al. (2023) developed a production model that combines Lean and Industry 4.0 principles to enhance productivity in small and medium-sized enterprises (SMEs) in Peru. This study demonstrates how ML can improve demand forecasting and inventory management, which are critical for maintaining strong relationships with retailers in the region. Maintenance and repair are support services for products after purchase, and manufacturers gain valuable insights into the cost of customer services. These services also include maintenance for manufacturing equipment and machinery, ensuring a smooth workflow in production processes and reducing quality issues and defects. ML is being applied to optimize maintenance and repair in Latin America, for example, Saucedo-Dorantes et al. (2020) used self-organizing maps (SOMs) for condition assessment of rotating machinery, demonstrating how ML can enhance predictive maintenance. Similarly, Montoya Herrera (2020) employed long short-term memory (LSTM) neural networks for remaining useful life estimation of turbofan engines, showcasing how ML can improve maintenance planning. These studies highlight the potential of ML to reduce downtime and improve equipment reliability in the region's manufacturing sector. Recycling is now a competitive advantage, as customers prefer sustainable companies, and organizations aim to reduce costs, waste, and preserve natural resources. Some businesses even use recycled materials for production, reducing the carbon footprint of manufacturing and promoting a sustainable future. As an example, in LATAM, Maldonado-Romo and Aldape-Pérez (2021) used ML for sustainable path planning, demonstrating how ML can enhance the efficiency of recycling logistics. This study highlights the potential of ML to support sustainable practices in the region's manufacturing value chain.

MANUFACTURING PROCESSES

This section explores the key elements, types, stages, and technologies involved in manufacturing processes, along with their benefits and challenges. It is important to understand that the digital transformation related to manufacturing in Latin America is still in early stages. The procedures and actions required to convert raw materials into completed goods are all included in the manufacturing process. It is essential to industrial production and plays a key role in satisfying consumer demand and promoting economic expansion. The cost, quality, and delivery of products are all directly impacted by how well and efficiently manufacturing processes work (Singha, S. K., & Singh, S. J., 2015). In Latin America, ML is increasingly being applied to optimize these processes. Rojas-Rodríguez et al. (2023) developed an ML model to predict failures in injection molding processes, a key manufacturing process in the plastics industry. Similarly, Saldaña et al. (2023) used ML to optimize grinding and milling processes in the Chilean copper mining industry, demonstrating how ML can enhance efficiency in resource-intensive sectors. In the energy sector, Petri et al. (2020) applied ML to evaluate pollution levels in transmission line insulators, showcasing the potential of ML to improve quality control and maintenance in critical infrastructure. Additionally, Cruz et al. (2022) demonstrated the use of ML in smart strawberry farming, highlighting how ML can optimize agricultural processes, which are vital to many LATAM economies.

Elements of Manufacturing Processes

- **Raw Materials**: The basic substances from which products are made. They can be natural (e.g., wood, minerals) or synthetic (e.g., plastics, alloys).
- **Labor**: Human resources involved in the manufacturing process, including operators, technicians, engineers, and managers.
- **Machinery and Equipment**: Tools, machines, and technology used to convert raw materials into finished products.
- **Methods**: The techniques, procedures, and workflows used in production.
- **Management**: Planning, organizing, directing, and controlling the manufacturing activities to ensure efficiency and effectiveness.

TYPES OF MANUFACTURING AND PROCESSES IN ENGINEERING

1. Casting and Molding

Casting and molding processes have a long-established history, deeply interwoven with the development of human society. Metal casting emerged as an innovative technique for the mass production of metal objects, revolutionizing industries in ancient Egypt, China, Greece, and Rome. In modern times, ML is being applied to optimize these processes. For example, Rojas-Rodríguez et al. (2023) developed an ML model to predict failures in injection molding processes, a key manufacturing process in the plastics industry. This study demonstrates how ML can enhance quality control and reduce defects in molding processes, which are critical for producing high-quality products in Latin America. The dawn of the Industrial Revolution and the introduction of iron and steel manufacturing propelled the mass production of durable components essential for locomotive engines, steamers, and warships. However, with the growing population and pollution concerns, the need for eco-friendly materials such as plastic, clay, wood, and gypsum ceramics emerged. This pushed the boundaries of casting technologies further into domains comprising low and high melting point metals and redox processes. The exploration of casting and molding technologies revolves around the mitigation of cast structural defects and shrinkage, modeling of the fluid flow and solidification of casting, sustainable casting through semi-solid state processing, and investment processes. The machinery available for casting processes and investment materials for molds and cores are also addressed (Lei, Z., & Su, W., 2019).

Molding is a common primary manufacturing process, as well as foundry terminology, in modern manufacturing. However, machinery to produce the mold is also considered an essential topic and is included in these titles. The simplest interpretation of casting is pouring of metals into a mold cavity, creation of the shape by solidification and removal of the part from the mold. Generally, shapes are internal and external casted surfaces, and it is a net shape operation, but it is also possible to imprint shape difference from the casting to the mold. There are many casting processes such as sand, lost foam, investment or metal die; and molds can be made using plaster, metal, high-speed machining, replicating or high precision milling. It includes a variety of processes according to the types of materials.

General classification of casting and molding processes primarily considers the following attributes: process category (sand, shell molding, investment, etc.), binder (none, resin, plaster, metal), dimensional accuracy (for investment, mold making) precision, molding method (manual, semi-automatic, fully automatic), thermal stability (low thermal, up to 600°C, normal, over 600°C), pattern return (metal expansion first casting, all metal mass casting), gating type (top gating, bottom gating), or using ceramic particles (also called koffl casting). It should also be noted that on the market products from casting materials go as Lost Foams in Inductotherm Co., Catamold in BASF, and DCM in 3M.

Casting and molding processes are fundamental in manufacturing for creating everything from small intricate parts to large, complex structures. Each process has its own advantages and is selected based on the material, product design, production volume, and cost considerations. Understanding these processes is essential for designing parts and selecting appropriate manufacturing methods (Callister & Rethwisch, 2020).

2. Forming

Forging is another critical forming process in modern engineering, especially for producing high-strength components. Forging involves deforming metal through the application of compressive forces, typically through hammering or pressing, to shape it into the desired form. The use of ML to improve these procedures is growing. Saldaña et al. (2023), for example, used ML to optimize the grinding and milling procedures used in the Chilean copper mining sector, which entails turning metal into goods that can be used. This study demonstrates how ML can increase productivity and decrease waste in metal-forming operations, which are essential to the mining industry in Latin America. The process can be carried out at various temperatures, with hot forging performed at high temperatures to reduce the metal's resistance to deformation, and cold forging done at or near room temperature for better dimensional accuracy and surface finish.

Two common types of forging are open-die forging and closed-die forging. In open-die forging, the metal is shaped between a hammer and anvil, allowing for greater flexibility in the size and shape of the part. This method is typically used for larger components, such as shafts and rods. In contrast, closed-die forging involves placing the metal between two dies that completely enclose the material, allowing to produce more complex shapes. This process is commonly used for smaller, high-strength parts like automotive crankshafts, gears, and aerospace components.

The advantages of forging include improved mechanical properties, such as increased strength and toughness, as well as the ability to produce parts with superior fatigue resistance. Modern forging processes have also benefited from advancements in simulation technology, such as finite element analysis (FEA), which allows engineers to predict material flow, optimize die design, and improve the quality of forged components.

3. Extrusion

Extrusion is another widely used forming process in modern engineering, where material, typically metal or plastic, is forced through a die to create a continuous shape with a fixed cross-section. Machine learning is being applied to optimize extrusion processes in Latin America. For example, Petri et al. (2020) used ML to evaluate pollution levels in transmission line insulators, which are often produced through extrusion processes. This study demonstrates how ML can improve quality control and ensure the reliability of extruded products, which are critical for the region's energy infrastructure. The extrusion process can be classified into hot extrusion and cold extrusion, depending on the temperature at which it is performed. In hot extrusion, the material is heated to reduce its resistance to deformation, while in cold extrusion, the material is shaped at room temperature, offering better surface finish and dimensional accuracy.

Extrusion is used to produce long, straight parts with uniform cross-sections, such as pipes, tubes, beams, and frames. It is particularly useful for materials like aluminum, which can be easily shaped into complex profiles. In the plastics industry, plastic extrusion is a common method for producing pipes, tubing, and sheeting. The versatility of the extrusion process, combined with its ability to produce continuous shapes with minimal waste, makes it an important technique in modern manufacturing.

The extrusion process has seen several advancements in recent years, including the development of multi-material extrusion, where different materials can be combined in a single extrusion process to create composite parts with tailored properties. Additionally, the use of 3D extrusion in additive manufacturing has enabled the creation of more complex geometries, expanding the range of applications for this process.

4. Injection Molding and Powder Metallurgy

In plastics manufacturing, injection molding is one of the most common forming processes. It involves injecting molten plastic into a mold cavity, where it cools and hardens into the desired shape. In Latin America, these procedures are being optimized through the use of ML. To illustrate how ML can improve quality control and lower faults in plastic manufacturing, Rojas-Rodríguez et al. (2023) created an ML model to forecast injection molding process failures. Cruz et al. (2022) also used ML to optimize agricultural processes, which frequently include injection-molded components. These studies demonstrate how ML can enhance injection molding and associated industries' productivity and product quality. Injection molding is highly efficient for mass production and is used to manufacture a wide variety of products, from automotive components to consumer electronics and medical devices.

Powder metallurgy is another important forming process, particularly for metals. It involves compressing metal powders into a desired shape and then heating (sintering) them to fuse the particles. This process is ideal for producing complex parts that would be difficult or costly to manufacture using traditional methods. Powder metallurgy is used in the production of gears, bearings, and other precision components (Torralba et al., 2013)

Forming processes are integral to engineering, enabling the efficient production of components with specific shapes, sizes, and mechanical properties. With advancements in technology, these processes have become more precise, efficient, and sustainable, meeting the demands of today's industries. Whether it is stamping, forging, rolling, extrusion, or more specialized techniques like injection molding and powder metallurgy, forming processes continue to evolve, providing manufacturers with the tools needed to produce high-quality products at scale.

5. Machining Process in Manufacturing

The machining operations are fundamental to engineering, serving as essential methods for shaping materials into finished products or components with high precision and tight tolerances. In Latin America, ML is being used more and more to streamline these procedures. To illustrate how ML may increase productivity and decrease waste in machining operations, Saldaña et al. (2023) employed ML to optimize grinding and milling procedures in the Chilean copper mining sector. Petri et al. (2020) also used ML to assess the pollution levels in transmission line insulators, which are frequently made by machining operations. These studies demonstrate how ML can improve machining accuracy and quality control, two crucial aspects of the region's manufacturing industry. These processes involve the removal of material, typically metals, but also plastics, ceramics, and composites, to create parts with specific geometries and dimensions. The ability to achieve exacting standards of accuracy and surface finish is what sets machining apart from other manufacturing processes such as forming or casting. In industries like aerospace, automotive, electronics, medical devices, and more, machining processes are indispensable for producing high-quality, intricate parts.

As technology has advanced, so too have machining processes. Traditional manual methods have largely been supplanted by automated, computer-controlled techniques such as CNC (Computer Numerical Control) machining, which has revolutionized precision manufacturing. Alongside CNC, other innovations such as additive machining, multi-axis machining, and the use of cutting-edge materials have elevated machining to new levels of efficiency and sophistication (Deshpande et al., 2023)

This introduction delves into the key machining processes used in modern engineering, their technological advancements, and their critical importance in today's industrial landscape.

Machining is indispensable engineering, enabling the production of parts with high precision, complex geometries, and excellent surface finishes. As technology continues to advance, machining processes are becoming more efficient, automated, and capable of producing components from a broader range of materials. From traditional processes like turning and milling to advanced techniques like EDM and laser cutting, machining is at the heart of modern manufacturing, driving innovation and supporting industries ranging from aerospace to consumer electronics. The future of machining promises even greater levels of sophistication, making it a critical area of focus for engineers and manufacturers alike.

6. Additive Manufacturing

Additive manufacturing (AM) is a technology that produces parts by joining layers of material together following the instructions of a computer model. AM has several advantages over traditional manufacturing methods due to its lack of tooling and its ability to create complex geometries that would otherwise be impossible to produce. In Latin America, for example, Montoya Herrera (2020) estimated the remaining useful life of turbofan engines—which are frequently made using AM—using long short-term memory (LSTM) neural networks. This study shows how ML can enhance the performance and dependability of AM-produced parts, which are essential for the aerospace sector. Furthermore, Escobar-Naranjo et al. (2023) demonstrated how ML might improve AM procedures in Latin America by using deep Q-networks (DQN) for autonomous robot navigation. These features allow AM to accelerate the progression from model to manufacturing, complementing rapid prototyping and small- and medium-volume production.

The application of 3D printing is a revolutionary process that builds objects layer by layer using digital models. Unlike traditional subtractive methods, where material is removed from a solid block, additive manufacturing selectively adds material, offering greater design flexibility, minimal waste, and the ability to create complex geometries that are difficult or impossible to achieve using conventional methods. AM has found applications in industries such as aerospace, automotive, healthcare, and consumer goods, driving innovation in product development and manufacturing. AM involves several steps, starting with the creation of a 3D digital model using computer-aided design (CAD) software.

The integration of ML with AM is revolutionizing how parts are designed, produced, and optimized. Machine learning algorithms are used to analyze and predict various aspects of the AM process, improving efficiency, quality, and reducing defects. By leveraging data from sensors, cameras, and other monitoring systems, ML enhances decision-making and process control, offering significant advantages across multiple stages of AM.

The model is then sliced into thin layers, which guide the printer on how to build the object layer by layer. Materials used in AM vary widely, including plastics, metals, ceramics, and composites. The specific AM technique employed depends on the material and the required properties of the final product. Common AM technologies include:

Fused Deposition Modeling (FDM): In this method, thermoplastic material is heated and extruded through a nozzle, depositing it layer by layer. FDM is widely used for prototyping and small-scale production due to its simplicity and low cost.

Stereolithography (SLA): SLA uses a UV laser to cure liquid resin into a solid form. This method produces highly detailed parts with excellent surface finish, making it suitable for applications like medical models and casting molds.

Selective Laser Sintering (SLS): SLS employs a laser to fuse powdered material (such as nylon or metal) into solid layers. This technique is valued for its ability to create durable parts with complex geometries without the need for support structures.

Despite its advantages, AM faces several challenges. The high cost of advanced materials, limited material options, slower production rates compared to traditional manufacturing, and post-processing requirements can hinder widespread adoption. Additionally, quality control and ensuring the structural integrity of printed parts remain significant concerns, especially for critical applications in aerospace and medical devices.

Additive manufacturing represents a transformative shift in how products are designed and manufactured. While challenges remain, ongoing advancements in materials, process efficiency, and post-processing technologies are likely to expand the use of AM across industries. As technology matures, it holds the potential to revolutionize traditional manufacturing, offering greater customization, efficiency, and design possibilities. Machine learning enhances the capabilities of Additive Manufacturing by improving process control, detection defect, and material performance. As these technologies continue to advance, the integration of ML in AM will unlock new levels of automation, efficiency, and innovation, making AM more reliable and scalable for industrial use. (Jiang, X.; Lu, J.; Zhao, N.; Chen, Z.; Zhao, 2024)

MACHINE LEARNING

Machine learning (ML) is a branch of artificial intelligence that allows computers by using algorithms to model based on experiences and accurately predict future events (Dogan & Birant, 2021). Machine learning is fundamentally about creating systems that can learn from data and make decisions or predictions based on that data. In Latin America, a variety of ML techniques are being applied to optimize manufacturing processes. Maldonado-Romo and Aldape-Pérez (2021) demonstrated the use of cutting-edge machine learning approaches in the area by using generative adversarial networks (GANs) for sustainable path design. Self-organizing maps (SOMs) were used in the predictive maintenance area by Saucedo-Dorantes et al. (2020) to evaluate the state of rotating machinery, demonstrating the application of unsupervised learning in industrial settings. Further demonstrating the application of deep learning in predictive maintenance, Montoya Herrera (2020) used long short-term memory (LSTM) neural networks for estimating the remaining usable life of turbofan engines. Furthermore, to demonstrate the use of reinforcement learning in the manufacturing sector of Latin America, Escobar-Naranjo et al. (2023) employed deep Q-networks (DQN) for autonomous robot navigation. These examples demonstrate the diverse range of ML techniques being applied to address specific challenges in the region's industrial processes. Depending on the available data and the problem to attend, the major ML approaches can be classified into three main categories: supervised learning, unsupervised learning and reinforcement learning (Janiesch, Zschech & Heinrich, 2021).

According to Liu (2011), supervised learning is called classification or inductive learning that is very similar to the way the human brain learns in the sense that from past experiences gain knowledge and increase the ability to perform tasks in daily life. To substitute "experiences", computers via machine learning uses past data that represents situations in the real-world to learn. Supervised learning is used in all types of situations, including text and Web, and has had very good results in applications of the real-world, involving training a model on a labeled dataset, meaning the data comes with input-output pairs (Janiesch, Zschech & Heinrich, 2021). Typical problems in supervised learning are classification and regression. This is particularly useful for tasks like predictive maintenance where historical data can be used to predict future equipment failures. Examples of this learnings are SVMs, neural nets, logistic regression, naive bayes, memory-based learning, random forests, decision trees, bagged trees, boosted trees, and boosted stumps.

Important amount of research studies in ML has enquired classification problems where the expected result is a categorical class affiliation of data, whereas some manufacturing problems are clustering type which is the task of finding groups of elements that share common properties, those groups are clusters (Dogan & Birant, 2021). Some algorithms used for classification include support vector machines and decision trees (Sarker, 2021); while k-means is the most widely extended clustering technique (Ahmad & Dey, 2007).

In contrast with supervised learning, unsupervised learning uses unstructured raw data releasing from the need of labeling data, that is done manually by engineering techniques, so it simplifies the machine learning methods making them flexible, general, and automated. Because of that, it has applications in computer vision, natural language processing, speech recognition, and optimal control (e.g., for developing autonomous self-driving cars), and by using network data, to improve network performance and provide services, such as traffic engineering, anomaly detection, Internet traffic classification, and quality of service optimization (Usama et al., 2019). So that, unsupervised learning deals with unlabeled data and is used for identifying patterns or clusters within the data, is commonly used in clustering and dimensionality reduction problems, which is essential for anomaly detection and process improvement

Unsupervised learning and supervised learning used in conjunctions are called semi-supervised learning, and its use is very common, by doing a preliminary process of the data and then performing the analysis, in that way can achieve a good feature representation and in finding patterns and structures in unlabeled data (Usama et al., 2019).

Reinforcement learning is an area of machine learning concerned with how software agents ought to take actions in an environment to maximize some notion of cumulative reward (Mahesh, 2018), involves training models to make a sequence of decisions by rewarding desired behaviors and punishing undesired ones, is used to achieve a goal in systems maximizing rewards (Janiesch, Zschech & Heinrich, 2021). Reinforcement learning is one of three basic machine learning paradigms, alongside supervised learning and unsupervised learning (Mahesh, 2018). This technique is especially valuable in dynamic and complex industrial environments where real-time adjustments are crucial.

A separate mention is required for artificial neural networks (ANN) and deep learning, a complex and advanced neural network. ANN is a flexible structure consisting of mathematical representation of connected processing units called artificial neurons, that can be modified to be used in all three categories of ML (Janiesch, Zschech & Heinrich, 2021). By applying algorithms that mimic the processes of real neurons, we can make the network 'learn' to solve many types of problems. Ranging from speech recognition to prediction of protein secondary structure, classification of cancers and gene prediction.

Deep learning (deep structured learning, hierarchical learning or deep machine learning), a subset of ML, that model complex patterns in large datasets, making it powerful for tasks such as image and speech recognition, which can be applied in quality control and monitoring systems (Janiesch, Zschech & Heinrich, 2021). Deep learning is replacing, just like unsupervised or semi-supervised learning, the handmade steps of characteristics learning and hierarchical classification of characteristics extraction, and it is a set of algorithms that model high-level abstractions in data by the usage of multiple layers with complex structures or integrated of multiple non-linear transformations (Benuwa, 2016).

METHODOLOGY

A systematic literature review was executed following a methodical approach to localize, obtain and synthesis research products. The process started with a rigorous search in Google Scholar, to identify a preliminary group of journal papers and chapters of books related to machine learning, manufacturing, case studies and only studies published by Latin American authors between 2020 to 2024 were considered to create a dataset. Secondly, duplicates and documents without full access were removed, so the list of possible matches to the research objectives and questions was ready for a further revision of the complete text article. Once the research findings were consciously read, it was decided which ones fulfilled the criteria of eligibility to be used in the content of the review. Finally, we obtained a set of documents that allow us to understand how the advancement of the applications of machine learning techniques in manufacturing in Latin America is. By sticking to the methodology presented, this work constitutes a rigorous examination of the existing literature for the following discussion.

Identification

In the initial search 840 documents were identified in the search engine Google Scholar.

Screening

It was necessary to remove 432 repeated documents and 10 unavailable studies, leaving 398 research findings retained. A revision of the author's affiliation, title, abstract and keyword, it turned out to exclude 357 documents, leaving 41 documents.

Eligibility

Of the remaining 41 documents, 14 were excluded because there was no access to the full text. This process resulted in 27 documents being eligible.

Final selection

The final selection emphasized Machine Learning models and activities related to manufacturing value chain cases of study, leaving aside literature reviews, developed by authors affiliated in Latin American institutions.

FINDINGS

About the first research question, there are insufficient observations to declare a specific industrial sector in Latin America that has adopted the ML more extensively. This is expected mainly due to confidentiality and competitive reasons. Companies do not tend to share information that would be required for a more extensive analysis. Only 12 of the 27 research documents are applied in a specific industry, none of them declare the name of the organization. Others test their proposals in laboratories, in virtual scenarios or take databases from several industries to probe the algorithms (Figure 1 and Table 1). Of the 12 studies developed in specific industries as diverse as, just agro-industrial is repeated twice, the others with one mention are steel, medical, mining, food, electrical, gas and oil, appliances, automotive, aerospace and construction.

Figure 1. Type of case of study or test developed in the document

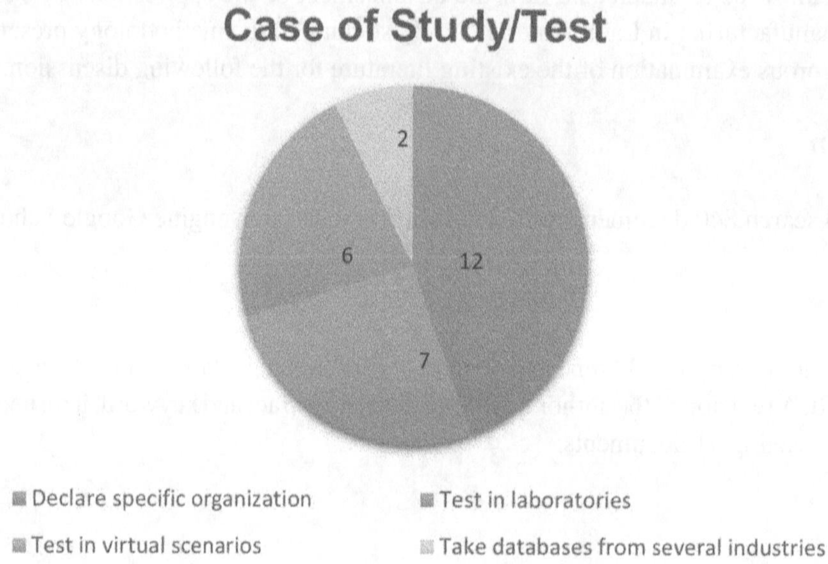

(Self-made)

Table 1. References for type of case of study or test developed

Case	References
Declare specific organization	(Ferreira et al., 2021), (Cruz et al., 2022), (Komori-Zevallos et al., 2023), (Paredes Rodríguez et al., 2023), (Rodríguez-Gallegos et al., 2023), (Shafi et al, 2023), (Petri et al, 2020), (Saldaña et al, 2023), (Gómez-Ramírez and Soler Anguiano, 2022), (Orellana Cordero, 2020), (Reséndiz-Flores et al, 2022), (Reyes et al., 2023).
Test in laboratories	(Saucedo-Dorantes et al., 2023), (Hernández- Barragan et al., 2020), (Escobar-Naranjo et al., 2023), (Parra, 2020), (Ruiz et al., 2023), (Rojas-Rodríguez et al., 2023), (Fortoul-Diaz et al., 2023).
Test in virtual scenarios	(Maldonado-Romo and Aldape-Pérez, 2021), (Ferreira de Rezende et al., 2023), (Lefrenc, 2023), (Rojas et al., 2022), (Montoya Herrera, 2020), (Ponce et al., 2024).
Take databases from several industries	(Aguilar-Lasserre et al., 2020), (Metri et al., 2021).

Source: self-made

The second research question about the type of manufacturing value chain processes problems are solved using ML techniques in Latin America. Within the manufacturing value chain process attended in Latin America stand out transportation in production for load and storage (Maldonado-Romo et al., 2021; Hernandez-Barragan et al., 2020; Escobar-Naranjo et al., 2023; Rojas et al., 2022). Additionally, Ponce et al. (2024) included in their study quality control to detect occurrences for faulty conditions besides transport in production. In all those works, the authors use ML for programming of robots and vehicles for path planning for autonomous systems, however, in Maldonado-Romo et al. (2021), they substitute the physical element by virtualization for sustainability. Another type of process included in the studies is transportation for production via extraction of oil in Ferreira de Rezende et al. (2023) by substitution of physical elements by virtualization. In assembly processes, the ML applications are used for quality control in detection of occurrences for faulty conditions in Rodriguez-Gallegos et al. (2023), Shafi et al. (2023) and Orellana Cordero (2020). Specifically, in Orellana Cordero also considers the stops for maintenance as occurrences for faulty conditions. Some of the studies address the condition-based maintenance for predicting maintenance plans in different machinery, like rotating machinery (Saucedo-Dorantes et al., 2020), hammers for shredding (Ferrerira Neto et al., 2021) and turbofan engines (Montoya Herrera, 2020); also, for evaluating the degree of damage in pollution line insulators of facility maintenance and quality control of the production of line insulators (Petri et al., 2020). In quality control, Cruz et al. (2022) verified in the food sector the occurrence for faulty conditions in the products and Parra (2020) cared about the properties of polymer nanocomposites and electric fields. In addition, some of the studies verified the occurrences for faulty conditions in a complete production line involving load, quality control and storage (Ruiz et al., 2023; Fortoul-Diaz et al., 2023), even include processes such as cut and weld (Lefrenc, 2023). Rojas-Rodriguez et al. (2023) used ML to predict failures in an injection molding processes, Reséndiz-Flores et al. (2022) performed smart fault detection in foam injection and welding processes, and Saldaña et al. (2023) carried out an optimization of productive indicators in grinding and milling processes. Whereas Aguilar-Lasserre et al. (2020) did functional optimization of a packing for packaging processes. Customer satisfaction is addressed in product validation and design processes in the study of Metri et al. (2021). Decision support system for demand forecast and path planning processes of value chain in manufacture is treated for small and medium-sized enterprises (Komori-Zevallos et al., 2023), for inventory classification (Paredes Rodríguez et al., 2023), for equipment logistics (Gómez-Ramírez and Soler Anguiano, 2023), and for supplier selection (Reyes et al., 2023).

The third research question inquiries about the machine learning methods commonly used to handle manufacturing value chain tasks in Latin America. Firstly, 13 of the studies used one ML algorithm, 10 used two ML algorithms, two used three algorithms, one used four algorithms and five algorithms, in Figure 2. It is important to point out that Rodriguez-Gallegos et al. do not declare specifically which algorithm was used, only indicate that it used one ML technique of Azure Cloud Services.

Figure 2. Number of ML algorithms used in the studies. In dark gray appear the number of studies that used only one algorithm in the research, in light gray appear the number of documents that used two algorithms in their research, in medium gray appear the ones that used three algorithms, in black and the lightest gray appear those that used four and five algorithms, respectively

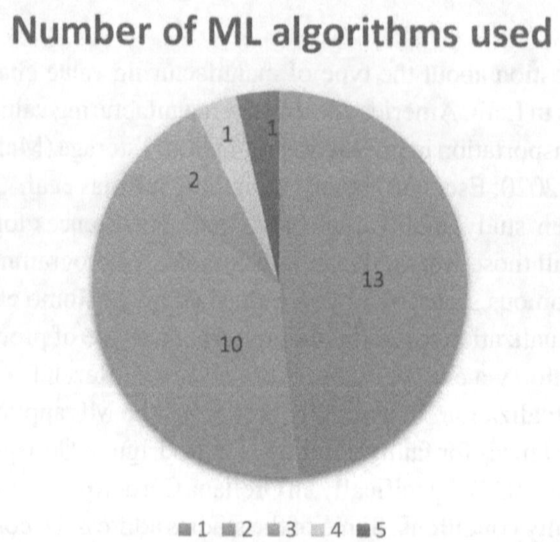

(Self-made)

In the studies, 34 different types of ML methods, more of the half are classified as supervised learning, a fifth are classified as unsupervised learning, the rest are other types of methods (Figure 3).

Figure 3. Classification of ML methods utilized in research studies. The darkest gray represents the number of studies that used the supervised approach, in black with white dots are the unsupervised approach research, in solid black are the reinforcement approach studies, in white with black dots and light gray are the studies with semi-supervised and supervised and unsupervised approach, finally in medium gray is shown the studies that didn´t specify which algorithm used

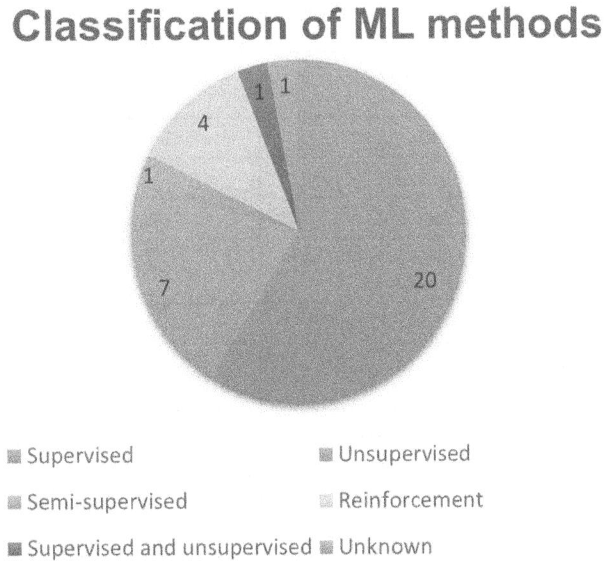

(Self-made)

Table 2. ML methods used in Latin America research studies during the period 2020-2024

Classification of ML methods	ML methods and references	References
Supervised	Genetic algorithms (GA)	Aguilar-Lasserre et al. 2020
	Adaptative Single Neuron (ASN)	Hernandez-Barragan et al., 2020
	Long-short term memory-Neural Networks (LSTM)	Ferreira de Rezende et al., 2023 ; Montoya Herrera, 2020
	Multilayer Perceptron-Neural Networks	Ferreira de Rezende et al., 2023
	Convolutional Neural Networks (CNN)	Parra Garcés, 2020; Fortoul-Díaz et al., 2023 ; Shafi et al., 2023 ; Ponce et al., 2024 ; Petri et al., 2020
	Neural Networks (NN)	Parra Garcés, 2020; Ruiz et al., 2023 ; Saldaña et al., 2023
	Multi-headed Neural Networks	Parra Garcés, 2020
	Random Forest	Parra Garcés, 2020; Orelllana Cordero, 2020
	Autoregressive Integrated Moving Average (ARIMA)	Komori-Zevallos et al., 2023
	Seasonal ARIMA (SARIMA)	Komori-Zevallos et al., 2023
	Multivariate Linear Regression Model (MLRM)	Metri et al., 2021
	Adaptive Neuro-Fuzzy Inference System	Metri et al., 2021
	Support Vector Machine	Ruiz et al., 2023
	Decision trees	Ruiz et al., 2023; Saldaña et al., 2023; Orellana Cordero, 2020
	Assembly methods Adaboost	Ruiz et al., 2023
	Assembly methods Random Forest	Ruiz et al., 2023
	Agent Based Simulation (ABS)	Gómez Ramírez and Soler Anguiano, 2023
	Regression	Saldaña et al., 2023
	Lineal algorithm	Orellana Cordero, 2020
	Sequential feature selection (SFS)	Reséndez-Flores et al., 2022
Unsupervised	Generative adversarial networks (GAN)	Maldonado-Romo and Adape-Pérez, 2021
	Self-organizing maps (SOM) Neural networks	Saucedo-Dorantes et al., 2020
	Isolation Forest model	Cruz et al., 2022
	AHP Fuzzy methodology	Paredes Rodriguez et al., 2023
	Remaining Useful Life – Deep learning	Montoya Herrera, 2020
	Machine learning	Rojas-Rodríguez et al., 2023
	K-means grouping	Reyes et al., 2023
Semi-supervised	YOLO V5 model	Cruz et al., 2022
Supervised and unsupervised	Fuzzy inference system	Petri et al., 2020; Ponce et al., 2024

continued on following page

Table 2. Continued

Classification of ML methods	ML methods and references	References
Reinforcement	Deep Reinforcement Learning	Ferreira Neto et al., 2021; Rojas et al., 2022
	Deep Q-Network (DQN)	Escobar-Naranjo et al., 2023
	Learning	Lefranc, 2023
	Q-Learning	Lefranc, 2023
No specified	Azure Cloud System	Rodríguez-Gallegos et al., 2023

Source: self-made

The fourth research question asks about the manufacturing value chain processes datasets that are popular and used as a benchmark in Latin America. Datasets depend on the application, in those whose purpose is transportation, the datasets are the characteristics of vehicle, environment and the interaction between the two, simulated or experimental (Maldonado-Romo and Aldape-Pérez, 2021; Hernadez-Barragan et al., 2020; Escobar-Naranjo et al., 2023; Lefranc, 2023). For applications of design, product validation and decision support systems, the datasets are actual characteristics of the product or process accompanied with de ponderation of desired requirements (Aguilar-Lasserre et al., 2020; Metri et al., 2021; Reyes et al., 2023). In maintenance and quality control processes, the data is in real time about the system condition (Saucedo-Dorantes et al., 2020; Ferreira de Rezende et al., 2023; Ferreira Neto et al., 2021; Montoya Herrera, 2020; Rojas-Rodriguez et al., 2023; Fortoul-Diaz et al., 2023; Petri et al., 2020; Saldaña et al., 2023; Gómez-Ramírez and Soler Anguiano, 2023; Orellana Cordero, 2020; Reséndiz-Flores et al., 2022). Datasets of real historical data is used in applications of forecasting (Komori-Zevallos et al., 2023; Paredes Rodríguez et al., 2023). Images are used as datasets in applications of quality control (Cruz et al., 2022; Parra Garcés, 2020; Rodriguez-Gallegos et al., 2023; Shafi et al., 2023; Ponce et al., 2024; Ruiz et al., 2023). Finally, Rojas et al., 2022 decided to use several libraries to facilitate the training of algorithm for applications of transportation using robots.

For the fifth research, the question about the performance metrics are generally used to evaluate ML models constructed in manufacturing studies, has a similar answer than the previous question, depend on the application. For transportation uses, avoidance of obstacles (Ponce et al., 2024), efficient movement via minimizing cost and traversing the shortest distance with limited timeframe (Rojas-Rodriguez et al., 2023; Escobar-Naranjo et al., 2023; Hernadez-Barragan et al., 2020) and, in addition, sustainability (Maldonado-Romo and Aldape-Pérez, 2021). For quality control purposes, fault detection is the main performance metric (Reséndiz-Flores et al., 2023; Petri et al., 2020; Shafi et al., 2020; Ruiz et al., 2023; Rodriguez-Gallegos et al., 2023; Cruz et al., 2022) and for maintenance applications (Saucedo-Dorantes et al., 2020; Ferreira Neto et al., 2020; Montoya Herrera, 2020). For purposes of a variety of production and assembly processes, metrics are productivity, efficiency, and other characteristics of the different processes (Lefrenc, 2023; Parra Garcés, 2020; Fortoul-Días et al., 2023, Petri et al., 2020; Saldaña et al., 2023; Orellana Cordero, 2020). In applications for forecasting and decision support systems, the metrics are reduced time and cost of delivery (Gómez Rodríguez and Soler Anguiano, 2020), minimum purchase cost (Reyes et al, 2023) and productivity and reduce waste and obsolete products (Komori-Zevallos et al., 2023). In design and product validation, the metrics are the desired characteristics of packing (Aguilar-Lasserre et al. 2020) and acceptability (Metri et al., 2021).

The last three research questions are difficult to fully answer considering the literature review made previously. For the sixth question about a comparison with Latin America, North America, Europe and Asia advances in the adoption of ML techniques in manufacturing value chain processes in the Industrial Development Report 2020: Industrializing in the digital age (UNIDO, 2019) pointed out that only United States, Japan, Germany, China, Taiwan Province of China, France, Switzerland, the United Kingdom, the Republic of Korea and the Netherlands, 10 countries possessed 100 or more global patent family applications in advanced digital production (ADP) technologies, that includes artificial intelligence, big data analytics, cloud computing, Internet of Things (IoT), advanced robotics and additive manufacturing, among others; what it implied that those countries had the 90 percent of patents and 70 percent of sells in goods with these technologies. According to the same report, of the nations of Latin America and the Caribbean only Brazil was indicated as a follower in production, Mexico, Colombia and Argentina as followers in use, Chile, Venezuela and Dominican Republic as latecomers in production and El Salvador, Ecuador and Costa Rica as latecomers in use. The followers' economies represent 8 percent of patents and a little less than half of the imports of the goods with ADP technologies. The latecomers are economies with some patenting activity, exporting and importing ADP technologies. In the literature review completed in the 27 research works only six countries appeared: Brazil, Chile, Colombia, Ecuador, Mexico and Peru (Figure 4).

Figure 4. Research works for Latin American countries founded during literature review

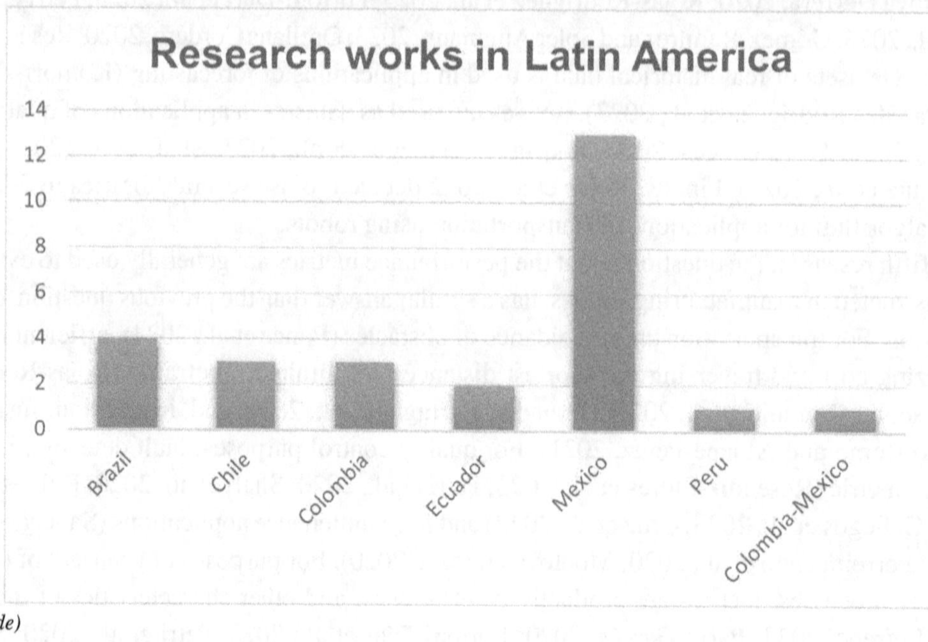

(Self-made)

It noted that all the economies mentioned in IDR2020 appeared in the review, but Peru has one contribution. In the IDR2024, there isn't a straightforward mention of the industry 4.0 and the ADP technologies, the information comes in terms of the achievement of sustainable development global objectives, particularly the ninth about Industry, Innovation and Infrastructure (SDG9). The performance in innovation is 7 percent below the developed countries average, the Caribbean and Central

America subregions have the lower performance, while South America has better innovation index, however, there is a need to foment policies, collaborative environments, and industrial clusters that allow an advance in technological innovations. Industry index is closer to the average of developed economies, nonetheless, this index presents a decrease since the pandemia times, increasing the disparity with other regions (UNIDO, 2024). The Central America subregion has 13 percent higher than the average of Latin America and the Caribbean in the industry index, this percentage is strongly influenced by the strong industry of Mexico (Idem). It is the strong industry in Mexico that has caused the number of investigations found in the review to be three times greater than that of any other country.

In the seventh research question, it is address about the main challenges of ML that are faced in manufacturing value chain in Latin America, according with IDR2024, the region has a delay in generation and adoption of advanced digital production technologies, hence is a need the creation and implementation of public policies that cause the establish of innovation environments and entrepreneurship, sustain research and development, improve education required for the smart factories technologies, and generated robust infrastructure for industry (UNIDO, 2024). On the other hand, micro, small, and medium enterprises (MSMEs) in Latin America and the Caribbean are the 90 percent of businesses and employ more than a half of the population, but only contribute to the 25 percent of gross domestic product (GDP), in comparison to 50 percent in OECD economies), because there are many informal small business that don't have access to appropriate financial instruments necessary to invest in growth, innovation, digitalization, and expansion, to create quality jobs for hiring qualified personnel (Diaz-Granados, 2023).

For the eighth research question about the possible research directions for ML studies in manufacturing value chain shortly, Metri et al. (2021) and Parra Garcés (2020) proposed more data for the same ML algorithms and the same application; also, Rojas et al. (2023) proposed more data in same application but captured in real time. While Maldonado-Romo and Aldape Pérez (2021) pointed out the use of the same methodology but applied to other situations; in the same line of ideas, Saucedo-Dorantes et al (2020) proposed the repetition of the methodology in other kind of applications with the consideration of more variables. Maldonado et al (2023) and Paredes-Rodriguez et al (2023) considered as future work the rerun of the research just expanding the variables. Likewise, Cruz et al (2020) propounded the repeat of the study considering more variables but including autonomous systems for the capture of data. However, some authors contemplated the application of other ML algorithms and different variables (Aguilar-Lasserre et al., 2020; Gómez-Ramírez and Soler Anguiano, 2023), inclusive the implementation in the real world (Ponce et al., 2024); or the ampliation of the application environments or situations (Hernandez Barragan et al, 2020; Ferreira Neto et al., 2021; Fortoul-Díaz et al., 2023). But not all the authors proposed increment of the variables, just propounded the use of more ML algorithms and other training environments (Escobar-Naranjo et al., 2023; Rojas et al., 2023; Rodriguez-Gallegos et al., 2023). Lefranc (2023) and Montoya Herrera (2020) only suggested the use of other ML algorithms to the same situation and variables. While Shafi et al. (2023) proposed to improve the model used. Komori-Zevallos et al. (2023) proposes improvements to the process not related to machine learning. Some studies that didn't include future work (Ferreira de Rezende et al. 2023; Rojas-Rodríguez et al., 2023; Orellana Cordero, 2020).

CONCLUSIONS

Although there are a great number of published papers and book chapters about the applications of Machine Learning, there are very few research studies about Machine Learning applied to manufacturing value chain processes in Latin America during 2020 to 2024. The reasons are diverse and include the disparity of industrialization of the region, the type of enterprises installed (SMEs) and their characteristics, the secrecy of the industries in their advanced digital production technologies used for competitiveness, the lack of public policies that promote the inversion in digitalization and innovation. In the research papers review, Mexico and Brazil lead the production. The industrial sectors are as diverse as agro-industrial, steel, medical, mining, food, electrical, gas and oil, appliances, automotive, aerospace and construction. The supervised learning is the most used in the papers, specifically Artificial Neural Networks. The datasets are as diverse as the situations treated and solved by Machine Learning, including images, information of the processes, a variety of paths, historical production data, and others, addressing situations of path planning of autonomous systems, decision support systems, quality control, description and control of production processes, forecasting of sales and condition-based maintenance. This review allows the visualization of the status of ML adoption in Latin America and the Caribbean, so that it is possible to know how to keep advancing in this subject. In the future it will be desirable to extend this study for searching in specified scientific databases to have access to more research works.

REFERENCES

Jiang, X., Lu, J., Zhao, N., Chen, Z., & Zhao, Z. (2024). A Review of Wear in Additive Manufacturing: Wear Mechanism, Materials, and Process. *Lubricants (Basel, Switzerland)*, *12*(9), 321.

Aguilar-Lasserre, A. A., Torres-Sanchez, V. E., Fernandez-Lambert, G., Azzaro-Pantel, C., Cortes-Robles, G., & Roman-del Valle, M. A. (2020). Functional optimization of a Persian lime packing using TRIZ and multi-objective genetic algorithms. Computers & Industrial Engineering, 139. 105558. ISSN 0360-8352 2.

Ahmad, A., & Dey, L. (2007). A k-mean clustering algorithm for mixed numeric and categorical data. *Data & Knowledge Engineering*, *63*(2), 503–527. DOI: 10.1016/j.datak.2007.03.016

Aytug, H., Bhattacharyya, S., Koehler, G. J., & Snowdon, J. L. (1994). A review of machine learning in scheduling. *IEEE Transactions on Engineering Management*, *41*(2), 165–171. DOI: 10.1109/17.293383

Babu, S. S., Mourad, A. H. I., Harib, K. H., & Vijayavenkataraman, S. (2022). Recent developments in the application of machine-learning towards accelerated predictive multiscale design and additive manufacturing. *Virtual and Physical Prototyping*, *18*(1), e2141653. Advance online publication. DOI: 10.1080/17452759.2022.2141653

Benuwa, B. B., Zhan, Y. Z., Ghansah, B., Wornyo, D. K., & Banaseka Kataka, F. (2016). A review of deep machine learning. *International Journal of Engineering Research in Africa*, *24*, 124–136. DOI: 10.4028/www.scientific.net/JERA.24.124

Callister, W. D., & Rethwisch, D. G. (2020). Callister's Materials Science and Engineering. Wiley. https://books.google.com.mx/books?id=IUjODwAAQBAJ

Canz, T., & Jagdale, S. (1995). Decision support for manufacturing using artificial neural-networks. Journal of Materials Processing Technology, 52(1), 9-26. ISSN 0924-0136. DOI: 10.1016/0924-0136(94)01439-8

Cheepu, M. (2023). Machine Learning Approach for the Prediction of Defect Characteristics in Wire Arc Additive Manufacturing. *Transactions of the Indian Institute of Metals*, *76*(2), 447–455. DOI: 10.1007/s12666-022-02715-1

Cica, D., Sredanovic, B., Tesic, S., & Kramar, D. (2024). Predictive modeling of turning operations under different cooling/lubricating conditions for sustainable manufacturing with machine learning techniques. *Applied Computing and Informatics*, *20*(1/2), 162–180. DOI: 10.1016/j.aci.2020.02.001

Ciccarelli, M., Corradini, F., Germani, M., Menchi, G., Mostarda, L., Papetti, A., & Piangerelli, M. (2023). SPECTRE: A deep learning network for posture recognition in manufacturing. *Journal of Intelligent Manufacturing*, *34*(8), 3469–3481. DOI: 10.1007/s10845-022-02014-y

Cruz, M., Mafra, S., Teixeira, E., & Figueiredo, F. (2022). Smart Strawberry Farming Using Edge Computing and IoT. *Sensors (Basel)*, *22*(15), 5866. DOI: 10.3390/s22155866 PMID: 35957425

Cser, L., Geiger, M., Greska, W., & Hoffmann, M. (1991). Three kinds of case-based learning in sheet metal manufacturing. Computers in Industry, 17(2–3), 195-206. ISSN 0166-3615. DOI: 10.1016/0166-3615(91)90032-5

Deshpande, Y., Ayer, S., Agrawal, T., Agrawal, V., Madankar, T., & Barve, P. (2023). Application of Smart Strategies for Sustainable Manufacturing of Conventional Machining Process: A Review. Journal of The Institution of Engineers (India): Series C, 104. DOI: 10.1007/s40032-023-00995-0

Díaz-Granados, S. (2023). The green and digital path for Latin American SMEs. In Development Bank of Latin America and the Caribbean. Access https://www.caf.com/en/knowledge/views/2023/11/the-green-and-digital-path-for-latin-american-smes/

Dogan, A., & Birant, D. (2021). Machine learning and data mining in manufacturing. Expert Systems with Applications, 166. 114060. ISSN 0957-4174. DOI: 10.1016/j.eswa.2020.114060

Duquesnoy, M., Liu, C., Zapata Dominguez, D., Kumar, V., Ayerbe, E., & Franco, A. A. (2023). Machine learning-assisted multi-objective optimization of battery manufacturing from synthetic data generated by physics-based simulations. Energy Storage Materials, 56, 50-61. ISSN 2405-8297. DOI: 10.1016/j.ensm.2022.12.040

Escobar-Naranjo, J., Caiza, G., Ayala, P., Jordan, E., Garcia, C. A., & Garcia, M. V. (2023). Autonomous navigation of robots: Optimization with DQN. *Applied Sciences (Basel, Switzerland)*, *13*(12), 7202. DOI: 10.3390/app13127202

Ferreira de Rezende, S. W., Soares Silva, L., Cavalini Jr, A. A., & Vieira de Moura Jr, J. R., & Steffen Jr., V. (2023). Digital Twin of an Offshore Riser Systems for Time Series Prediction Using Deep Learning Models. At A.S. De Paula, A.T. Fabro & M.R. Machado (Eds) Proceedings of the XIX International Symposium on Dynamic Problems of Mechanics. DINAME 2023. Brazil. DOI: 10.26678/ABCM.DINAME2023.DIN2023-0069

Ferreira Neto, W., Cavalcante, C., & Do, P. (2021). Deep reinforcement learning-based maintenance decision-making for a steel production line. In *Proceedings of the Proceedings of the 31st European Safety and Reliability Conference*; Castanier, B., Cepin, M., Bigaud, D., & Berenguer, C., Eds (pp. 2611-2618). DOI: 10.3850/978-981-18-2016-8_600-cd

Fortoul-Diaz, J. A., Carrillo-Martinez, L. A., Centeno-Tellez, A., Cortes-Santacruz, F., Olmos-Pineda, I., & Flores-Quintero, R. R. (2023). A smart factory architecture based on industry 4.0 technologies: Open-source software implementation. *IEEE Access : Practical Innovations, Open Solutions*, *11*, 101727–101749. DOI: 10.1109/ACCESS.2023.3316116

Ghasemkhani, B., Aktas, O., & Birant, D. (2023). Balanced K-Star: An Explainable Machine Learning Method for Internet-of-Things-Enabled Predictive Maintenance in Manufacturing. *Machines (Basel)*, *11*(3), 322. DOI: 10.3390/machines11030322

Gómez-Ramírez, A., & Soler Anguiano, F. I. (2023). Simulation based optimization of drilling equipment logistics: a case of study (Vol. 217). Procedia Computer Science., https://www.sciencedirect.com/science/article/pii/S1877050922023614, , ISSN 1877-0509.DOI: 10.1016/j.procs.2022.12.283

Grosman, N., Braude, H., Rovira, S., & Patiño, S. (2021). “Made in Latam: how smart manufacturing can give Latin America new hope for industrialization”, Project Documents (LC/TS.2021/111), Santiago, Economic Commission for Latin America and the Caribbean (ECLAC).

Hernandez-Barragan, J., Rios, J. D., Alanis, A. Y., Lopez-Franco, C., Gomez-Avila, J., & Arana-Daniel, N. (2020). Adaptative Single Neuron Anti-Windup PID Controller Based on the Extended Kalman Filter Algorithm. *Electronics (Basel)*, *9*(636), 636. Advance online publication. DOI: 10.3390/electronics9040636

Herzog, T., Brandt, M., Trinchi, A., Sola, A., & Molotnikov, A. (2024). Process monitoring and machine learning for defect detection in laser-based metal additive manufacturing. *Journal of Intelligent Manufacturing*, *35*(4), 1407–1437. DOI: 10.1007/s10845-023-02119-y

Horňas, J., Běhal, J., Homola, P., Senck, S., Holzleitner, M., Godja, N., Pásztor, Z., Hegedüs, B., Doubrava, R., Růžek, R., & Petrusová, L. (2023). Modelling fatigue life prediction of additively manufactured Ti-6Al-4V samples using machine learning approach. International Journal of Fatigue, 169, 107483. ISSN 0142-1123. DOI: 10.1016/j.ijfatigue.2022.107483

Horvath, L., & Rudas, I. J. (1993). A machine learning based approach to manufacturing process planning. ISIE '93 - Budapest: IEEE International Symposium on Industrial Electronics Conference Proceedings, Budapest, Hungary, pp. 429-433, DOI: 10.1109/ISIE.1993.268767

Institute for Manufacturing. (2023). Porter's Value Chain. In IfM Management Technology Policy of University of Cambridge. https://www.ifm.eng.cam.ac.uk/research/dstools/value-chain-/

International Monetary Fund – IMF. (2023). Country Composition of WEO Groups. World Economic Outlook Database. Groups and Aggregates Information. In https://www.imf.org/en/Publications/WEO/weo-database/2023/April/groups-and-aggregates

Irani, K. B., Cheng, J., Fayyad, U. M., & Qian, Z. (1993). Applying machine learning to semiconductor manufacturing. *IEEE Expert*, *8*(1), 41–47. DOI: 10.1109/64.193054

Janiesch, C., Zschech, P., & Heinrich, K. (2021). Machine learning and deep learning. *Electronic Markets*, *31*(3), 685–695. DOI: 10.1007/s12525-021-00475-2

Jiang, X., Lu, J., Zhao, N., Chen, Z., & Zhao, Z. (2024). A Review of Wear in Additive Manufacturing: Wear Mechanism, Materials, and Process. *Lubricants (Basel, Switzerland)*, *12*(9), 321. DOI: 10.3390/lubricants12090321

Ko, H., Lu, Y., Yang, Z., Ndiaye, N. Y., & Witherell, P. (2023). A framework driven by physics-guided machine learning for process-structure-property causal analytics in additive manufacturing. Journal of Manufacturing Systems, 67, 213-228. ISSN 0278-6125. DOI: 10.1016/j.jmsy.2022.09.010

Komori-Zevallos, A. R., Montedoro-Garay, F. M., Garcia-Lopez, Y. J., & Quiroz-Flores, J. C. (2023). A Production Model That Combines Lean And Industry 4.0 Principles To Enhance The Productivity Of Small And Medium-Sized Enterprises (SMEs) In Peru's Food Manufacturing Sector [Conference]. 5th Conference On Production Systems and Logistics CPSL 2023-2.Hannover, Germany. DOI: DOI: 10.15488/1525

Krogh, A. (2008). What are artificial neural networks? *Nature Biotechnology*, *26*(2), 195–197. DOI: 10.1038/nbt1386 PMID: 18259176

Lee, J. (1995). Machine performance monitoring and proactive maintenance in computer-integrated manufacturing: Review and perspective. *International Journal of Computer Integrated Manufacturing*, *8*(5), 370–380. DOI: 10.1080/09511929508944664

Lefranc, G. (2023). Multirobot Allocation In A Flexible Manufacturing System, Using Reinforcement Learning For Decision-Making, Case of Study. *Procedia Computer Science*, *221*, 41–48. DOI: 10.1016/j.procs.2023.07.006

Lei, Z., & Su, W. (2019). Research and application of a rolling gap prediction model in continuous casting. *Metals*, *9*(3), 380. DOI: 10.3390/met9030380

Liu, B. (2011). Supervised Learning. In *Web Data Mining. Data-Centric Systems and Applications*. Springer., DOI: 10.1007/978-3-642-19460-3_3

Liu, J., Ye, J., Silva Izquierdo, D., Vinel, A., Shamsaei, N., & Shao, S. (2023). A review of machine learning techniques for process and performance optimization in laser beam powder bed fusion additive manufacturing. *Journal of Intelligent Manufacturing*, *34*(8), 3249–3275. DOI: 10.1007/s10845-022-02012-0

Luther, D. (2023). Manufacturing Supply Chain Explained. In Oracle NetSuite. https://www.netsuite.com/portal/resource/articles/erp/manufacturing-supply-chains.shtml

Mahesh, B. (2020). Machine learning algorithms-a review. International Journal of Science and Research (IJSR).[Internet], 9(1), 381-386.

Maldonado-Romo, J., & Aldape-Pérez, M. (2021). Sustainable circular micro index for evaluating virtual substitution using machine learning with the path planning problem as a case study. *Sustainability (Basel)*, *13*(23), 13436. DOI: 10.3390/su132313436

Metri, J., Ramirez, N., Ramirez, M., & Baigts, D. (2021). Artificial Intelligence for Sensory Acceptability Prediction. At L.R.Hernandez & M.A. Serrano-Meneses (Eds.), Technology, Science and Culture: A Global Vision, Volume IV 2021 (pp. 329-344). UNESCO Hydrometeorological Risks.

Montoya Herrera, L. F. (2020). LSTM Neural Networks for Remaining Useful Life Estimation of Turbofan Engines.

Nuhu, A. A., Zeeshan, Q., Safaei, B., & Shahzad, M. A. (2023). Machine learning-based techniques for fault diagnosis in the semiconductor manufacturing process: A comparative study. *The Journal of Supercomputing*, *79*(2), 2031–2081. DOI: 10.1007/s11227-022-04730-x

Orellana Cordero, F. R. (2020). Desarrollo de un modelo de análisis de improductividad en una línea de ensamble de refrigeración para empresas de manufactura en el Ecuador.

Paredes Rodríguez, A. M., Bravo Bastidas, J. J., Osorio Gómez, J. C., Peña Orozco, D. L., & González Feliu, J. (2023). Fuzzy AHP TOPSIS Methodology for Multicriteria ABC Inventory Classification. *Journal of Engineering*, *2023*(1), 7661628. DOI: 10.1155/2023/7661628

Parra Garcés, V. (2020). Nanocomposites analysis through machine learning. Universidad de los Andes. Disponible en: http://hdl.handle.net/1992/48944

Parsazadeh, M., Sharma, S., & Dahotre, N. (2023). Towards the next generation of machine learning models in additive manufacturing: A review of process dependent material evolution. Progress in Materials Science, 135, 101102. ISSN 0079-6425. DOI: 10.1016/j.pmatsci.2023.101102

Perez, R.A.; Lilkendey, J.T.; and, Koh, S.W. (1994). Machine learning for a dynamic manufacturing environment. ACM SIGICE Bulleting, 19(3). pp. 5.9. DOI: 10.1145/182063.182067

Petri, L. D. P. S., Moutinho, E. A., Silva, R. P., Capelini, R. M., Salustiano, R., Ferraz, G. M. F., Neto, E. T. W., Villibor, J. P., & Pinto, S. S. (2020). A portable system for the evaluation of the degree of pollution of transmission line insulators. *Energies*, *13*(24), 6625. DOI: 10.3390/en13246625

Ponce, P., Anthony, B., Bradley, R., Maldonado-Romo, J., Méndez, J. I., Montesinos, L., & Molina, A. (2024). Developing a virtual reality and AI-based framework for advanced digital manufacturing and nearshoring opportunities in Mexico. *Scientific Reports*, *14*(1), 11214. DOI: 10.1038/s41598-024-61514-4 PMID: 38755242

Porter, M. E. (1985). *Competitive advantage*. The Free Press.

Rai, R., Tiwari, M. K., Ivanov, D., & Dolgui, A. (2021). Machine learning in manufacturing and industry 4.0 applications. *International Journal of Production Research*, *59*(16), 4773–4778. DOI: 10.1080/00207543.2021.1956675

Ranjan, N., Kumar, R., Kumar, R., Kaur, R., & Singh, S. (2023). Investigation of Fused Filament Fabrication-Based Manufacturing of ABS-Al Composite Structures: Prediction by Machine Learning and Optimization. *Journal of Materials Engineering and Performance*, *32*(10), 4555–4574. DOI: 10.1007/s11665-022-07431-x

Reséndiz-Flores, E. O., Navarro-Acosta, J. A., & García-Calvillo, I. D. (2022). Smart fault detection and optimal variables identification using Kernel Mahalanobis Distance for industrial manufacturing processes. *International Journal of Computer Integrated Manufacturing*, *35*(9), 942–950. DOI: 10.1080/0951192X.2022.2027019

Reyes, A., Escobar, J. W., & Londoño, J. C. (2023). Methodology for the supplier selection and valuation problem by using a combined grouping and multi-criteria approach. *International Journal of Logistics Systems and Management*, *46*(2), 174–205. DOI: 10.1504/IJLSM.2023.134401

Riddle, P., Segal, R., & Etzioni, O. (1994). Representation design and brute-force induction in a boeing manufacturing domain. *Applied Artificial Intelligence*, *8*(1), 125–147. DOI: 10.1080/08839519408945435

Rodriguez-Gallegos, A. Y., Mendez-Gonzalez, L. C., Hernandez-Holguin, A. I., & Rodriguez-Picon, L. A. (2023). Failure Detection System Controlled by a Mixed Reality Interface. At L. C. Méndez-González et al. (eds.), Innovation and Competitiveness in Industry 4.0 Based on Intelligent Systems, (pp. 21-42), EAI/Springer Innovations in Communication and Computing, DOI: 10.1007/978-3-031-29775-5_2

Rojas, M., Hermosilla, G., Yunge, D., & Farias, G. (2022). An Easy to Use Deep Reinforcement Learning Library for AI Mobile Robots in Isaac Sim. *Applied Sciences (Basel, Switzerland)*, *12*(17), 8429. DOI: 10.3390/app12178429

Rojas-Rodriguez, A., Chiwo, F. S., Arcos-Gutierrez, H., Ovando-Vazquez, C., & Garduño, L. E. (2023). Development and Evaluation of a Machine Learning Model for the Prediction of Failures in an Injection Moulding Process. At L. C. Méndez-González et al. (eds.), Innovation and Competitiveness in Industry 4.0 Based on Intelligent Systems, (pp. 101-121). EAI/Springer Innovations in Communication and Computing, DOI: 10.1007/978-3-031-29775-5_5

Romero Bravo, G. J., Espinoza Maza, J., Macgluf Issasi, A., Suárez Álvarez, Á., & Rodríguez Rodríguez, L. A. (2021). Aplicación de Machine Learning en la Industria 4.0 en tiempos de pandemia. *Interconectando Saberes*, (11). Advance online publication. DOI: 10.25009/is.v0i11.2692

Romero Bravo, G. J., Espinoza Maza, J., Macgluf Issasi, A., Suárez Álvarez, Á., & Rodríguez Rodríguez, L. A. (2021). Aplicación de Machine Learning en la Industria 4.0 en tiempos de pandemia. *Interconectando Saberes*, (11). Advance online publication. DOI: 10.25009/is.v0i11.2692

Ruiz, S. A., Beltrán, E. R., Olivares, J. A. P., Mora, J. L. O., & Valadez, J. O. V. (2023). DIAGNÓSTICO DE FALLAS INTERMITENTES EN SISTEMAS AUTOMÁTICOS DE MANUFACTURA USANDO MACHINE LEARNING (DIAGNOSIS OF INTERMITTENT FAILURES IN AUTOMATIC MANUFACTURING SYSTEMS USING MACHINE LEARNING). Pistas Educativas, 44(144).

Saldaña, M., Gálvez, E., Navarra, A., Toro, N., & Cisternas, L. A. (2023). Optimization of the SAG Grinding Process Using Statistical Analysis and Machine Learning: A Case Study of the Chilean Copper Mining Industry. *Materials (Basel)*, *16*(8), 3220. DOI: 10.3390/ma16083220 PMID: 37110055

Sarker, I. H. (2021). Machine Learning: Algorithms, Real-World Applications and Research Directions. *SN Computer Science*, *2*(3), 160. DOI: 10.1007/s42979-021-00592-x PMID: 33778771

Saucedo-Dorantes, J. J., Jaen-Cuellar, A. Y., & Elvira-Ortiz, D. A. (2020). Smart Monitoring Based on Novelty Detection and Artificial Intelligence Applied to the Condition Assessment of Rotating Machinery in the Industry 4.0. At L. Romeral-Martinez; R.A. Osornio-Rios; and M. Delgado-Prieto (Eds.), New Trends in the Use of Artificial Intelligence for the Industry 4.0,(pp. 63-76). IntechOpen. ISBN 978-1-83880-141-0

Scallan, P. (2003). *Process planning: the design/manufacture interface*. Elsevier. DOI: 10.1016/B978-075065129-5/50003-9

Seabra Lopes, L., & Camarinha-Matos, L. M. (1995). A machine learning approach to error detection and recovery in assembly. Proceedings 1995 IEEE/RSJ International Conference on Intelligent Robots and Systems. Human Robot Interaction and Cooperative Robots, Pittsburgh, PA, USA, 3, 197-203. DOI: 10.1109/IROS.1995.525884

Shafi, I., Mazhar, M. F., Fatima, A., Alvarez, R. M., Miró, Y., Espinosa, J. C. M., & Ashraf, I. (2023). Deep learning-based real time defect detection for optimization of aircraft manufacturing and control performance. *Drones (Basel)*, *7*(1), 31. DOI: 10.3390/drones7010031

Shaw, M. J., Park, S., & Raman, N. (1992). Intelligent Scheduling with Machine Learning Capabilities: The Induction of Scheduling Knowledge. *IIE Transactions*, *24*(2), 156–168. DOI: 10.1080/07408179208964213

Singha, S. K., & Singh, S. J. (2015). Analysis and optimization of sand casting defects with the help of artificial neural network. *International Journal of Research in Engineering and Technology*, *4*(5), 24–29. DOI: 10.15623/ijret.2015.0405005

Suarez, F. F., Cusumano, M. A., & Fine, C. H. (1995). An empirical study of flexibility in manufacturing. *MIT Sloan Management Review*.

Suleiman, Z., Shaikholla, S., Dikhanbayeva, D., Shehab, E., & Turkyilmaz, A. (2022). Industry 4.0: Clustering of concepts and characteristics. *Cogent Engineering*, *9*(1), 2034264. DOI: 10.1080/23311916.2022.2034264

Torralba, J., Hidalgo, J., & Jiménez-Morales, A. (2013). Powder injection moulding: Processing of small parts of complex shape. *International Journal of Microstructure and Materials Properties*, *8*(1/2), 87–96. DOI: 10.1504/IJMMP.2013.052648

United Nations Industrial Development Organization. (2019). Industrial Development Report 2020. Industrializing in the digital age. Vienna.

United Nations Industrial Development Organization. (2024). *Industrial Development Report 2024*. Turning Challenges into Sustainable Solutions. The New Era of Industrial Policy.

Usama, M., Qadir, J., Raza, A., Arif, H., Alvin Yau, K. L., Elkhatib, Y., Hussain, A., & Al-Fuqaha, A. (2019). Unsupervised Machine Learning for Networking: Techniques, Applications and Research Challenges. *IEEE Access : Practical Innovations, Open Solutions*, *7*, 65579–65615. DOI: 10.1109/ACCESS.2019.2916648

Vergara Villegas, O. O., Nandayapa, M., Sossa Azuela, J. H., Cossio Franco, E. G., & Rubin Linares, G. T. (2021). Artificial Intelligence for Industry 4.0 in Iberoamerica. *Computación y Sistemas*, *25*(4), 761–773. DOI: 10.13053/cys-25-4-4056

Zhu, X., Jiang, F., Guo, C., Wang, Z., Dong, T., & Li, H. (2023). Prediction of melt pool shape in additive manufacturing based on machine learning methods. *Optics & Laser Technology*, *159*, 108964.

Chapter 17
Machine Learning Algorithms Applied to Machine Vision and Inertial Systems

Gabriel Trujillo-Hernández
https://orcid.org/0000-0003-1556-387X
Universidad Politécnica de Baja California, Mexico

Luis Roberto Ramírez-Hernández
https://orcid.org/0000-0001-5578-8745
Universidad Autónoma de Baja California, Mexico

Jorge Alejandro Valdez-Rodríguez
Cetys Universidad, Mexico

Julio C. Rodríguez-Quiñonez
Universidad Autónoma de Baja California, Mexico

Wendy Flores-Fuentes
https://orcid.org/0000-0002-1477-7449
Universidad Autónoma de Baja California, Mexico

Oleg Sergiyenko
https://orcid.org/0000-0003-4270-6872
Universidad Autonóma de Baja California, Mexico

Jonathan J. Sanchez-Castro
https://orcid.org/0000-0001-6868-6959
Universidad Autónoma de Baja California, Mexico

ABSTRACT

This chapter presents machine learning algorithms applied to enhance the accuracy of the systems based on laser scanning, monovision, stereovision, and inertial systems. Furthermore, it outlines the principles of supervised algorithm operation, such as artificial neural networks, nearest neighbors, and support vector regression. Then, these algorithms are employed to reduce the RMSE of the variables required to apply triangulation in laser scanning systems, improving the measurements of 3D coordinates. On the other hand, the machine learning algorithms applied in monovision and stereovision reduce errors in tasks such as depth estimation, feature extraction, ad analysis, and stereo matching, among others. Lastly, the chapter presents machine learning algorithms that minimize errors in systems such as accelerometers and gyroscopes.

DOI: 10.4018/979-8-3693-6910-4.ch017

INTRODUCTION

Artificial intelligence (AI) is part of our daily lives, revolutionizing various sectors and changing how people interact with technology. Machine learning is a key component within the field of AI, as it enables systems to learn from data and improve their performance over time. Some machine learning algorithms widely used are artificial neural networks (ANN), K-nearest neighbors (KNN), and support vector regression (SVR) are often applied in machine vision and inertial systems (Ibrahim & Abdulazeez, 2021), (Ngeljaratan & Bas, 2024), (Jia et al., 2023), (Jwo & Mir, 2023).

In machine vision, the camera system and laser scanner allow 2D and 3D data to produce a point cloud to create a digital model. However, the measurement errors complicate the analysis of the experimental data. Therefore, algorithms such as ANN and KNN have been implemented on laser scanners based on dynamic triangulation to reduce the measurement errors caused by the vibrations produced by mechanical components and the energy center deformation, as well as ambient conditions (Real et al., 2017), (Trujillo et al., 2019). These algorithms have reduced errors in calculating the position of objects within the field of view. On the other hand, the reduction of errors due to beam laser deformation is achieved using the SVM algorithm. However, the SVM algorithm requires more computational resources than KNN and ANN, which can be a limitation in real-time systems, such as indoor and outdoor navigation (Flores et al., 2014).

In systems based on monovision, machine-learning algorithms can be trained with ground truth to improve the accuracy of spatial measurements after the calibration procedure. Furthermore, the SVM, KNN, and ANN can be used to perform tasks such as object detection and recognition, depth estimation, image enhancement, and preprocessing. In image enhancement and preprocessing, the ANN can map the relation between a noisy image and a noise-free one, creating a noisy pixel to achieve accurate reconstruction. In object detection, the SVM maps training targets as points in space, creating a plane that separates the categories with the maximum possible margin; this procedure enables an accurate classification. In depth estimation, although monovision lacks depth perception, the algorithm can be used to estimate the depth of a single image using techniques like depth estimation leverage neural networks (Zhang et al., 2023), (Okuno et al., 2024).

In stereovision systems, machine learning algorithms reduce measurement errors in applications of stereo matching, outlier detection, removal, depth estimation, and adaptive learning. In stereo matching, CNN determines the minimum function cost to find the disparity and create accurate point clouds. In applications of outlier detection and removal, the left-right checking algorithm (LRC) can be used to detect outliners caused by occlusions or mismatches (Qian et al., 2023). In applications of depth estimation, neural networks can predict depth information from the stereo image, achieving an accuracy of the 3D reconstruction and reducing measurement errors. Finally, adaptative learning algorithms based on neural networks incorporate an online adaptation step directly into the learning objective, obtaining a base set of weights that are more conducive to unsupervised online adaptation. These machine-learning algorithms can continuously learn from new data and adapt to changes in the environment (Duran et al., 2024).

In inertial systems, an approach to dimming the influence of drift is the usage of machine learning algorithms to provide an adaptative solution that seeks to compensate for the increments in the error over time. In this task, neural networks are implemented to identify parameters that increment drift occurs through certain moments. The algorithms' forecasted data can reduce error, aiding in angular velocity estimation. As the algorithms are complemented with other regression analyses, the use of neural networks reduces the error through time (Halitim et al., 2024).

This chapter offers a comprehensive overview of machine learning algorithms applied in machine vision and inertial systems to reduce measurement errors. The chapter presents the operation principle of machine learning algorithms in applications of machine vision systems such as laser scanners, monovision, and stereovision. Furthermore, it presents these algorithms applied in accelerometers and gyroscopes.

PRINCIPLE OF MACHINE LEARNING ALGORITHMS

The machine learning algorithms employed in the task of regression and classification applications are linear regression (LR), k-nearest neighbor (KNN), support vector machines (SVM), and artificial neural networks(ANN). Machine learning includes deep learning, which simulates how the brain makes decisions using neural networks. An ANN is classified as deep learning when it has more than one hidden layer. This section will explain basic machine learning algorithms like K-Nearest Neighbors (KNN), Support Vector Machines (SVM), and simple artificial neural networks (ANNs). These algorithms are used to reduce measurement errors.

Artificial Neural Networks

Artificial neural networks are algorithms employed for learning based on data training and specific targets; these are known as supervised algorithms. These algorithms are widely used to solve tasks of machine vision and inertial systems, such as reducing measurement errors, object detection, and autonomous navigation, among others.

Biological Neuron

According to Levitan et al. (2015), biological neurons are cells that process information via electrical impulse-based stimuli. Figure 1 shows the components of a biological neuron, which include:

- **Dendrites:** These are branching fibers that receive signals from other neurons and relay them to the nucleus.
- **Axons:** These long, slender fibers transmit action potentials through one or more synaptic junctions.
- **Synapses:** These are the junctions where communication is performed between two or more neurons.
- **Nucleus:** This central component integrates incoming signals and generates the appropriate response.

Figure 1. Biological neurons are composed of dendrites, nuclei, axons, and synapses

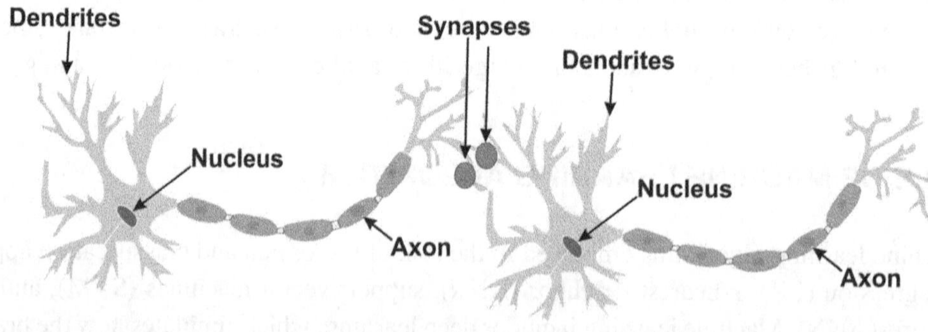

Perceptron Model

A perceptron is similar to a biological neuron; it receives external signals from other neurons. A perceptron comprises weights, inputs, outputs, and an activation function as shown in Figure 2.

Weights (W_n): These are adjusted to reduce the discrepancy between the output and the actual value, a method utilized by artificial neurons to facilitate learning.

Inputs (X_n): These are categorized into two types: binary or continuous. Binary neurons accept only two possible values, often (0,1) or (-1,1). Continuous neurons, in contrast handle values within a defined range, usually from (-1 to 1).

Activation Function (F(x)): This function determines the neuron's output based on the input values X and, in some cases, the neuron's previous output.

Output (Y_p): This indicates the neuron's output value that is calculated based on the activation function F_n of the neuron.

Bias (B_n): This parameter added to the neuron input to adjust the activation. They work as an adjustment term, allowing the model to learn complex and flexible patterns.

Figure 2. Perceptron model used in artificial neural networks

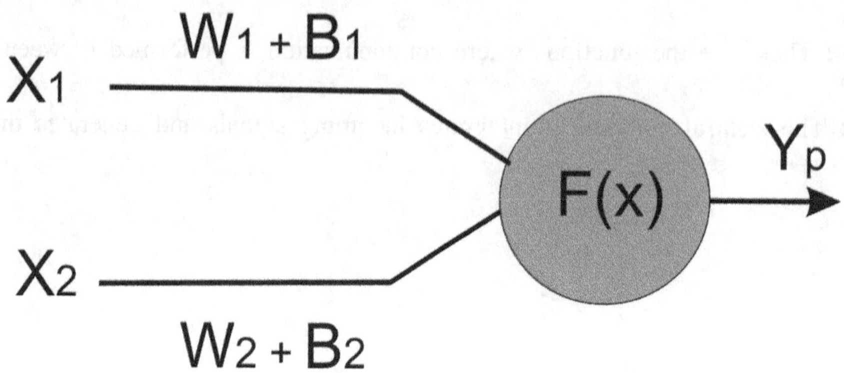

The perceptron model can be denoted as shown in the Equation 1. Where each input (x_i) is multiplied by the weight (w_i) and a bias (b_n) is added. The output ($\hat{y}_p$) consist of the sum of the results of this procedure for each branch (Huang et al., 2019).

$$\hat{y}_p = \sum_{i=1}^{n} x_i w_i + b_n \quad (1)$$

Multilayer Perceptron Model

One perceptron is not enough to implement a learning algorithm and solve complex problems; for these tasks, a multilayer perceptron model is needed. In this model, the perceptron's output is used as the input for other perceptron allowing the network to learn relationships between variables in each iteration by adjusting their parameters to learn from the data and make more accurate predictions. A multilayer perceptron model comprises an input, hidden, and output layer. The input layer is the first layer of the neural network, which receives the inputs of the dataset. It is responsible for taking raw data and passing it to the subsequent layers of the network. The hidden layer is a layer between the input layer and the output layer. It uses the input data to apply transformations through weights and activation functions to extract important features. The output layer is the final layer of the neural network, which produces the model's output and predictions based on the processed data. The data from the output layer can be used for regression or classification applications. Figure 3 shows an artificial neural network composed of an input layer, a hidden layer, and an output layer. The green circles represent the neurons (n), while x_1 and x_2 denote the inputs. The blue, yellow, and orange circles correspond to the weights w_1, w_2, and w_3, respectively. Additionally, b_1, b_2, and b_3 represent the biases of the network (Vang et al., 2020).

Figure 3. Example of an artificial neural network composed of an input, hidden, and output layer

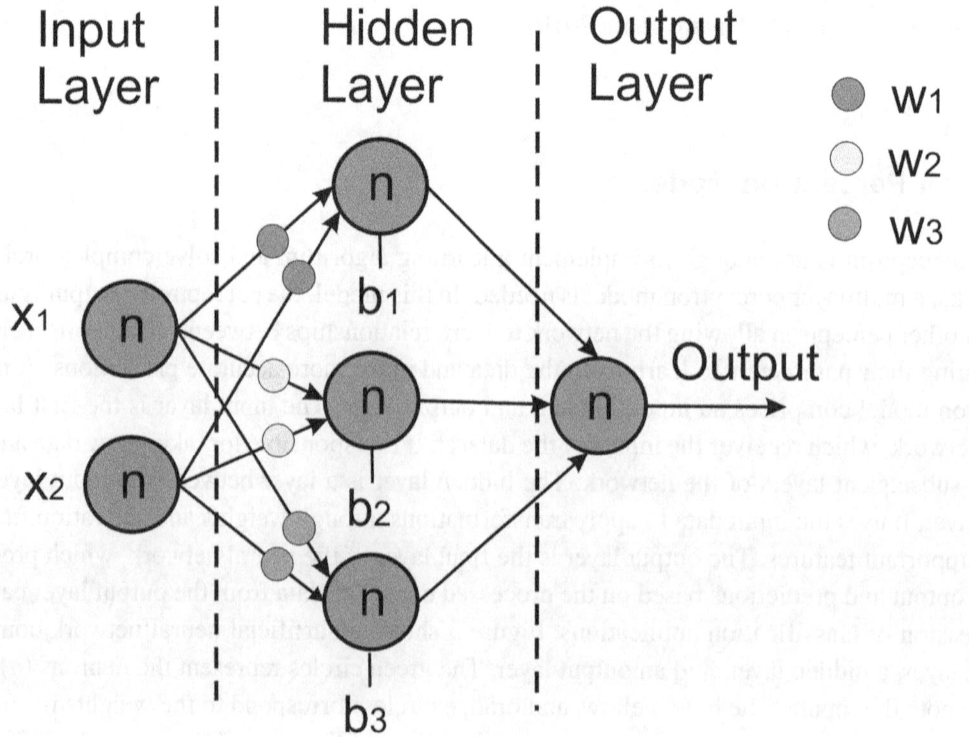

Feed Forward and Backpropagation

Feed Forward refers to the mathematical process where input data is introduced into an artificial neural network, and outputs are computed through a series of transformations. Operations in neural networks are performed from left to right, i.e., the process flows from the input layer to the hidden layers and then to the output layer (Shekhovtsov et al., 2018). Afterward, it compares the prediction value $\hat{y}_i$ and real value y_i using a metric R^2, where μ_i represents the mean of the observations as shown in Equation 2. This function indicates the error (loss value - loss).

$$R^2 = 1 - \frac{\sum(y_i - \hat{y}_i)^2}{\sum(y_i - \mu_i)} \tag{2}$$

Once the loss values are computed, the technique of gradient descent is used to define parameters that enable the minimization of the error utilizing the backpropagation technique. The error (difference between the predicted output and the actual target) is calculated and propagated backward through the network to update the weights. This method uses gradient descent to optimize the weights and improve the network's performance over time.

Support Vector Machine Regression

The Support Vector Machine (SVM) is a supervised machine learning technique designed to address classification issues involving both linear and non-linear data. Support Vector Machines (SVM) can also be utilized for solving regression issues. However, in this specific context, numerous articles make reference to a variant called Support Vector Regression (SVR).

Principle of Operation

In SVR algorithms, there is a hyperplane that is used to fit the data, unlike SVM classification, where the hyperplane separates different classes (Ahmed et al., 2023), (Shataee et al., 2012), (Specht et al., 2023). In SVR, the hyperplane represents the best regression function that intends to approximate the target values within a tolerance margin. The adjustment of the data fit depends on several parameters, such as hyperplane, margin of tolerance, kernel, and support vectors, which are explained below and shown in Figure 4.

- **Hyperplane:** In SVR algorithms, the hyperplane is a line used to fit the data inward to the margin of tolerance.
- **Margin of Tolerance (ε)**: This parameter defines a threshold within which errors are considered acceptable. Data points outward of this margin do not contribute to the loss function. The approach is to define a hyperplane that fits as many points as possible within this ε-tube.
- **Kernel:** This is a function that transforms input data into a higher-dimensional space to capture complex relationships between variables.
- **Support Vectors:** These are the data points that lie on the boundary of the ε-tube or outside it. They are crucial because they define the position of the hyperplane.

Figure 4. Support vector regression principle, the green circle indicates the fit data, and the blue circles are the support vectors

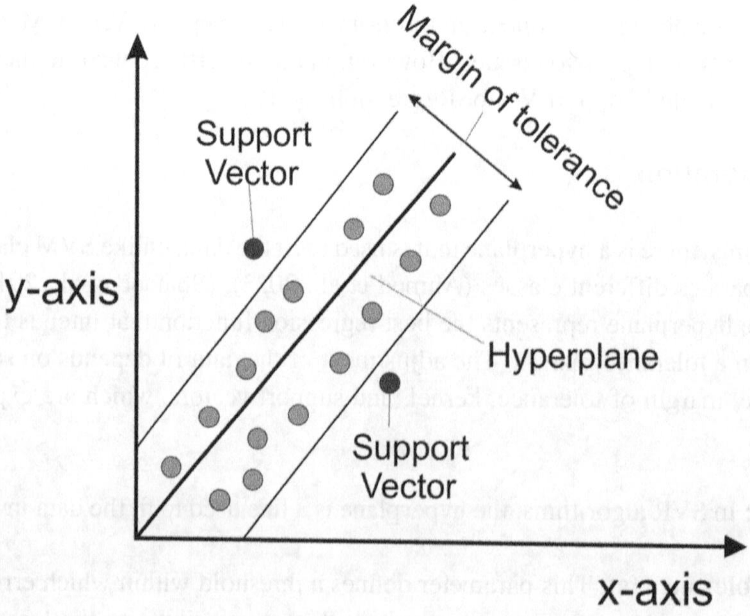

The SVR aims to define a function with the simplest form to reduce the risk of overfitting (Jayadeva et al., 2018). For linear functions, the hyperplane is denoted by the Equation 3.

$$y = wx + b \tag{3}$$

Where w represents the slope of the hyperplane, b is the point when the hyperplane intersects the y-axis, x is the feature vector used as input for prediction, and y is the predicted value (Xu et al., 2014). For univariate linear SVR (zero tolerance error), the w parameter can be minimized by applying the minimization Equation 4. Where $\|w\|^2$ is the regularization term that penalizes the complexity of the model (for a linear kernel, w is the weight vector that controls the slope of the line or the hyperplane).

$$Min = \frac{1}{2}\|w\|^2 \tag{4}$$

The constraints equation is denoted as follows:

$$constrains\begin{cases} y_i - wx_i - b \leq \varepsilon \\ wx_i + b - y \leq \varepsilon \end{cases} \tag{5}$$

In SVR algorithms, the term b is a constant that adjusts the vertical position of the hyperplane, and the parameters ε^+ and ε^- are the lower and upper bounds of the tolerance margin, respectively (Figure 5).

Figure 5. Support Vector Regression (SVR) operates without support vectors, enforcing a strict zero-tolerance for error within the margin

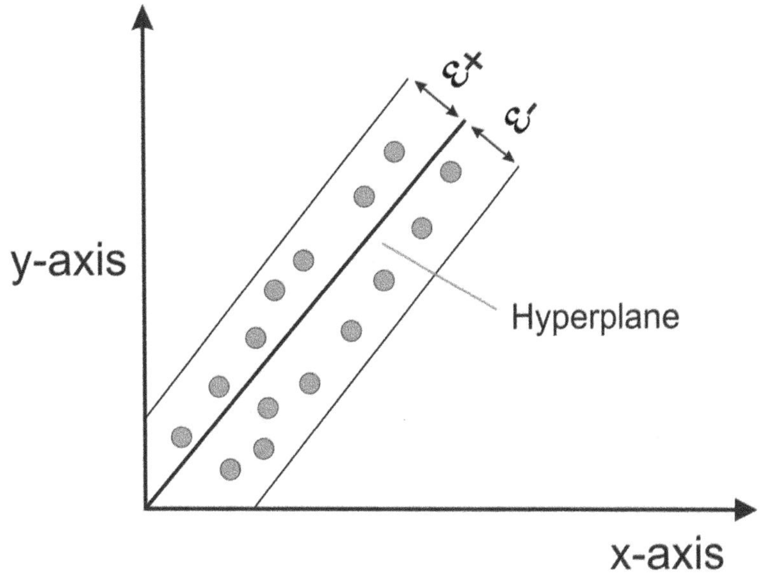

On the other hand, the SVR algorithm can allow a certain amount of error, ensuring that the model can effectively handle real-world data, avoid overfitting, and achieve better generalization. To allow these errors, it is necessary to introduce support vectors, which are data points that lie outside the margin of tolerance, specifically beyond ε^+ and ε^- boundaries. The distances of support vector variables from the ε-insensitive tube boundary are represented by ξ_i and ξ_i^* as shown in Figure 6. On the other hand, to balance model complexity with deviations outside the ε-insensitive tube, a regularization parameter $C > 0$ is included (Zhang et al., 2022). The SVR algorithm assigns zero prediction errors to the data inside the ε-insensitive tube and penalizes errors outside the ε-insensitive tube depending on the distance of support vector variables ξ. This approach allows for more effective handling of overfitting than traditional regression models. For univariate linear SVR (with tolerance error), the w parameter can be minimized by applying the minimization Equation 6.

$$Min = \frac{1}{2}\|w\|^2 + C \sum_{i=1}^{l} \left(\xi_i - \xi_i^* \right) \quad (6)$$

The constrain equations are denoted as follows:

$$constrains \begin{cases} y_i - wx_i - b \leq \varepsilon + \xi_i \\ wx_i + b - y_i \leq \varepsilon + \xi_i^* \\ \xi_i, \xi_i^* \geq 0 \end{cases} \quad (7)$$

Figure 6. SVR operates with support vectors blue points with distances ξ_i and ξ^*

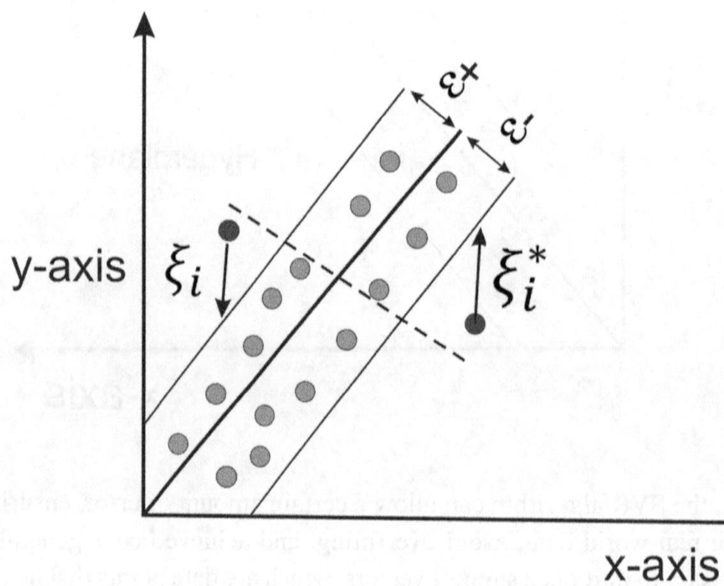

K-nearest neighbors Regression

The k-nearest neighbor is a supervised algorithm employed to solve regression or classification tasks. This algorithm is widely used because it is less complex, unlike other machine learning algorithms. However, it is considered lazy because it does not use data training to create a function to predict new data. Otherwise, keep the training data in the testing phase to predict new data. Therefore, the KNN algorithm is used more in applications that do not require processing data in real-time. This section focuses on the operation principle of KNN in regression applications.

Principle of Operation

In regression, the KNN algorithm estimates the value of a continuous variable using k neighbors and the distances from every training data to calculate the predicted value. The steps to apply the KNN regression algorithm are explained below (Zhang et al., 2021), (Cosenza et al., 2021).

1. Calculating the distances between the prediction point and all points of the training data. There are different metric distances that can be used; however, Euclidean and Mahalanobis are explained in this section. Euclidean distance is calculated by applying Equation 8, where *d(x,y)* is the distance

between a predicted value and a training data point, (x_1,y_1), (x_2, y_2) are the first point and second point, respectively. (x_n,y_n) is the coordinate of the n point.

$$d(x,y) = \sqrt{(x_2 - x_1)^2 + (y_2 - y_1)^2 + \ldots + (x_n,y_n)^2} \quad (8)$$

Mahalanobis distance is a measure that considers the correlation between variables and dispersion between data. This distance between two points x and y in a space n can be computed as shown in equation (9). Where, x and y are points in a space of n dimension. S^{-1} is an inverse covariance matrix, which captures the correlations between the variables. $(x - y)^T$ is the transpose vector of the differences of the points. $D_M(x,y)$ is the calculated distance.

$$D_M(x,y) = \sqrt{(x - y)^T S^{-1}(x - y)} \quad (9)$$

In Figure 7, the blue point is the value to predict, and the green points are the training points. d_1, d_2, d_3, d_4, d_5 values are the distances used to calculated a predicted value depending on the k neighbors selection.

Figure 7. Distances of the point to predict regarding each training point

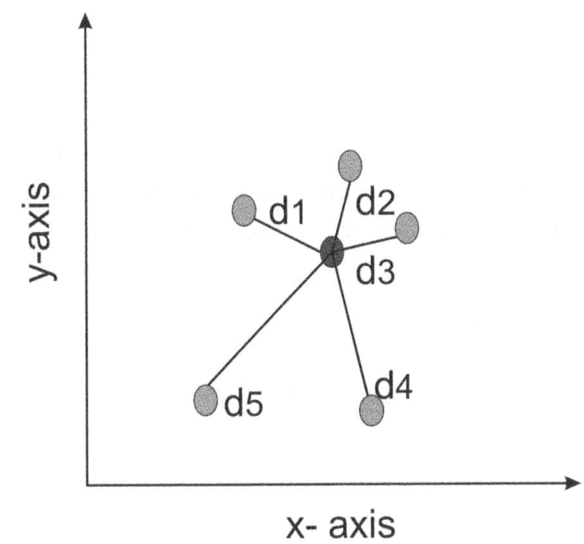

2. Selecting a *k-value* positive integer to determine the number of neighbors used to calculate the predicted value. These values can be adjusted to enhance the accuracy and performance of the algorithm. Figure 8 shows an example using k=3, where three green points inside the red circle are the k values due to are located to the distances nearest (d_1, d_2, d_3) regarding the predicted value (blue point).

Figure 8. Example using the three nearest neighbors.

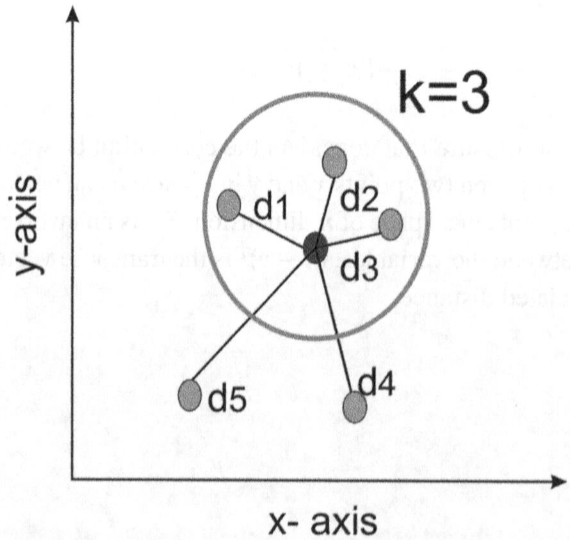

3. Calculating the mean of the values k nearest neighbors to determine the predicted values. In some cases, it is possible to apply the median to obtain the predicted value. According to the example presented in Figure 8, the predicted value (P) is calculated as shown in Equation 10.

$$P = \frac{d_1 + d_2 + d_3}{3} \tag{10}$$

4. The algorithm is evaluated using metrics such as mean square error (MSE), mean absolute error (MAE), or the root mean square error (RMSE). Here y_i represents the real value, $\hat{y}_i$ represents the predicted value, and n represents the number of observations (Chicco et al., 2021).

$$MSE = \frac{1}{n}\sum_{i=1}^{n}(y_i - \hat{y}_i)^2 \tag{11}$$

$$MAE = \frac{1}{n}\sum_{i=1}^{n}|y_i - \hat{y}_i| \tag{12}$$

$$RMSE = \sqrt{\frac{1}{n}\sum_{i=1}^{n}(y_i - \hat{y}_i)^2} \tag{13}$$

These metrics enable a comparison between predicted values and real values to calculate the algorithm error. However, the RMSE is the most useful for validation, as it penalizes the severe errors (outliers) more than the MSE and MAE.

MACHINE LEARNING ALGORITHMS TO REDUCE MEASUREMENT ERRORS IN 3D LASER SCANNER.

Vision systems are widely used to capture a 3D scene to determine object positions or dimensions for performing specific tasks. These systems are utilized in medical fields to create customized prostheses, in structural health monitoring to assess building damage, and in both indoor and outdoor navigation for autonomous robots (Grigorescu et al., 2023), (Castro et al., 2020), (Trujillo et al, 2022). This chapter presents systems based on laser scanners, monovision, and stereovision, which involve determining 3D points for several applications. However, the machine vision system presents measurement errors due to multiple factors. The laser scanner is a system composed of optoelectronic and mechanical elements, and it is sensitive to environmental lighting. The stereovision and monovision systems also present measurement errors due to lighting and lens distortion. Therefore, machine learning algorithms are trained to enhance the accuracy in the measurements. The regression algorithms used in this chapter are nearest neighbor (KNN), artificial neural networks (ANN), and support vector machine (SVM) (Boumedine& Bentaieb et al., 2023), (Vezilić et al., 2018), (Pinto et al., 2013).

3D Laser Scanner and Algorithms to Reduce Measurement Errors

There are various types of laser scanners that use techniques like triangulation, time of flight, and phase shift to measure surfaces (Wang et al., 2024). The next section introduces a laser scanner based on a variant of triangulation known as dynamic triangulation, where the angles of the emitter and receptor are modified, unlike in simple triangulation, where the angles remain fixed. This method is implemented in a Technical Vision System (TVS) with machine learning algorithms to minimize measurement errors in specific work areas.

Principle of triangulation

The principle of triangulation in the TVS consists of three elements: a scanning aperture (SA) and a positioning laser (PL) separated by a fixed distance (a), as shown in Figure 9. The positioning laser projects a beam toward the surface then reflected the scanning apertures. The angle C_{ij} is formed between the projected laser beam and the horizontal line of a, while the angle B_{ij} is formed between the horizontal line of a and the laser beam reflected toward the scanning aperture. This information is used with trigonometric function to calculate the 3D coordinates (x, y, z) of the beam laser on a surface; the Equations 14, 15, and 16, are applied. Where β angle is used to calculate the coordinate z. The measurements can be acquired in real time to create 3D cloud points, which could be used to create a 3D model through reconstruction (Trujillo et al., 2024).

$$x_{ij} = a\left(\frac{\sin B_{ij} . \sin C_{ij}}{\sin\left(B_{ij} + C_{ij}\right)}\right) \tag{14}$$

$$y_{ij} = a\left(\frac{1}{2} - \frac{\cos B_{ij} . \sin C_{ij}}{\sin\left(B_{ij} + C_{ij}\right)}\right) \tag{15}$$

$$z_{ij} = a\left(\frac{\sin B_{ij}.\sin C_{ij}.\tan \beta_{ij}}{\sin\left(B_{ij}+C_{ij}\right)}\right) \tag{16}$$

Figure 9. Principle of triangulation in the technical vision system

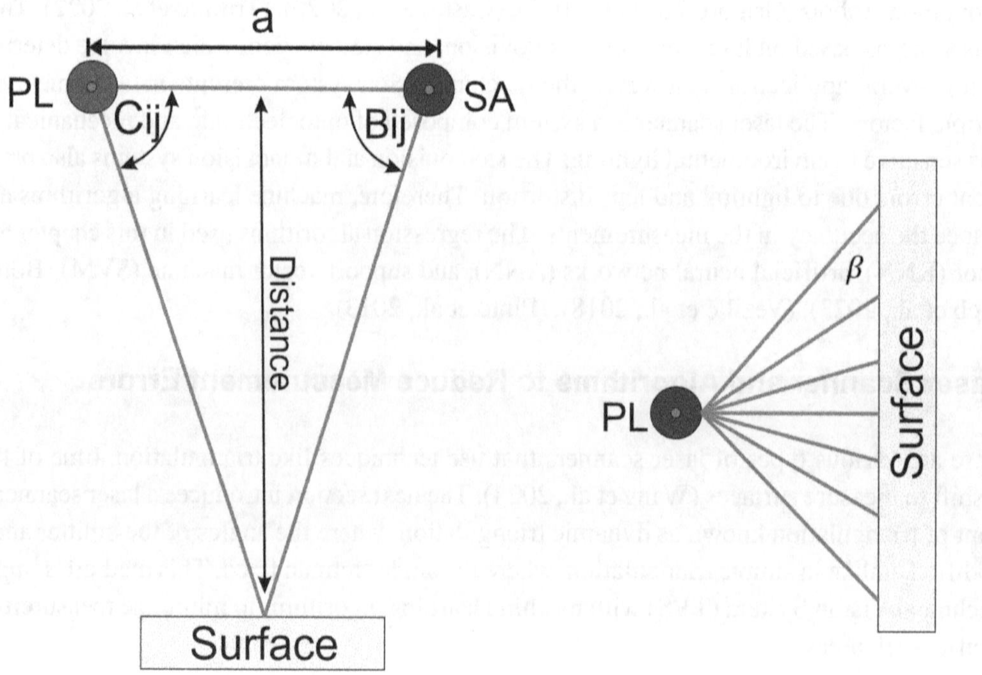

Mechanical and Optoelectronic Components

The positioning laser consists of a laser that projects a beam toward a 45-degree mirror joint connected to a DC motor. The motor rotates, causing the mirror to deflect the laser beam orthogonally onto the surface. On the other hand, the scanning aperture consists of a DC motor with an encoder connected to a 45-degree joint, which redirects the beam vertically through two pairs of biconvex lenses. These lenses cause the laser to form parallel lines, and the second lens focuses these parallel lines onto a photoreceptor. Additionally, the scanning aperture includes a photoreceptor that measures the revolutions per minute of the DC motor. The angle C_{ij} and β are angles obtained through PID controllers with excellent accuracy and a distance. However, the principal cause of inaccuracy in the systems is related to the calculus of the B_{ij} angles. This angle is calculated by means of pulse trains of the photoreceptor, encoders, optoswitch, and generated square signal, as shown in Figure. Optoswich signal is activated for each revolution of the DC motor; then, the system counts the pulses of the encoder and generates a square signal. If, after receiving another pulse of the encoder, the laser beam is not detected, the value of the generated square signal is set to 0. However, if the laser beam is detected, the number of pulses from the generated square signal is counted. This pulse count is then used to calculate the angle B_{ij} until the next pulse from encoder, as shown in Equation 17.

$$B_{ij} = P_A\left(Q_A + \frac{Q_S}{Q_T}\right) \tag{17}$$

$$P_A = \frac{P_{360}}{360^o} \tag{18}$$

Where, Q_A represents the number of pulses of the encoder before the detection photoreceptor pulse, Q_s is the number of pulses counted from the generated square signal of Q_A to the photoreceptor, and Q_t is the number of pulses from generated square signal of Q_A to another pulse. P_A are the degrees per pulses of the encoder signal, which can be calculated using the Equation 18. Where, P_{360} is the number of pulses per revolution (Real et al, 2017).

Machine Learning Algorithms Applied to Reduce Measurement Errors

To reduce the error, artificial neural networks (ANN) and K-nearest neighbors (KNN) algorithms have been implemented in the TVS, achieving a low root mean square error (RMSE) in the B_{ij} measurement. Before training the ANN algorithm, a dataset of 1000 samples based on the inputs B_{ij}, C_{ij}, and Tick are used. According to Trujillo et al. (2019), the samples are captured on a grid (Figure 10) with distances from 15 cm to 35 cm from the center of the 45-degree mirrors of the scanning aperture and positioning laser. Moreover, the grid is graduated with angles ranging from 30° to 130°, which are used to position the laser beam. This obtained dataset is used to train the ANN and KNN algorithms.

Figure 10. Grid design used to train the ANN and KNN algorithms

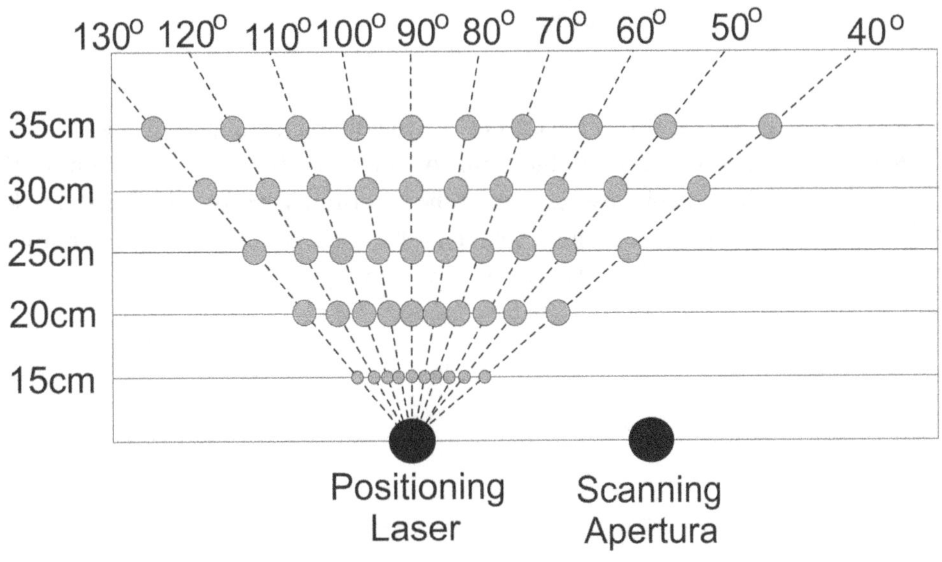

The design ANN consists of three inputs (B_{ij}, C_{ij}, Tick) and one output (offset). The ANN design includes one hidden layer with seven neurons, each with a bias and a tanh activation function, and one output layer with a single neuron and a purely linear function. Figure 11 shows a representation of the ANN design.

Figure 11. Design of the ANN employed for the technical vision system

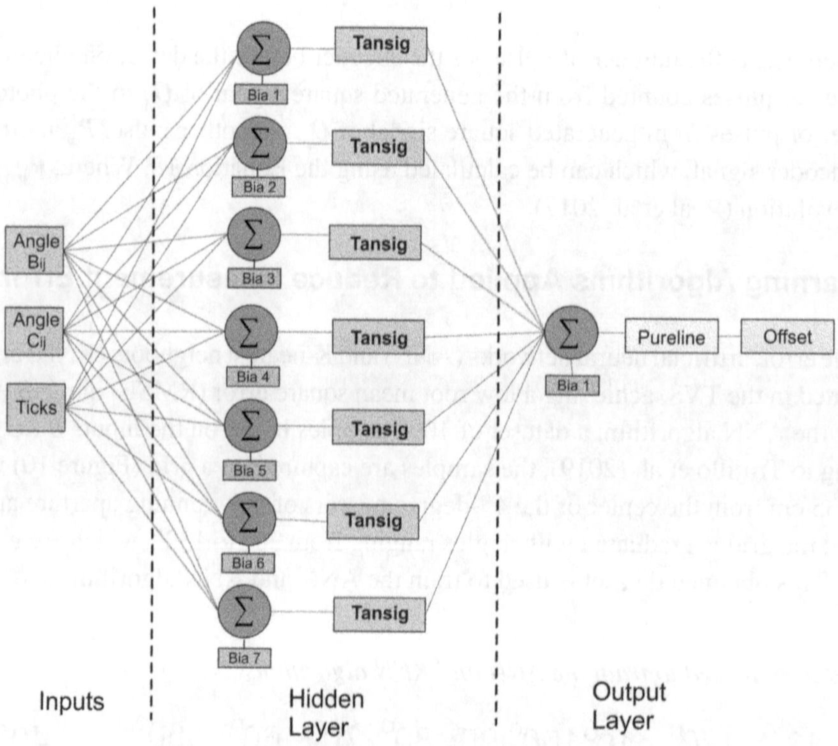

The training process involves applying the Levenberg-Marquardt method for ANN and utilizing three neighbors for KNN. However, to evaluate the algorithms' performance, the cross-validation (k-fold) method can be employed. This method enables assessment of the algorithms' performance without dependency on the partitioning of the training and testing data. i.e., 90% of the dataset is used for training and 10% for test traversing in each iteration of the dataset, as shown in Figure 12.

Figure 12. Cross-validation method used to evaluate the algorithms

	Fold1	Fold2	Fold3	Fold4	Fold5	Fold6	Fold7	Fold8	Fold9	Fold10
n=1	Test	Train	Train	Train	Train	Train	Train	Train	Train	Train
n=2	Train	Test	Train	Train	Train	Train	Train	Train	Train	Train
n=3	Train	Train	Test	Train	Train	Train	Train	Train	Train	Train
n=4	Train	Train	Train	Test	Train	Train	Train	Train	Train	Train
n=5	Train	Train	Train	Train	Test	Train	Train	Train	Train	Train
n=6	Train	Train	Train	Train	Train	Test	Train	Train	Train	Train
n=7	Train	Train	Train	Train	Train	Train	Test	Train	Train	Train
n=8	Train	Train	Train	Train	Train	Train	Train	Test	Train	Train
n=9	Train	Train	Train	Train	Train	Train	Train	Train	Test	Train
n=10	Train	Train	Train	Train	Train	Train	Train	Train	Train	Test

The algorithm training is performed considering the parameters of epochs=500, goal=0, mu=0.001, and max-fail=6. In each iteration, it calculates a root mean square error (RMSE); these values can also be used to calculate the average RMSE. Once the algorithms are trained, the ANN and KNN are implemented in the TVS. Subsequently, the TVS is tested at 100 points on the grid to analyze the performance of the system using both algorithms. According to Trujillo et al. (2019) and Real et al. (2016), the obtained B_{ij} RMSEs for KNN and ANN are 0.55° and 0.38° within the workspace. On the other hand, there are several algorithms that can be tested in the TVS, such as support vector machines (SVM) and random forest, which present good results in applications with laser scanners. In the publication by Monnet et al. (2011), a support vector machine (SVM) algorithm is implemented in a laser scanner. The obtained measurements are then used to compare multiple regression models with least squares for height, basal area, mean diameter, and stem density in plot volume prediction applications. In the publication by Urbas et al. (2021), algorithms are tested using a laser scanner based on the principle of triangulation. These algorithms include linear regression, support vector machines, neural networks, k-nearest neighbors, AdaBoost, and random forests. The results indicate that random forests exhibit lower deviation compared to the other methods. Random forests algorithm can be used for classification and regression tasks, and it is based on multiple decision trees, which operate like a flow chart that represents a series of decisions based on the data features to improve the model's accuracy and robustness. Additionally, these algorithms can be used to estimate the rapidly growing forest plantation yield based on statistical vegetation metrics obtained from an Airborne Laser Scanning (ALS) survey (Görgens et al., 2015).

MACHINE LEARNING ALGORITHMS FOR REDUCING MEASUREMENT ERRORS IN VISION MONOVISION AND STEREO VISION.

Machine learning has revolutionized many things in the last few years and that includes computer vision as well. Sophisticated learning algorithms have greatly upped the accuracy and efficiency of vision systems. A key application is in reducing measurement errors, a flash demonstration of some usage cases being for vision systems needed across various industries ranging from manufacture to autonomous vehicles. In this section, we illustrate the adoption of machine learning algorithms in compensating for measurement errors embedded within two vision system categories: monovision and stereo vision. A monovision system uses a single camera to capture the images and extract the information, while stereo vision involves multiple cameras for depth perception and 3D reconstruction. Using machine learning, the vision systems are able to achieve far better precision and reliability. This chapter presents different machine learning algorithms and the challenges they face in optimizing overall performance and accuracy, which is discussed further below.

Monovision Systems

Systems based on monovision (using just one camera), due to their simplicity and the fact that they are usually cheaper, can be found in many applications. Those who apply this technology to manufacturing for inspection and quality control (Di Leo et al., 2017), ensuring the characteristics of products meet specified standards, as in medicine to aid diagnostic procedures through essentially non-invasive detailed images of the human body (Chang et al., 2020) or enable autonomous vehicles by directing lane detection, object recognition and traffic sign reading algorithms based on input from visual media sensors (Hsu et al., 2018). It can be used on robotics applications requiring cameras instead of sensors embedded at various positions onto a mobile entity that must autonomously navigate within its complex environment, particularly in scenarios where deploying multiple cameras is impractical (Al-Jarrah, 2016). Moreover, a Monovision system can be employed for surveillance to monitor and analyze environments (Noor et al., 2017). While this offers some advantages, monovision systems are also severely constrained with respect to accuracy and reliability in both depth perception (they do not account for it) and measurement precision. Figure 13 shows a schematic monovision system with camera coordinates (x_c, y_c, z_c) and real-world coordinates (x_w, y_w, z_w).

Figure 13. Schematic monovision system

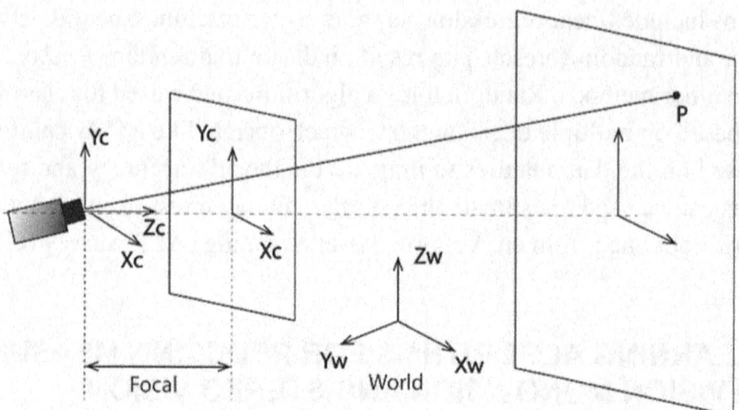

Machine learning models can be trained to identify and correct measurement errors by comparing predicted measurements with known ground truth data. This continuous learning process allows for incremental improvements in accuracy over time. Here are some ways in which these algorithms are applied.

Image Enhancement and Preprocessing: Machine learning models can improve image quality through noise reduction, contrast enhancement, and resolution upscaling, which are essential for accurate measurement and analysis (Lu et al., 2021). In this case is used a fully connected neural network (FCNN) to map a relation between a noisy image and a noise-free one, creating noisy pixel for achieve accuracy reconstruction. Figure 14 shows an example of a denoised image comparison.

Figure 14. Denoised image comparison. a) original image; b) noisy image with 80% noise density; c) deep-learning FCNN mean filter

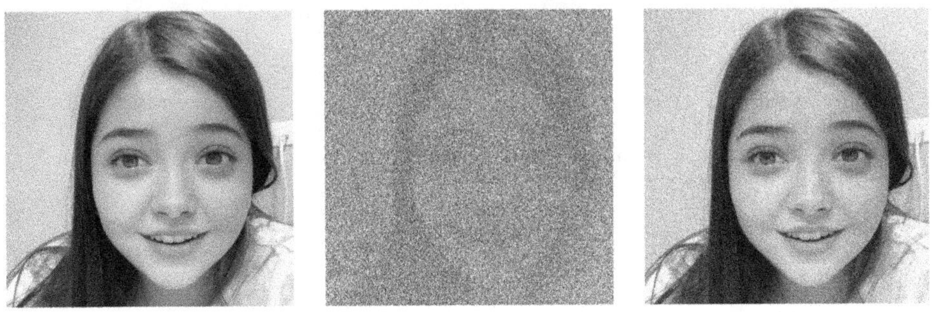

(a) (b) (c)

The FCNN targets the selection of neighboring pixels that are noise-free and share similar characteristics with the central pixel, considering factors such as spatial distances, pixel deviation, and overall similarity. Accordingly, the feature vector Ψ_{ij} is expressed by Equation 20.

$$\Psi_{ij} = \left[\eta_{ij}^{T} k_{ij}^{T} r_{ij}^{T} \sigma_{ij}^{T}\right] \quad (20)$$

where k_{ij}^{T} and σ_{ij}^{T} represent the transpose of measures similarities, and the pixels free of noise, respectively. η_{ij}^{T} and r_{ij}^{T} denote the transpose of noise free flag and the transpose of spatial distance, respectively. Each pixel with noise is examined using a 3 × 3 window.

Subsequently, central pixel noise is determined by the central position of pixels free of noise with the smallest gray-level differences. The deep algorithm FCNN is more accuracy than other mean filters, even in low noise levels conditions. This is primarily because the algorithm identifies the most similar neighboring pixels in the search scene considered the noisy image and the pixels free of noise. As a result, it achieves better restoration quality than employing the neighboring pixels with lowest variation.

Object Detection and Recognition: Advanced algorithms, such as SVM, are used to detect and recognize objects within images. This capability is vital for applications like autonomous driving and industrial inspection, where precise identification is crucial (Zhang et al., 2022). Figure 15 shows an example of the object detection and recognition using SVM.

Figure 15. Object detection and recognition using SVM

SVM algorithm maps targets identifying them as points in space, creating a plane that separates the categories with the maximum possible margin. The algorithm starts by creating a population $\{x_i(0) | x^L_{ij} \leq x_{ij}(0) \leq x^U_{ij}, i = 1,2,\cdots,NP\ ;1,2,\cdots .D\}$ randomly generated by applying Equation 21.

$$x_{ij} = (0,1)*\left(x^U_{ij} - x^L_{ij}\right) \tag{21}$$

$x_i(0)$ denotes the i-th individual of the 0-th generation in the population, and $x_{ij}(0)$ denotes the j-th "gene" of the i-th individual in the 0-th generation. Moreover, NP denotes the size of the population, and rand (0,1) denotes a number with interval (0,1). When new targets are introduced, they are mapped into this same space, and their categories are predicted based on their position relative to the margin. The main objective of SVM is create a separating hyperplane within the data which it is going be used to reduce classification errors. This process involves first identifying the support vectors, then maximizing the distance between these vectors and the separation plane, ultimately finding the most suitable hyperplane.

Depth Estimation: Although monovision systems inherently lack depth perception, machine learning algorithms can infer depth from one image. Techniques such as monocular depth estimation leverage neural networks to predict depth information, thereby improving measurement accuracy (Wang et al., 2023). Monocular depth estimation is classified by using parametric or non-parametric methods. Parametric methods involve techniques with parameters obtained by energy function which they are focused on determining the parameters themselves. On the other hand, non-parametric methods leverage the data obtained in similarity process to estimate depth without relying on parameter previously learning. These depth estimation algorithms predict the depth by combining the depths of images with others RGB datasets. Models with monocular depth estimation requires one reference RGB image and their corresponding depth map obtained by cameras or laser radar sensors. This reference RGB image is called ground truth (GT), and it is employed for training. Unlike earlier methods, supervised models learn by using a pixel data without needing specific features. The general workflow of a monocular supervised model to estimate the depth employing neural networks is illustrated in Figure 16.

Figure 16. Supervised model for monocular deep estimation employing neural networks

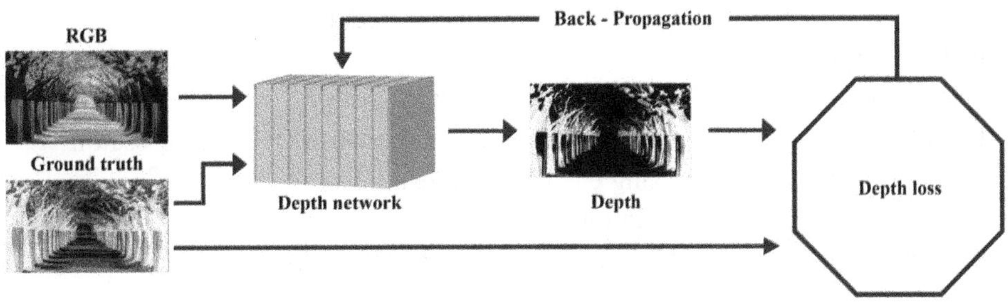

The method is evaluated by using root mean square error (RMSE), logarithm root means square error (RMSE log), absolute relative error (Abs Rel), and the square relative error (Sq Rel), which they are formulated as follows.

$$\text{RMSE} = \sqrt{\frac{1}{|N|}\sum_{i\in N}\|d_i - d_i^*\|^2} \tag{21}$$

$$\text{RMSE log} = \sqrt{\frac{1}{|N|}\sum_{i\in N}\|\log(d_i) - \log(d_i^*)\|^2} \tag{22}$$

$$\text{Abs Rel} = \frac{1}{|N|}\sum_{i\in N}\frac{|d_i - d_i^*|}{d_i^*} \tag{23}$$

$$\text{Sq Rel} = \frac{1}{|N|}\sum_{i\in N}\frac{\|d_i - d_i^*\|^2}{d_i^*} \tag{24}$$

where N denotes the total number of pixels with real depth values. d_i and d_i^* denotes the prediction depth value of pixel and the GT of depth, respectively.

Neural network recovers depth based on the local plane assumption, progressively refining the depth map layer by layer to achieve high accuracy. Backpropagation is used to efficiently train a neural network by applying the chain rule. Simply put, after each forward pass through the network, backpropagation performs a backward pass to adjust the model's parameters, seeking to reduce the error percentage to a minimum.

Feature Extraction and Analysis: Algorithms can automatically extract relevant features from images, which are then used to make precise measurements. These features can include edges, textures, and specific patterns relevant to the application (Joseph et al., 2021). The method employed for feature extraction is the K nearest neighbor (KNN), which is considered a non-parametric method as it does not require parameter tuning or estimation. It works by dynamically finding the k most similar observations in a dataset to a new observation. To determine these similarities, KNN employs a distance metric, such as Euclidean distance, Minkowski distance, or Mahalanobis distance. The process for feature extraction with KNN considers a training phase and testing phase, which can be described as follows:

Training Phase:

- Select and place the training images into a designated folder.
- Load the training images.
- Preprocess each image and apply a corner detection algorithm.
- Extract local features from the preprocessed image to construct its feature vector.
- Create a feature matrix by organizing each image's feature vector as a row.
- Repeat steps 2 through 5 for all training images.
- Train and tune the KNN method in preparation for the testing phase.

Testing Phase:

- Load the test images.
- Apply the KNN algorithm to identify the nearest neighbors using the Euclidean distance function based on the training data.
- If the K nearest neighbors have identical labels, assign the same label to the query and exit; otherwise, calculate pairwise distances among the K neighbors and create a distance matrix.
- Convert the distance matrix into a kernel matrix using the kernel trick method.
- Display the classified object and its label.

The KNN algorithm's role in feature extraction is to leverage the extracted features to classify new data points by comparing their feature vectors to those of the training data. The effectiveness of the KNN classification depends on the quality of feature extraction and the choice of distance metrics. By integrating these machine learning techniques, monovision systems can overcome their inherent limitations, resulting in more accurate and reliable measurements. This integration not only enhances the overall performance of these systems but also expands their applicability across various fields.

Stereovision Systems

These systems are based on the principle of stereovision in which two or more cameras capture images from a slightly different perspective, analogous to human eyes. This configuration helps the system in sensing depth and creating 3D structures from 2D images. Stereovision systems are employed in scenarios where depth information is important to obtain since they allow a better understanding of the environment than monovision systems (Ramirez et al., 2020). Figure 17 shows a schematic of the stereovision system.

Figure 17. Schematic stereovision system

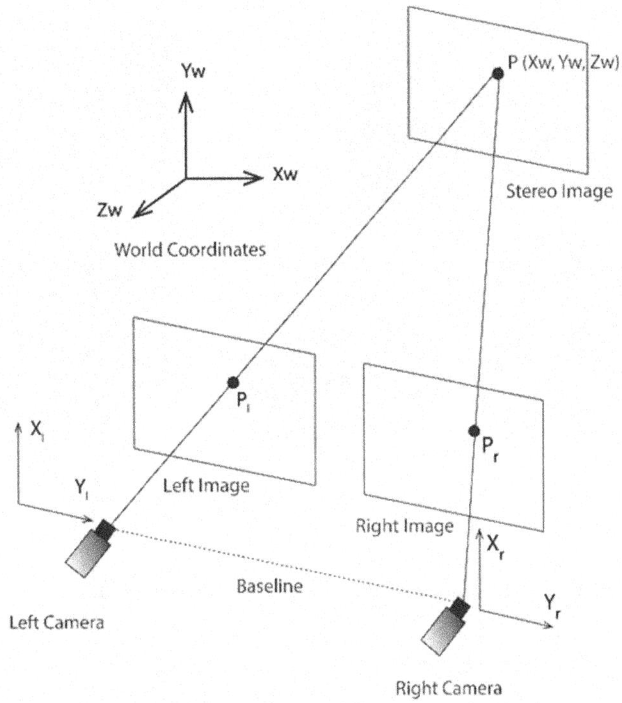

Applications for the Stereovision system can be found in different fields: In robotics is used to navigate through complex environments, perform precise manipulation tasks and interact safely with a human; in autonomous vehicles to detect obstacles, estimate distances, and understand the spatial layout of the surroundings (Emani et al., 2019), in medical to create detailed 3D models of anatomical structures, Creating useful 3D visualizations that can be used in the medical field to increase the precision of surgical planning and diagnostic procedures (Gonzalez et al., 2020), in virtual and augmented reality, augmenting real-world models rendering for a realistic experience, and providing an immersive virtual environment (Smiles et al., 2023), or in security, having a higher level monitoring capabilities due to depth perception allowing human mind assess situations properly fairly quickly and assess threats really fast (Hozyń et al., 2021). There are many algorithms where machine learning was employed to enhance the accuracy and reliability of stereo vision systems by addressing various sources of measurement errors.

Disparity Map Refinement: The core of stereo vision lies in computing disparity maps, which represent the differences between corresponding points in the stereo images. Machine learning models, particularly deep learning techniques, can refine these disparity maps, reducing errors and improving depth estimation (Jospin et al., 2022). The disparity map can be obtained by using the Bayesian probabilistic model, where the sole meta-parameter to adjust is the desired reconstruction accuracy. Figure 18 shows a disparity map refinement of a determined stereo image.

Figure 18. Stereo disparity map. a) original image; b) stereo images; c) disparity map: d) disparity map refinement

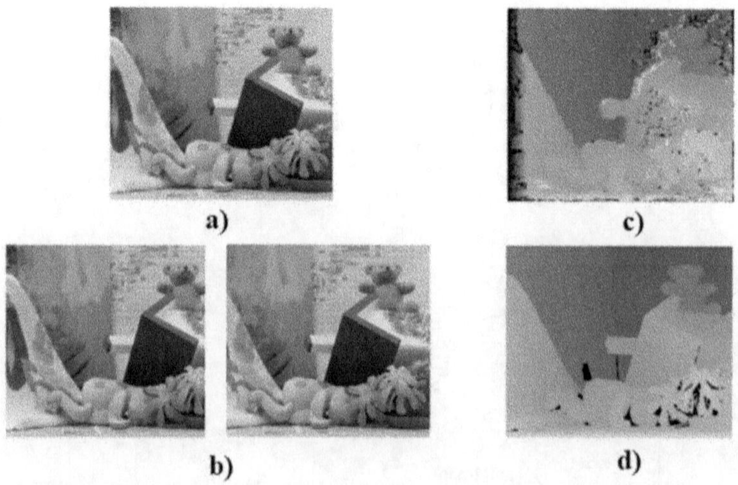

To minimize the amount of data, an equation is denoted as follows:

$$V^T\left(k\right) = \begin{cases} V(d_{raw} + k) \; if \, d_{raw} + k \in \left[o, d_{max}\right], \\ 0 \; otherwise, \end{cases} \tag{25}$$

where $V^T$is the cost volume, which is defined at pixel *k=(i,j)*.

Let also X be a tensor defined at pixel (*i ,j*) as:

$$X\left(k\right) = \begin{cases} 1 \; if \, d_{raw} + k \in \left[o, d_{max}\right], \\ 0 \; otherwise, \end{cases} \tag{26}$$

Here, d_{raw} id the maximum disparity under consideration.

Using a Bayesian model has the benefit of allowing one to consider both epistemic uncertainty (the uncertainty resulting from insufficient information during model training) and aleatoric uncertainty (the intrinsic uncertainty in the data). The model uses a deep neural network architecture with an outlier detection branch, trained unsupervised, and a refinement branch trained supervised. The function of the outlier detection branch is to invalidate outlier pixels during training so that the refinement network may focus on the remaining inlier pixels, sometimes known as validated pixels, which are simpler to rebuild. Acting as a kind of meta-learning, this outlier detection branch influences the loss obtained from a Bayesian Probabilistic Graphical Model (PGM) applied to train the refinement branch. The branch on outlier identification keeps polishing the differences in the model forecasts during testing. Comparatively to a similar refinement network that does not discover and invalidate outliers, the Bayesian probabilistic model doubles the accuracy of disparity refinement at validated pixels. The proposed design achieves errors under 0.2 pixels in verified areas despite its small scale. Furthermore, the model can extend to passive stereo; however, this yields a noticeably lower fraction of inliers. This indicates that the architecture helps the network to identify when it can generate a safe prediction, which is a very desired quality of the model.

Stereo Matching: Reliable depth perception requires accurate stereo matching. Convolutional neural networks (CNNs) learn complicated patterns and correlations between image pairs to improve matching (Liu et al., 2020). The sum of absolute differences in image blocks can be used to determine D(x,y) by comparing block similarities to find pixel points:

$$D\left(x,y\right) = \sum_{s=1}^{M}\sum_{t=1}^{N}|P(x+s-1,y+t-1)-Q(s,t) \tag{27}$$

Where P and Q represent one pixel in the corresponding pixel point

Figure 19 shows a Typical Network architecture. The sub-networks consist of several convolutional layers, each followed by rectified linear units except for the final layer. Each sub-network generates a vector that encapsulates the characteristics of the input patch. These two vectors are then compared using the cosine similarity measure to generate the network's final output.

Figure 19. Network architecture for stereo matching

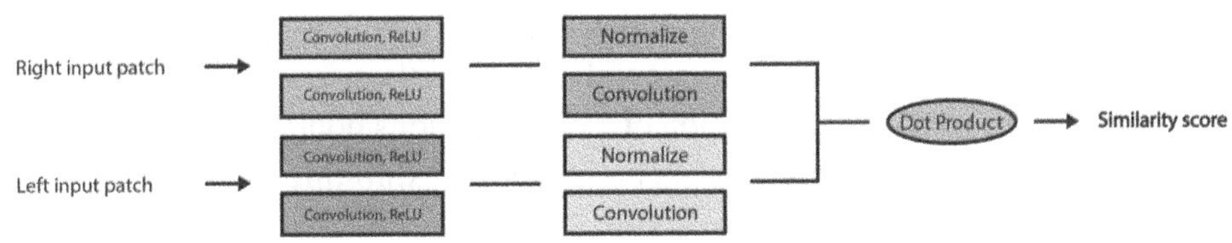

The layer is employed to derive various feature details from the left and right image blocks. The network undergoes training through the minimization of a hinge loss. This loss is calculated by evaluating pairs of examples that are centered on the same image position, where one example is from the positive class and the other from the negative class. After each convolutional layer, a batch normalization layer is incorporated to accelerate the network training convergence. Additionally, batch normalization is utilized to enhance the smoothness of the training process, allowing for a higher learning rate and reduced dependency on initialization. The cost function is formulated to be inversely proportional to the similarity score. The higher the similarity between image blocks, the lower the cost. Subsequently, the cost function undergoes post-processing through cross-cost aggregation and semi-global matching, selecting the minimum cost to determine the disparity.

Outlier Detection and Removal: Machine learning models can identify and remove outliers from the data, ensuring that the measurements are based on accurate and reliable information (Dong et al., 2019). The left-right checking algorithm (LRC) is used to detect outliners caused by occlusions or mismatches. Effective stereo matching is achieved by accurately detecting outliers and then using several disparity refinement methods to analyze occlusion and mismatch outliers. Similar to confidence measure detection in stereo vision, outlier detection determines if a point in the image is likely to be an outlier based on how low its confidence value is. An outliner identification example is shown in Figure 20.

Figure 20. Outlier Identification using LRC algorithm. a) original image; b) stereo images; c) outlier map classified by LRC, the green and red points represent occlusion and mismatch outliers, respectively

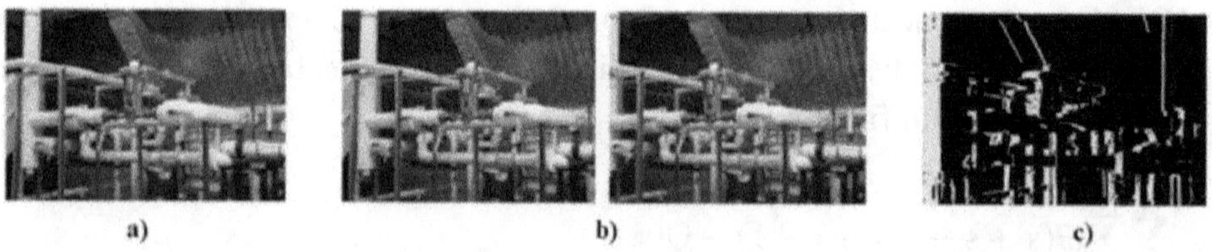

The identification of mismatch and occlusion outliers through LRC is not consistently precise. In certain occlusion areas, outliers are mistakenly identified as mismatches, negatively affecting subsequent disparity refinement. The Occlusion points usually cluster together instead of being isolated. Any mismatch outlier is reclassified as an occlusion outlier if the count of occlusion points within a specific window exceeds a set threshold. During this process, each occlusion point receives a reliable 3D label to improve stereo-matching efficacy. For each occlusion point, the nearest dependable points are located along horizontal and vertical axes. One of these dependable points is chosen to provide a reliable 3D label. The search is confined to nearby areas since points that are distant often belong to different planes. The search is halted upon encountering a discontinuous boundary, which can be identified by texture or disparity edges. The search for 3D labels in horizontal and vertical orientations concludes with the discovery of a credible 3D label or the identification of a discontinuous boundary. Consequently, disparity is established by designating the accurate 3D label to the point exhibiting the minimal disparity value, as occlusion spots typically manifest in the backdrop.

Depth Estimation and Reconstruction: Advanced neural networks can directly predict depth information from stereo image pairs, enhancing the accuracy of 3D reconstruction and reducing measurement errors (Tosi et al., 2019).

The training loss is defined as follows:

$$_{total} = \sum_{s=1}^{n_i} {}_{init} + \sum_{s=1}^{n_r} {}_{ref} \tag{28}$$

where s indicates the output resolution, n_i and n_r the number of considered scales during loss computation. $_{init}$ and $_{ref}$ are formalised as:

$$_{init} = \alpha_{ap}\left({}^{l}_{ap} + {}^{r}_{ap} \right) + \alpha_{ds}\left({}^{l}_{ds} + {}^{r}_{ds} \right) + \alpha_{ps}\left({}^{l}_{ps} + {}^{r}_{ps} \right) \tag{29}$$

$$_{ref} = \alpha_{ap}\, {}^{l}_{ap} + \alpha_{ds}\, {}^{l}_{ds} + \alpha_{ps}\, {}^{l}_{ps} \tag{30}$$

where $_{ap}$ is an image reconstruction loss, $_{ds}$ is a smothness term and $_{ps}$ is a proxy supervised loss. Figure 21 shows the general diagram of 3D reconstruction using stereo vision.

Figure 21. General diagram of 3D reconstruction using stereo vision

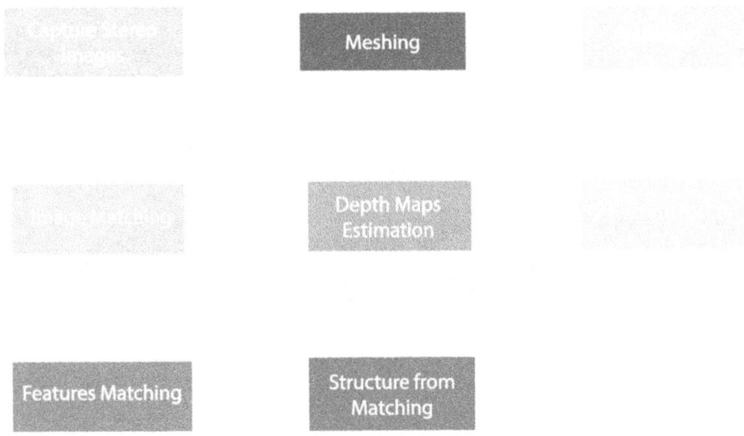

Starting with an initial disparity estimate at each scale obtained from the network, often marked by errors at depth discontinuities and occluded areas, this stage predicts multi-scale residual signals using a few stacked nonlinear layers. These signals are then employed to compute the final left-view aligned disparity map. This method simplifies the end-to-end learning process of the entire network. Furthermore, the geometrical constraints are crucial for enhancing final depth accuracy. In the process, a matching cost in feature space is computed via a horizontal correlation layer typically employed in deep stereo methods. A computed right-view disparity map is utilized to generate right-view features from the left features via a differentiable bilinear sampler. The network additionally obtains an error term defined as the absolute difference between the left and virtual right features at input resolution, with the latter mapped back to the same locations as the former. The system generates a synthetic right view along with its feature representation and computes the final disparity map based on stereo principles. This approach creates a unified end-to-end architecture that efficiently executes stereo operations from a singular input view. Moreover, trials conducted using this strategy validated the precision of the entirely self-supervised, end-to-end algorithm.

Adaptive Learning: Machine learning algorithms can continuously learn from new data, adapting to changes in the environment and improving their accuracy over time (Tonioni et al., 2019). These algorithms incorporate an online adaptation step directly into the learning objective, obtaining a base set of weights that are more conducive to unsupervised online adaptation. Figure 22 shows adaptive learning using stereo vision and CNN.

Figure 22. Adaptive learning process with stereo vision and CNN

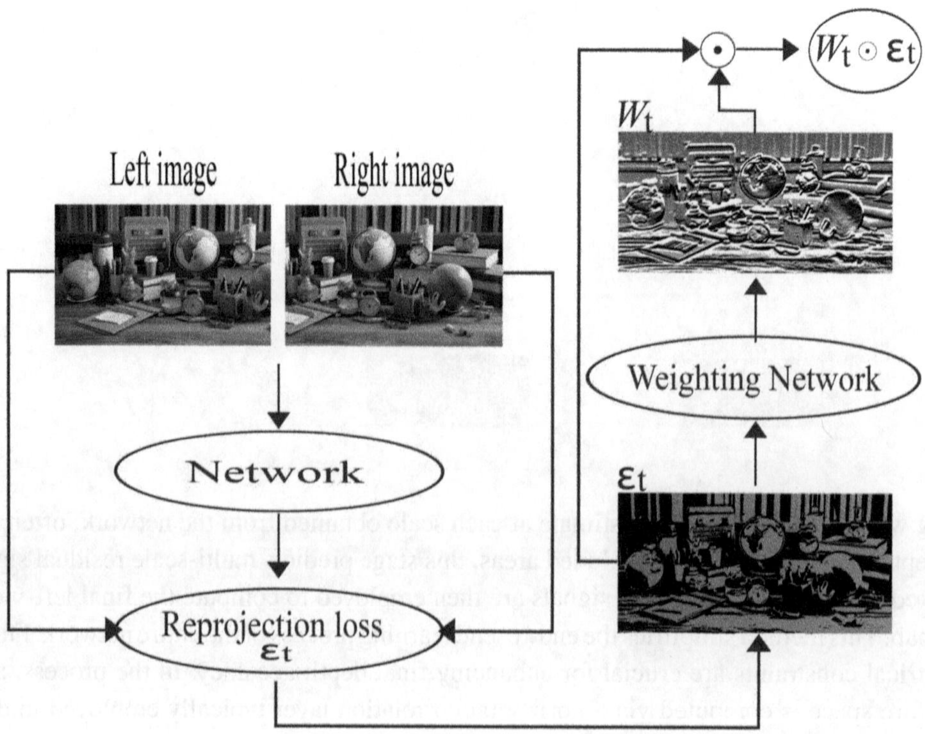

The adaptation is inherently chaotic and has a detrimental effect on the overall algorithm, as it occurs in an unsupervised manner. To prevent this problem, a confidence measure for the unsupervised loss has been devised, and a confidence-weighted loss is applied to update the network parameters. This prevents harmful parameter updates by reducing the noise in the adaptation phase. In the implementation, the confidence measures are predicted using a small CNN that is integrated into the meta-learning framework. This enables the network to be trained end-to-end without the need for additional supervision. Stereo vision systems and machine learning techniques can be implemented to enhance precision and dependability, rendering them appropriate for a diverse array of demanding applications. The system's performance is not only improved but its capabilities are also expanded by the combination of stereo vision and machine learning, which allows for more precise and sophisticated measurements.

MACHINE LEARNING ALGORITHM APPLIED TO REDUCE MEASUREMENT ERRORS IN INERTIAL SYSTEMS

Inertial systems are a suitable option when a machine or device is in motion and measure the changes in the movements performed. Hence, it is expected to find arrangements where two or more inertial sensors are combined to measure in different axes, providing information on the movements; even so, most sensors are contained in one device called an "Inertial Measurement Unit" or IMU. The IMU mentioned above can provide data when the system is in motion; this makes the inertial systems a suitable option when it is required to obtain data as the movements are performed (Ku, 2019), (Xuan et al., 2024). Besides the benefits proportioned by an IMU, they present an inherent error called "drift". This error originated due to voltage and temperature variations in one of the sensors or the IMU, differences in the measurement when sensor fusion of two or more instruments is implemented, and an increment of the error through the dead-reckoning process, among other similar errors.

As different approaches exist to diminish the influence of drift in inertial systems, machine learning algorithms provide an adaptative solution that seeks to compensate for the increments in the error over time. In some works, neural networks are implemented to describe transient behaviors in an inertial system. As some disturbances that increment drift occur through certain moments, the algorithm seeks to identify parameters such as temperature variations, as in Tu research (Tu & Peng, 2021). The forecasted data can diminish the error, aiding the estimation of angular velocity. As the algorithm is complemented with other regression analyses, the use of neural networks reduces the error through time. Additionally, inertial systems contain neural networks to increment the precision in estimating parameters, which is required as odometry. This implementation reduces the error when they are estimated vehicle trajectories or a body navigating. As wearable gadgets increase their demand, the drift error in these devices is still present and subjected to giving data with an error that adheres to the measured motion. Figure 23 shows an example of IMU data processed by Neural Network.

Figure 23. Inertial system measurement error reduction diagram. In an inertial system, the IMU data is processed by a neural network, which diminishes the drift influence and performs the fusion with the additional instruments if they exist

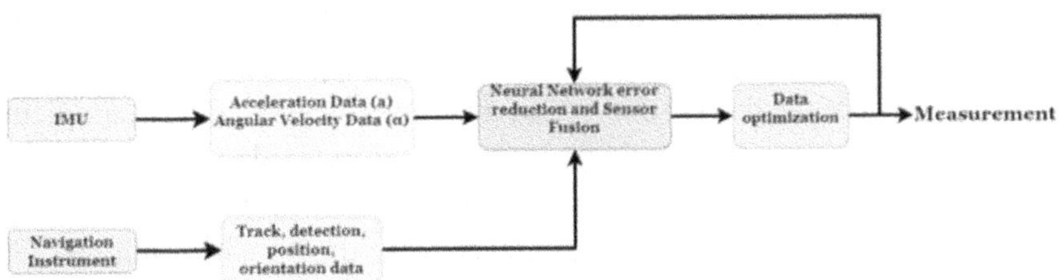

This section discusses the implementation of an algorithm applied to reduce error in accelerometers and gyroscopes, two inertial sensors that are sometimes combined to improve the precision of the measurements performed by inertial systems. As one of the most common applications is navigation, this section is not limited to techniques that apply to trajectory estimation.

Algorithms Applied to Reduce the Error in Accelerometers

An accelerometer is a mechanical sensor widely implemented in vibration measurements and through calculation to obtain data such as velocity and position as an aid in inertial navigation. These instruments respond to the inertial forces as a mass-spring-damper configuration, where the mass's displacement depends on the applied force. Additionally, accelerometers in electronic systems are mainly presented as Micro Electro Mechanical Systems (MEMS), incrementing the measurement variation (Zhang et al.,

2021). This is due to the sensitivity of an accelerometer increasing when it is in the size of micro accelerometers to external forces, even when it is intended to align the force to the instrument axis.

Figure 24. Accelerometer free body diagram. Where the accelerometer is represented as a mass-damping-spring system, which, as a force is applied, the measured result is the acceleration

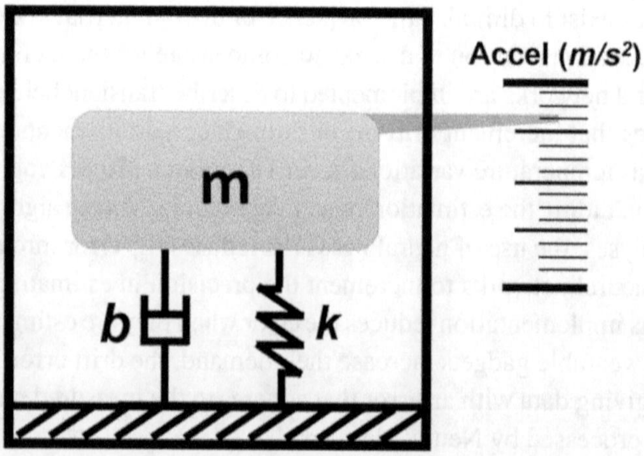

Figure 24 presents the free-body diagram of an accelerometer. The accelerometer is composed of a mass m that moves when the sensor is in motion and is linked to a damper b and a spring k (Zhao et al., 2022). As different types of accelerometers exist, the influence of external forces in the instrument increments the measurement error. Hence, as some techniques directly diminish the error in the data of accelerometers, it is important to note that some other methods apply sensor fusion to reduce the error. Hence, to diminish the effect of the drift in the data of an accelerometer, could be implemented a neural network of three layers. In the first layer, an input layer x is considered the acceleration of the body in the three axes. Therefore, a hidden layer could be implemented that considers bias b and the relationship between the patterns in the data. Finally, an output layer takes the classification prediction function f to determine the acceleration Y with the error diminished (Farahani et al., 2023). Figure 25 shows a diagram of how to reduce the drift in an accelerometer.

Figure 25. Deep neural network diagram to reduce drift influence in an accelerometer

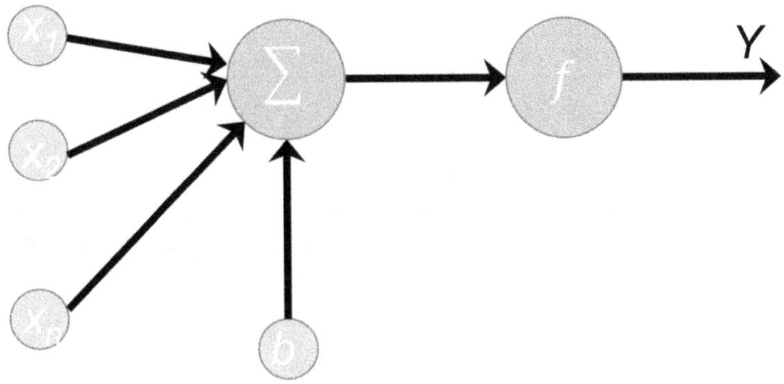

The model is expressed as:

$$Y(t+1) = \sum_{i=1}^{n} x_i w_i(t) \tag{31}$$

Where w_i is the input weighting factor. It is important to note that the bias b is adjusted during training. Finally, the activation function is a binary function $\varphi(x)$ as shown below:

$$\varphi(x) = \begin{cases} 1 & wx + b > threshold \\ 0 & Otherwise \end{cases} \tag{32}$$

Algorithms Applied to Reduce the Error in Gyroscopes

A gyroscope is a spinning mass that responds to angular velocity movements. As these devices could be presented in different forms according to their size, it is possible to measure the rate of degrees per second. These inertial sensors are commonly used to measure orientation by integrating angular velocity. Hence, as an inertial device, they are subjected to the angular velocities in different axes, which could add noise to the data obtained with a gyroscope (Dell'Olio et al., 2023), (Miao et al., 2022).

Figure 26. Free body diagram of a gyroscope. The spinning mass rotating in an axis orthogonal to Earth's gravity

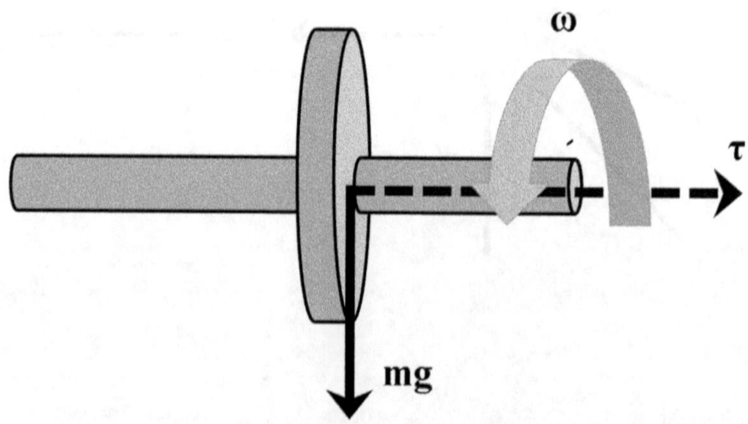

Figure 26 presents a mass spinning with an angular velocity ω, in an axis orthogonal to Earth's gravity and with torque τ applied in that axis. These sensors are susceptible to torque disturbances and external vibrations that add noise to the angular velocity signal of the gyroscope. Hence, to diminish the influence of disturbances with a machine learning technique in a gyroscope is taken in consideration the dynamic of the device to develop a neural network to reduce the error (Gao et al., 2023), (Li & Liu, 2021).

A recurrent neural network (RNN) model is composed as shown below:

$$x_k = f\left(\sum_{i=1}^{I} V_i g_i(k) + c\right) \tag{33}$$

Where the output g_i is:

$$g_i(k) = f\left(\sum_{j=1}^{p} w_{ij} x_{k-j} + \sum_{d=1}^{q}\sum_{i=1}^{I} w_{iid} g_i(k-d) + b_i\right) \tag{34}$$

Where p and I are the number of input and hidden layers in the network (Shao & Shi, 2020). The farthest step to k is denoted by p, and w_{ij} represents the weight, V_i is the connection between the weight and the hidden layer, b_i is the threshold, and f is the nonlinear function. As the inputs of the networks are the dynamic measured by the gyroscope, the presented equation considers the angular velocity of the gyroscope, which could be influenced by external forces

CONCLUSIONS

This chapter provides an overview of machine learning algorithms used to reduce measurement errors in machine vision and inertial systems. The algorithms presented are nearest neighbor (KNN), support vector regression (SVR), and artificial neural networks (ANN). These algorithms are applied in systems based on laser scanning, monovision, stereovision, or inertial measurement units (IMU).

Laser scanning uses the ANN and KNN to reduce measurement errors in the B_{ij} angle, enhancing coordinate calculation accuracy and enabling the creation of a 3D point cloud. One input layer, one hidden input, and one output layer form the ANN to crate the model prediction, while the KNN algorithm uses three neighbors to calculate the predicted value. The Root Mean Square Error (RMSE) values for KNN and ANN within the workspace are 0.55° and 0.38°, respectively, indicating an improvement when using an ANN. Furthermore, a disadvantage of the KNN is that it requires more computational resources; therefore, it is necessary to decrease the scanning velocity to process all data in real-time.

In monovision systems, image enhancement and preprocessing using FCNN is useful for noise reduction, identifying the most similar neighboring pixels, and achieving better results than using pixels with the smallest variation. In other applications using monovision systems, SVM detects and recognizes objects in images, establishing a hyperplane to minimize classification errors. For depth estimation using monovision systems, neural networks are used, evaluating the error results with RMSE, RMSE log, absolutive relative error, and square relative error, achieving high accuracy. In other applications using monovision systems, KNN is employed for feature extraction, showing a high classification that depends on the chosen distance metrics and the quality of feature extraction.

In stereovision systems, a Bayesian probabilistic model is used for disparity map refinement, achieving errors around 0.2 pixels in validated areas. In other applications, CNNs are employed for stereo-matching tasks, determining the minimum function cost to find the disparity. For outliner detection on stereovision systems, machine learning models like LRC are used to detect outliners caused by occlusions or mismatches. Afterward, the detected points receive a 3D label to improve the stereo-matching process. For depth estimation and reconstruction applications with stereo vision, neural networks can predict the depth information, enhancing the high accuracy of 3D reconstruction and reducing measurement errors. Moreover, CNN is employed for adaptive learning, allowing to be trained without additional supervision, achieving better error results as it trains itself.

Finally, in accelerometers, external forces influence the instrument, increasing measurement errors. The neural network can be applied to diminish the effect of the drift in the data of an accelerometer. On the other hand, the gyroscope is subjected to the angular velocities in different axes, which could add noise to the obtained data. Therefore, to diminish the influence of disturbances, a machine learning technique in a gyroscope is taken into consideration, and the dynamic of the device is used to develop a neural network to reduce the error.

REFERENCES

Ahmed, H. U., Mostafa, R. R., Mohammed, A., Sihag, P., & Qadir, A. (2023). Support vector regression (SVR) and grey wolf optimization (GWO) to predict the compressive strength of GGBFS-based geopolymer concrete. *Neural Computing & Applications*, *35*(3), 2909–2926. DOI: 10.1007/s00521-022-07724-1

Al-Jarrah, M. A. (2016, July). Developing 3D model for mobile robot environment using mono-vision system. In 2016 7th International Conference on Computer Science and Information Technology (CSIT) (pp. 1-4). IEEE. DOI: 10.1109/CSIT.2016.7549466

Boumedine, A. Y., Bentaieb, S., & Ouamri, A. (2023). An improved KNN classifier for 3D face recognition based on SURF descriptors. *Journal of Applied Security Research*, *18*(4), 808–826. DOI: 10.1080/19361610.2022.2099688

Castro-Toscano, M. J., Rodríguez-Quiñonez, J. C., Sergiyenko, O., Flores-Fuentes, W., Ramirez-Hernandez, L. R., Hernández-Balbuena, D., Lindner, L., & Rascón, R. (2020). Novel sensing approaches for structural deformation monitoring and 3D measurements. *IEEE Sensors Journal*, *21*(10), 11318–11328. DOI: 10.1109/JSEN.2020.3031882

Chang, J. S., Liu, S. C., Ng, J. C., & Ma, P. L. (2020). Monovision with a bifocal diffractive multifocal intraocular lens in presbyopic patients: A prospective, observational case series. *American Journal of Ophthalmology*, *212*, 105–115. DOI: 10.1016/j.ajo.2019.11.010 PMID: 31765626

Chicco, D., Warrens, M. J., & Jurman, G. (2021). The coefficient of determination R-squared is more informative than SMAPE, MAE, MAPE, MSE and RMSE in regression analysis evaluation. *PeerJ. Computer Science*, *7*, e623. DOI: 10.7717/peerj-cs.623 PMID: 34307865

Cosenza, D. N., Korhonen, L., Maltamo, M., Packalen, P., Strunk, J. L., Næsset, E., Gobakken, T., Soares, P., & Tomé, M. (2021). Comparison of linear regression, k-nearest neighbour and random forest methods in airborne laser-scanning-based prediction of growing stock. Forestry. *Forestry*, *94*(2), 311–323. DOI: 10.1093/forestry/cpaa034

Dell'Olio, F., Natale, T., Wang, Y.-C., & Hung, Y.-J. (2023). Miniaturization of interferometric optical gyroscopes: A review. *IEEE Sensors Journal*, *23*(24), 29948–29968. DOI: 10.1109/JSEN.2023.3327217

Di Leo, G., Liguori, C., Pietrosanto, A., & Sommella, P. (2017). A vision system for the online quality monitoring of industrial manufacturing. *Optics and Lasers in Engineering*, *89*, 162–168. DOI: 10.1016/j.optlaseng.2016.05.007

Dong, Q., & Feng, J. (2019). Outlier detection and disparity refinement in stereo matching. *Journal of Visual Communication and Image Representation*, *60*, 380–390. DOI: 10.1016/j.jvcir.2019.03.007

Duran, O., Turan, B., & Kaya, M. (2024). Machine-learning-based ensemble regression for vehicle-to-vehicle distance estimation using a toe-in style stereo camera. *Measurement*, *115540*. Advance online publication. DOI: 10.1016/j.measurement.2024.115540

Emani, S., Soman, K. P., Sajith Variyar, V. V., & Adarsh, S. (2019). Obstacle detection and distance estimation for autonomous electric vehicle using stereo vision and DNN. In Soft Computing and Signal Processing: Proceedings of ICSCSP 2018, Volume 2 (pp. 639-648). Springer Singapore. DOI: 10.1007/978-981-13-3393-4_65

Farahani, H., Blagojević, M., Azadfallah, P., Watson, P., Esrafilian, F., & Saljoughi, S. (2023). Deep neural network. In Springer eBooks (pp. 145–185). DOI: 10.1007/978-3-031-31172-7_6

Flores-Fuentes, W., Rivas-Lopez, M., Sergiyenko, O., Rodríguez-Quiñonez, J. C., Hernández-Balbuena, D., & Rivera-Castillo, J. (2014). Energy center detection in light scanning sensors for structural health monitoring accuracy enhancement. *IEEE Sensors Journal*, *14*(7), 2355–2361. DOI: 10.1109/JSEN.2014.2310224

Gao, Y., Shi, D., Li, R., Liu, Z., & Sun, W. (2023). Gyro-net: IMU gyroscopes random errors compensation method based on deep learning. *IEEE Robotics and Automation Letters*, *8*(3), 1471–1478. DOI: 10.1109/LRA.2022.3230594

González Izard, S., Sánchez Torres, R., Alonso Plaza, O., Juanes Mendez, J. A., & García-Peñalvo, F. J. (2020). Nextmed: Automatic imaging segmentation, 3D reconstruction, and 3D model visualization platform using augmented and virtual reality. *Sensors (Basel)*, *20*(10), 2962. DOI: 10.3390/s20102962 PMID: 32456194

Görgens, E. B., Montaghi, A., & Rodriguez, L. C. E. (2015). A performance comparison of machine learning methods to estimate the fast-growing forest plantation yield based on laser scanning metrics. *Computers and Electronics in Agriculture*, *116*, 221–227. DOI: 10.1016/j.compag.2015.07.004

Grigorescu, S. D., Seritan, G. C., Enache, B. A., Argatu, F. C., Adochiei, F. C., & Stanciu, V. (2023). Indoor 3d scanning and navigation system for an automated guided vehicle. Revue Roumaine Des Sciences Techniques—Série Électrotechnique Et Énergétique, 68(2), 232-235.

Halitim, A. M., Bouhedda, M., Tchoketch-Kebir, S., & Rebouh, S. (2024). Artificial neural network for tilt compensation in yaw estimation. *Transactions of the Institute of Measurement and Control*, *46*(11), 2087–2096. DOI: 10.1177/01423312231214832

Hożyń, S., & Żak, B. (2021). Stereo Vision System for Vision-Based Control of Inspection-Class ROVs. *Remote Sensing (Basel)*, *13*(24), 5075. DOI: 10.3390/rs13245075

Hsu, Y. W., Zhong, K. Q., Perng, J. W., Yin, T. K., & Chen, C. Y. (2018, November). Developing an on-road obstacle detection system using monovision. In *2018 International Conference on Image and Vision Computing New Zealand (IVCNZ)* (pp. 1-9). IEEE. DOI: 10.1109/IVCNZ.2018.8634799

Huang, K., Hussain, A., Wang, Q. F., & Zhang, R. (Eds.). (2019). Deep learning: fundamentals, theory and applications (Vol. 2). springer. DOI: 10.1007/978-3-030-06073-2

Ibrahim, I., & Abdulazeez, A. (2021). The role of machine learning algorithms for diagnosing diseases. *Journal of Applied Science and Technology Trends*, *2*(1), 10–19. DOI: 10.38094/jastt20179

Jayadeva Khemchandani, R., & Chandra, S. (2018). *Twin support vector machines: models, extensions and applications*. Springer Publishing Company, Incorporated.

Jia, X., Ma, P., Tarwa, K., & Wang, Q. (2023). Machine vision-based colorimetric sensor systems for food applications. *Journal of Agriculture and Food Research, 11*, 100503. DOI: 10.1016/j.jafr.2023.100503

Joseph, S., & George, J. (2021). Convolutional autoencoder based feature extraction and KNN classifier for handwritten MODI script character recognition. In Data Science and Security: Proceedings of IDSCS 2021 (pp. 142-149). Springer Singapore. DOI: 10.1007/978-981-16-4486-3_15

Jospin, L. V., Laga, H., Boussaid, F., & Bennamoun, M. (2022). Bayesian learning for disparity map refinement for semi-dense active stereo vision. arXiv preprint arXiv:2209.05082. https://doi.org//arXiv .2209.05082.DOI: 10.48550

Jwo, D. J., Biswal, A., & Mir, I. A. (2023). Artificial neural networks for navigation systems: A review of recent research. *Applied Sciences (Basel, Switzerland), 13*(7), 4475. DOI: 10.3390/app13074475

Ku, D. Y., & Patel, A. (2020). Kinematic state estimation using multiple DGPS/MEMS-IMU sensors. *IEEE Sensors Letters, 4*(12), 1–4. Advance online publication. DOI: 10.1109/LSENS.2020.3040661

Levitan, I. B., & Kaczmarek, L. K. (2015). *The neuron: cell and molecular biology*. Oxford University Press. DOI: 10.1093/med/9780199773893.001.0001

Li, D., Zhou, J., & Liu, Y. (2021). Recurrent-neural-network-based unscented Kalman filter for estimating and compensating the random drift of MEMS gyroscopes in real time. *Mechanical Systems and Signal Processing, 147*(107057), 107057. DOI: 10.1016/j.ymssp.2020.107057

Liu, Y., Lv, B., Wang, Y., & Huang, W. (2020). An end-to-end stereo matching algorithm based on improved convolutional neural network. *Mathematical Biosciences and Engineering, 17*(6), 7787–7803. DOI: 10.3934/mbe.2020396 PMID: 33378920

Lu, C. T., Wang, L. L., Shen, J. H., & Lin, J. A. (2021). Image enhancement using deep-learning fully connected neural network mean filter. *The Journal of Supercomputing, 77*(3), 3144–3164. DOI: 10.1007/s11227-020-03389-6

Miao, T., Li, Q., Hu, X., Wu, X., Wu, W., & Xiao, D. (2022). Virtual rotating MEMS gyrocompassing with honeycomb disk resonator gyroscope. *IEEE Electron Device Letters, 43*(8), 1331–1334. DOI: 10.1109/LED.2022.3185985

Monnet, J. M., Chanussot, J., & Berger, F. (2011). Support vector regression for the estimation of forest stand parameters using airborne laser scanning. *IEEE Geoscience and Remote Sensing Letters, 8*(3), 580–584. DOI: 10.1109/LGRS.2010.2094179

Ngeljaratan, L., Bas, E. E., & Moustafa, M. A. (2024). Unmanned Aerial Vehicle-Based Structural Health Monitoring and Computer Vision-Aided Procedure for Seismic Safety Measures of Linear Infrastructures. *Sensors (Basel), 24*(5), 1450. DOI: 10.3390/s24051450 PMID: 38474985

Noor, K., Siddiquee, E. A., Sarma, D., Nandi, A., Akhter, S., Hossain, S., . . . Hossain, M. S. (2017, June). Performance analysis of a surveillance system to detect and track vehicles using Haar cascaded classifiers and optical flow method. In 2017 12th IEEE Conference on Industrial Electronics and Applications (ICIEA) (pp. 258-263). IEEE. DOI: 10.1109/ICIEA.2017.8282853

Okuno, F., Yoshimoto, K., Takahashi, H., & Hamagishi, G. (2024, March). Evaluation method of matching points for improving accuracy of position estimation using monocular endoscope. In *Endoscopic Microscopy XIX* (Vol. 12820, pp. 49–55). SPIE., DOI: 10.1117/12.3001311

Pinto, A. M., Rocha, L. F., & Moreira, A. P. (2013). Object recognition using laser range finder and machine learning techniques. *Robotics and Computer-integrated Manufacturing*, *29*(1), 12–22. DOI: 10.1016/j.rcim.2012.06.002

Qian, R., Zhao, Y., Feng, R., Yang, W., & Zhang, Z. (2023). An algorithm for single view occlusion area detection in binocular stereo matching. *International Journal of Pattern Recognition and Artificial Intelligence*, *37*(2), 2350003. DOI: 10.1142/S0218001423500039

Ramírez-Hernández, L. R., Rodríguez-Quinoñez, J. C., Castro-Toscano, M. J., Hernández-Balbuena, D., Flores-Fuentes, W., Rascón-Carmona, R., Lindner, L., & Sergiyenko, O. (2020). Improve three-dimensional point localization accuracy in stereo vision systems using a novel camera calibration method. *International Journal of Advanced Robotic Systems*, *17*(1), 1729881419896717. DOI: 10.1177/1729881419896717

Real-Moreno, O., Castro-Toscano, M. J., Rodrízuez-Ouiñonez, J. C., Hernández-Balbuena, D., Flores-Fuentes, W., & Rivas-Lopez, M. (2018, October). Implementing k-nearest neighbor algorithm on scanning aperture for accuracy improvement. In IECON 2018-44th Annual Conference of the IEEE Industrial Electronics Society (pp. 3182-3186). IEEE. DOI: 10.1109/IECON.2018.8592804

Real-Moreno, O., Rodriguez-Quiñonez, J. C., Sergiyenko, O., Basaca-Preciado, L. C., Hernandez-Balbuena, D., Rivas-Lopez, M., & Flores-Fuentes, W. (2017, June). Accuracy improvement in 3D laser scanner based on dynamic triangulation for autonomous navigation system. In 2017 IEEE 26th International Symposium on Industrial Electronics (ISIE) (pp. 1602-1608). IEEE. DOI: 10.1109/ISIE.2017.8001486

Risi, N., Aimar, A., Donati, E., Solinas, S., & Indiveri, G. (2020). A spike-based neuromorphic architecture of stereo vision. *Frontiers in Neurorobotics*, *14*, 568283. DOI: 10.3389/fnbot.2020.568283 PMID: 33304262

Shao, X., & Shi, Y. (2020). Neural adaptive control for MEMS gyroscope with full-state constraints and quantized input. *IEEE Transactions on Industrial Informatics*, •••, 1–1. DOI: 10.1109/TII.2020.2968345

Shataee, S., Kalbi, S., Fallah, A., & Pelz, D. (2012). Forest attribute imputation using machine-learning methods and ASTER data: Comparison of k-NN, SVR and random forest regression algorithms. *International Journal of Remote Sensing*, *33*(19), 6254–6280. DOI: 10.1080/01431161.2012.682661

Shekhovtsov, A., & Flach, B. (2018). Feed-forward propagation in probabilistic neural networks with categorical and max layers. In *International conference on learning representations*.

Smiles, A. L., Chavanakunakorn, K. D., Omarali, B., Oh, C., & Farkhatdinov, I. (2023, September). Implementation of a Stereo Vision System for a Mixed Reality Robot Teleoperation Simulator. In Annual Conference Towards Autonomous Robotic Systems (pp. 494-502). Cham: Springer Nature Switzerland. DOI: 10.1007/978-3-031-43360-3_40

Specht, M., Szostak, B., Lewicka, O., Stateczny, A., & Specht, C. (2023). Method for determining of shallow water depths based on data recorded by UAV/USV vehicles and processed using the SVR algorithm. *Measurement*, *221*, 113437. DOI: 10.1016/j.measurement.2023.113437

Tonioni, A., Rahnama, O., Joy, T., Stefano, L. D., Ajanthan, T., & Torr, P. H. (2019). Learning to adapt for stereo. In *Proceedings of the IEEE/CVF Conference on Computer Vision and Pattern Recognition* (pp. 9661-9670).

Tosi, F., Aleotti, F., Poggi, M., & Mattoccia, S. (2019). Learning monocular depth estimation infusing traditional stereo knowledge. In *Proceedings of the IEEE/CVF Conference on Computer Vision and Pattern Recognition* (pp. 9799-9809). DOI: 10.1109/CVPR.2019.01003

Trujillo-Hernández, G., Flores-Fuentes, W., Rodríguez-Quiñonez, J. C., Hernández-Balbuena, D., Real-Moreno, O., Miranda-Vega, J. E., & Bhateja, V. (2022). 3D reconstruction of human body biometry. In *Optoelectronic Devices in Robotic Systems* (pp. 195–225). Springer International Publishing., DOI: 10.1007/978-3-031-09791-1_8

Trujillo-Hernández, G., Rodríguez-Quiñonez, J. C., Flores-Fuentes, W., Sanchez-Castro, J. J., Real-Moreno, O., Castro-Toscano, M. J., & Hernández-Balbuena, D. (2024). Optical 3D Scanning System in Medical Applications. In Scanning Technologies for Autonomous Systems (pp. 159-196). Cham: Springer Nature Switzerland. DOI: 10.1007/978-3-031-59531-8_6

Trujillo-Hernández, G., Rodríguez-Quiñonez, J. C., Ramírez-Hernández, L. R., Castro-Toscano, M. J., Hernández-Balbuena, D., Flores-Fuentes, W., . . . Mercorelli, P. (2019, October). Accuracy improvement by artificial neural networks in technical vision system. In IECON 2019-45th Annual Conference of the IEEE Industrial Electronics Society (Vol. 1, pp. 5572-5577). IEEE. DOI: 10.1109/IECON.2019.8927596

Tu, Y., & Peng, C. (2021). An ARMA-Based digital twin for MEMS gyroscope drift dynamics modeling and Real-Time compensation. *IEEE Sensors Journal, 21*(3), 2712–2724. DOI: 10.1109/JSEN.2020.3028140

Urbas, U., Vlah, D., & Vukašinović, N. (2021). Machine learning method for predicting the influence of scanning parameters on random measurement error. *Measurement Science & Technology, 32*(6), 065201. DOI: 10.1088/1361-6501/abd57a

Vang-Mata, R. (Ed.). (2020). *Multilayer perceptrons: theory and applications*. Nova Science Publishers.

Vezilić, B., Gajić, D. B., Dragan, D., Petrović, V., Mihić, S., Anišić, Z., & Puhalac, V. (2018). Binary classification of images for applications in intelligent 3D scanning. *Intelligent Distributed Computing, XI*, 199–209. DOI: 10.1007/978-3-319-66379-1_18

Wang, D., Liu, Z., Shao, S., Wu, X., Chen, W., & Li, Z. (2023, October). Monocular Depth Estimation: A Survey. In IECON 2023-49th Annual Conference of the IEEE Industrial Electronics Society (pp. 1-7). IEEE. DOI: 10.1109/IECON51785.2023.10311687

Wang, Y., Zhou, P., Yao, C. W., Wang, H. Y., & Lin, B. (2024). Development of a novel multi-line laser triangulation scanning system based on the rotary diffraction grating. *Measurement, 225*, 113988. DOI: 10.1016/j.measurement.2023.113988

Xu, Z., Gao, Y., & Jin, Y. (2014). Application of an optimized SVR model of machine learning. *International Journal of Multimedia and Ubiquitous Engineering, 9*(6), 67–80. DOI: 10.14257/ijmue.2014.9.6.08

Xuan, L., He, X., Yi, Y., Shen, A., Yang, X., Dong, J., & Dong, S. (2024). Research on IMU-Based Motion Attitude Acquisition and Motion Recognition. *IEEE Sensors Journal*, *24*(13), 20786–20793. DOI: 10.1109/JSEN.2024.3394903

Zhang, H., Sobreviela, G., Pandit, M., Chen, D., Sun, J., Parajuli, M., Zhao, C., & Seshia, A. A. (2021). A low-noise high-order mode-localized MEMS accelerometer. Journal of Microelectromechanical Systems: A Joint IEEE and ASME Publication on Microstructures. *Journal of Microelectromechanical Systems*, *30*(2), 178–180. DOI: 10.1109/JMEMS.2021.3057260

Zhang, S. (2021). Challenges in KNN classification. *IEEE Transactions on Knowledge and Data Engineering*, *34*(10), 4663–4675. DOI: 10.1109/TKDE.2021.3049250

Zhang, T., Zheng, J., & Zou, Y. (2023). Weighted voting ensemble method for predicting workpiece imaging dimensional deviation based on monocular vision systems. *Optics & Laser Technology*, *159*, 109012. DOI: 10.1016/j.optlastec.2022.109012

Zhang, X., Li, S., Yang, J., Wang, Y., Huang, Z., & Zhang, J. (2022). Tactile Perception Object Recognition Based on an Improved Support Vector Machine. *Micromachines*, *13*(9), 1538. DOI: 10.3390/mi13091538 PMID: 36144161

Zhang, Y., Wang, Q., Chen, X., Yan, Y., Yang, R., Liu, Z., & Fu, J. (2022). The prediction of spark-ignition engine performance and emissions based on the SVR algorithm. *Processes (Basel, Switzerland)*, *10*(2), 312. DOI: 10.3390/pr10020312

Zhao, M., Qi, Y., Zhang, X., Li, B., Jiang, Z., & Wei, X. (2022). Noise Analysis and Performance Improvement of a MEMS Fabry-Pérot Seismic Accelerometer. *IEEE Sensors Journal*, *22*(1), 365–372. DOI: 10.1109/JSEN.2021.3128534

Chapter 18
AI and the Creative Process:
Balancing Innovation, Copyright, and Artistic Authenticity

Victor Manuel Gonzalez Davila
Universidad Politécnica de Baja California, Mexico

Martha Elena González Correa
https://orcid.org/0009-0000-6073-9585
Universidad Politécnica de Baja California, Mexico

E. Ivette Cota-Rivera
https://orcid.org/0009-0002-6240-9269
Universidad Politécnica de Baja California, Mexico

Martha Angélica Martínez Martínez
Universidad Politécnica de Baja California, Mexico

Luis Antonio Bernal Marín
Universidad Politécnica de Baja California, Mexico

María Ysabel Márquez Montenegro
Universidad Politécnica de Baja California, Mexico

ABSTRACT

In this article the ethical implications and the risk that the work of artist suffers with the use of AI are discussed. The use of artificial intelligence is framed by the internet that stores intangible cultural assets as part of a cyberculture marked by the appropriation, the optimistic speech of CEO Sam Altman and the behavior of early adopters accept it in the midst of a culture of plagiarism and appropriation practices within digital art. Ethical and fair suggestions are established in favor of digital artists and the implementation of the glaze tool to protect them artwork, while the authorities don't take charge.

DOI: 10.4018/979-8-3693-6910-4.ch018

INTRODUCTION

A digital artist, as opposed to a traditional analog artist, is someone who works directly with drawing, editing, and computer-aided modeling software. They typically use social media to showcase their work, attract commission clients, and secure both temporary and indefinite employment opportunities.

It is important to introduce concepts such as heritage, cultural assets, economic assets, and the creative industry to determine the exchange and service of digital business models on the web. In this context, it is crucial to define the type of asset to which the set or sets of images, valued for their artistic components created by humans, belong.

To facilitate this research, the artworks presumably used to train AI were divided into two groups. The first group comprises works with a physical version outside the web, considered to be of lower risk: paintings, sculptures, installations, and performances. The second group, deemed to be of higher risk, includes works created by humans that have a more intangible presence: photographs, graphic design, digital illustrations, 3D models, cinema, videos, literature, screenplays, and storyboards.

In this context, without a doubt, a significant portion of both tangible and intangible works were used to train artificial intelligence. However, it is currently difficult to determine which authors were used more extensively. Although the potential future repercussions are concerning, they are beyond the scope of this book. Instead, we establish that the first group pertains to all areas of human activity related to the visual arts, including photography, while the second group encompasses video.

The cultural industry is part of the creative industry; there is no difference if we consider the aim to promote audiovisual products that are part of cinema, television, and the internet. From Theodor Adorno's perspective, the cultural industry, specifically cinema and television, serves the function of manipulating information and creating new needs (Adorno & Horkheimer,1947). Considering Adorno's critical thoughts, we can attribute the same negative purposes to the internet as an essential component of a capitalist economic model. The new needs of AI can be regarded within the creative industry as reducing production times and costs.

To fully understand the landscape of the arts and the creative industries, it can be defined that the former predominantly encompasses all cultural assets conceived and managed by human communities, extending to major museums. In contrast, the latter pertains to a network of collaboration between state participation and private enterprise, led by entrepreneurs for the management and development of creative production, including cinema, 3D animation, and video games.

In this regard, it is important to note that it is specifically the second group, the one comprised of intangible materials mentioned earlier, that concerns us the most due to its high potential for AI training. Considering digital illustrations, 3D models, storyboards, and scripts as crucial for the development of the creative industry, their intangible nature exposes the works of authors to significant risk of being readily appropriated by AI. This raises concerns about the potential impact on other activities, businesses, and services that have previously derived benefits from these creative endeavors, without foreseeing how they might be affected.

Once the above is understood, it is important to recognize that while the web serves as a public good intended for the exchange of primarily intangible information derived from real-world models—both physical and digital—it is pertinent to investigate whether the web constitutes both a cultural and economic asset. This inquiry becomes crucial in light of the AI's increasing utilization, often disregarding human agreements to uphold copyright and promoting a culture of appropriation among users for creative purposes.

Appropriation within visual arts has been a highly controversial practice involving the unauthorized use of another artist's work and presenting it as one's own. While it does not entail physical theft of the original piece, claiming someone else's work as your own raises' questions about creative contribution. Needless to say, this has sparked criticism from the public and, at times, legal issues for those involved. Such actions appear to be increasingly facilitated by the use of AI and data scraping techniques (Jones, 2023)

Therefore, through a qualitative methodology, an analysis is conducted on the discourse of key representatives from Meta, OpenAI, and Google regarding their arguments legitimizing the use of artificial intelligence to democratize talent (Francisco, Santilli, & Demetrios, 2024). This analysis aims not to discourage its use, as AI serves as a valuable tool for enhancing processes across various activities. However, it is evident that this democratizing stance appears to overlook job opportunities and the dignity of digital artists.

Furthermore, the discourse of optimistic legitimization towards new technologies by social media users is compared to understand whether it aligns with a discourse of late modernism in relation to postmodern values in 2024. This comparison also serves to assess the different stages of progress and democratization in the history of visual arts, highlighting the differences and similarities with the current supposed democratization facilitated, in part, by the use of AI among digital artists and users.

Finally, a study was conducted on the creative process of a digital artist in the editing, modification of images, and digital collage to determine how the use of AI can enhance their work without entirely appropriating the work of others. In this context, the use of AI is crucial to the final outcome, depending on the creative stage. For this purpose, the following phases are utilized linearly: brainstorming or definition, the research phase considering what other artists have already done, the analysis phase where selections or sketches are made, the execution phase where the artist works on the image, and the editing and correction phases of the image.

The intersection of AI and the creative industry presents both opportunities and challenges. While AI has the potential to enhance creative processes and broaden access to artistic tools, it also poses risks to the rights and livelihoods of artists. By critically examining the implications of AI and advocating for ethical practices, a future can be envisioned where technology and creativity coexist harmoniously, enriching the cultural landscape and supporting the diverse talents of digital artists. The goal of this document is to explore whether the use of AI can enhance the creative process of digital artists without infringing on copyright laws and undermining the economic livelihood that artists rely on to continue entertaining and astonishing us.

APPROPRIATION AS ART TECHNIQUE A PERSPECTIVE FROM THE ART WORLD

In the 1980s, the action of taking a work created by another artist as one's own became institutionalized in the art world. The visual artist Sherrie Levine stands out as the most prominent figure in what we now call appropriationism (Foster, 1983). It is important to remember that this would not be the first time an artist has taken an object not made by their own hands as a piece of art. In the history of 20th-century art we have the Dada Baroness (Gammel, 2002), who enjoyed collecting industrial objects not for their use or historical value, but for their unique components and physical qualities. Similarly, perhaps the most notable artist in terms of selecting domestic objects to incorporate into art spaces and discussions is Marcel Duchamp (Guasch,2009)., who is recognized as the father of object art and a forerunner of conceptual art, among other art trends, and is remembered for his famous urinal.

In this regard, it is very important to differentiate between the act of choosing an everyday object, the readymade, and the piece of art that results from the act of manipulating or transforming an object. In the first case, it consists of absolutely respecting the physical properties of the object, and for this reason, we can understand that in no way could we apply any other technique such as: painting it, making a copy at a larger or smaller scale, cutting it, using it as a base for another object, exhibiting some of its parts, etc.

On the contrary, in the second case, the original object can undergo all sorts of interventions; it can be painted, cut, altered in various parts or in its entirety, combined with other objects, and so on. In any case, at least within the art world, there is a code of permissibility that does not respect or acknowledge industrial and intellectual property and the original purposes of the object because, likewise, the purpose of art and artists lies in the appreciation of objects and not in their use.

However, the unmarked boundary between appreciating an object as the final purpose and not using the object itself to justify taking an object not created by the artist and exhibiting it as a work of art, a work by the artist who chose it, becomes illegible when it comes to photography. Whether it is a photograph for a commercial, for the sale of a product, as a documentary of an event of historical relevance or not, photography as reportage or editorial news, and even more so, photography for artistic purposes, regardless of whether its function is to communicate or not, its main objective lies in being appreciated. This is why artists from the aforementioned appropriationist movement have been sanctioned for the improper use of photographs (Brown, 2019)., even when they are used as references for a painting or for a silkscreen that is later altered with various artistic techniques.

In the world of art, it has never been so transparent when it comes to offering explanations about whether there are commercial agreements between artists and major brands for the use of industrial property owned by those mass consumer brands, including poster designs, packaging design, product design, etc. In this sense, the permissibility does not always seem to be consensual or pre-established between the parties: between the owner of the original product or object and the artist who takes it as their own. It is worth mentioning that the appropriate sanctions do not apply in the same way, at least not as they do outside the art world. There are rare exceptions where photographers dedicated to photojournalism and documentary photography have managed to win lawsuits against visual artists who used their photos without consent (Bowley, 2015)

APPROPRIATION AS USER BEHAVIOR A PERSPECTIVE FROM THE WEB

In our days, we cannot deny how we are participating in appropriation, no longer as an artistic act but as a culture. If we consider culture as this human manifestation resulting from agreements on ways of thinking and learned and acquired behaviors, we must analyze with perspective how we use the content we find mainly on the internet, a space where we have left our marks as appropriators. Whether for artistic purposes or not, in most cases not, it has become part of our culture to take images of ourselves or strangers, to share images without knowing their origin or whether they have copyright. At the same time, while we practice copy and paste, take and share, there are websites dedicated to the buying and selling of images that are photographs of public events, such as Getty Images.

Just as the culture of appropriation has been practiced on the web, it is perhaps from this that artists and creatives in various fields have sought to protect their digital works, such as photographs from Getty Images, through watermarks and author signatures. However, those familiar with manipulation and digital retouching software are well aware of the few guarantees available, as they can easily remove or disguise these marks and signatures. In this regard, it goes without saying that the artificial intelligences that are already part of such software allow for even greater ease in eliminating these marks and author signatures., (Appel, Neelbauer, & Schweidel, 2023).

Although images are available to everyone, any business that does not want to engage in illegal activity must pay for copyright rights for their commercial use, sale, and exhibition. However, this applies to the web; it does not mean that a physical, local business has the same caution and willingness to use images for its exploitation and commerce. This depends on the laws of each country. We know that in Japan, they are very protective when it comes to safeguarding their intellectual properties (U.S. Department of Commerce, 2024) that are part of the cultural entertainment industry. Similarly, the laws in the United States are equally strict in protecting their cinema from websites that display licensed movies without authorization (United States Congress, 1998).

Another type of appropriationism that exists is the one we authorize through the use of social networks. Facebook and Instagram, in their usage policies, grant permission to manage the images uploaded by users in a convenient manner. In the same vein, it seems that the social network X has taken a similar approach, even allowing the use of material for training AI (Morales, 2025).

Among the memes and images that web users identify with, visual artists have always tried to represent part of contemporary life, and suddenly we can see those same memes painted or printed in museums. Whether this is ethical or not, just as we know that trademarks are insufficient to protect images, we also know that those who appropriate images, photographs, and films for profit have been doing so for a long time. It goes without saying that they do not think about an art museum to appreciate the original work from another perspective.

UNESCO, for example, is taking the issue of artificial intelligence more seriously because they recognize the appropriative capabilities of individuals and the potential misuse of this technology (UNESCO, 2021). We know that how this situation is addressed depends on each country, but to counteract the culture of taking what belongs to others as one's own, it is also necessary to promote a culture of protecting the digital works of artists

THE OPTIMISTIC SPEECH OF LATE MODERNITY PROMOTE THE USE OF AIS: EARLIER ADOPTERS OF THE NEW FROM THE PAST TO OUR DAYS.

At the end of the nineteenth century, photography was part of the culture of Parisians and the Impressionist painters had undertaken a new way of painting to remain relevant within artistic creation (Honnef, 2003). Its popular the speech that artists at that time were at odds with photography is popular, however, we know thanks to the art historians that this was not entirely true and we even know that painters used photography. Therefore, this does not mean that technology is seen as an enemy for the visual artist. And it is even known that artists such as Pablo Picasso used photography as part of his creative process.

without forgetting to mention what it has meant for video artists, internet art artists, among many others the use of technologies for the realization of art.

Likewise, the use of technology since the early years of the twentieth century has been encouraged by the values of modernity, by confidence in the future. In those years there was very little concern about the environmental implications that would come from the creation of flying cars, the human rights implications of the establishment of fascist and communist regimes. And all this is understood by the tendency of a utopian thought, full of illusion and self-confidence. A self-confidence conferred on change, change as a currency of exchange posed as a social, cultural and artistic dynamic (Honnef, 2003).

At that time, it was a time full of great and eloquent speeches in the name of change. These speeches were in favor of the new man, in masculine, in the name of the whole of humanity circumscribed to a nation, to the hegemonic power, to the working masses and as part of the rupturing mechanics in art. Whether with the aim of social justice or not, the speeches served to promote participation in the First World War and, later, the Second World War, to justify and encourage artistic creation in the very diverse artistic movements that were aimed at establishing novelty in art. It is as if in every sense, cultural and human, society or a part of it has proposed to go against the forms of the past as a guarantee of a splendid future.

Therefore, everything that represented the past in the academic, political and social spheres was seen as in bad taste and was rejected by the artistic avant-garde (Honnef, 2003). This cultural change led to accusations and even alignments for those artists who did not move under the same interests, the same aesthetic concerns or artistic approaches. We are talking about being against everything related to the new times: the pass of the time and its relativity, the distortion of reality, the promotion of machines such as the automobile, abstraction as the original language of art for art's sake, the veneration of the working class and communist leaders, life itself as a subject for art, and so on (Honnef, 2003). This attitude would be repeated as new forms of art continue to be discovered or revised well into the twenty-first century. However, such an attitude has more to do with supply and demand, with the inherent practices of the market and its need to offer new products, trends and new work of art for new users or new collectors.

However, there were several blows that connected modern thought head-on and, which could not be replaced, prefigured by the self-confidence of humanity itself towards its own technological, creative creations and new ways of doing politics and leading nations: the destruction that the two world wars of the twentieth century meant at the psychological, social and economic level. All this mixture of challenges that ended up diminishing the utopian self-confidence of humanity, of that new man, ended up leading to increasingly individualistic desires and increasingly complex thoughts and a society more disoriented in what we call postmodern thinking (Bauman, 1992).

Postmodernity, the postmodern, is proposed by authors such as Zigmund Bauman as an opposing position to modernity. And we can define it as denial. Postmodernity denies the values of modernity. Therefore, history can no longer be understood in a linear way, instead we are standing on a shaky thought that reviews the past without respecting its original principles (Bauman, 1992). In this sense, aesthetic values are often above ethical principles. Therefore, the new self-confidence acquired by the postmodern subject is the one that dictates his own self-construction as that set of knowledge, lifestyles, ideological and cultural interests acquired in the name of his individual freedom, in the name of his own person and not in the name of humanity. The emergence of collectives, urban tribes grouped by common interests and belonging to a certain skin color and social class are configured as ghosts of the social masses of the early 20th century, are examples of cultural diversity in the midst of postmodernity as opposed to the idea of a modern hegemonic culture that was easier to understand.

All of the above allows us to more or less analyze the speech of a late modernity from the commercialization of the first PC or MAC in the eighties to the present day as part of a propaganda from the technological establishment that has been normalizing among the technological giants, their CEOs and the earlier adopters who push the rest of the people to join in the use of technological novelty. Once again, the speech of acceptance of change is promoted, in this case, in the use of artificial intelligence

Meanwhile, the socially accepted speech is that opponents of the use of artificial intelligences are understood as people who want to live in the past or who oppose change. Change being seen as a natural practice, not as a human practice. This implies, of course, a return to modern thought based on the future, in that splendid future nourished by utopia that needs to be weighed by postmodern thought that questions forms; and that he sees them, certainly, from a less naïve perspective. Let's not forget that flying cars were also proposed without contemplating the consequences of use in the past and we also know how polluted our world is.

This is not to say that the use of artificial intelligence represents a danger to society, but neither should it mean that any criticism wishes to be a setback for our society, as the early adopters wrongly point out. This means that we no longer live in a unidirectional world or with a hegemonic culture or a humanity given over to what a grand modern narrative dictates. If we learned anything from postmodernity, it is that great stories are no longer trustworthy, they are more trustworthy as we seek our individual well-being within a society.

In this sense, we cannot ignore that these speeches are established in our current era in which large companies have expanded due to the free market and the reduction of participation in the national project that are part of globalization itself. Therefore, we should not be surprised by the freedom of CEOs in the construction of a speech that represents the interests of their company and that of the early adopters but that influences our lives and the culture in which we participate, whether Western or Eastern.

If we consider the newspaper headlines, the posts of the creators of artificial intelligence themselves that position it as a technological before and after at the height of the use of the PC or Mac, of the arrival of the web, we can only critically analyze what the impact of freedom in capital letters implies by CEO Sam Altman in his speech that says that artificial intelligence*: "Generative AI has the potential to democratize access to creative tools and empower people to express themselves in new and exciting ways"* (Altman,2023).

However, this has resulted in polarization among groups of digital artists and illustrators who oppose the non-consensual use of images and styles by users of artificial intelligence to the extent that sites such as ArtStation, a well-known site for conceptual art artists and 3D modeling artists, and Devianart, another site aimed at uploading digital illustrations, were the center of heated discussions, not to mention those Instagram artists who claimed to have been plagiarized by users who took their styles.And as we mentioned in a previous section, the culture of appropriationism within the web could not fall worse into the hands of artificial intelligence users. The phrase artificial intelligence does not democratize art was a trending topic in various websites. While digital artists with a solid career and millions of followers on social media have taken legal action against the use of the Artificial intelligence.

The omission in terms of use policies seems incredible, the excessive carelessness when it comes to the training of artificial intelligences. Although the web was born to share and build through data within a strictly scientific environment, the theft of intellectual material and non-consensual exposure is part of the culture of use today and there are legal agreements that prohibit such practices Although they have wanted to popularize the idea that AI was not trained with previous creations, at first that was part of the speech, there is an open investigation into the improper use of data for its training. Social networks such as Instagram, Facebook's Meta make use of all the images stored on their sites to trained (Growcoot, 2023)

It is too much to say that an artist invents or that the human being imitates out of nothing. Like learned language, humans learn from their surroundings and from their peers. Imitation for Aristotle meant learning, when an actor enters a role or when an artist tries with all his intellect to represent that model of reality is when he learns, he learns its texture, its shape, its size, its appearance, etc (Aristotle,1996). On the other hand, taking a photograph of that model and putting it in a photocopier is not learning and there is no way to teach others any representation technique. In that sense, using a machine to copy another artist's style, not to mention without their consent, can be considered a form of expression but has nothing to do with being creative. If we consider an author in the area of industrial design such as Bruno Munari he would explain that creativity is part of a process of design methodology that by the way is quite long and that has to do with solving a problem and even creativity is not the most important part (Munari,2016). In the future, if there are AIs more capable than the current ones, the appropriation of the creative and human activities that were previously used for their training must be recognized.

The same spirit with which the use of artificial intelligence is promoted by the Open AI CEO Sam Altman, deployed politician, when he says the purpose of use is the democratization of art. Such a provocation concerns us to analyze the history of art as when the United States became the center of art in the mid-fifties of the twentieth century (Honnef, 2003). There is much we can say about this if we consider the main ideas of authors such as Clement Greenberg who proposed as a democratizing project of art: the negation of an autobiographical art in favor of a universal art that promoted art for art's sake (Greenberg,1971). If we connect both discourses we can prefigure that:

1. The development of art is not limited to digital artist who have received an academic training, if they ask it has never been, therefore it promotes its universality. Universality is not defined by the subject of the works, but by the artistic talents conferred through the use of artificial intelligences on a large number of people; it reminds us a visual artist like Joseph Beuys once said that anyone is an artist (Beuys, 1993), at least potentially, is now possible thanks to the use of artificial intelligences.
2. The discourse rests on self-confidence towards the future typical of modern utopianism, while it is also a libertarian gesture of postmodernity itself, as a global enterprise, which rests ethical responsibility on the users themselves.
3. Using artificial intelligence is related to the democratic as opposed to the authoritarian, to the unique. While works of art history exist in many cases as unique pieces.

CULTURAL ASSETS AS PART OF THE CONTENTS OF THE WEBSITE.

So far we have made clear the appropriationist behaviors on the part of artists within the art world or not and the appropriationist behaviors on the part of the rest of the users of the web. And in the midst of all this, we have a discourse that promotes the use of AIs without neglecting the polarization it has generated between digital artists and early adopters. We have also pointed out that this discourse masks a certain lack of concern and omission for what could happen, that is, the owners leave the ethical responsibility to the consideration of the users themselves. However, we also know that the results have not been as positive as his discourse, as he has annoyed web artists because they have plagiarized his work and his styles and pushed some to take legal action; an author like Noam Chomsky is not so patient to denounce that artificial intelligences are not intelligent but are machines dedicated to stealing (Chomsky,2023).

Well, in the midst of all this reflection, several questions arise, among which I consider the most important is the nature of the cultural good, because I think it is an issue that we overlook and that we forget that they are part of a real physical world. And if this should be part of and change the turn of the discussion regarding the use of AI and if this contributes to the idea of the importance of having more protection measures for digital and traditional artists.

It is therefore very important that we understand the existence of cultural asset in two categories: tangible and intangible (Linaki,2020). Regarding tangible goods, we understand that they are those that are part of the real world, that physical world in which we live. These same goods also become intangible goods when they become part of the digital world of the web in the form of information: as text, as video, as image, etc. In this way, we have handmade works (tangible goods) that are then digitized and that are now part of the web (intangible goods) (Linaki,2020). But there are also intangible assets that do not have their physical form in the real world and that were created by computer, which could have their physical presentation by means of an impression if we think of 3D illustrators and modelers and even a projection if we think of animators or filmmakers, without forgetting the musicians who still record their music on vinyl or CDs.

The point here is that cultural assets are part of communities in the physical world and the digital world, as referred to by What is different about cyberculture and Internet-based information resources from traditional materials? Essentially two kinds of information exist in the net-based environment: (a) information that we can classify as a by-product of 'real communities'; and, (b) information that can be classified as a by-product of virtual communities.

Without forgetting that these materials (cultural assets) are created by humans for the use, consumption and enjoyment of humans although, some enthusiasts attribute feelings to, AIs they are not part of humanity. Considering this, artificial intelligence not only could not and should not be considered for the protection of its creations, as the Supreme Court of Justice has already openly stated, while the creations derived from its use are not human, cannot enjoy protection as a cultural asset. This opens up other issues to discuss within the culture of appropriation, because if we consider that they cannot be protected (U.S. Supreme Court, 2021). how could a user report plagiarism to another user for using their prompt or for example that someone copies and paste a script made by AI for a movie. It is interesting to consider that although artificial intelligence is a cultural asset because it was created by humans but its creations, on the other hand, cannot be considered cultural asset.

THE CREATIVE PROCESS AND THE USE OF ARTIFICIAL INTELLIGENCE

If there is one thing that real and virtual communities have in common, it is a lack of interest, not to say a contempt for investigating how things are produced, the objects they use (Baudrillard,1968), the clothes they put on top, or at least that is how authors such as Jean Baudrillard and scholars of culture and consumption refer to it. In this sense, the culture of apropriationism (taking the cultural assets of others without their consent or in an unjustified way) without forgetting that there are also artists who use these cultural assets of others to transform and enrich them, has little positive value in the use of artificial intelligence. On the contrary, it feeds more the prejudice that things are already given, everything is already done for my personal satisfaction, as if the other did not exist: that human who dedicated his time and money to create that cultural asset; It is such a ridiculous scenario in which there are many people who believe that paying the monthly internet fee gives you the right to profit from the work of others. In that sense, there is no room for naivety, such social positions are part of the problem and contribute to the discredit of AIs in addition to not dignifying the work of the artist.

Figure 1, Figure 2 and Figure 3. As a result of the AI training the users can't imitate whatever kind of art styles o digital techniques. As an example, I have decide to write prompts for Copilot AI to try to imitate my own work of Figure 1. Figure 2 and Figure 3 are AI works, and we can say that it had made a good job in terms of color palette and design. At this point it is a good example of what we are proposing, that no one cares about how things are done or the creative process to carry them out.

Figure 1. Prompts for Copilot AI

Figure 2. AI work

Figure 3. AI work

For this reason, we explain what each of the design stages or methodologies consists of to create a design or an artistic material. Design and art are creative areas with very different and diverse objectives, but it must be very clear that, although both can be used for the entertainment or distribution of an audience, the former seeks utility and its appreciation and questioning, whether visual or not. Art and design are based on brainstorming, use of materials, sketches, proofs and corrections. At least visual artists are not thinking about users, they think more about audiences, but they do not have them in mind because they are more aware of the various current issues of our culture: violence, injustice, global warming, consumerism, the end of the world, etc.. Understanding this, the question arises as to what kind of artists are the illustrators and digital modelers who are more dedicated to the design of characters and backgrounds because it is clear that they are not visual artists and they are not graphic and industrial designers. We could say that it is a branch of artists that emerged between the film industry and prolonged animation to the video game industry. In that sense, digital artists on the web are independent artists from the same industry.

Whether they are a visual artist, a designer or an independent digital artist or not, they are all part of a real community of a virtual community and their works and creations are exposed to the use of artificial intelligence. For this reason, let us continue with what each of the stages consists of and how we could take advantage of artificial intelligences in a more appropriate way and dignifying human artistic work. We know that it is very difficult to match each creative stage of each area, so they are not going to be listed in a linear way but this will facilitate the explanation of how you can benefit from the use of AIs.

- **Brainstorming.** It consists of taking a piece of paper and listing all the possible ideas to start a project Many times, creatives are limited by their own ideas and create teams from different areas to enrich the possibilities but sometimes they remain isolated, working alone, so they use the internet. In that case, artificial intelligence can be used as a search engine to expand the topics and their possibilities.

- **Research.** It consists of investigating if there is already a project, product or art similar to the one you are planning to create, whether physical or virtual. And if so, creative investigates enough in the materials as in the results to make something that makes it different from what exists Again, using artificial intelligence can be used to speed up search time and even to generate questions for surveys or for market research of a certain age group and social class.
- **Sketching.** It consists of making exploratory drawings as visual representations of our first ideas as previsualizations of the product or art This is up to each creative, whether or not to provide the first sketches of an intellectual property to an artificial intelligence with the purpose of offering other variants, if we consider that with this we continue to train it for the use of other users.
- **Tests**. It consists of using digital and physical materials to subject our art or product to tests to evaluate its resistance, its ability to survive in the environment, its duration in physical and virtual space, its spatial execution, etc. (Munari,2016). Artificial intelligence could be used in repetitive tasks, provide possible solutions and proposals. For example, if an image requires you to have many dots distributed geometrically or in a certain order and there are thousands, or finish filling a space with the same color, copy and paste the same shape over and over again. But not that their use does not involve decision-making because that decision-making is inherent to human creation. And of course, to prohibit ourselves from the improper use of another artist's art or without their consent.
- **Corrections.** It consists of adjusting, repairing, adapting, adapting our product or art for one's own satisfaction, that of the customer, the public or the user (Munari,2016). Similar to testing and sketching, we can propose to artificial intelligence that it makes improvements to our own creations. If we are really ethical, we could limit the use between our final result and the final result that the machine suggests to us; do not forget that this also contributes to the training of AIs.

Of course, our considerations will not be so well received by a wider group, for those early adopters and their promoters of IAs, nor by those companies and industries that seek to limit human participation to minimize production costs. However, we believe that with this we contribute to the use of AIs, for a fairer use for human artists and creatives.

We have mentioned that artificial intelligences in one way or another must be limited for human benefit and to be used in repetitive tasks. In this case we would not want to be misunderstood because the truth is that behind any repetitive task there is a human who can be replaced, whether or not he is an artist. And with this we do not want to subscribe to the idea that humans must adapt to change. We know how long this speech has been coming, we know that it is not new. But we must take precautions to use it like any product. AI is a product, which must provide the opportunity for users to refrain from supporting without any type of labor stigma. Not to mention the environmental toll that the use of intelligence represents in the midst of a policy in favor of the environment, we believe that at least a less optimistic consideration is necessary.

THE PLAN B FOR ARTISTS AND CREATIVES

Without a doubt, limiting ourselves in the use of artificial intelligence is insufficient to protect digital art from its improper and non-consensual use. On the other hand, we know that it will take time for agreements to be established between the various nations in the use of AI, which also implies local agreements related to each culture. Nor does it ensure the protection of the group that we consider most vulnerable from the beginning of this document: digital artists. And to be more precise, those 2D digital artists and 3D digital artists who work, outside the industry, independently. These digital artists have been part of cyberculture for a few years now, they have become influencers and have resorted to the exposure of their image and their person in order to get more followers. This also helps to counteract and make public any disagreements or disputes between the parties.

However, if there is one thing we all agree on, it is that we cannot and should not limit the actions of others. In other words, we can choose how to use artificial intelligence so that it does not affect the economy of independent artists, but we do not have legal resources to limit the use of artificial intelligence for the rest of the people. In this case, we must thank a group of academics and scientists from the University of Chicago who developed software to limit the capabilities of AIs for their empathy towards the artistic cause. This ingenious solution prevents us from falling into disputes with other users and wasting our energies in convincing local congresses of the importance of establishing laws that regulate AIs.

The name of the tool is Glaze and it was developed by computer science engineers: Shawn Shan, Jenna Cryan, Emily Wenger, Haitao Zheng, Rana Hanocka and Ben Y. Zhao. Glaze consists of protecting one of the greatest demands on the part of artists, which the justice system itself cannot protect and the action that is most exploited by users and promoters of AIs: the art style (Shan, Cryan, Wenger, Zheng, Hanocka, & Zhao, 2023). For example, if we think of a painting to explain this situation:

We all visually recognize the unmistakable style of this piece of art. We know that it was painted by myself and the style of the art work it's an abstract painting. We also recognize that it is a painting by the quality of the brushes and the marks. If someone in any way exploits it as their own, the law protects the artist from that outrage. However, if I don't want no one to paint in my style, the law cannot protect artistic styles.

Figure 4. A painting that I made for local art exhibition as an example of abstract painting with a lot relates with the style as a main thing

And it is precisely what makes an artist famous is their style, digital artists thanks to their style have gained many followers. This is why we consider Glaze to be a marvel because it protects artists where the other parties left them defenseless.

CONCLUSION

None of the sentences of this text were intended to disqualify the use of artificial intelligences nor have we dedicated a single paragraph to statistically address the economic risk it represents for independent digital artists. Instead, we have distributed responsibilities among the users themselves, the speeches and behaviors that are part of our cyberculture. We have also agreed that artificial intelligences can be used appropriately so as not to hinder creative activities in a way that is fair to all parties.

On the one hand, we believe that leading authorities can do more to protect cultural property from artificial intelligence. On the other hand, CEO's do what is expected, they promote their latest developments as if the history of humanity depended on it, while early adopters feel that FOMO (Fear Of missing Out) is a race in which they do not want to be left behind. The result of all this situation has been that the authorities are lagging behind to speak out for the good of all parties. We can also realize that despite what we thought, there is a very poor culture regarding copyright, although we should not limit human technological development, the authorities have done very little to protect the use of copyrighted works that were used for the initial training of IAs. Even the information we have, it is incredible to us that it is only thanks to a group of engineers from the University of Chicago that digital artists feel that they are doing something to protect their works, their economy and their prestige. And contrary to what is thought: no, AI is not just a camera. And no, we are not still living in the XIX century, our world is more complex.

REFERENCES

Altman, S. (2023). Generative AI has the potential to democratize access to creative tools and empower people to express themselves in new and exciting ways. Retrieved from https://openai.com

Anantrasirichai, N., & Bull, D. (2022). Artificial intelligence in the creative industries: A review. *Artificial Intelligence Review, 55*(1), 589–656. DOI: 10.1007/s10462-021-10039-7

Aristotle, . (1996). *Poetics* (Butcher, S. H., Trans.). Dover Publications. (Original work published 335 BCE)

Baudrillard, J. (1968). *El sistema de los objetos.* Siglo XXI Editores.

Bauman, Z. (1992). *Intimations of Postmodernity.* Routledge.

Beuys, J. (1993). *Joseph Beuys: The Revolution is Us.* Thames & Hudson.

Bowley, G. (2015, January 22). The verdict that flies in the face of art history: Luc Tuymans guilty of copyright infringement. The Art Newspaper. Retrieved from https://www.theartnewspaper.com/news/the-verdict-that-flies-in-the-face-of-art-history-luc-tuymans-guilty-of-copyright-infringement

Brown, M. (2019). Richard Prince ordered to pay damages to photographers in copyright infringement lawsuits over Instagram portraits. The Art Newspaper. Retrieved from https://www.theartnewspaper.com/news/richard-prince-ordered-to-pay-damages-to-photographers-in-copyright-infringement-lawsuits-over-instagram-portraits

Buchloh, B. H. D. (2015). *Formalism and Historicity: Models and Methods in Twentieth-Century Art.* MIT Press.

Chikarkova, M., & Fedkovych, Y. (2023). Artificial intelligence and digital art: current state and development prospects. Retrieved from https://www.researchgate.net/publication/376881413

Chomsky, N. (2023). The False Promise of ChatGPT. The New York Times. Retrieved from https://www.nytimes.com/2023/03/08/opinion/noam-chomsky-chatgpt.html

Christou, E. (2017). The digital time of internet art. Retrieved from https://www.researchgate.net/publication/320623021

Foster, H. (Ed.). (1983). *The Anti-Aesthetic: Essays on Postmodern Culture.* New Press.

Francisco, J., Santilli, M., & Demetrios, D. (2024). Rare Sam Altman Interview, Constructing "Dreams" with GenAI, and Sora & Marvel. AI Minds. Retrieved from https://www.aiminds.com/p/ai-art-dreams-and-hollywood

Gaiki. (2020, 13 de febrero). Los algoritmos de la creatividad. Retrieved from https://gaiki.org/los-algoritmos-de-la-creatividad/#:~:text=El%20algoritmo%20de%20la%20Creatividad%20por%20Requerimiento%20es%20aquel%20que,para%20resolver%20el%20problema%20dado

Gammel, I. (2002). *Baroness Elsa: Gender, Dada, and Everyday Modernity: A Cultural Biography.* MIT Press. DOI: 10.7551/mitpress/1517.001.0001

Greenberg, C. (1971). *Art and Culture: Critical Essays.* Beacon Press.

Growcoot, M. (2023). Meta is Using Your Instagram Photos to Train its 'Amazing' AI Image Generator. Retrieved from https://petapixel.com/2023/03/15/meta-is-using-your-instagram-photos-to-train-its-amazing-ai-image-generator/

Guasch, A. M. (2009). *El arte del siglo XX en sus exposiciones: 1945-2007*. Akal.

Harris, J. (2008). *American Art: A Cultural History*. Oxford University Press.

Honnef, K. (2003). *Arte Del Siglo XX*. Editorial Océano de México.

Jones, L. (2023). OpenAI and Microsoft accused of scraping and exploiting private data for AI training. WinBuzzer. Retrieved from https://winbuzzer.com/2023/06/29/openai-accused-of-scraping-and-exploiting-private-data-for-ai-training-xcxwbn/

Jones, R. (2023). How Sam Altman got it wrong on a key part of AI: 'Creativity has been easier for AI than people thought'. Fortune. Retrieved from https://finance.yahoo.com/news/sam-altman-got-wrong-key-210436142.htm

Linaki, E. (2020). The definition of intangible and tangible cultural assets in digital technology. Retrieved from https://www.researchgate.net/publication/344396747

Morales, J. (2025). Elon Musk's xAI launches Grok 3, expanding AI capabilities with deep search and reasoning. Computerworld. Retrieved from https://www.computerworld.com/article/3826765/musks-xai-launches-grok-3-expanding-ai-capabilities-with-deep-search-and-reasoning.html

Munari, B. (2016). *Cómo nacen los objetos?: Apuntes para una metodología proyectual*. Editorial GG.

Schröter, J. (2019). Artificial Intelligence and the Democratization of Art. In A. Sudmann (Ed.), *The democratization of artificial intelligence. Net politics in the era of learning algorithms* (pp. 297-311). transcript. Retrieved from https://mediarep.org/server/api/core/bitstreams/1c616f3d-1cfa-4d99-87cc-d0282770c540/content

Shawn, S., Cryan, J., Wenger, E., Zheng, H., Hanocka, R., & Zhao, B. Y. (2023). Glaze: Protecting Artists from Style Mimicry by Text-to-Image Models. Retrieved from https://glaze.cs.uchicago.edu

UNESCO. (2021). Recommendation on the Ethics of Artificial Intelligence. Retrieved from https://unesdoc.unesco.org/ark:/48223/pf0000380455

U.S. Department of Commerce. (2024). Japan - Protecting Intellectual Property. U.S. Department of Commerce. Retrieved from https://www.trade.gov/country-commercial-guides/japan-protecting-intellectual-property

U.S. Supreme Court. (2021). Supreme Court ruling on copyright eligibility for AI-generated works. Retrieved from https://www.supremecourt.gov

ADDITIONAL READING

Cota-Rivera, E. I., Correa, M. E. G., Marín, L. A. B., Montenegro, M. Y. M., Herrera, A. M., & Martinez, M. A. A. M. (1970, January 1). *Transforming education with the power of Artificial Intelligence: Case Studies*. IGI Global.

KEY TERMS AND DEFINITIONS

Art: Is a broad term encompassing a variety of creative expressions, including visual, auditory, and performing arts. It can be a product of human imagination and skill, often intended to evoke emotion or thought.

Digital Art: Refers to artwork created using digital technology such as computers, tablets, or smartphones. It can include various forms, including digital painting, graphic design, animation, and interactive art.

Cultural Assets: Are tangible and intangible resources that contribute to a society's cultural identity and heritage.

Cultural Industry: Refers to economic activities related to the production, distribution, and consumption of cultural products and services.

Creative Industry: Are a broader category that encompasses a range of economic activities involving creativity, innovation, and intellectual property. They often include the cultural industries, as well as sectors like design, architecture, and advertising.

Chapter 19
Ethical Implications of Students Using Artificial Intelligence

Ma. Rocío Mendoza-Belmonte
https://orcid.org/0009-0001-6671-8969
Universidad Politécnica de Baja California, Mexico

Laura Edith Cabanillas-Vázquez
Universidad Politécnica de Baja California, Mexico

Arturo López-Velasco
https://orcid.org/0009-0009-4912-7735
Universidad Politécnica de Baja California, Mexico

Marcela Barreras-Hernández
Universidad Politécnica de Baja California, Mexico

ABSTRACT

Currently, artificial intelligence (AI) has taken on great relevance in the daily lives of people, companies, and university students. This tool seeks to process large amounts of data and information to perform various tasks autonomously such as creating reports and/or reports, analyzing and classifying information. The purpose of this tool is that the AI reduce data processing time and let humans be the decision makers. In education, the use of AI in carrying out academic tasks or activities is increasingly common. This raises questions and discussions about the use of AI, to what level its use is ethical or not. Should it be regulated? Should we train teachers and students to obtain the greatest benefit from the use of AI? In this chapter, the authors analyze the use, the lack of regulation of the use of AI, the challenges that we will face in the future, and how the application of ethics can stop the indiscriminate use of this tool.

INTRODUCTION

Currently the University students have access to new tools to do their class assignments quickly and creatively without much effort, to solve mathematical problems, to do quality research in less time, etc. Shabbir (2018), in his book "AI and its role in the near future", tells us that this tool can improve

DOI: 10.4018/979-8-3693-6910-4.ch019

learning, reasoning, autonomous work and deep analysis in topics of interest to the user, allowing them to improve their academic performance.

Another definition of AI mentions the Spanish Program Recovery, Transformation and Resilience Plan "Artificial intelligence is a field of computing that focuses on creating systems that can perform tasks that normally require human intelligence, such as learning, reasoning and perception" (España, 2023). As we see, AI helps us process thousands of data automatically to make correct decisions in large companies. It is used to increase productivity, efficiency, learn about customer opinions, learn about new trends or needs, in the educational field it is not the exception is that AI is used to improve learning, accelerate academic research, bring closer ties of collaboration in other languages, bridge distances between students, teachers and researchers from anywhere in the world, among other new functions that emerge according to the needs or creativity of each user. However, Ramos (2024) tells us at the Global Forum on Ethics in AI "This tool creates a new challenge, that of regulating the use of AI, since it requires mutual learning based on the lessons and good practices that emerge from different jurisdictions" around the world.

This new way of doing things through AI is not one hundred percent regulated, so there is a big gap between the ways of working and the user's ethics, which can cause problems, since each student, professor or university acts based on their interests and may engage in unethical practices, putting personal information collected at risk, discrimination in the use of technology since not all of us have access to the internet or an electronic device, among other challenges. The objective of this delivery is to be able to understand what AI is, how it originated, its current academic use, the regulations that will mark the standards of conduct regarding the use of this tool and the ethics in its application, use and protection of information.

ETHICS AND AI

As mentioned above, AI is a computer tool that helps people process data in increasingly shorter times, resulting in automatic problem solving, thereby avoiding wasted time that helps us focus on other activities such as ever more relevant. With AI we can carry out any university activity such as texts, essays, research, presentations and an endless number of tasks within the university work, with the aim of saving time and improving the quality of the work, however, this is where we question the ethical use of AI. The increase in the use of this tool has contributed to students making less and less effort to carry out their tasks or activities that were traditionally done, so these days the effort, the curiosity of the student to investigate and learn about a topic has been diminished, the student's reasoning with respect to solving problems of all kinds such as social problems, mathematical problems, economic problems, labor problems has been replaced by AI. However, this tool has its positive side in its use and application, with the use of AI, work on translating books, research, among others, for example, is faster, breaking down communication and collaboration barriers with other peers. Other countries and languages, these jobs are carried out with fewer personnel and in a short time. "While AI powered language translation tools break down language barriers, facilitating collaboration and global understanding. AI-guided interventions can identify and support students with learning difficulties, making education more inclusive and equitable" (Cota-Rivera et al. 2024). This is what is expected from the use of auxiliary AI in quality educational processes, reducing time, distances, but always prevailing ethics as a standard of conduct in the use of this tool.

ORIGIN OF ARTIFICIAL INTELLIGENCE

The origin of AI is very uncertain, as there were several events in the last century that gave shape to what we now call AI. One of the first records of the concept was in the thirties, when the English mathematician Alan Turing proposed the idea of creating a machine that could solve problems by itself. In 1950, Alan Turing asked himself a question: "Can machines think? In fact, this simple question and the Turing Test would change the world and lay the foundation for AI (Scientist, 2022). More than seventy years have passed since Turing's famous 1950 proposal of an imitation game or test of AI, and interpretations of it still vary widely. We still ask the question: Can machines think? (Gonçalves, 2023).

A second known event where the concept of AI first appeared was at a conference that brought together the best scientists of the time. "In 1956 during the Dartmouth Conference, a historic event that brought together some of the best scientists of the time to discuss the possibility of creating a machine that could think like a human being" (España, 2023). In the 1950s and 1960s, the first AI programs were developed, the most famous of which was Claude Shannon's chess program. The first experimental test of AI was in the city of Los Alamos in 1956 with the experiment called Maniac I, which consisted of a chess game, man against machine, according to Alberto Portillo (2019). Another event that marked the beginning of AI was the "Geometry Theorem Prover" by Allen Newell and Herbert Simon. "In the sixties, they gave rise to the computer model theory of the General Problem Solver; a human information processing system with the ability to learn and adapt to the demands of a problematic context, through the generation of decision-oriented programs and finding solutions from their interaction with a wide set of tasks. Since the nineties and until today, AI has evolved and to know how it works, it is divided into two aspects, one called Machine Learning and the other called Deep Learning. Let's review each of them for further study. A review of machine learning and deep learning applications (Shinde, 2018).

- **Machine Learning** is a discipline of AI that provides machines with the ability to learn like humans and execute predictive analyzes based on certain mathematical algorithms (APD, 2023).

This type of AI has many uses or applications in different disciplines of daily life, of which we can mention some such as the use of autonomous vehicles, which have the purpose of allowing cars to drive themselves without the need for a driver, in genetics. To detect possible diseases, detection of malicious software to prevent identity theft or fraud in the banking system, voice recognition and natural language processing as we know them today as SIRI or Alexa, face detection to identify people for various uses, generation of content for all kinds of presentations or investigations such as the GPT chat, personalized content suggestions widely used on platforms such as Netflix or Spotify that, according to history, suggest a movie, series or song. In education, this discipline of AI is used in the development of academic research, use of technology such as GPT chat to carry out activities related to the subjects taught at the higher level, obtain data and information from students and it processes them in an almost instantaneous, among other benefits in the use of machine learning.

- **Deep Learning,** in the strictest sense, would be translated as 'deep learning or machine learning,' meaning that within Deep Learning there is a group of algorithms that imitate the human brain 'learning' to recognize repetition patterns, specific words, behaviors. Frequent (APD,2018). According to the Royal Spanish Academy, these algorithms are defined as an "ordered and finite set of operations that allows finding the solution to a problem" (Royal Spanish Academy, 2023). In other words, this learning tries to imitate the neural work of the brain.

A very clear example of how Deep Learning works is the case of Facebook, when we are talking about something or we write on our wall, advertising appears for articles we were talking about, or when we write on our smartphone and suggested words appear for the word used in it that moment. In education, it allows us to know, according to scenarios, the possible candidates to enter, where they are located, what their social and economic position is, what are the areas where the graduates work, what positions they have, what would be the trend in the future for the university, among other topics of interest for education.

To understand these concepts, we will see the difference between a learning machine and deep machine see Figure 1.

Figure 1. Differences between machine learning and deep learning

Machine Learning	Deep Learning
• Subfield of artificial intelligence that allows machines to learn from data without being explicitly programmed. • Uses algorithms to identify patterns in data and make predictions. • Requires humans to design features and select algorithms. • Generally requires less data and computational power than deep learning. • Examples: linear regression, support vector machines, decision trees.	• Subfield of machine learning that uses artificial neural networks with multiple layers. • Learns complex representations of data in a hierarchical way. • It can learn features automatically without human intervention. • Generally requires large amounts of data and computing power. • Examples: convolutional neural networks (CNN), recurrent neural networks (RNN).

Nowadays, the use of AI is increasingly common since it can process thousands of data in seconds, which facilitates the rational use of information, makes our lives easier since it can identify us in seconds our human voice or face, we can investigate any topic in minutes, it can store a large amount of data which makes it easier for us to use and process it, among other things.

AI has become popular in practically all business areas, in the financial area, in marketing and in educational and research areas, since with this tool data can be processed, stored and analyzed in less time.

Today the use of this tool has increased in educational establishments at almost all levels from basic education to higher levels, which has resulted in its indiscriminate and unethical use in carrying out educational activities.

In the field of education, the use of AI has increased, Jauregui (2024), tells us in areas such as intelligent tutoring, tutoring, learning tool/companion and advisor in policy formulation.

Ethics plays an important role in the use of AI since it involves, in the first instance, personal data, teaching processes, learning processes, activities related to knowledge which must be handled ethically to avoid misuse of this tool and that may somehow fall into excesses such as discrimination, disclosure of personal data, security in the handling of this information.

However, students and teachers in the constant use of AI in educational establishments face new challenges in ethical matters, new dilemmas in the use and behaviors of this computer tool, Jauregui (2024), comments that based on the United Nations Educational, Scientific and Cultural Organization (UNESCO) report related to Al, she points out six challenges to avoid falling into unethical practices:

1. Make public policies that are inclusive and equitable: to date there is no comprehensive regulation regarding the use of AI, this regulation must be designed for all users regardless of sex, sexual preferences, religious beliefs or policies, race or ethnicity, in short, without difference or discrimination.
2. Prepare teachers to teach with AI: the same public policies must integrate teachers in training in the proper use of this tool with the aim of giving it effective, formal and ethical use.
3. Training AI in education, with regulation and training for teachers, they can download information on the use of AI to all students for its proper use, which is strictly to facilitate learning and research.
4. Develop comprehensive data systems with quality: data systems must include the entire society where they are used; they must not discriminate against marginalized communities with little access to technologies or even the Internet.
5. Ensure ethics and transparency in the collection, use and dissemination of data: ethics in the use of AI must be monitored and regulated so that high school and higher education students do not misuse the data collected, disclosure of said information. Remember that ethics is the application of standards of conduct for good coexistence and the well-being of all people.
6. Avoid discrimination, inequality, xenophobia, risks to privacy, identity, integrity, etc.: this challenge is one of the most important since it must be regulated very carefully with values and ethics so as not to incur in acts considered unethical, discriminatory, management of inappropriate personal information, avoid xenophobic acts, always taking care of the integrity of the student body, the teachers and the institution itself. Include the entire community in the use of technology that is available to everyone (Jauregui, 2024).

In conclusion, the use of AI in education must be regulated with ethical policies in which the privacy and integrity of the user, their identity, the new forms of teaching and learning of the student and using this tool safely can be safeguarded. in which teachers support this tool to facilitate its use but in a responsible and ethical manner in order to make the most of this powerful tool that is revolutionizing the way of doing things, priority must be given to learning and teaching in order to be better students and teachers. Melissa Guerra Jauregui tells us, "public policies for AI must be developed in a responsible and transparent manner, therefore, the priority is not the technical aspects but the ethical implications of its use" (Jauregui, 2024).

THE USE OF AI IN TEACHING AND LEARNING PROCESSES

In the teaching-learning process, according to the new needs of this world in constant transformation, the student has been placed as the center of the educational process. The objective is for the student to acquire the necessary competencies for life. The teacher becomes the facilitator of this knowledge so that the student acquires it, becoming a facilitator and guide for learning. Therefore, it is proposed that the use of AI in education can be useful, since it can be used at any time inside and outside the classroom through an electronic device such as cellular, tablets and computer, which are increasingly present in the hands of people. This has allowed the expansion of AI into the daily lives of both children and adults. However, as everything has its advantages and disadvantages, this section will present its main aspects.

It has changed the trends of education and schooling by introducing personalization, where educating someone depends upon the knowledge level of the student, his/her speed of learning, and goals to be obtained from education or the course. Unlike the traditional manner, the learning histories of each student are analyzed continuously to assess the weaknesses and offer courses of interest and improvement (Ahmad et al., 2021).

AI in education is driven by the principle of novelty being that it is very attractive to use because it is already present in various aspects of life such as smartphones and computers that has allowed simplifying the use of this tool that is AI as in the use of photography, creating images, tables, improving presentations, consulting information, etc. Technologies play a fundamental role in achieving this objective. In addition, in companies its use is more common, and they are betting on investing budget in the incorporation of this type of tool for productivity. Therefore, training centers and universities should incorporate the use of AI in various fields of knowledge into their curricula.

It is claimed that AI is playing an increasing role in the research of educational technology, management sciences and operational research areas. Intelligence is commonly considered as the ability to collect knowledge to solve complex problems. In the near future intelligent machines will replace human capabilities in many areas (Verma, 2018).

Nowadays with the use of Information and Communication Technologies (ICT) in Education it is increasingly common to use them in the teaching-learning process from basic to higher levels. The demand for their use has led teachers to incorporate them in their classes to be at the forefront and generate attractive and meaningful learning for new generations of students who are increasingly involved with technology at younger ages. In fact, it is a reality, as students are active beings of their learning, they by their own initiative look for AI tools to simplify the realization of presentations, documents, videos or any type of task assigned within the teaching-learning process. AI in education is very attractive for students as there are options that allow them to learn through games, which makes it a novel tool for meaningful learning. In addition, its functionality to prepare academic content remotely and face-to-face for educators has made it a viable option to include in the academic planning of teachers, as it allows to speed up the preparation and application times of thematic content.

Applications of ICT and AI in Modern Education. Representation of the main areas where artificial intelligence and information technologies impact education see Figure 2.

Figure 2. How ICT and AI are implemented in educational processes, highlighting their impact on teaching and learning

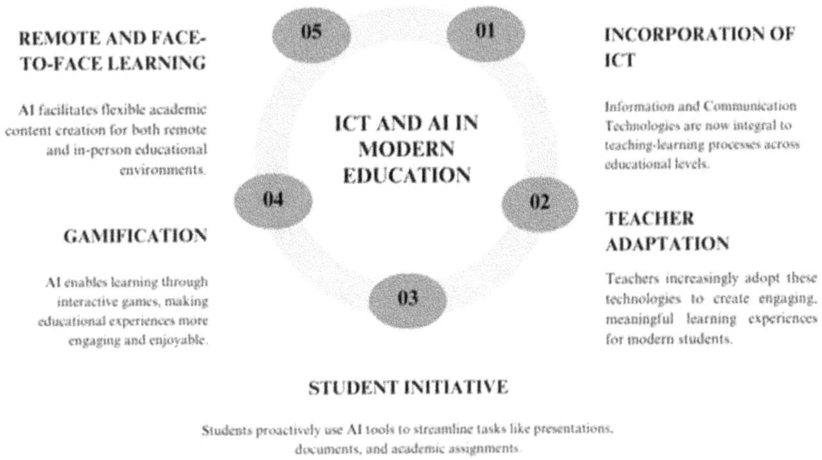

For example, students can access learning from a location where they have no access to physical classrooms in a real environment, they merely need the required AI technology to teach them or communicate according to the teaching or learning objectives. Distance and geography are no more hurdles in education if AI is being used. Similarly, for schools and colleges, AI can make it easy to register as many students as they need regardless of their location, etc. (Ahmad et al., 2021).

There is some mistrust about the use that can be given to AI in education with respect to how the role of the teacher will be when using this tool, it is being generalized that it will partly replace the relevance of this in the teaching-learning process. This may have negative repercussions in its use due to the fear of being replaced in the near future. The use of AI is becoming more and more common in the educational field where new generations of students are becoming more familiar with the tool and see it as a viable option that facilitates their academic activities. Teachers have an infinite number of tools at their disposal to incorporate into their academic planning. The use of AI in education does not replace the role of the teacher in the educational process because it is only a utility that facilitates the teaching-learning process.

Teachers who are starting to use AI in education may at first feel a little lost in choosing which tools with AI can be used since it is increasingly present in everyday life and there are many options to consider in the market, so it can be based on testing tools either by recommendation by other teachers, articles read on the network and video tutorials found on video platforms. Some tools are free and others are paid, which allows the user to decide according to their features, accessibility and costs to choose the most appropriate to their needs. In some countries the economic resources for education are limited so teachers have to allocate their own budget to acquire materials and supplies for their classes, being the AI considering the options that have cost, teachers who are attracted to implement the latest developments in AI in education that are generally paid will have to invest from their pocket, in some cases the AI tools have payment options at lower cost for students or educational institutions and even offer as part of corporate social responsibility free premium options for education. However, it should not be lost sight of the fact that these companies are for-profit and continue to be a business that is becoming more profitable every day and there are customers who pay for their use, which is why the range of AI that are free is limited.

Challenges in Implementing AI in Education. A diagram showing the main challenges teachers face when integrating artificial intelligence into their teaching, see Figure 3.

Figure 3. Barriers such as tool selection, resource exploration, and budget constraints are identified, among others

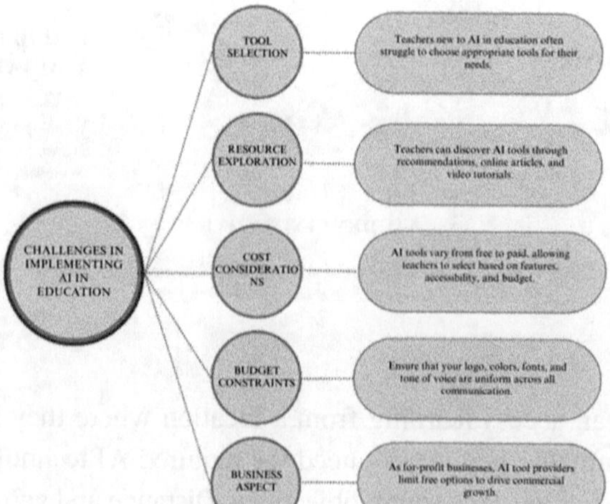

Teachers should be prepared to incorporate AI in their classes, avoiding resistance to change and adapting to its use because it can become a useful tool, the directors of educational institutions should provide as an agenda within the continuous training. Even teachers can look for tutorials on the Internet or courses that can be offered on the network on how to use AI in education. In addition, its use worldwide is growing steadily as international organizations such as the United Nations Educational, Scientific and Cultural Organization (UNESCO), known in short as UNESCO encourages its use in schools around the world.

UNESCO encourages its use in the field of education. The advent of AI is raising a fundamental question regarding how education can prepare people to live and work effectively with AI and to build sustainable societies (UNESCO, 2019).

We live in a changing and increasingly demanding world with the demands that society requires for the betterment of our society, so the main role of education is to train individuals who are increasingly committed to the common good and environmentally conscious. It is not unknown that world organizations are concerned about the future and that the resources we have at our disposal are sufficient for the new generations that are and will be on their way. Education plays a fundamental role, which is why the 2030 Agenda of the United Nations (UN) proposes 17 Sustainable Development Goals (SDGs), where the 4th Goal is a quality education.

With regard to education and learning, there is consensus that AI has a dual impact, with advantages and risks. With a view to rethinking education systems for the AI era, policymakers will need to analyze the benefits and potential risks of AI in the context of achieving Sustainable Development Goal 4 (SDG 4) Education 2030 agenda on ensuring inclusive and equitable quality education and promoting lifelong learning opportunities (UNESCO, 2019).

So what does AI have to do with SDG 4? Being a tool that is having an impact on its use on individuals in aspects of their lives, so it should be sought to have a favorable impact on society. A changing society that requires developing skills and abilities for the improvement of a sustainable society.

Forming active and competent citizens with technologies requires not only the skills to handle different digital tools. The fundamental challenge is to generate critical thinking, which allows analyzing and evaluating information, promoting an ethical stance on the use of technology, as well as stimulating creative processes through the generation of new ideas (*Universidad ORT Uruguay*, s.f.).

It is important that all participants in the educational process are familiar with the growing use of AI and that it is part of the requirements demanded by society, it can no longer be seen as a simple fashion, but as a tool that comes to simplify social, educational and productive aspects.

Developing Critical Thinking with AI. The diagram illustrates key elements in fostering critical thinking through AI, including digital tool handling, critical thinking, ethical stance, and creative processes, see Figure 4.

Figure 4. The image highlights the importance of equipping individuals with digital skills, promoting ethical use of technology, and encouraging creative and analytical thinking

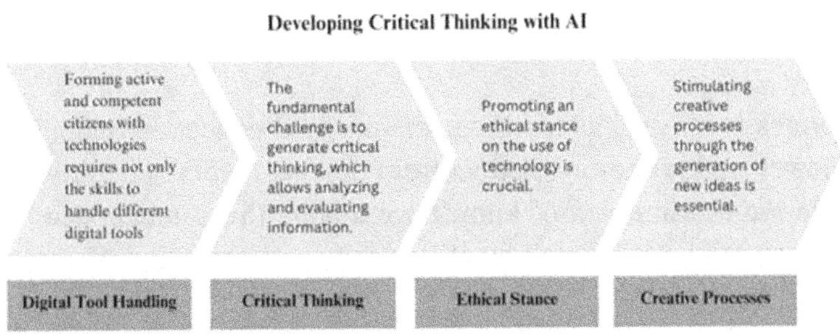

Since UNESCO adopted the 2030, Agenda for Sustainable Development, whose fundamental objectives include ensuring equitable access to quality education worldwide SDG4, technological development and evolution has changed the role of teachers in the transmission of knowledge (Vivar, J. M. F., & Peñalvo, F. J. G, 2023).

The use of AI in education can be a disadvantage that is not inclusive and equitable for many because its use also involves the economic and technological aspect where not everyone will have access to these tools, because there is no access to internet and equipment for its operation. This requires planning and allocation of the necessary resources by those responsible or educational authorities to bring AI to the classroom.

It is essential to ensure that the integration of AI in higher education is inclusive and equitable, ensuring that all our students have the opportunity to access and take advantage of the benefits of technology (Vera, F., 2023).

What is better within the classroom under the supervision of teachers to teach the student the proper use, with the aim of making it a fruitful tool for the teaching-learning process.

In particular, it is essential that humans are protected from becoming victims of AI tools, so we must understand that AI should be used to augment and amplify human capabilities, not to replace them. And this understanding begins in education (Vivar, J. M. F., & Peñalvo, F. J. G, 2023).

There are also the ethical implications with the use of AI in education since it can lend itself to the storage of information and private data of users, in this case teachers and students, which can be an attack on privacy and the access that private companies can have of information to a database that can be used by third parties for profit-making purposes.

The integration of AI in higher education offers a wide range of opportunities to improve teaching and learning, as well as to optimize institutional management. However, it also poses challenges and ethical dilemmas that must be carefully addressed (Vera F., 2023).

The use of AI to automate routine administrative tasks, diagnose student competencies, and provide learning content and feedback tailored to individual progress (Vivar, J. M. M. F., & Peñalvo, F. J. G, 2023).

What may worry teachers the most is that they overexploit the use of AI when performing their learning activities, or even expect them to perform the activity completely through the use of AI, limiting the learning of students because they do not apply their knowledge or expected learning, apart from the ethical implications that could arise similar to the fact that it is literally "copying" a work not properly done by the student.

Thus is effective in educational technology to make the teaching learning process more effective and concrete with the help of various AI teaching techniques (Verma, M., 2018).

The teacher's duty is to adequately guide their learners the use of AI within the classroom, so it is important as previously mentioned that the teacher is familiar with its use, avoiding resistance to change and seeing it as a utility.

However, many professions, such as teaching, training and education undoubtedly require the human mind, not AI. However, this does not detract from the fact that the cognitive capacity of teachers, the continuous flow in the transmission of knowledge among their students and the tasks of content management and creation can be supported by algorithmic assistants and AI tools (Vivar, J. M. F., & Peñalvo, F. J. G., 2023).

The collection and use of students' and scholars' personal data, such as their online browsing patterns or academic performance, raises ethical and legal issues, such as privacy protection, data ownership, and informed consent. Therefore, it is critical to establish appropriate policies and regulations to ensure the security and privacy of personal data in the integration of AI in higher education (Vera, 2023).

AI's role in education is through its intelligent methods for tutoring, communication, analysis, assessment and evaluation of the student or learner in addition to supervision, process control and optimization. AI technology performs all the tasks needed for a teacher and a student. The technology must be capable of communication such that the language etc. from the user side is understandable and makes sense (Ahmad et al., 2021).

It is interesting to study or repercussions of AI in the use in education as it is increasingly present in this field, it may prove to be a useful tool for the teaching-learning process in educational institutions. At the time the use of ICT impacted to adapt to the changes that society demanded, the use of technological tools for everyday life and work. Therefore, it is to be assumed that AI will be increasingly present in the various aspects of our lives and it is important to know how it works correctly. Those involved in education from the governmental side should allocate relevant resources for the incorporation of AI because as previously mentioned they are tools that require economic costs to access them, agreements can be sought with AI companies to agree to provide them at more moderate costs and even provide them for free as part of corporate social responsibility. Educators should get involved with the familiarization of their use to be able to adapt to incorporate them in their teaching process, especially to overcome resistance to change, and students should make responsible use of them and be aware of the proper use of AI tools without neglecting their learning processes and engaging in unethical practices such as allowing AI alone to perform an activity without using their knowledge. This is one of the most discussed dilemmas due to resistance to change or fear of using AI in education, so teachers must guide the learner in its correct use.

ETHICAL IMPLICATIONS OF THE USE OF AI IN UNIVERSITY HIGHER EDUCATION AND ITS REGULATORY FRAMEWORK

As discussed in the background of this chapter, the history of AI dates back to the late nineteenth century, with the mathematician Ada Lovelace, who was a pioneer in computer science, she programmed the first algorithm to be processed by a machine; but it was not until the middle of the last century that the concept of AI was formalized as a new field of scientific study, and this was thanks to the contributions of Lovelace, who made visible the work of women in science, technology, engineering and mathematics. In addition to other great contributions to science, such as the example of the visionary names of Alan Newell and Hebert Simon, who published the first AI program "Logic Theory Machine", helping to devise a heuristic approach that has influenced computer science in various aspects of human problem solving. Already in the first decade of this century, advances in specialized hardware accelerated the training and performance of neural networks by reducing energy consumption, in addition to the consumption of open data available online and low-cost services, which boosted the development of AI internationally (Abeliuk et al., 2021).

AI has undergone a dizzying development over the past few decades, making its way into higher education. Since the 1960s, scientists have been developing teaching systems based on AI, which soon encountered several difficulties, such as the lack of flexibility to adapt to different learning styles, as well as the lack of emotional capacity in the process. Currently, AI has a greater reach and impact among students, creating new forms of social and virtual interaction in university teaching and learning environments. However, today we are witnessing a major problem in such environments, such as plagiarism, omission of citations, copying, pasting, and such actions contribute to weaken meaningful learning and create a digital dependency of students on technology, hindering the development of critical thinking skills and problem solving (Moran-Ortega et al., 2024).

The phenomenon associated with AI and its disruptive development is creating a new social order in which this technology is empowering in the political, economic and cultural spheres. It is also worth mentioning that this event is unfolding with little regulation of its use, since AI is also present in all dimensions related to behavior, thinking, reasoning, and even the capacity of the human brain; we are facing what authors usually call a quantum leap, a phenomenon that puts us in a state of vulnerability to technological advances. For this reason, respect for human dignity, decision-making capacity and user autonomy should be guaranteed. Let us be aware that the main characteristic of AI is its relatively autonomous operation, which can lead to confusion among social actors who think that it is correct to give AI the power to make decisions, and this forces us to ask ourselves what should be done in such a situation, since in the international legal context there is almost no legal regulation, let alone ethical or moral limitations. Therefore, we are irrevocably heading towards a future in which AI will be the protagonist and decision- maker, shaking the foundations of humanity and the values that govern us today, and facing imposing ethical challenges such as the impact on human autonomy, socio- economic inequality, privacy, data protection, legal and regulatory challenges, changes in professional ethics, and technological dependence, to name a few (Dávila et al., 2023).

Notwithstanding the above, it is imperative to recognize that AI has positively influenced our students by making them more insightful and curious as technology plays a very important role. According to UNESCO and the SDG 4, which states that to address the major challenges in education today, AI has the potential to address these formidable challenges as well as the progress of education and therefore the fulfillment of this goal, which aims to seek and ensure inclusive, equitable and quality education to promote learning opportunities throughout life. And therefore, the care of thinking, analysis and the reflective nature of the university student should be privileged as a valuable tool, making positive use of the innumerable benefits that AI offers for the generation of scientific knowledge, but of course supported by ethics, process in which the human being takes responsibility for the consequences of his decisions, despite the prejudices that may arise along the way, since the use of AI implies being aware that ethics is not a new complement to science, but is rooted in the generation of scientific knowledge, it is natural to clarify the moral values and social responsibility of science, its public discussion, its governance, and its use and application. The ethical researcher always seeks to maximize good outcomes and minimize risks and harms as part of an intelligent and sensitive planning, treatment and evaluation process (Labrador-Fernandez et al., 2023).

Thus, we can see that AI has arrived to impact our society with new challenges and incidents derived from its use, and this movement has alerted international communities to generate a relevant and appropriate regulatory framework for the reality we live in. Technologies pose new challenges, and even if specific legal frameworks already exist in some countries, they are very restrictive and insufficient in the face of the growing and overflowing technological expansion; therefore, governments are overwhelmed and powerless in the face of continuous changes in technology, and to complicate matters further, it is known that for legal logic, AI is not a natural thing, but is the product of human intelligence created to perform some kind of function, this makes it a non-biological artifact and cannot opt for another purpose for which it was programmed, that is, AI can be considered an artifact according to traditional regulations.

Already some countries have regulations in place for the safety and protection of people in relation to legal issues around the use of AI, but the reality is that enforcement and legal compliance is proving to be very complicated and this is largely because the nature of AI is such a dynamic form that one solution could be to have very active regulations, however, the good development and application of the instruments of law have the potential to guide the development of AI for the social good, ethics and legal norms, but unfortunately in the current reality the rigid and traditional structure of legal systems and families, do not solve the regulatory challenges in the field of AI. It is possible to recognize that science seeks the truth, while legal regulation seeks the ought to be; then the main challenge is to seek a balance that does not stifle technological progress and that in turn evolves in harmony with the legal and ethical balance of its application, creating a legal value that comes to guarantee the use of AI as a new dynamic form of regulation (Llamas et al., 2022).

In a relatively short period of time, AI has generated an uninterrupted controversy in the field of education, especially in university educational spaces, but in an ambivalent way, it is assumed that AI could be an effective tool to achieve knowledge, as long as it does not become a carrier of information and excessive agglomeration of data. Since a positive aspect of its use would be the reduction of social gaps in education, it becomes rigorous and a priority to carry out a transversal literacy of these tools in educational spaces, to understand their application and usefulness, strengthening scientific integrity, ethical efforts in research, as well as respect for the rights of all those involved. AI could be subordinated, or in its case homologated, to the ethical and legal system of the international community and the European Union, since these organizations are at the forefront in this field, as can be seen in the different regulatory frameworks and normative systems. In this way, it can be considered that AI can become a reliable tool for a more supportive, inclusive and effective education in accordance with a social purpose of transformation and integration of knowledge in the face of the objectives of sustainable development (Cornejo-Plaza et al., 2023).

The issue of the ethical implications of the use of AI in university higher education is constantly debated, since it presents great advantages, but also marked disadvantages. For example, AI can personalize learning thanks to its ability to analyze data, preferences and skills, allowing it to adapt to specific needs and adjust to the learning pace of each student. Another example would be individualized tutoring, in which immediate feedback can be provided, as it identifies areas for improvement and even offers relevant support resources, attending to the individuality of the student and evaluating their progress. AI can also analyze tasks and tests performed, identifying weaknesses and strengths individually, allowing the pedagogical approach to be adjusted. However, conflicts arise regarding the security and privacy of student data and ensuring that such information is collected in an ethical and confidential manner. The use of AI has great potential to transform the teaching-learning process, although it also poses great challenges in terms of confidentiality, teacher training and the necessary regulation for the use and application of AI, the key is in prevention and regulation for the best use of the new tools offered by science and technology, this can help our education to be of high quality, ensuring that AI is used in a responsible, legal and inclusive way for our society (Ruíz, 2024).

While the majority of countries have yet to implement regulations governing the use of AI, the European Union has taken the lead in this area, establishing two resolutions in 2020. The first resolution addresses the technical aspects of AI, robotics, and related technologies, while the second focuses on a civil liability regime in the field of AI. In 2022, ethical guidelines on the use of AI and data in teaching and learning for educators were established, and a draft law on AI has recently been presented with the objective of establishing a common regulatory and legal framework in the European Union that will be effective and valid before 2026. In a document published in 2022, the Organization for Economic Cooperation and Development (OECD) set forth a series of intergovernmental policy guidelines on AI. Furthermore, UNESCO drafted the Beijing Consensus on AI in 2021, comprising 44 recommendations that address the challenges posed by AI in education. This consensus approved the Recommendation on the Ethics of AI, which is the first international standard on AI ethics adopted by the 193 member states of the European Union. Furthermore, the ethical implications of its use were examined from a variety of perspectives, with a particular focus on education and research. This highlighted the need for collaboration to reduce digital divides and inequality in access to digital technology. In 2023, UNESCO published a guide on ChatGPT and AI in higher education. This guide addresses some of the main challenges and ethical implications of AI in educational settings and offers practical steps that universities can take. The objective of these guidelines is to address the ethical and legal challenges posed by the use of this technology in higher education, with the aim of minimizing its potential risks, It is imperative to consider the potential impact of this technology on academic integrity in teaching and learning processes. There is an urgent need to foster an ethical commitment within the university community to preserve the fundamental values of higher education. This can be achieved by preventing new forms of plagiarism and academic fraud that seek to imitate or impersonate authorship (Gallent-Torres et al., 2023).

An increasing number of experts are highlighting the intricacy of the challenges associated with the utilization of AI, particularly within the context of university education. There is a growing discourse surrounding the potential for bias to be embedded within AI driven decisions, whether due to the information provided by humans or the algorithms developed by biological persons for their operation. This eliminates the necessity for humans to assume both the responsibility and the control of its use, thereby reinforcing the ethical implications identified by researchers regarding intelligent systems that make decisions and act based on this information. This raises a series of questions regarding the ethical and moral principles that should regulate the use of AI in education, as well as the tasks that should be programmed for the AI to perform autonomously. One potential response is that the ethical code should be assumed by the AI programmer, however, it is unclear whether this is the most appropriate approach. The proposal is that this could be achieved through interdisciplinary work at the university level, specifically in subjects related to AI learning, including ethical and philosophical technological understanding, through inclusive plans and flexible strategies; it is evident that in the teaching-learning process, the empathy of a human teacher plays a pivotal role. Consequently, it is erroneous to assume that AI can serve as the primary source of inspiration for learning. Furthermore, it is misguided to believe that ethics can be effectively implemented by the programmers themselves or their creators. This is because the evolution of AI must be approached and developed from a multidisciplinary perspective. Such an approach must be interdisciplinary, involving the input of social scientists, philosophers, sociologists, anthropologists, and behavioral scientists. This will facilitate the development of regulatory frameworks and consensual policies to order and raise awareness about the ethical use of AI (Flores, 2023).

CONCLUSION

As demonstrated, the researchers emphasize that the integration of AI in the teaching- learning process, particularly in the university educational context, presents both opportunities and challenges, as well as potential disadvantages, however, they also suggest that it is possible to achieve a balance between the advantages of AI use and the risks associated with technological dependence. Furthermore, they acknowledge the difficulties and disadvantages inherent in its use, the necessity of ensuring the responsible use of AI in education is underscored, emphasizing the importance of ethical and technical training to guarantee the improvement and progress of education. This entails developing effective strategies that maximize the power of AI while also establishing a concern regarding the authenticity of knowledge and the ethical risks generated by its use. While there is a clear positive perception of AI, it is imperative that its use be standardized or legally regulated. This should entail the establishment of clear guidelines, as well as the promotion of good practices and ethical awareness among students. Furthermore, the necessity to cultivate analytical, reflective, and critical thinking abilities is underscored, ensuring effective and ethical integration. This enables students to refrain from substituting their own efforts with those of others and to engage in questioning, analysis, and independent, ethical, and well-founded information processing with regard to AI, which guarantees and facilitates the acquisition of authentic and valuable knowledge within the context of university education.

REFERENCES

Abeliuk, A., & Gutiérrez, C. (2021). Historia y evolución de la inteligencia artificial. Revista Bits de Ciencia, (21), 14-21.

Ahmad, S. F., Rahmat, M. K., Mubarik, M. S., Alam, M. M., & Hyder, S. I. (2021). Artificial Intelligence and Its Role in Education. *Sustainability (Basel), 13*(22), 22. Advance online publication. DOI: 10.3390/su132212902

APD. (2018). *Redacción APD.* Retrieved from Que es el deep learning y por qué es clave para la inteligencia artificial? https://www.apd.es/que-es-deep-learning/

APD, R. (2023). ¿Qué es machine learning y para qué sirve? *APD.*

Beyoud, L. (2023, julio 18). *Blooberg.* Retrieved from https://www.elfinanciero.com.mx/economia/2023/07/17/sec-advierte-sobre-la- inteligencia-artificial-pone-en-peligro-la-estabilidad-financiera/

CII.IA. (2021, septiembre 20). *La prueba de Turing.* Retrieved from https://www.ciiia.mx/noticiasciiia/la-prueba-de-turing

Cornejo-Plaza, I., & Cippitani, R. (2023). Consideraciones éticas y jurídicas de la Inteligencia Artificial en Educación Superior: desafíos y perspectivas. Revista De Educación Y Derecho, (28). DOI: 10.1344/REYD2023.28.43935

DataScientest. (2022). *Inteligencia Artificial: definición, historia, usos y peligros.* Retrieved from https://datascientest.com/es/inteligencia-artificial-definicion

Davila Morán, R., & Agüero Corzo, E. (2023). Desafíos éticos de la inteligencia artificial: implicaciones para la sociedad y la economía. Conrado, 19(94), 137-144. http://scielo.sld.cu/scielo.php?script=sci_arttext&pid=S1990-86442023000500137&lng=es&tlng=es

España, G. d. (2023). *Que es la inteligencia artificial.* Retrieved from https://planderecuperacion.gob.es/noticias/que-es-inteligencia-artificial-ia- prtr#:~:text=La%20inteligencia%20artificial%20(IA)%20es,el%20razonamiento%20y%20la%20percepci%C3%B3n

Flores, J. (2023). ¿Están claros los límites de la ética en educación e inteligencia artificial? Universidad Complutense de Madrid, Unidad de Cultura Científica y de Innovación.https://www.ucm.es/otri/noticias-etica-inteligencia-artificial-ucm

Gallent-Torres, C., Zapata-González, A., & Ortego-Hernando, J. (2023) El impacto de la inteligencia artificial generativa en educación superior: una mirada desde la ética y la integridad académica/RELIEVE. Revista Electrónica de Investigación y Evaluación Educativa, 29(2). https://www.redalyc.org/journal/916/91676028011/html/

Gonçalves, B. (2023). *Can machines think? The controversy that led to the Turing test.* Retrieved from Springer link: https://link.springer.com/article/10.1007/s00146-021-01318-6

Gonzalez, F. (2023, mayo 23). *Wired.* Retrieved from Propuesta de Ley para regular la IA en México: https://es.wired.com/articulos/diputado-presenta-propuesta-de-ley-para-regula-la-ia-en-mexico

Iberdrola. (2022). *Innovación, Que es la Inteligencia Artificial.* Retrieved from https://www.iberdrola.com/innovacion/que-es-inteligencia-artificial

IG. (2022). *Trading Algoritmico.*

Jauregui, M. G. (2024, febrero 12). *Principios éticos de la educación con inteligencia artificial.* Retrieved from https://observatorio.tec.mx/edu-news/principios-eticos- de-la-educacion-con-inteligencia-artificial-ia/

Labrador-Fernández, J. (2023). Implicaciones éticas de la Inteligencia Artificial en las Ciencias de la Educación. Revista Arbitrada Interdisciplinaria Koinonía, 8(16), 1-3. DOI: 10.35381/r.k.v8i16.2545

Llamas, J., Mendoza, O., & Graff, M. (2022). Enfoques regulatorios para la inteligencia artificial (IA). Revista chilena de derecho, 49(3), 31-62. DOI: 10.7764/R.493.2

Morán-Ortega, S.-A., Ruiz-Tirado, S.-G., Simental-López, L.-M., & Tirado-López, A.-B. (2024). Barreras de la Inteligencia Artificial generativa en estudiantes de educación superior. Percepción docente. *Revista De Investigación En Tecnologías De La Información, 12*(25), 26–37. DOI: 10.36825/RITI.12.25.003

Olmos, F. G. (2023, Junio 2). *Expansión.* México avanza en la adopción de la IA y así se está usando en las empresas.

RAE. (2023). *Diccionario de la lengua española.* Retrieved from https://dle.rae.es

Ramos, G. (2024, Febrero). *Foro Global sobre la ética de la IA2024*. Retrieved from https://www.unesco.org/es/artificial-intelligence/recommendation-ethics

Raya, A. P. (2019). *Fundamentos de la IA para el ajedrez.* Retrieved from https://idus.us.es/bitstream/handle/11441/93632/1/Portillo%20Raya%20Alberto%20TFG.pdf?sequence=1

Roldán, P. N. (2017, febrero 17). *Economipedia*. Retrieved from https://economipedia.com/definiciones/finanzas.html

Ruíz, E. (2024). Las implicaciones del uso de la inteligencia artificial en la academia. Gaceta Universidad Abierta y a Distancia UNAD. Gaceta Extraordinaria. https://gaceta.unadmexico.mx/historico-anual/95-2023/julio-septiembre- 2023/educacion/205-las-implicaciones-del-uso-de-la-inteligencia-artificial-en-la-academia

Shabbir, J. (2018). Artificial intelligence and its role in near future. Retrieved from https://arxiv.org/abs/1804.01396

Shinde, P. P. (2018, AGOSTO 18). *A review of machine learning and deep learning applications.*https://ieeexplore.ieee.org/abstract/document/8697857

The impact of artificial intelligence in education and teaching - Universidad ORT Uruguay. (n.d.). https://ie.ort.edu.uy/blog/el-impacto-de-la-inteligencia-artificial-en-la-educacion-y-enla-docencia

UNESCO. (2019). Review of the potential of artificial intelligence to accelerate progress towards SDG 4-Education 2030. In *UNESDOC Digital Library* (206 EX/44 + REV.CORR.). United Nations Educational, Scientific and Cultural Organization. Retrieved July 29, 2024, from https://unesdoc.unesco.org/ark:/48223/pf0000367373_spa

Universidad, O. R. T. Uruguay. (n.d.). El impacto de la inteligencia artificial en la educación y en la docencia. https://ie.ort.edu.uy/blog/el-impacto-de-la-inteligencia-artificial-en-la-educacion-y-en-la-docencia

Vera, F. (2023). Integrating Artificial Intelligence in Higher Education: Challenges and opportunities. *Transform*, *4*(1), 17–34.

Verma, M. (2018). Artificial intelligence and its scope in different areas with special reference to the field of education. *International Journal of Advanced Educational Research*, *3*(1), 5–10. https://files.eric.ed.gov/fulltext/ED604401.pdf

Vivar, J. M. F., & Peñalvo, F. J. G. (2023). Reflections on the ethics, potentials and challenges of Artificial Intelligence in the framework of Quality Education (SDG4). Comunicar: Revista científica de comunicación y educación, (74), 37-47.

Work.r. (2023). *Work.r.* Retrieved from Enterprise IA platform: https://irp- cdn.multiscreensite.com/1e240500/files/uploaded/Workr-Aplicaciones-Cartera- %20Brochure.pdf

ADDITIONAL READING

Cota-Rivera, E. I., González Correa, M. E., Bernal Marín, L. A., Márquez Montenegro, M. Y., Herrera, A. M., & Martínez Martínez, M. A. (2024). Transforming Education With the Power of Artificial Intelligence: Case Studies. In Demir, Ş., & Demir, M. (Eds.), *Enhancing Higher Education and Research With Open AI Models* (pp. 113–140). IGI Global. DOI: 10.4018/979-8-3693-1666-5.ch006

Crawford, J., Cowling, M., & Allen, K. (2023). Leadership is needed for ethical ChatGPT: Character, assessment, and learning using artificial intelligence (AI). Journal of University Teaching & Learning Practice, 20(3). .DOI: 10.53761/1.20.3.02

Delgado, N., Campo Carrasco, L., Sainz de la Maza, M., & Etxabe-Urbieta, J. M. (2024). Application of Artificial Intelligence (AI) in Education: The benefits and limitations of AI as perceived by primary education, secondary education and higher education teachers. *Interuniversity Electronic Journal of Teacher Education*, *27*(1), 207–224. DOI: 10.6018/reifop.577211

Gallent-Torres, C., Zapata-González, A., & Ortego-Hernando, J. L. (2023). El impacto de la inteligencia artificial generativa en educación superior: Una mirada desde la ética y la integridad académica. *RELIEVE*, *29*(2), M5. Advance online publication. DOI: 10.30827/relieve.v29i2.29134

Janiesch, C., Zschech, P., & Heinrich, K. (2021). Machine learning and deep learning. *Electronic Markets*, *31*(3), 685–695. DOI: 10.1007/s12525-021-00475-2

Jáuregui, M. G. (2024, 5 August). Glossary of Artificial Intelligence (AI) for education. Observatory / Institute for the Future of Education. https://observatorio.tec.mx/edunews/glosario-de-inteligencia-artificial-ia-para-la- educacion/

Magallanes Ronquillo, K. K., Plúas Pérez, L. del R., Aguas Veloz, J. F., & Freire Solís, R. L. (2023). Artificial intelligence applied to educational innovation in the teaching and learning process: Artificial intelligence applied to educational innovation in the teaching and learning process. LATAM Revista Latinoamericana De Ciencias Sociales Y Humanidades, 4(2), 1597-1613. DOI: 10.56712/latam.v4i2.706

Mujica-Sequera, R. M. (2024). Classification of Artificial Intelligence Tools in Education. Docentes 2.0 Journal, 17(1), 31-40. DOI: 10.37843/rted.v17i1.513

Salman, K. (2024). *Brave New Words: How AI Will Revolutionize Education (and Why That's a Good Thing)*. Viking.

Yébenes, J. A. V. (2024). Alan Turing y el origen de la inteligencia artificial: La superación de la intuición. *Naturaleza y Libertad. Revista de Estudios Interdisciplinares*, (18).

KEY TERMS AND DEFINITIONS

Artificial Intelligence: Field of computer science that is dedicated to solving cognitive problems associated with human intelligence such as learning.

Digital Dependence: This is when a person uses technological devices more and more, losing the power to manage both time and how to use them.

Educational Space: These are specific places prepared to carry out teaching and learning activities in the best conditions.

Ethics: It is a philosophical discipline that studies human behavior and its relationship with the notions of good and evil.

Learning: It is the transmission of knowledge, values and ideas between people.

Meaningful Learning: Type of learning in which the student associates new information with what he already has, building new coherent and structured knowledge.

Regulatory Framework: It is a set of laws, regulations and rules in general that regulate a certain area of the state.

Science: It is the knowledge obtained through a set of specific techniques and methods.

Teaching: It is the transmission of knowledge, values and ideas between people.

Technology: It is the application of notions, skills or scientific knowledge that human beings use to achieve objectives or solutions.

Compilation of References

Aafrein, R., & Chinnaiyan, D. R. (2023). Detection of pneumonia using chest X-ray images with deep learning techniques-review. *International Journal of Scientific Research in Science and Technology*, •••, 793–800. DOI: 10.32628/IJSRST523103147

Abad-Segura, E., & González-Zamar, M.-D. (2019). Effects of financial education and financial literacy on creative entrepreneurship: A worldwide research. *Education Sciences*, *9* (3), 238. DOI: 10.3390/educsci9030238

Abdalla, R., & Esmail, M. (2018). *WebGIS for disaster management and emergency response*. Springer.

Abdullah, S. M. S. A., Ameen, S. Y. A., Sadeeq, M. A., & Zeebaree, S. (2021). Multimodal emotion recognition using deep learning. *Journal of Applied Science and Technology Trends*, *2*(01), 73–79. DOI: 10.38094/jastt20291

Abdullah, T., & Ahmet, A. (2022). Deep learning in sentiment analysis: Recent architectures. *ACM Computing Surveys*, *55*(8), 1–37. DOI: 10.1145/3548772

Abdulrahman, A., & Baykara, M. (2020). Fake news detection using machine learning and deep learning algorithms. *2020 International Conference on Advanced Science and Engineering (ICOASE)*, 18–23. DOI: 10.1109/ICOASE51841.2020.9436605

Abedalla, A., Al-Sadi, A., & Abdullah, M. (2019). A closer look at fake news detection: A deep learning perspective. *Proceedings of the 3rd International Conference on Advances in Artificial Intelligence*, 24–28. DOI: 10.1145/3369114.3369149

Abeliuk, A., & Gutiérrez, C. (2021). Historia y evolución de la inteligencia artificial. Revista Bits de Ciencia, (21), 14-21.

Abirami, T., & Gayathri, R. (2016). A Survey on Efficient Power allocation for OFDM–Based Cognitive Radio Systems. *on Journal of Chemical and Pharmaceutical Sciences (ICPS), 8*, 83-87.

Abrams, Z. (2023). AI is changing every aspect of psychology. Here's what to watch for Psychologists and their skills are irreplaceable, but thoughtful and strategic implementation of AI is crucial. *American Psychological Association*, Vol. 54 No. 5, p46.

Acolita. (n.d.). *Digitalización automática en QGIS usando inteligencia artificial*. Acolita. https://acolita.com/digitalizacion-automatica-en-qgis-usando-inteligencia-artificial/

Adebola, O. O., Onyekwelu, B., & Orogun, A. (2019). Predicting Consumer Behaviour in Digital Market: A Machine Learning Approach. *International Journal of Innovative Research in Science, Engineering and Technology*, *8*(8), 1–13. DOI: 10.15680/IJIRSET.2019.0808006

Agarwal, S., Kumar, S., & Goel, U. (2021). Social media and the stock markets: An emerging market perspective. *Journal of Business Economics and Management*, *22*(6), 1614–1632. DOI: 10.3846/jbem.2021.15619

Aggarwal, A., Tam, C. C., Wu, D., Li, X., & Qiao, S. (2023). Artificial Intelligence–Based Chatbots for Promoting Health Behavioral Changes: Systematic review. *Journal of Medical Internet Research, 25*, e40789. DOI: 10.2196/40789 PMID: 36826990

Aguilar-Lasserre, A. A., Torres-Sanchez, V. E., Fernandez-Lambert, G., Azzaro-Pantel, C., Cortes-Robles, G., & Roman-del Valle, M. A. (2020). Functional optimization of a Persian lime packing using TRIZ and multi-objective genetic algorithms. Computers & Industrial Engineering, 139. 105558. ISSN 0360-8352 2.

Ahmad, A., & Dey, L. (2007). A k-mean clustering algorithm for mixed numeric and categorical data. *Data & Knowledge Engineering*, *63*(2), 503–527. DOI: 10.1016/j.datak.2007.03.016

Ahmad, I., Yousaf, M., Yousaf, S., & Ahmad, M. O. (2020). Fake news detection using machine learning ensemble methods. *Complexity*, *2020*(1), 8885861. DOI: 10.1155/2020/8885861

Ahmad, S. F., Rahmat, M. K., Mubarik, M. S., Alam, M. M., & Hyder, S. I. (2021). Artificial Intelligence and Its Role in Education. *Sustainability (Basel)*, *13*(22), 22. Advance online publication. DOI: 10.3390/su132212902

Ahmed, A. A. A., Aljabouh, A., Donepudi, P. K., & Choi, M. S. (2021). Detecting fake news using machine learning: A systematic literature review. *arXiv Preprint arXiv:2102.04458*.

Ahmed, A. A. A., Agarwal, S., Kurniawan, I. G. A., Anantadjaya, S. P., & Krishnan, C. (2022). Business boosting through sentiment analysis using Artificial Intelligence approach. *International Journal of System Assurance Engineering and Management*, *13*(S1, Suppl 1), 699–709. DOI: 10.1007/s13198-021-01594-x

Ahmed, H. U., Mostafa, R. R., Mohammed, A., Sihag, P., & Qadir, A. (2023). Support vector regression (SVR) and grey wolf optimization (GWO) to predict the compressive strength of GGBFS-based geopolymer concrete. *Neural Computing & Applications*, *35*(3), 2909–2926. DOI: 10.1007/s00521-022-07724-1

Alamoudi, E., & Alghamdi, N. (2021). Sentiment classification and aspect-based sentiment analysis on yelp reviews using deep learning and word embeddings. *Journal of Decision Systems*, *30*(2–3), 259–281. DOI: 10.1080/12460125.2020.1864106

Alghamdi, J., Luo, S., & Lin, Y. (2024). A comprehensive survey on machine learning approaches for fake news detection. *Multimedia Tools and Applications*, *83*(17), 51009–51067. DOI: 10.1007/s11042-023-17470-8

Alhuwaydi, A. M. (2024). Exploring the Role of Artificial Intelligence in Mental Healthcare: Current Trends and Future Directions – A Narrative Review for a Comprehensive Insight. *Risk Management and Healthcare Policy*, *17*, 1339–1348. DOI: 10.2147/RMHP.S461562 PMID: 38799612

Al-Jarrah, M. A. (2016, July). Developing 3D model for mobile robot environment using mono-vision system. In 2016 7th International Conference on Computer Science and Information Technology (CSIT) (pp. 1-4). IEEE. DOI: 10.1109/CSIT.2016.7549466

Allen, K. S., Hood, D. R., Cummins, J., Kasturi, S., Mendonca, E. A., & Vest, J. R. (2023). Natural language processing-driven state machines to extract social factors from unstructured clinical documentation. *JAMIA Open*, *6*(2), ooad024. Advance online publication. DOI: 10.1093/jamiaopen/ooad024 PMID: 37081945

Al-qablan, T. A. (2023). A survey on sentiment analysis and its applications. In *Neural Computing and Applications* (Issue August). Springer London. DOI: 10.1007/s00521-023-08941-y

Alqahtan, T., Badreldin, H. A., & Mohammed, A. (2023). The emergent role of artificial intelligence, natural learning processing, and large language models in higher education and research. *Research in Social & Administrative Pharmacy*, *19*(8), 1236–1242. DOI: 10.1016/j.sapharm.2023.05.016 PMID: 37321925

Alslaity, A., & Orji, R. (2024). Machine learning techniques for emotion detection and sentiment analysis: Current state, challenges, and future directions. *Behaviour & Information Technology*, *43*(1), 139–164. DOI: 10.1080/0144929X.2022.2156387

Altman, S. (2023). Generative AI has the potential to democratize access to creative tools and empower people to express themselves in new and exciting ways. Retrieved from https://openai.com

Al-Utaibi, K. A., & El-Alfy, E. S. M. (2018). Intrusion detection taxonomy and data preprocessing mechanisms. *Journal of Intelligent & Fuzzy Systems*, *34*(3), 1369–1383. DOI: 10.3233/JIFS-169432

Amazon Web Services. (n.d.). *What is an API?* Amazon. https://aws.amazon.com/es/what-is/api/

American Psychiatric Association. (2000). *Diagnostic and statistical manual of mental disorders.* Fifth EditionText revision.

Amidon, A. F. M., Yusoff, Z. M., Ismail, N., Abd Almisreb, A., & Taib, M. N. (2024). kNN Euclidean Distance Model Performance on Aquilaria Malaccensis Oil Qualities. *Journal of Advanced Research in Applied Sciences and Engineering Technology*, *48*(2), 16–28. DOI: 10.37934/araset.48.2.1628

Anantrasirichai, N., & Bull, D. (2022). Artificial intelligence in the creative industries: A review. *Artificial Intelligence Review*, *55*(1), 589–656. DOI: 10.1007/s10462-021-10039-7

Anderson, N. E., & Kiehl, K. A. (2012). The psychopath magnetized: Insights from brain imaging. *Trends in Cognitive Sciences*, *16*(1), 52–60. DOI: 10.1016/j.tics.2011.11.008 PMID: 22177031

Anjaria, M., & Guddeti, R. M. R. (2014). A novel sentiment analysis of social networks using supervised learning. *Social Network Analysis and Mining*, *4*(1), 1–15. DOI: 10.1007/s13278-014-0181-9

APD, R. (2023). ¿Qué es machine learning y para qué sirve? *APD.*

APD. (2018). *Redacción APD.* Retrieved from Que es el deep learning y por qué es clave para la inteligencia artificial? https://www.apd.es/que-es-deep-learning/

Arabameri, A., Chandra Pal, S., Rezaie, F., Asadi Nalivan, O., Chowdhuri, I., Saha, A., Lee, S., & Moayedi, H. (2021). Modeling Groundwater Potential Using Novel GIS-Based Machine-Learning Ensemble Techniques. *Journal of Hydrology. Regional Studies*, *36*, 100848. Advance online publication. DOI: 10.1016/j.ejrh.2021.100848

Araújo, M., Pereira, A., & Benevenuto, F. (2020). A comparative study of machine translation for multilingual sentence-level sentiment analysis. *Information Sciences, 512*, 1078–1102. DOI: 10.1016/j.ins.2019.10.031

Ardissono, L., & Goy, A. (2000). Tailoring the interaction with users in Web stores. *User Modeling and User-Adapted Interaction, 10*(4), 251–303. DOI: 10.1023/A:1008372122567

Ardissono, L., Goy, A., Petrone, G., & Segnan, M. (2002). Personalization in business-to-customer interaction. *Communications of the ACM, 45*(5), 52–53. DOI: 10.1145/506218.506245

Ariffin, N. H. M., Iqbal, M. I. M., Yusoff, M., & Zulkefli, N. A. M. (2024). A Study on the Best Classification Method for an Intelligent Phishing Website Detection System. *Journal of Advanced Research in Applied Sciences and Engineering Technology, 48*(2), 197–210. DOI: 10.37934/araset.48.2.197210

Aristotle, . (1996). *Poetics* (Butcher, S. H., Trans.). Dover Publications. (Original work published 335 BCE)

Asghar, J., Akba, S., & Zubair, M. A. (2021). Detection and Classification of Psychopathic Personality Trait from Social Media Text Using Deep Learning Model. *Computational and Mathematical Methods in Medicine*, Volume 2021, Article ID 5512241, 10 pages DOI: 10.1155/2021/5512241

Aslan, S., Durham, L. M., Alyuz, N., Okur, E., Sharma, S., Savur, C., & Nachman, L. (2024). Immersive multi-modal pedagogical conversational artificial intelligence for early childhood education: An exploratory case study in the wild. *Computers and Education: Artificial Intelligence, 6*(April), 100220. DOI: 10.1016/j.caeai.2024.100220

Atef, I., Ahmed, W., & Abdel-Maguid, R. H. (2023). Modelling of land use land cover changes using machine learning and GIS techniques: A case study in El-Fayoum Governorate, Egypt. *Environmental Monitoring and Assessment, 195*(6), 637. DOI: 10.1007/s10661-023-11224-7 PMID: 37133528

Aytug, H., Bhattacharyya, S., Koehler, G. J., & Snowdon, J. L. (1994). A review of machine learning in scheduling. *IEEE Transactions on Engineering Management, 41*(2), 165–171. DOI: 10.1109/17.293383

Azzouza, N., Akli-Astouati, K., Oussalah, A., & Ait Bachir, S. (2017). A Real-Time Twitter Sentiment Analysis using an unsupervised Method. *ACM International Conference Proceeding Series, Part F1294*. DOI: 10.1145/3102254.3102282

Baarir, N. F., & Djeffal, A. (2021). Fake news detection using machine learning. *2020 2nd International Workshop on Human-Centric Smart Environments for Health and Well-Being (IHSH)*, 125–130.

Baba, A., & Bunji, K. (2023). Prediction of mental health problem using annual student health survey: Machine learning approach. *JMIR Mental Health, 10*, e42420. DOI: 10.2196/42420 PMID: 37163323

Babu, S. S., Mourad, A. H. I., Harib, K. H., & Vijayavenkataraman, S. (2022). Recent developments in the application of machine-learning towards accelerated predictive multiscale design and additive manufacturing. *Virtual and Physical Prototyping, 18*(1), e2141653. Advance online publication. DOI: 10.1080/17452759.2022.2141653

Bakliwal, A., Foster, J., van der Puil, J., O'Brien, R., Tounsi, L., & Hughes, M. (2013, June). Sentiment analysis of political tweets: Towards an accurate classifier. Association for Computational Linguistics.

Bala, A., Zakariyya Shuaibu, M., Karami Lawal, Z., & Zakari Yusuf, R. (2016). Performance Analysis of Apriori and FP-Growth Algorithms (Association Rule Mining). *International Journal Of Engineering And Computer Science, 7*(2), 279–293. https://www.researchgate.net/publication/297898164

Balaji, P., Nagaraju, O., & Haritha, D. (2017, March). Levels of sentiment analysis and its challenges: A literature review. In *2017 International Conference on Big Data Analytics and Computational Intelligence (ICBDAC)* (pp. 436-439). IEEE. DOI: 10.1109/ICBDACI.2017.8070879

Balaji, V., Acharjee, P. B., Elangovan, M., Kalnoor, G., Rastogi, R., & Patidar, V. (2023). Developing a semantic framework for categorizing IoT agriculture sensor data: A machine learning and web semantics approach. *The Scientific Temper, 14*(4), 1332–1338. DOI: 10.58414/SCIENTIFICTEMPER.2023.14.4.40

Banaee, H., Ahmed, M. U., & Loutfi, A. (2013). Data mining for wearable sensors in health monitoring systems: A review of recent trends and challenges. *Sensors (Basel), 13*(12), 17472–17500. DOI: 10.3390/s131217472 PMID: 24351646

Bansal, M., Goyal, A., & Choudhary, A. (2022). A comparative analysis of K-Nearest Neighbor, Genetic, Support Vector Machine, Decision Tree, and Long Short Term Memory algorithms in machine learning. *Decision Analytics Journal, 3*(100071), 100071. DOI: 10.1016/j.dajour.2022.100071

Barba, P. (2020). *Machine Learning (ML) for Natural Language Processing (NLP)*. Lexalytics.

Barnes, B., Dunn, S., & Wilkinson, S. (2019). Natural hazards, disaster management and simulation: A bibliometric analysis of keyword searches. *Natural Hazards, 97*(2), 813–840. DOI: 10.1007/s11069-019-03677-2

Barreto, B. B., Luz, M., & Gusmao-Flores, D. (2020). Nonpharmacologic interventions to prevent or mitigate adverse long-term outcomes among ICU survivors: Is the ICU diary really important? *Critical Care Medicine, 48*(3), e265–e265. DOI: 10.1097/CCM.0000000000004136 PMID: 32058409

Bastani, H., Bastani, O., & Sinchaisri, W. P. (2021). Learning best practices: Can machine learning improve human decision-making. *Proceedings - Academy of Management, 1* (1), 14006. DOI: 10.5465/AMBPP.2021.14006abstract

Baudrillard, J. (1968). *El sistema de los objetos*. Siglo XXI Editores.

Bauer, M., Glenn, T., Monteith, S., Bauer, R., Whybrow, P. C., & Geddes, J. (2017). Ethical perspectives on recommending digital technology for patients with mental illness. *International Journal of Bipolar Disorders, 5*(1), 6–12. DOI: 10.1186/s40345-017-0073-9 PMID: 28155206

Bauman, Z. (1992). *Intimations of Postmodernity*. Routledge.

Bayuk, J., & Altobello, S. A. (2019). Can gamification improve financial behavior? The moderating role of app expertise. *International Journal of Bank Marketing, 37* (4), 951–975. DOI: 10.1108/IJBM-04-2018-0086

Beck, J. S. (2020). *Cognitive Behavior Therapy: Basics and Beyond* (3rd ed.). Guilford Press.

Behdenna, S., Barigou, F., & Belalem, G. (2018). Document Level Sentiment Analysis: A survey. *EAI Endorsed Transactions on Context-Aware Systems and Applications*, *4* (13), 154339. DOI: 10.4108/eai.14-3-2018.154339

Beidel, D. C., Frueh, B. C., Neer, S. M., Bowers, C. A., Trachik, B., Uhde, T. W., & Grubaugh, A. (2019). Trauma management therapy with virtual-reality augmented exposure therapy for combat-related PTSD: A randomized controlled trial. *Journal of Anxiety Disorders*, *61*(1), 64–74. DOI: 10.1016/j.janxdis.2017.08.005 PMID: 28865911

Benamara, N. K., Val-Calvo, M., Alvarez-Sanchez, J. R., Diaz-Morcillo, A., Ferrandez-Vicente, J. M., Fernandez-Jover, E., & Stambouli, T. B. (2021). Real-time facial expression recognition using smoothed deep neural network ensemble. *Integrated Computer-Aided Engineering*, *28*(1), 97–111. DOI: 10.3233/ICA-200643

Benuwa, B. B., Zhan, Y. Z., Ghansah, B., Wornyo, D. K., & Banaseka Kataka, F. (2016). A review of deep machine learning. *International Journal of Engineering Research in Africa*, *24*, 124–136. DOI: 10.4028/www.scientific.net/JERA.24.124

Beuys, J. (1993). *Joseph Beuys: The Revolution is Us*. Thames & Hudson.

Beyoud, L. (2023, julio 18). *Blooberg*. Retrieved from https://www.elfinanciero.com.mx/economia/2023/07/17/sec-advierte-sobre-la- inteligencia-artificial-pone-en-peligro-la-estabilidad-financiera/

Bhanumurthy, V., Rao, K. V., Rao, S. S., Rao, K. R. M., Chandra, P. S., Vidhyasagar, J., Diwakar, P. G., & Dadhwal, V. K. (2014). Enabling heterogenous multi-scale database for emergency service functions through geoinformation technologies. *International Archives of the Photogrammetry, Remote Sensing and Spatial Information Sciences - ISPRS Archives, XL–8*(1), 7–14. DOI: 10.5194/isprsarchives-XL-8-7-2014

Bhaskar, J., Sruthi, K., & Nedungadi, P. (2014). Enhanced sentiment analysis of informal textual communication in social media by considering objective words and intensifiers. In *International Conference on Recent Advances and Innovations in Engineering (ICRAIE-2014)* (pp. 1–6). IEEE. DOI: 10.1109/ICRAIE.2014.6909220

Bhatia, P., Ji, Y., & Eisenstein, J. (2007). *Better Document-level Sentiment Analysis from RST Discourse Parsing*.

Bhonde, R., Bhagwat, B., Ingulkar, S., & Pande, A. (2015). Sentiment analysis based on dictionary approach. *International Journal of Emerging Engineering Research and Technology*, *3*(1), 51–55.

Bickmore, T., & Picard, R. W. (2020). Establishing and maintaining long-term human-computer relationships. *ACM Transactions on Computer-Human Interaction*, *7*(2), 95–132. DOI: 10.1145/1057237

Bilgin, M., & Şentürk, İ. F. 2017. Sentiment Analysis on Twitter Data with Semi-Supervised Doc2Vec. 2nd International Conference on Computer Science and Engineering, UBMK 2017: 661–666. . 2017.8093492. DOI: 10.1109/UBMK.2017.8093492

Binetti, M. S., Massarelli, C., & Uricchio, V. F. (2024). Machine Learning in Geosciences: A Review of Complex Environmental Monitoring Applications. *Machine Learning and Knowledge Extraction*, *6*(2), 1263–1280. DOI: 10.3390/make6020059

Binns, R. (2018). Fairness in Machine Learning: Lessons from Political Philosophy. Proceedings of the 1st Conference on Fairness, Accountability and Transparency, in Proceedings of Machine Learning Research 81:149-159 Available from https://proceedings.mlr.press/v81/binns18a.html.

Bishop, C. M. (2006). *Pattern Recognition and Machine Learning*. Springer.

Black, M. H., Helander, J., Segers, J., Ingard, C., Bervoets, J., de Puget, V. G., & Bölte, S. (2024). Resilience in the face of neurodivergence: A scoping review of resilience and factors promoting positive outcomes. *Clinical Psychology Review, 113*(May), 102487. Advance online publication. DOI: 10.1016/j.cpr.2024.102487 PMID: 39178757

Bockting, C. L., Hollon, S. D., Jarrett, R. B., Kuyken, W., & Dobson, K. (2015). A lifetime approach to major depressive disorder: The contributions of psychological interventions in preventing relapse and recurrence. *Clinical Psychology Review, 41*, 16–26. DOI: 10.1016/j.cpr.2015.02.003 PMID: 25754289

Boeldt, D., McMahon, E., McFaul, M., & Greenleaf, W. (2019). Using virtual reality exposure therapy to enhance treatment of anxiety disorders: Identifying areas of clinical adoption and potential obstacles. *Frontiers in Psychiatry, 10*(1), 1–6. DOI: 10.3389/fpsyt.2019.00773 PMID: 31708821

Bolanle, O. I. (2014). Impact of Customer Satisfaction and Customer Retention on Customer Loyalty: A Case Study of Enterprise Bank in OYO. *International Journal of Education and Research, 2*(9), 427–450.

Bollmer, G. (2017). Empathy machines. *Media International Australia, Incorporating Culture & Policy, 165*(1), 63–76. DOI: 10.1177/1329878X17726794

Bond, R. R., Mulvenna, M. D., Potts, C., Siobhan O'Neill, Ennis, E., Torous, J. (2023). Digital transformation of mental health services. NJP *Mental Health Research*, 2(1):13. DOI: 10.1038/s44184-023-00033-y

Boress, K., Gaasedelen, O. J., Croghan, A., Johnson, M. K., Caraher, K., Basso, M. R., & Whiteside, D. M. (2022). Validation of the Personality Assessment Inventory (PAI) scale of scales in a mixed clinical sample. *The Clinical Neuropsychologist, 36*(7), 1844–1859. DOI: 10.1080/13854046.2021.1900400 PMID: 33730975

Bormah, A. M. (2023). *Productivity Forecasting of Employees Performance Using Machine Learning and Adaptive Neuro-Fuzzy Inference System* [PhD Thesis]. KING ABDULAZIZ UNIVERSITY.

Boukes, M., Van de Velde, B., Araujo, T., & Vliegenthart, R. (2020). What's the tone? Easy doesn't do it: Analyzing performance and agreement between off-the-shelf sentiment analysis tools. *Communication Methods and Measures, 14*(2), 83–104. DOI: 10.1080/19312458.2019.1671966

Boumedine, A. Y., Bentaieb, S., & Ouamri, A. (2023). An improved KNN classifier for 3D face recognition based on SURF descriptors. *Journal of Applied Security Research, 18*(4), 808–826. DOI: 10.1080/19361610.2022.2099688

Bowley, G. (2015, January 22). The verdict that flies in the face of art history: Luc Tuymans guilty of copyright infringement. The Art Newspaper. Retrieved from https://www.theartnewspaper.com/news/the-verdict-that-flies-in-the-face-of-art-history-luc-tuymans-guilty-of-copyright-infringement

Boyd, D., & Crawford, K. (2012). Critical questions for big data: Provocations for a cultural, technological, and scholarly phenomenon. *Information Communication and Society, 15*(5), 662–679. DOI: 10.1080/1369118X.2012.678878

Bradlow, E. T., Gangwar, M., Kopalle, P., & Voleti, S. (2017). The Role of Big Data and Predictive Analytics in Retailing. *Journal of Retailing, 93*(1), 79–95. DOI: 10.1016/j.jretai.2016.12.004

Braun, V., & Clarke, V. (2006). Using thematic analysis in psychology. *Qualitative Research in Psychology, 3*(2), 77–101. DOI: 10.1191/1478088706qp063oa

Bray, J. H. (2010). The future of psychology practice and science. *The American Psychologist, 65*(5), 355–369. DOI: 10.1037/a0020273 PMID: 20642298

Breiman, L. (2001). Random forests. *Machine Learning, 45*(1), 5–32. DOI: 10.1023/A:1010933404324

Breiter, G., & Naik, V. K. (2013, March). *A framework for controlling and managing hybrid cloud service integration. In 2013 IEEE international conference on Cloud engineering (ic2e)* (pp. 217-224). IEEE.

Brezinka, V. (2012). Computer games supporting cognitive behavior therapy in children. *Clinical Child Psychology and Psychiatry, 19*(1), 100–110. DOI: 10.1177/1359104512468288 PMID: 23258925

Brown, M. (2019). Richard Prince ordered to pay damages to photographers in copyright infringement lawsuits over Instagram portraits. The Art Newspaper. Retrieved from https://www.theartnewspaper.com/news/richard-prince-ordered-to-pay-damages-to-photographers-in-copyright-infringement-lawsuits-over-instagram-portraits

Brown, M. P. S., Grundy, W. N., Lin, D., Cristianini, N., Sugnet, C. W., Furey, T. S., Ares, M.Jr, & Haussler, D. (2000). Knowledge-based analysis of microarray gene expression data by using support vector machines. *Proceedings of the National Academy of Sciences of the United States of America, 97*(1), 262–267. DOI: 10.1073/pnas.97.1.262 PMID: 10618406

Brown, T., Mann, B., Ryder, N., Subbiah, M., Kaplan, J. D., Dhariwal, P., Neelakantan, A., Shyam, P., Sastry, G., Askell, A., & Agarwal, S. (2020). *Language models are few-shot learners*. Advances in Neural Information Processing.

Bryan, K. N., & George, R. (2003). *Geographic information systems. The methods and materials of demography,* 733.

Buchloh, B. H. D. (2015). *Formalism and Historicity: Models and Methods in Twentieth-Century Art*. MIT Press.

Burges, C. J. C. (1998). A Tutorial on Support Vector Machines for Pattern Recognition. *Data Mining and Knowledge Discovery, 2*(2), 121–167. DOI: 10.1023/A:1009715923555

Burr, C., Taddeo, M., & Floridi, L. (2020). The ethics of digital well-being: A thematic review. *Science and Engineering Ethics, 26* (4), 2313–2343. DOI: 10.1007/s11948-020-00175-8 PMID: 31933119

Burrell, J. (2016). How the machine 'thinks': Understanding opacity in machine learning algorithms. *Big Data & Society, 3*(1), 2053951715622512. DOI: 10.1177/2053951715622512

Byravan, A., & Ramanaiah, N. V. (2002). On the incremental validity of MMPI-2 Psychopathology-5 scales over the Revised Neo Personality Inventory scales for predicting personality disorders. *Psychological Reports*, *90*(3_part_2), 1084–1090. DOI: 10.1177/003329410209000305.2 PMID: 12150388

Cai, J., Luo, J., Wang, S., & Yang, S. (2018). Feature selection in machine learning: A new perspective. *Neurocomputing*, *300*, 70–79. DOI: 10.1016/j.neucom.2017.11.077

Cai, Y., Cai, H., & Wan, X. (2019, July). Multi-modal sarcasm detection in twitter with hierarchical fusion model. In *Proceedings of the 57th annual meeting of the association for computational linguistics* (pp. 2506-2515). DOI: 10.18653/v1/P19-1239

Callister, W. D., & Rethwisch, D. G. (2020). Callister's Materials Science and Engineering. Wiley. https://books.google.com.mx/books?id=IUjODwAAQBAJ

Cambria, E., Das, D., Bandyopadhyay, S., & Feraco, A. (2017). Affective computing and sentiment analysis. A practical guide to sentiment analysis, 1-10.

Campolo, M., Andreussi, P., & Soldati, A. (1999). River flood forecasting with a neural network model. *Water Resources Research*, *35*(4), 1191–1197. DOI: 10.1029/1998WR900086

Canz, T., & Jagdale, S. (1995). Decision support for manufacturing using artificial neural-networks. Journal of Materials Processing Technology, 52(1), 9-26. ISSN 0924-0136. DOI: 10.1016/0924-0136(94)01439-8

Capuano, N., Fenza, G., Loia, V., & Nota, F. D. (2023). Content-based fake news detection with machine and deep learning: A systematic review. *Neurocomputing*, *530*, 91–103. DOI: 10.1016/j.neucom.2023.02.005

Cashin, P., Mohaddes, K., & Raissi, M. (2017). Fair weather or foul? The macroeconomic effects of El Niño. *Journal of International Economics*, *106*, 37–54. DOI: 10.1016/j.jinteco.2017.01.010

Castro, S., Hazarika, D., Pérez-Rosas, V., Zimmermann, R., Mihalcea, R., & Poria, S. (2019). Towards multimodal sarcasm detection (an _obviously_ perfect paper). *arXiv preprint arXiv:1906.01815*. DOI: 10.18653/v1/P19-1455

Castro-Toscano, M. J., Rodríguez-Quiñonez, J. C., Sergiyenko, O., Flores-Fuentes, W., Ramirez-Hernandez, L. R., Hernández-Balbuena, D., Lindner, L., & Rascón, R. (2020). Novel sensing approaches for structural deformation monitoring and 3D measurements. *IEEE Sensors Journal*, *21*(10), 11318–11328. DOI: 10.1109/JSEN.2020.3031882

Cataldo, I., Lepri, B., Neoh, M. J. Y., & Esposito, G. (2021). Social media usage and development of psychiatric disorders in childhood and adolescence: A review. *Frontiers in Psychiatry*, *11*(1), 1–15. DOI: 10.3389/fpsyt.2020.508595 PMID: 33519535

Chambers, F., & Sandford, R. (2019). Learning to be human in a digital world: A model of values fluency education for physical education. *Sport Education and Society*, *24* (9), 925–938. DOI: 10.1080/13573322.2018.1515071

Chandana, C., Sarkar, A., Deshmukh, K., Kulkarni, P., Acharjee, P. B., & Lourens, M. (2024). Analysing Employee Management Using Machine Learning Techniques and Solutions in Human Resource Management. *2024 4th International Conference on Innovative Practices in Technology and Management (ICIPTM)*, 1–6.

Chang, K. T. (2016). *Geographic information system. International encyclopedia of geography: people, the earth, environment and technology*, 1-10.

Chang, J. S., Liu, S. C., Ng, J. C., & Ma, P. L. (2020). Monovision with a bifocal diffractive multifocal intraocular lens in presbyopic patients: A prospective, observational case series. *American Journal of Ophthalmology, 212*, 105–115. DOI: 10.1016/j.ajo.2019.11.010 PMID: 31765626

Chang, S., & Pyland, K. M. (2022). The Psychopath's Brain: Is it Structurally Different from the Normal Brain? *Journal of Student Research, 11*(2). Advance online publication. DOI: 10.47611/jsrhs.v11i2.2515

Chan, S. W., & Chong, M. W. (2017). Sentiment analysis in financial texts. *Decision Support Systems, 94*, 53–64. DOI: 10.1016/j.dss.2016.10.006

Chansa Chanda, T., Edwin Phiri, V., Chitondo, L., Chansa Thelma, C., & Vinandi Phiri, E. (2024). Ensuring National Food Security in Southern Africa: Challenges and Solutions. *International Journal of Novel Research in Interdisciplinary Studies, 11*(2), 1–18. DOI: 10.5281/zenodo.10785042

Chase, G. E., Brown, M. T., & Jensen, M. (2022). Emerging adults' digital technology engagement and mental health during the COVID-19 pandemic. *Frontiers in Psychology, 13*, 1023514. Advance online publication. DOI: 10.3389/fpsyg.2022.1023514 PMID: 36467177

Chaubey, G., Gavhane, P. R., Bisen, D., & Arjaria, S. K. (2023). Customer purchasing behavior prediction using machine learning classification techniques. *Journal of Ambient Intelligence and Humanized Computing, 14*(12), 16133–16157. DOI: 10.1007/s12652-022-03837-6

Chauhan, D. S., Akhtar, M. S., Ekbal, A., & Bhattacharyya, P. (2019). Context-aware interactive attention for multi-modal sentiment and emotion analysis. *EMNLP-IJCNLP 2019 - 2019 Conference on Empirical Methods in Natural Language Processing and 9th International Joint Conference on Natural Language Processing,Proceedings of the Conference*, 5647–5657. DOI: 10.18653/v1/D19-1566

Chauhan, D. S., Dhanush, S. R., Ekbal, A., & Bhattacharyya, P. (2020, July). Sentiment and emotion help sarcasm? A multi-task learning framework for multi-modal sarcasm, sentiment and emotion analysis. In *Proceedings of the 58th annual meeting of the association for computational linguistics* (pp. 4351-4360). DOI: 10.18653/v1/2020.acl-main.401

Chauhan, D. S., Ghosal, D., Poria, S., Ekbal, A., & Bhattacharyya, P. (2016). Multi-task Learning for Multi-modal Emotion Recognition and Sentiment Analysis. *ArXiv:1905.05812v1*.

Cheepu, M. (2023). Machine Learning Approach for the Prediction of Defect Characteristics in Wire Arc Additive Manufacturing. *Transactions of the Indian Institute of Metals, 76*(2), 447–455. DOI: 10.1007/s12666-022-02715-1

Chekroud, A. (2018). T107. Why validation matters: A demonstration predicting antipsychotic response using 5 RCTs. *Schizophrenia Bulletin, 44*(suppl_1), S157–S157. DOI: 10.1093/schbul/sby016.383

Chen, R., Zhang, W., & Wang, X. (2020). Machine learning in tropical cyclone forecast modeling: A review. *Atmosphere (Basel), 11*(7), 1–29. DOI: 10.3390/atmos11070676

Chiauzzi, E., Clayton, A., & Huh-Yoo, J. (2020). Video conferencing-based telemental health: Important questions for the COVID-19 era from clinical and patient-centered perspectives. *JMIR Mental Health, 7*(12), 1–12. DOI: 10.2196/24021 PMID: 33180739

Chicco, D., Warrens, M. J., & Jurman, G. (2021). The coefficient of determination R-squared is more informative than SMAPE, MAE, MAPE, MSE and RMSE in regression analysis evaluation. *PeerJ. Computer Science, 7*, e623. DOI: 10.7717/peerj-cs.623 PMID: 34307865

Chikarkova, M., & Fedkovych, Y. (2023). Artificial intelligence and digital art: current state and development prospects. Retrieved from https://www.researchgate.net/publication/376881413

Choi, Y. (2023). GeoAI: Integration of artificial intelligence, machine learning, and deep learning with GIS. *Applied Sciences (Basel, Switzerland), 13*(6), 3895. DOI: 10.3390/app13063895

Chomsky, N. (2023). The False Promise of ChatGPT. The New York Times. Retrieved from https://www.nytimes.com/2023/03/08/opinion/noam-chomsky-chatgpt.html

Choudhary, M., Jha, S., Saxena, D., Singh, A. K., & Associates. (2021). A review of fake news detection methods using machine learning. *2021 2nd International Conference for Emerging Technology (INCET)*, 1–5.

Choudhury, D., & Acharjee, T. (2023). A novel approach to fake news detection in social networks using genetic algorithm applying machine learning classifiers. *Multimedia Tools and Applications, 82*(6), 9029–9045. DOI: 10.1007/s11042-022-12788-1

Christou, E. (2017). The digital time of internet art. Retrieved from https://www.researchgate.net/publication/320623021

Chrysochoou, M., Zaghi, A. E., & Syharat, C. M. (2022). Reframing neurodiversity in engineering education. *Frontiers in Education, 7*(November), 1–12. DOI: 10.3389/feduc.2022.995865

Chu, E., & Roy, D. (2017, November). Audio-visual sentiment analysis for learning emotional arcs in movies. In *2017 IEEE International Conference on Data Mining (ICDM)* (pp. 829-834). IEEE. DOI: 10.1109/ICDM.2017.100

Chujfi, S., & Meinel, C. (2019). Machine learning and human cognition combined to enhance knowledge discovery fidelity. *2019 IEEE First International Conference on Cognitive Machine Intelligence (CogMI)*, 1–7. DOI: 10.1109/CogMI48466.2019.00010

Chung, J., Gulcehre, C., Cho, K., & Bengio, Y. (2014). Empirical evaluation of gated recurrent neural networks on sequence modelling. In *Deep learning and representation learning workshop*. Neural Information Processing Systems.

Cica, D., Sredanovic, B., Tesic, S., & Kramar, D. (2024). Predictive modeling of turning operations under different cooling/lubricating conditions for sustainable manufacturing with machine learning techniques. *Applied Computing and Informatics, 20*(1/2), 162–180. DOI: 10.1016/j.aci.2020.02.001

Ciccarelli, M., Corradini, F., Germani, M., Menchi, G., Mostarda, L., Papetti, A., & Piangerelli, M. (2023). SPECTRE: A deep learning network for posture recognition in manufacturing. *Journal of Intelligent Manufacturing, 34*(8), 3469–3481. DOI: 10.1007/s10845-022-02014-y

CII.IA. (2021, septiembre 20). *La prueba de Turing.* Retrieved from https://www.ciiia.mx/noticiasciiia/la-prueba-de-turing

Cleary, M., Visentin, D., West, S., Lopez, V., & Kornhaber, R. (2018). Promoting emotional intelligence and resilience in undergraduate nursing students: An integrative review. *Nurse Education Today*, *68*(1), 112–120. DOI: 10.1016/j.nedt.2018.05.018 PMID: 29902740

Clouder, L., Karakus, M., Cinotti, A., Ferreyra, M. V., Fierros, G. A., & Rojo, P. (2020). Neurodiversity in higher education: A narrative synthesis. *Higher Education*, *80*(4), 757–778. DOI: 10.1007/s10734-020-00513-6

CONAPRED. (2021). *Situación de las personas con discapacidad en México.*

Contreras-Rodríguez, O., Pujol, J., Batalla, I., Harrison, B. J., Bosque, J., Ibern-Regàs, I., Hernández-Ribas, R., Soriano-Mas, C., Deus, J., López-Solà, M., Pifarré, J., Menchón, J. M., & Cardoner, N. (2014). Disrupted neural processing of emotional faces in psychopathy. *Social Cognitive and Affective Neuroscience*, *9*(4), 505–512. DOI: 10.1093/scan/nst014 PMID: 23386739

Coppersmith, G., Dredze, M., & Harman, C. (2015). Quantifying mental health signals in Twitter. *Proceedings of the Fifth International Conference on Social Media Technologies, Communication, and Informatics*, 51-60

Cordeiro, J. V. (2021). Digital technologies and data science as health enablers: An outline of appealing promises and compelling ethical, legal, and social challenges. *Frontiers in Medicine*, *8*(1), 1–13. DOI: 10.3389/fmed.2021.647897 PMID: 34307394

Cornejo-Plaza, I., & Cippitani, R. (2023). Consideraciones éticas y jurídicas de la Inteligencia Artificial en Educación Superior: desafíos y perspectivas. Revista De Educación Y Derecho, (28). DOI: 10.1344/REYD2023.28.43935

Cosenza, D. N., Korhonen, L., Maltamo, M., Packalen, P., Strunk, J. L., Næsset, E., Gobakken, T., Soares, P., & Tomé, M. (2021). Comparison of linear regression, k-nearest neighbour and random forest methods in airborne laser-scanning-based prediction of growing stock. Forestry. *Forestry*, *94*(2), 311–323. DOI: 10.1093/forestry/cpaa034

Cota-Rivera, E. I., & Gonzalez Correa, M. E. Martinez, M., Bernal Marin, L. A., & Marquez Montenegro, M. Y. (2024). Transforming Education With the Power of Artificial Intelligence: Case Studies. In *Enhancing Higher Education and Research With OpenAI Models* (pp. 113–140). IGI Global.

Cotton, M., Smith, L., & Johnson, T. (2022). Empathy in Education: The Role of Human Interaction. *Educational Leadership*, *79*(5), 23–29.

Coussement, K., & Poel, D. Van Den. (2008). *Churn prediction in subscription services : An application of support vector machines while comparing two parameter-selection techniques. 34*, 313–327. DOI: 10.1016/j.eswa.2006.09.038

Crawford, K. (2013, April 1). *The Hidden Biases in Big Data.* Harvard Business Review, https://hbr.org/2013/04/the-hidden-biases-in-big-data

Creswell, J. W., & Poth, C. N. (2018). *Qualitative Inquiry and Research Design Choosing among Five Approaches* (4th ed.). SAGE Publications, Inc.

Cruz, M., Mafra, S., Teixeira, E., & Figueiredo, F. (2022). Smart Strawberry Farming Using Edge Computing and IoT. *Sensors (Basel), 22*(15), 5866. DOI: 10.3390/s22155866 PMID: 35957425

Crysel, L. C., Crosier, B. S., & Webster, G. D. (2013). The Dark Triad and risk behavior. *Personality and Individual Differences, 54*(1), 35–40. DOI: 10.1016/j.paid.2012.07.029

Cser, L., Geiger, M., Greska, W., & Hoffmann, M. (1991). Three kinds of case-based learning in sheet metal manufacturing. Computers in Industry, 17(2–3), 195-206. ISSN 0166-3615. DOI: 10.1016/0166-3615(91)90032-5

Cyr, L., & Francis, K. (1992). Measures of clinical agreement for nominal and categorical data: The kappa coefficient. *Computers in Biology and Medicine, 22*(4), 239–246. DOI: 10.1016/0010-4825(92)90063-S PMID: 1643847

Dair, Z., Donovan, R., & O'Reilly, R. (2021). Classification of emotive expression using verbal and non-verbal components of speech. In 2021 32nd Irish Signals and Systems Conference (ISSC) (pp. 1–8). IEEE. DOI: 10.1109/ISSC52156.2021.9467869

Dalvi, C., Rathod, M., Patil, S., Gite, S., & Kotecha, K. (2021). A survey of ai-based facial emotion recognition: Features, ml & dl techniques, age-wise datasets and future directions. *IEEE Access : Practical Innovations, Open Solutions, 9*, 165806–165840. DOI: 10.1109/ACCESS.2021.3131733

Daradkeh, T. K., Ghubash, R., El-Rufaie, O. E. F., & Abou-Saleh, M. T. (1999). The rationale, development and reliability of a new screening psychiatric instrument. *Social Psychiatry and Psychiatric Epidemiology, 34*(4), 223–228. DOI: 10.1007/s001270050137 PMID: 10365629

Das, R., & Singh, T. D. (2023). Multimodal sentiment analysis: A survey of methods, trends, and challenges. *ACM Computing Surveys, 55*(13s), 1–38. DOI: 10.1145/3586075

Dass, A., & Boopathi, S. (2016). Experimental Study of Eco-friendly Wire-Cut Electrical Discharge Machining Processes. *International Journal of Innovative Research in Science, Engineering and Technology, 5* (1), 666–675.

Das, T. K. (2016). A customer classification prediction model based on machine learning techniques. *Proceedings of the 2015 International Conference on Applied and Theoretical Computing and Communication Technology, ICATccT 2015*, 321–326. DOI: 10.1109/ICATCCT.2015.7456903

DataScientest. (2022). *Inteligencia Artificial: definición, historia, usos y peligros*. Retrieved from https://datascientest.com/es/inteligencia-artificial-definicion

Davenport, T., & Kalakota, R. (2019). The potential for artificial intelligence in healthcare. *Future Healthcare Journal, 6*(2), 94–98. DOI: 10.7861/futurehosp.6-2-94 PMID: 31363513

Davila Morán, R., & Agüero Corzo, E. (2023). Desafíos éticos de la inteligencia artificial: implicaciones para la sociedad y la economía. Conrado, 19(94), 137-144. http://scielo.sld.cu/scielo.php?script=sci_arttext&pid=S1990-86442023000500137&lng=es&tlng=es

De Luca, G. N., Silva, J. M., Cerasoli, S., Araújo, J., Campos, J., Di Fazio, S., & Modica, G. (2019). Object-based land cover classification of cork oak woodlands using UAV imagery and Orfeo ToolBox. *Remote Sensing*, *11*(10), 1238. DOI: 10.3390/rs11101238

de Medeiros, K., Robert, P., Gauthier, S., Stella, F., Politis, A., Leoutsakos, J., Taragano, F., Kremer, J., Brugnolo, A., Porsteinsson, A. P., Geda, Y. E., Brodaty, H., Gazdag, G., Cummings, J., & Lyketsos, C. (2010). The Neuropsychiatric Inventory-Clinician rating scale (NPI-C): Reliability and validity of a revised assessment of neuropsychiatric symptoms in dementia. *International Psychogeriatrics*, *22*(6), 984–994. DOI: 10.1017/S1041610210000876 PMID: 20594384

Del-Ben, C. M., Rodrigues, C. R., & Zuardi, A. W. (1996). Reliability of the Portuguese version of the structured clinical interview for DSM-III-R (SCID) in a Brazilian sample of psychiatric outpatients. *Brazilian Journal of Medical and Biological Research = Revista Brasileira de Pesquisas Médicas e Biológicas*, *29*(12). https://pubmed.ncbi.nlm.nih.gov/9222432/ PMID: 9222432

Del-Ben, C. M., Vilela, J. A. A., Crippa, J. A. D. S., Hallak, J. E. C., Labate, C. M., & Zuardi, A. W. (2001). Confiabilidade da" Entrevista Clínica Estruturada para o DSM-IV-Versão Clínica" traduzida para o português. *The British Journal of Psychiatry*, *23*, 156–159.

Dell'Olio, F., Natale, T., Wang, Y.-C., & Hung, Y.-J. (2023). Miniaturization of interferometric optical gyroscopes: A review. *IEEE Sensors Journal*, *23*(24), 29948–29968. DOI: 10.1109/JSEN.2023.3327217

Deng, X., & Yang, Z. (2021). Digital proficiency and psychological well-being in online learning: Experiences of first-generation college students and their peers. *Social Sciences (Basel, Switzerland)*, *10*(6), 1–16. DOI: 10.3390/socsci10060192

Deshpande, Y., Ayer, S., Agrawal, T., Agrawal, V., Madankar, T., & Barve, P. (2023). Application of Smart Strategies for Sustainable Manufacturing of Conventional Machining Process: A Review. Journal of The Institution of Engineers (India): Series C, 104. DOI: 10.1007/s40032-023-00995-0

Devika, M. D., Sunitha, C., & Ganesh, A. (2016). Sentiment analysis: A comparative study on different approaches. *Procedia Computer Science*, *87*, 44–49. DOI: 10.1016/j.procs.2016.05.124

Devlin, J., Chang, M. W., Lee, K., & Toutanova, K. (2019). Bert: Pre-training of deep bidirectional Transformers for language understanding. arXiv preprint arXiv:1810.04805.

Dey, L., Chakraborty, S., Biswas, A., Bose, B., & Tiwari, S. (2016). Sentiment analysis of review datasets using naive bayes and k-nn classifier. *arXiv preprint arXiv:1610.09982*.

Di Leo, G., Liguori, C., Pietrosanto, A., & Sommella, P. (2017). A vision system for the online quality monitoring of industrial manufacturing. *Optics and Lasers in Engineering*, *89*, 162–168. DOI: 10.1016/j.optlaseng.2016.05.007

Diakopoulos, N. (2016). Accountability in algorithmic decision making. *Communications of the ACM*, *59*(2), 56–62. DOI: 10.1145/2844110

Díaz-Granados, S. (2023). The green and digital path for Latin American SMEs. In Development Bank of Latin America and the Caribbean. Access https://www.caf.com/en/knowledge/views/2023/11/the-green-and-digital-path-for-latin-american-smes/

Dienlin, T., & Johannes, N. (2020). The impact of digital technology use on adolescent well-being. *Dialogues in Clinical Neuroscience, 22* (2), 135–142. DOI: 10.31887/DCNS.2020.22.2/tdienlin PMID: 32699513

Diesner, J., & Evans, C. S. (2015, August). Little bad concerns: Using sentiment analysis to assess structural balance in communication networks. In Proceedings of the 2015 IEEE/ACM International Conference on Advances in Social Networks Analysis and Mining 2015 (pp. 342-348).

Ding, V. Y., Hubbard, R. A., Rutter, C. M., & Simon, G. E. (2013). Assessing the accuracy of profiling methods for identifying top providers: Performance of mental health care providers. *Health Services and Outcomes Research Methodology, 13*(1), 1–17. DOI: 10.1007/s10742-012-0099-5 PMID: 23565050

Dini, P., Elhanashi, A., Begni, A., Saponara, S., Zheng, Q., & Gasmi, K. (2023). Overview on intrusion detection systems design exploiting machine learning for networking cybersecurity. *Applied Sciences (Basel, Switzerland), 13*(13), 7507. DOI: 10.3390/app13137507

Dodda, R.. (2023). Precision in classification: A comparative study of logistic regression, Naive Bayes, LSTM, and CNN for spam email detection. *International Journal on Recent and Innovation Trends in Computing and Communication, 11*(9), 2276–2280. DOI: 10.17762/ijritcc.v11i9.9233

Dogan, A., & Birant, D. (2021). Machine learning and data mining in manufacturing. Expert Systems with Applications, 166. 114060. ISSN 0957-4174. DOI: 10.1016/j.eswa.2020.114060

Dongen, J. D. (2020). The Empathic Brain of Psychopaths: From Social Science to Neuroscience in Empathy. *Frontiers in Psychology, 11*, 695. DOI: 10.3389/fpsyg.2020.00695 PMID: 32477201

Dong, Q., & Feng, J. (2019). Outlier detection and disparity refinement in stereo matching. *Journal of Visual Communication and Image Representation, 60*, 380–390. DOI: 10.1016/j.jvcir.2019.03.007

Doran, J., Peterson, D., & Price, S. (2012). Earnings conference call content and stock price: The case of REITs. *The Journal of Real Estate Finance and Economics, 45*(2), 402–434. DOI: 10.1007/s11146-010-9266-z

dos Santos Rocha, R., Widera, A., van den Berg, R. P., de Albuquerque, J. P., & Helingrath, B. (2017). Improving the involvement of digital volunteers in disaster management. *IFIP Advances in Information and Communication Technology, 501*, 214–224. DOI: 10.1007/978-3-319-68486-4_17

Doshi-Velez, F., & Kim, B. (2017). *Towards a rigorous science of interpretable machine learning*. arXiv preprint arXiv:1702.08608

Douglas, P. K., Harris, S., Yuille, A., & Cohen, M. S. (2011). Performance comparison of machine learning algorithms and number of independent components used in fMRI decoding of belief vs. disbelief. *NeuroImage, 56*(2), 544–553. DOI: 10.1016/j.neuroimage.2010.11.002 PMID: 21073969

Dowgwillo, E. A., & Aaron, P. L. (2017). Differentiating Dark Triad Traits Within and Across Interpersonal Circumplex Surfaces. *Assessment, 24*(1), 24–44. Advance online publication. DOI: 10.1177/1073191116643161 PMID: 27084838

Dowson, B., Schneider, J., McDermott, O., & Orrell, M. (2023). Online singing groups for people with dementia: Adaptation and resilience in the face of the COVID-19 pandemic. *Dementia (London), 22*(7), 1348–1371. DOI: 10.1177/14713012231179262 PMID: 37294296

Duquesnoy, M., Liu, C., Zapata Dominguez, D., Kumar, V., Ayerbe, E., & Franco, A. A. (2023). Machine learning-assisted multi-objective optimization of battery manufacturing from synthetic data generated by physics-based simulations. Energy Storage Materials, 56, 50-61. ISSN 2405-8297. DOI: 10.1016/j.ensm.2022.12.040

Duran, O., Turan, B., & Kaya, M. (2024). Machine-learning-based ensemble regression for vehicle-to-vehicle distance estimation using a toe-in style stereo camera. *Measurement, 115540*. Advance online publication. DOI: 10.1016/j.measurement.2024.115540

Durant, K. T., & Smith, M. D. (2006). Mining sentiment classification from political web logs. In *Proceedings of Workshop on Web Mining and Web Usage Analysis of the 12th ACM SIGKDD International Conference on Knowledge Discovery and Data Mining (WebKDD-2006),* Philadelphia, PA (Vol. 16).

Du, Z., Sun, X., Zheng, S., Wang, S., Wu, L., An, Y., & Luo, Y. (2024). Optimal biochar selection for cadmium pollution remediation in Chinese agricultural soils via optimized machine learning. *Journal of Hazardous Materials, 476*(135065), 135065. DOI: 10.1016/j.jhazmat.2024.135065 PMID: 38943890

Dwijayanti, S., Iqbal, M., & Suprapto, B. Y. (2022). Real-time implementation of face recognition and emotion recognition in a humanoid robot using a convolutional neural network. *IEEE Access : Practical Innovations, Open Solutions, 10*, 89876–89886. DOI: 10.1109/ACCESS.2022.3200762

El Morr, C., Jammal, M., Ali-Hassan, H., & El-Hallak, W. (2022). K-Nearest Neighbors. In *International Series in Operations Research & Management Science* (pp. 301–318). Springer International Publishing.

El Naqa, I., & Murphy, M. J. (2015). *What is machine learning?* Springer International Publishing.

El-Haj, M., Rayson, P., Walker, M., Young, S., & Simaki, V. (2019). In search of meaning: Lessons, resources and next steps for computational analysis of financial discourse. *Journal of Business Finance & Accounting, 46*(3–4), 265–306. DOI: 10.1111/jbfa.12378

Emami, P., & Marzban, A. (2023). The synergy of artificial intelligence (ai) and geographic information systems (gis) for enhanced disaster management: Opportunities and challenges. *Disaster Medicine and Public Health Preparedness, 17*, 1–3. DOI: 10.1017/dmp.2023.174 PMID: 37694385

Emani, S., Soman, K. P., Sajith Variyar, V. V., & Adarsh, S. (2019). Obstacle detection and distance estimation for autonomous electric vehicle using stereo vision and DNN. In Soft Computing and Signal Processing: Proceedings of ICSCSP 2018, Volume 2 (pp. 639-648). Springer Singapore. DOI: 10.1007/978-981-13-3393-4_65

Escobar-Naranjo, J., Caiza, G., Ayala, P., Jordan, E., Garcia, C. A., & Garcia, M. V. (2023). Autonomous navigation of robots: Optimization with DQN. *Applied Sciences (Basel, Switzerland), 13*(12), 7202. DOI: 10.3390/app13127202

España, G. d. (2023). *Que es la inteligencia artificial.* Retrieved from https://planderecuperacion.gob.es/noticias/que-es-inteligencia-artificial-ia- prtr#:~:text=La%20inteligencia%20artificial%20(IA)%20es,el%20razonamiento%20y%20la%20percepci%C3%B3n

Esri. (n.d.). *ArcGIS API for Python.* Esri. https://www.esri.com/es-es/arcgis/products/arcgis-python-libraries/libraries/arcgis-api-python

Esri. (n.d.). *Get started with ArcGIS Pro.* ArcGIS Pro Documentation. https://pro.arcgis.com/es/pro-app/latest/get-started/get-started.htm

Fairburn, C. G., & Patel, V. (2017). The impact of digital technology on psychological treatments and their dissemination. *Behaviour Research and Therapy, 88*(1), 19–25. DOI: 10.1016/j.brat.2016.08.012 PMID: 28110672

Falahzadeh, M. R., Farokhi, F., Harimi, A., & Sabbaghi-Nadooshan, R. (2023). Deep convolutional neural network and gray wolf optimization algorithm for speech emotion recognition. *Circuits, Systems, and Signal Processing, 42*(1), 449–492. DOI: 10.1007/s00034-022-02130-3

Fan, X., Zhao, S., Zhang, X., & Meng, L. (2023). The impact of improving employee psychological empowerment and job performance based on deep learning and artificial intelligence. [JOEUC]. *Journal of Organizational and End User Computing, 35* (3), 1–14. DOI: 10.4018/JOEUC.321639

Farahani, H., Blagojević, M., Azadfallah, P., Watson, P., Esrafilian, F., & Saljoughi, S. (2023). Deep neural network. In Springer eBooks (pp. 145–185). DOI: 10.1007/978-3-031-31172-7_6

Fariborzi, M., Zeinalnezhad, M., & Saghaei, A. (2022). Accuracy improvement of mood disorders prediction using a combination of data mining and meta -heuristic algorithms. *Journal of Health and Biomedical Informatics, 9*(3), 106–119. DOI: 10.34172/jhbmi.2022.01

Fekadu, A., Soromessa, T., & Dullo, B. W. (2020). GIS-based assessment of climate change impacts on forest habitable Aframomum corrorima (Braun) in Southwest Ethiopia coffee forest. *Journal of Mountain Science, 17*(10), 2432–2446. DOI: 10.1007/s11629-019-5722-2

Ferdinandov, K. G. (2022). The accuracy distortions in self-assessment. *Psychological Thought, 15*(2), 114–147. DOI: 10.37708/psyct.v15i2.706

Ferreira de Rezende, S. W., Soares Silva, L., Cavalini Jr, A. A., & Vieira de Moura Jr, J. R., & Steffen Jr., V. (2023). Digital Twin of an Offshore Riser Systems for Time Series Prediction Using Deep Learning Models. At A.S. De Paula, A.T. Fabro & M.R. Machado (Eds) Proceedings of the XIX International Symposium on Dynamic Problems of Mechanics. DINAME 2023. Brazil. DOI: 10.26678/ABCM.DINAME2023.DIN2023-0069

Ferreira Neto, W., Cavalcante, C., & Do, P. (2021). Deep reinforcement learning-based maintenance decision-making for a steel production line. In *Proceedings of the Proceedings of the 31st European Safety and Reliability Conference*; Castanier, B., Cepin, M., Bigaud, D., & Berenguer, C., Eds (pp. 2611-2618). DOI: 10.3850/978-981-18-2016-8_600-cd

Flores, J. (2023). ¿Están claros los límites de la ética en educación e inteligencia artificial? Universidad Complutense de Madrid, Unidad de Cultura Científica y de Innovación.https://www.ucm.es/otri/noticias-etica-inteligencia-artificial-ucm

Flores-Fuentes, W., Rivas-Lopez, M., Sergiyenko, O., Rodríguez-Quiñonez, J. C., Hernández-Balbuena, D., & Rivera-Castillo, J. (2014). Energy center detection in light scanning sensors for structural health monitoring accuracy enhancement. *IEEE Sensors Journal, 14*(7), 2355–2361. DOI: 10.1109/JSEN.2014.2310224

Folini, A. (2020). *Integrating machine learning techniques into GIS software: development of a comprehensive and versatile QGIS plugin for cluster analysis on geospatial data*. GeoAI. (n.d). *ArcGIS-Pro*https://pro.arcgis.com/es/pro-app/latest/help/analysis/ai/geoai.htm

Fortoul-Diaz, J. A., Carrillo-Martinez, L. A., Centeno-Tellez, A., Cortes-Santacruz, F., Olmos-Pineda, I., & Flores-Quintero, R. R. (2023). A smart factory architecture based on industry 4.0 technologies: Open-source software implementation. *IEEE Access : Practical Innovations, Open Solutions, 11*, 101727–101749. DOI: 10.1109/ACCESS.2023.3316116

Foster, H. (Ed.). (1983). *The Anti-Aesthetic: Essays on Postmodern Culture*. New Press.

Francisco, J., Santilli, M., & Demetrios, D. (2024). Rare Sam Altman Interview, Constructing "Dreams" with GenAI, and Sora & Marvel. AI Minds. Retrieved from https://www.aiminds.com/p/ai-art-dreams-and-hollywood

Frankel, R., Johnson, M., & Skinner, D. J. (1999). An empirical examination of conference calls as a voluntary disclosure medium. *Journal of Accounting Research, 37*(1), 133–150. DOI: 10.2307/2491400

Freitas, J. R. S., Velosa, V. H. S., Abreu, L. T. N., Jardim, R. L., Santos, J. A. V., Peres, B., & Campos, P. F. (2021). Virtual reality exposure treatment in phobias: A systematic review. *The Psychiatric Quarterly, 92*(4), 1685–1710. DOI: 10.1007/s11126-021-09935-6 PMID: 34173160

Friedline, T., Chen, Z., & Morrow, S. (2021). Families' financial stress & well-being: The importance of the economy and economic environments. *Journal of Family and Economic Issues, 42* (S1), 34–51. DOI: 10.1007/s10834-020-09694-9 PMID: 32837140

Fukui, S., Wu, W., Greenfield, J., Salyers, M. P., Morse, G., Garabrant, J., Bass, E., Kyere, E., & Dell, N. (2023). Applying Machine Learning to Human Resources Data: Predicting Job Turnover among Community Mental Health Center Employees. *The Journal of Mental Health Policy and Economics, 26* (2), 63. PMID: 37357871

Furnham, A., & Grover, S. (2021). The moderating effects of emotional stability on the relationship between the Dark Triad and different measures of risk-taking. *Personality and Individual Differences, 171*, 110450. DOI: 10.1016/j.paid.2020.110450

Furtner, M. R., Maran, T., & Rauthmann, J. F. (2017). Dark Leadership: The Role of Leaders' Dark Triad Personality Traits. Leader Development Deconstructed. Annals of Theoretical Psychology, vol 15. Springer DOI: 10.1007/978-3-658-18189-5

Futrell, R. (2023). Validity, reliability, and significance: Empirical methods for NLP and data science. [Association for Computational Linguistics]. *Computational Linguistics*, *49*(1), 249–251. DOI: 10.1162/coli_r_00467

Gaasedelen, O., & Whiteside, D. (2016). Performance and symptom validity-4Evaluating the classification accuracy of the personality assessment inventory validity scales in detecting cognitive response bias. *Archives of Clinical Neuropsychology: The Official Journal of the National Academy of Neuropsychologists, 31*(6), 574.1-574. DOI: 10.1093/arclin/acw042.04

Gabor, D., & Brooks, S. (2020). The digital revolution in financial inclusion: International development in the fintech era. In *Material Cultures of Financialisation* (pp. 69–82). Routledge. DOI: 10.4324/9781351002103-6

Gaggioli, A., Villani, D., Serino, S., Baños, R. M., & Botella, C. (2019). Positive technology designing e-experiences for positive change. *Frontiers in Psychology*, *10*(1), 15–25. DOI: 10.3389/fpsyg.2019.01571 PMID: 31338053

Gaiki. (2020, 13 de febrero). Los algoritmos de la creatividad. Retrieved from https://gaiki.org/los -algoritmos-de-la-creatividad/#:~:text=El%20algoritmo%20de%20la%20Creatividad%20por%20 Requerimiento%20es%20aquel%20que,para%20resolver%20el%20problema%20dado

Gallent-Torres, C., Zapata-González, A., & Ortego-Hernando, J. (2023) El impacto de la inteligencia artificial generativa en educación superior: una mirada desde la ética y la integridad académica/RELIEVE. Revista Electrónica de Investigación y Evaluación Educativa, 29(2). https://www.redalyc.org/journal/916/91676028011/html/

Gammel, I. (2002). *Baroness Elsa: Gender, Dada, and Everyday Modernity: A Cultural Biography*. MIT Press. DOI: 10.7551/mitpress/1517.001.0001

Gao, Y., Shi, D., Li, R., Liu, Z., & Sun, W. (2023). Gyro-net: IMU gyroscopes random errors compensation method based on deep learning. *IEEE Robotics and Automation Letters*, *8*(3), 1471–1478. DOI: 10.1109/LRA.2022.3230594

García, S., Luengo, J., & Herrera, F. (2015). Data Preprocessing in Data Mining. In *Intelligent Systems Reference Library* (Vol. 72). DOI: 10.1007/978-3-319-10247-4

Garcia, D., & Rosenberg, P. (2016). he dark cube: Dark and light character profiles. *PeerJ*, *4*, e1675. DOI: 10.7717/peerj.1675 PMID: 26966650

Garcia-Palacios, A., Hoffman, H., Carlin, A., Furness, T.III, & Botella, C. (2002). Virtual reality in the treatment of spider phobia: A controlled study. *Behaviour Research and Therapy*, *40*(9), 983–993. DOI: 10.1016/S0005-7967(01)00068-7 PMID: 12296495

Ghasemkhani, B., Aktas, O., & Birant, D. (2023). Balanced K-Star: An Explainable Machine Learning Method for Internet-of-Things-Enabled Predictive Maintenance in Manufacturing. *Machines (Basel)*, *11*(3), 322. DOI: 10.3390/machines11030322

Ghosal, D., Akhtar, M. S., Chauhan, D., Poria, S., Ekbal, A., & Bhattacharyya, P. (2018). Contextual inter-modal attention for multi-modal sentiment analysis. *Proceedings of the 2018 Conference on Empirical Methods in Natural Language Processing, EMNLP 2018*, 3454–3466. DOI: 10.18653/v1/D18-1382

Ghosh, M., & Sanyal, G. (2018). An ensemble approach to stabilize the features for multi-domain sentiment analysis using supervised machine learning. *Journal of Big Data, 5* (1), 44. Advance online publication. DOI: 10.1186/s40537-018-0152-5

Giannopoulos, P., Perikos, I., & Hatzilygeroudis, I. (2018). Deep learning approaches for facial emotion recognition: A case study on FER-2013. *Advances in Hybridization of Intelligent Methods: Models, Systems and Applications*, 1–16.

Gidahatari. (n.d.). *Conozca Orfeo Toolbox (OTB) pàra procesamiento de imágenes satelitales*. Gidahatari. https://gidahatari.com/ih-es/conozca-orfeo-toolbox-otb-para-procesamiento-de-imagenes-satelitales

Gindl, S., Weichselbraun, A., & Scharl, A. (2010). Cross-domain contextualization of sentiment lexicons. In *ECAI 2010* (pp. 771–776). IOS Press.

Glanz, K., Rizzo, A., & Graap, K. (2003). Virtual reality for psychotherapy: Current reality and future possibilities. *Psychotherapy (Chicago, Ill.), 40*(1-2), 55–67. DOI: 10.1037/0033-3204.40.1-2.55

Glöckner, A., Michels, M., & Giersch, D. (2020). Predicting personality test scores with machine learning methodology: Investigation of a new approach to psychological assessment. In *PsyArXiv*. DOI: 10.31234/osf.io/ysd3f

Gogoulou, A., Salta, K., & Efstathiou, N. (2022). AI in Education: Supporting Personalized Learning for Neurodiverse Students. *Computers & Education, 176*, 104253. https://doi.org/https://doi.org/10.1016/j.compedu.2022.104253

Goldenberg, S. B., Gopalakrishnan, S. G., Tallapragada, V., Quirino, T., Marks, F., Trahan, S., Zhang, X., & Atlas, R. (2015). The 2012 triply nested, high-resolution operational version of the Hurricane Weather Research and Forecasting Model (HWRF): Track and intensity forecast verifications. *Weather and Forecasting, 30*(3), 710–729. DOI: 10.1175/WAF-D-14-00098.1

Gomes, N., Pato, M., Lourenço, A. R., & Datia, N. (2023). A Survey on Wearable Sensors for Mental Health Monitoring. *Sensors (Basel), 23*(3), 1330. DOI: 10.3390/s23031330 PMID: 36772370

Gómez-Cambronero, Á., Mann, A., Mira, A., Doherty, G., & Casteleyn, S. (2024). Smartphone-based serious games for mental health: A scoping review. *Multimedia Tools and Applications, 1*(2), 1–48. DOI: 10.1007/s11042-024-18971-w PMID: 39553422

Gómez-Ramírez, A., & Soler Anguiano, F. I. (2023). Simulation based optimization of drilling equipment logistics: a case of study (Vol. 217). Procedia Computer Science., https://www.sciencedirect.com/science/article/pii/S1877050922023614, , ISSN 1877-0509.DOI: 10.1016/j.procs.2022.12.283

Gon, S., Ma, H., Wan, Y., & Xu, A. (2019). Machine learning in human-computer nonverbal communication. In *NeuroManagement and Intelligent Computing Method on Multimodal Interaction* (pp. 1–7). DOI: 10.1145/3357160.3357670

Gonçalves, B. (2023). *Can machines think? The controversy that led to the Turing test.* Retrieved from Springer link: https://link.springer.com/article/10.1007/s00146-021-01318-6

Gönül, S., Namlı, T., Coşar, A., & Toroslu, İ. H. (2021). A reinforcement learning based algorithm for personalization of digital, just-in-time, adaptive interventions. *Artificial Intelligence in Medicine, 115*, 102062. DOI: 10.1016/j.artmed.2021.102062 PMID: 34001322

González Izard, S., Sánchez Torres, R., Alonso Plaza, O., Juanes Mendez, J. A., & García-Peñalvo, F. J. (2020). Nextmed: Automatic imaging segmentation, 3D reconstruction, and 3D model visualization platform using augmented and virtual reality. *Sensors (Basel), 20*(10), 2962. DOI: 10.3390/s20102962 PMID: 32456194

Gonzalez, F. (2023, mayo 23). *Wired.* Retrieved from Propuesta de Ley para regular la IA en México: https://es.wired.com/articulos/diputado-presenta-propuesta-de-ley-para-regula-la-ia-en-mexico

González-Bailón, S., & Paltoglou, G. (2015). Signals of public opinion in online communication: A comparison of methods and data sources. *The Annals of the American Academy of Political and Social Science, 659*(1), 95–107. DOI: 10.1177/0002716215569192

Goodfellow, I., Bengio, Y., & Aaron, C. (2016). *Deep Learning.* MIT Press.

Google Maps API Developer Guide. [Online]. Available: https://developers.google.com/maps/documentation [Accessed: Sep. 3, 2024].

Google. (n.d.). *Machine learning guide.* Google Earth Engine. https://developers.google.com/earth-engine/guides/machine-learning

Görgens, E. B., Montaghi, A., & Rodriguez, L. C. E. (2015). A performance comparison of machine learning methods to estimate the fast-growing forest plantation yield based on laser scanning metrics. *Computers and Electronics in Agriculture, 116*, 221–227. DOI: 10.1016/j.compag.2015.07.004

Goswami, T., & Sarma, H. (2020*). Intelligent computing for air pollution monitoring using gis, remote sensing and machine learning.* In Emerging Trends in Electrical, Communications, and Information Technologies: Proceedings of ICECIT-2018 (pp. 125-133). Springer Singapore."

Goy, A., Ardissono, L., & Petrone, G. (2007). Personalization in e-commerce applications. *Lecture Notes in Computer Science (Including Subseries Lecture Notes in Artificial Intelligence and Lecture Notes in Bioinformatics), 4321 LNCS*, 485–520. DOI: 10.1007/978-3-540-72079-9_16

Goyal, K., & Kumar, S. (2021). Financial literacy: A systematic review and bibliometric analysis. *International Journal of Consumer Studies, 45* (1), 80–105. DOI: 10.1111/ijcs.12605

Graham, S., Colin, D., & Lee, E. E. (2020). *Artificial Intelligence for Mental Health and Mental Illnesses: An Overview.* 21(11): 116 1-18 : Current psychiatry reports.DOI: 10.1007/s11920-019-1094-0

Granell, C., & Ostermann, F. O. (2016). Beyond data collection: Objectives and methods of research using VGI and geo-social media for disaster management. *Computers, Environment and Urban Systems, 59*, 231–243. DOI: 10.1016/j.compenvurbsys.2016.01.006

Greenberg, C. (1971). *Art and Culture: Critical Essays.* Beacon Press.

Grigorescu, S. D., Seritan, G. C., Enache, B. A., Argatu, F. C., Adochiei, F. C., & Stanciu, V. (2023). Indoor 3d scanning and navigation system for an automated guided vehicle. Revue Roumaine Des Sciences Techniques—Série Électrotechnique Et Énergétique, 68(2), 232-235.

Grimmer, J., & Stewart, B. M. (2013). Text as data: The promise and pitfalls of automatic content analysis methods for political texts. *Political Analysis*, *21*(3), 267–297. DOI: 10.1093/pan/mps028

Grosman, N., Braude, H., Rovira, S., & Patiño, S. (2021). "Made in Latam: how smart manufacturing can give Latin America new hope for industrialization", Project Documents (LC/TS.2021/111), Santiago, Economic Commission for Latin America and the Caribbean (ECLAC).

Growcoot, M. (2023). Meta is Using Your Instagram Photos to Train its 'Amazing' AI Image Generator. Retrieved from https://petapixel.com/2023/03/15/meta-is-using-your-instagram-photos-to-train-its-amazing-ai-image-generator/

Guasch, A. M. (2009). *El arte del siglo XX en sus exposiciones: 1945-2007*. Akal.

Guastello, S. J., Rieke, M. L., Guastello, D. D., & Billings, S. W. (1992). Assessing the validity of computer-based test interpretations: Rating reliability and individual differences among raters. *Journal of Personality Assessment*, *58*(1), 79–89. DOI: 10.1207/s15327752jpa5801_7 PMID: 16370874

Guo, J., Guo, Y., Kong, X., Zhang, M., & He, R. (2016, February). Discriminative analysis dictionary learning. In Proceedings of the AAAI conference on artificial intelligence (Vol. 30, No. 1). discussion sessions. *The Accounting Review*, *86*(4), 1383–1414. DOI: 10.2308/accr-10034

Hadden, J., Tiwari, A., Roy, R., & Ruta, D. (2007). Computer assisted customer churn management: State-of-the-art and future trends. *Computers & Operations Research*, *34*(10), 2902–2917. DOI: 10.1016/j.cor.2005.11.007

Halitim, A. M., Bouhedda, M., Tchoketch-Kebir, S., & Rebouh, S. (2024). Artificial neural network for tilt compensation in yaw estimation. *Transactions of the Institute of Measurement and Control*, *46*(11), 2087–2096. DOI: 10.1177/01423312231214832

Hall, O. P. (2021). Managing employee turnover: Machine learning to the rescue. *International Journal of Data Science*, *6* (1), 57–82. DOI: 10.1504/IJDS.2021.117472

Hall, R., Peterson, A., & Sandoval, J. (2024). Assistive Technology for Neurodivergent Students: Empowering Independence Through AI. *Learning and Instruction*, *55*, 120–133. https://doi.org/https://doi.org/10.1016/j.learninstruc.2024.101421

Hamilton, L. G., & Petty, S. (2023). Compassionate pedagogy for neurodiversity in higher education: A conceptual analysis. *Frontiers in Psychology*, *14*(February), 1–9. DOI: 10.3389/fpsyg.2023.1093290 PMID: 36874864

Hampson, S. E. (2012). Personality Processes: Mechanisms by which Personality Traits "Get Outside the Skin". *Annual Review of Psychology*, *63*(1), 315–339. DOI: 10.1146/annurev-psych-120710-100419 PMID: 21740225

Han, Z., Wu, J., Huang, C., Huang, Q., & Zhao, M. (2020). A Review on Sentiment Discovery and Analysis of Educational Big-Data. *Wiley Interdisciplinary Reviews. Data Mining and Knowledge Discovery, 10*(1), e1328. DOI: 10.1002/widm.1328

Haribalaji, V., Boopathi, S., & Balamurugan, S. (2014). Effect of Welding Processes on Mechanical and Metallurgical Properties of High Strength Low Alloy (HSLA) Steel Joints. *International Journal of Innovation and Scientific Research, 12* (1), 170–179.

Harris, J. (2008). *American Art: A Cultural History*. Oxford University Press.

Harrison, A., Summers, J., & Mennecke, B. (2016). The Effects of the Dark Triad on Unethical Behavior. *Journal of Business Ethics*, Volume 153, pages 53–77, (2018).

Hasselgren, B., Chrysoulas, C., Pitropakis, N., & Buchanan, W. J. (2022). Using social media & sentiment analysis to make investment decisions. *Future Internet, 15*(1), 5. DOI: 10.3390/fi15010005

Hassini, A., & Belbachir, A. H. (2014). Thermal method of remote sensing for prediction and monitoring earthquake. *2014 1st International Conference on Information and Communication Technologies for Disaster Management, ICT-DM 2014*. DOI: 10.1109/ICT-DM.2014.6917790

Hassouneh, A., Mutawa, A., & Murugappan, M. (2020). Development of a real-time emotion recognition system using facial expressions and EEG based on machine learning and deep neural network methods. *Informatics in Medicine Unlocked, 20*, 100372. DOI: 10.1016/j.imu.2020.100372

Hazarika, D., Gorantla, S., Poria, S., & Zimmermann, R. (2018). Self-Attentive Feature-level Fusion for Multimodal Emotion Detection. *2018 IEEE Conference on Multimedia Information Processing and Retrieval (MIPR)*, 196–201. DOI: 10.1109/MIPR.2018.00043

He, C., Levis, B., Riehm, K. E., Saadat, N., Levis, A. W., Azar, M., Rice, D. B., Krishnan, A., Wu, Y., Sun, Y., Imran, M., Boruff, J., Cuijpers, P., Gilbody, S., Ioannidis, J. P. A., Kloda, L. A., McMillan, D., Patten, S. B., Shrier, I., & Benedetti, A. (2020). The accuracy of the Patient Health Questionnaire-9 algorithm for screening to detect major depression: An individual participant data meta-analysis. *Psychotherapy and Psychosomatics, 89*(1), 25–37. DOI: 10.1159/000502294 PMID: 31593971

Henning, A. S. (2017). *Machine Learning And Natural Language Methods For Detecting*. 446: Electronic Theses and Dissertations.

Hernandez-Barragan, J., Rios, J. D., Alanis, A. Y., Lopez-Franco, C., Gomez-Avila, J., & Arana-Daniel, N. (2020). Adaptative Single Neuron Anti-Windup PID Controller Based on the Extended Kalman Filter Algorithm. *Electronics (Basel), 9*(636), 636. Advance online publication. DOI: 10.3390/electronics9040636

Herzog, T., Brandt, M., Trinchi, A., Sola, A., & Molotnikov, A. (2024). Process monitoring and machine learning for defect detection in laser-based metal additive manufacturing. *Journal of Intelligent Manufacturing, 35*(4), 1407–1437. DOI: 10.1007/s10845-023-02119-y

Hiramath, C. K., & Deshpande, G. (2019). Fake news detection using deep learning techniques. *2019 1st International Conference on Advances in Information Technology (ICAIT)*, 411–415.

Hizlisoy, S., Yildirim, S., & Tufekci, Z. (2021). Music emotion recognition using convolutional long short term memory deep neural networks. *Engineering Science and Technology, an International Journal, 24*(3), 760–767. DOI: 10.1016/j.jestch.2020.10.009

Hochreiter, S., & Schmidhuber, J. (1997). Long short-term memory. *Neural Computation, 9*(8), 1735–1780. DOI: 10.1162/neco.1997.9.8.1735 PMID: 9377276

Holden, M. K. (2005). Virtual environments for motor rehabilitation [review]. *Cyberpsychology & Behavior, 8*(3), 187–211. DOI: 10.1089/cpb.2005.8.187 PMID: 15971970

Holmes, W., Bialik, M., & Fadel, C. (2021). AI in education: Opportunities and risks. *International Journal of Artificial Intelligence in Education, 31*(1), 5–27. https://doi.org/https://doi.org/10.1007/s40593-021-00235-7

Honnef, K. (2003). *Arte Del Siglo XX*. Editorial Océano de México.

Horňas, J., Běhal, J., Homola, P., Senck, S., Holzleitner, M., Godja, N., Pásztor, Z., Hegedüs, B., Doubrava, R., Růžek, R., & Petrusová, L. (2023). Modelling fatigue life prediction of additively manufactured Ti-6Al-4V samples using machine learning approach. International Journal of Fatigue, 169, 107483. ISSN 0142-1123. DOI: 10.1016/j.ijfatigue.2022.107483

Horvath, L., & Rudas, I. J. (1993). A machine learning based approach to manufacturing process planning. ISIE '93 - Budapest: IEEE International Symposium on Industrial Electronics Conference Proceedings, Budapest, Hungary, pp. 429-433, DOI: 10.1109/ISIE.1993.268767

Hossain, M. S., & Muhammad, G. (2019). Emotion recognition using deep learning approach from audio–visual emotional big data. *Information Fusion, 49*, 69–78. DOI: 10.1016/j.inffus.2018.09.008

Houjeij, A., Hamieh, L., Mehdi, N., & Hajj, H. (2012). A novel approach for emotion classification based on fusion of text and speech. In 2012 19th International Conference on Telecommunications (ICT) (pp. 1–6). IEEE. DOI: 10.1109/ICTEL.2012.6221211

Hożyń, S., & Żak, B. (2021). Stereo Vision System for Vision-Based Control of Inspection-Class ROVs. *Remote Sensing (Basel), 13*(24), 5075. DOI: 10.3390/rs13245075

Hsu, Y. W., Zhong, K. Q., Perng, J. W., Yin, T. K., & Chen, C. Y. (2018, November). Developing an on-road obstacle detection system using monovision. In *2018 International Conference on Image and Vision Computing New Zealand (IVCNZ)* (pp. 1-9). IEEE. DOI: 10.1109/IVCNZ.2018.8634799

Hu, Y., Li, W., Wright, D., Aydin, O., Wilson, D., Maher, O., & Raad, M. (2019). *Artificial intelligence approaches.arXiv preprint arXiv:1908.10345.*

Huang, K., Hussain, A., Wang, Q. F., & Zhang, R. (Eds.). (2019). Deep learning: fundamentals, theory and applications (Vol. 2). springer. DOI: 10.1007/978-3-030-06073-2

Huang, D., Wang, S., & Liu, Z. (2021). A systematic review of prediction methods for emergency management. *International Journal of Disaster Risk Reduction, 62*, 102412. Advance online publication. DOI: 10.1016/j.ijdrr.2021.102412

Huang, H., Asemi, A., & Mustafa, M. B. (2023). Sentiment analysis in e-commerce platforms: A review of current techniques and future directions. *IEEE Access : Practical Innovations, Open Solutions, 11*, 90367–90382. DOI: 10.1109/ACCESS.2023.3307308

Hu, L., Wei, S., Zhao, Z., & Wu, B. (2022). Deep learning for fake news detection: A comprehensive survey. *AI Open, 3*, 133–155. DOI: 10.1016/j.aiopen.2022.09.001

Hunter, J. F., Walsh, L. C., Chan, C. K., & Schueller, S. M. (2023). Editorial: The good side of technology: how we can harness the positive potential of digital technology to maximize well-being. *Frontiers in Psychology, 14*(1), 1–3. DOI: 10.3389/fpsyg.2023.1304592 PMID: 37885744

Hussain, A., Arif, A., & Asad, K. M. (2020). Applying Deep Neural Networks for Predicting Dark Triad Personality Trait of Online Users. *International Conference on Information Networking (ICOIN)* (pp. pp. 102-105)). IEEE.

Hussain, H., Tamizharasan, P., & Rahul, C. (2022). Design possibilities and challenges of DNN models: A review on the perspective of end devices. *Artificial Intelligence Review, 55*(7), 1–59. DOI: 10.1007/s10462-022-10138-z

Iban, M. C., & Sekertekin, A. (2022). Machine learning based wildfire susceptibility mapping using remotely sensed fire data and GIS: A case study of Adana and Mersin provinces, Turkey. *Ecological Informatics, 69*, 101647. DOI: 10.1016/j.ecoinf.2022.101647

Iberdrola. (2022). *Innovación, Que es la Inteligencia Artificial.* Retrieved from https://www.iberdrola.com/innovacion/que-es-inteligencia-artificial

Ibrahim, I., & Abdulazeez, A. (2021). The role of machine learning algorithms for diagnosing diseases. *Journal of Applied Science and Technology Trends, 2*(1), 10–19. DOI: 10.38094/jastt20179

Ibrishimova, M. D., & Li, K. F. (2020). A machine learning approach to fake news detection using knowledge verification and natural language processing. *Advances in Intelligent Networking and Collaborative Systems: The 11th International Conference on Intelligent Networking and Collaborative Systems (INCoS-2019)*, 223–234.

IG. (2022). *Trading Algoritmico.*

INEGI. (2020). *Censo de Población y Vivienda.* https://www.inegi.org.mx

Ingale, K. K., & Paluri, R. A. (2022). Financial literacy and financial behaviour: A bibliometric analysis. *Review of Behavioral Finance, 14* (1), 130–154. DOI: 10.1108/RBF-06-2020-0141

Institute for Manufacturing. (2023). Porter's Value Chain. In IfM Management Technology Policy of University of Cambridge. https://www.ifm.eng.cam.ac.uk/research/dstools/value-chain-/

International Monetary Fund – IMF. (2023). Country Composition of WEO Groups. World Economic Outlook Database. Groups and Aggregates Information. In https://www.imf.org/en/Publications/WEO/weo-database/2023/April/groups-and-aggregates

Irani, K. B., Cheng, J., Fayyad, U. M., & Qian, Z. (1993). Applying machine learning to semiconductor manufacturing. *IEEE Expert, 8*(1), 41–47. DOI: 10.1109/64.193054

Issa, D., Demirci, M. F., & Yazici, A. (2020). Speech emotion recognition with deep convolutional neural networks. *Biomedical Signal Processing and Control*, *59*, 101894. DOI: 10.1016/j.bspc.2020.101894

Jadhav, S. D., & Channe, H. (2016). *Comparative study of K-NN, Naive Bayes and decision tree classification techniques*. DOI: 10.21275/v5i1.NOV153131

Jahangir, R., Teh, Y. W., Hanif, F., & Mujtaba, G. (2021). Deep learning approaches for speech emotion recognition: State of the art and research challenges. *Multimedia Tools and Applications*, *80*(16), 23745–23812. DOI: 10.1007/s11042-020-09874-7

Jain, K., & Kaushal, S. (2018). A comparative study of machine learning and deep learning techniques for sentiment analysis. In *2018 7th International conference on reliability, infocom technologies and optimization (Trends and Future Directions)(ICRITO)* (pp. 483-487). IEEE. DOI: 10.1109/ICRITO.2018.8748793

Jain, N., Kumar, S., Kumar, A., Shamsolmoali, P., & Zareapoor, M. (2018). Hybrid deep neural networks for face emotion recognition. *Pattern Recognition Letters*, *115*, 101–106. DOI: 10.1016/j.patrec.2018.04.010

Jain, P. K., Pamula, R., & Srivastava, G. (2021). A Systematic Literature Review on Machine Learning Applications for Consumer Sentiment Analysis Using Online Reviews. *Computer Science Review*, *41*, 100413. Advance online publication. DOI: 10.1016/j.cosrev.2021.100413

Jain, P., Coogan, S. C. P., Subramanian, S. G., Crowley, M., Taylor, S., & Flannigan, M. D. (2020). A review of machine learning applications in wildfire science and management. *Environmental Reviews*, *28*(4), 478–505. DOI: 10.1139/er-2020-0019

Jain, S., Kaur, N., Verma, S., Kavita, , Hosen, A. S. M. S., & Sehgal, S. S. (2022). Use of machine learning in air pollution research: A bibliographic perspective. *Electronics (Basel)*, *11*(21), 3621. DOI: 10.3390/electronics11213621

Jaisawal, R. (2022). Machine learning to evaluate important human capital (HC) determinants impacting IT compensation. *Ramanujan International Journal of Business and Research*, *7* (2), 16–25. DOI: 10.51245/rijbr.v7i2.2022.797

Jaiswal, A., Raju, A. K., & Deb, S. (2020). Facial emotion detection using deep learning. *2020 International Conference for Emerging Technology (INCET)*, 1–5.

Jang, J., Yoon, S., Son, G., Kang, M., Choeh, J. Y., & Choi, K.-H. (2022). Predicting personality and psychological distress using natural language processing: A study protocol. *Frontiers in Psychology*, *13*, 865541. Advance online publication. DOI: 10.3389/fpsyg.2022.865541 PMID: 35465529

Janiesch, C., Zschech, P., & Heinrich, K. (2021). Machine Learning and Deep Learning. *Electronic Markets*, *31*(3), 685–695. DOI: 10.1007/s12525-021-00475-2

Jauregui, M. G. (2024, febrero 12). *Principios éticos de la educación con inteligencia artificial*. Retrieved from https://observatorio.tec.mx/edu-news/principios-eticos- de-la-educacion-con-inteligencia-artificial-ia/

Jayadeva Khemchandani, R., & Chandra, S. (2018). *Twin support vector machines: models, extensions and applications*. Springer Publishing Company, Incorporated.

Jiang, X., Lu, J., Zhao, N., Chen, Z., & Zhao, Z. (2024). A Review of Wear in Additive Manufacturing: Wear Mechanism, Materials, and Process. *Lubricants (Basel, Switzerland)*, *12*(9), 321.

Jia, X., Ma, P., Tarwa, K., & Wang, Q. (2023). Machine vision-based colorimetric sensor systems for food applications. *Journal of Agriculture and Food Research*, *11*, 100503. DOI: 10.1016/j.jafr.2023.100503

Jim, J. R., Talukder, M. A. R., Malakar, P., Kabir, M. M., Nur, K., & Mridha, M. F. (2024). Recent advancements and challenges of NLP-based sentiment analysis: A state-of-the-art review. Natural Language Processing Journal, 100059.

Ji, M., Xie, W., Huang, R., & Qian, X. (2021). Forecasting the suitability of online mental health information for effective self-care developing machine learning classifiers using natural language features. *International Journal of Environmental Research and Public Health*, *18*(19), 10048. DOI: 10.3390/ijerph181910048 PMID: 34639348

Johnson, J. A., Sanghvi, P., & Mehrotra, S. (2021). Technology-based interventions to improve help-seeking for mental health concerns: A systematic review. *Indian Journal of Psychological Medicine*, *44*(4), 332–340. DOI: 10.1177/02537176211034578 PMID: 35949632

Johnson, J., Douze, M., & Jégou, H. (2019). Billion-scale similarity search with GPUs. *IEEE Transactions on Big Data*, *7*(3), 535–547. DOI: 10.1109/TBDATA.2019.2921572

Jones, L. (2023). OpenAI and Microsoft accused of scraping and exploiting private data for AI training. WinBuzzer. Retrieved from https://winbuzzer.com/2023/06/29/openai-accused-of-scraping-and-exploiting-private-data-for-ai-training-xcxwbn/

Jones, R. (2023). How Sam Altman got it wrong on a key part of AI: 'Creativity has been easier for AI than people thought'. Fortune. Retrieved from https://finance.yahoo.com/news/sam-altman-got-wrong-key-210436142.htm

Jones, F., & Müller, G. (2023). Collective Intelligence in Neurodiverse Learning Environments. *Journal of Creativity and Innovation*, *15*(1), 33–47. https://doi.org/https://doi.org/10.1057/s41117-023-00285-8

Jordan, M. I., & Mitchell, T. M. (2015). Machine learning: Trends, perspectives, and prospects. *Science*, *349*(6245), 255–260. DOI: 10.1126/science.aaa8415 PMID: 26185243

Joseph, S., & George, J. (2021). Convolutional autoencoder based feature extraction and KNN classifier for handwritten MODI script character recognition. In Data Science and Security: Proceedings of IDSCS 2021 (pp. 142-149). Springer Singapore. DOI: 10.1007/978-981-16-4486-3_15

Joshipura, K. (2008). Value of predictive values. *Journal of Dental Research*, *87*(10), e1–e1. DOI: 10.1177/154405910808701016 PMID: 18809740

Jospin, L. V., Laga, H., Boussaid, F., & Bennamoun, M. (2022). Bayesian learning for disparity map refinement for semi-dense active stereo vision. arXiv preprint arXiv:2209.05082. https://doi.org//arXiv.2209.05082.DOI: 10.48550

Jozefowicz, R., Vinyals, O., Schuster, M., Shazeer, N., & Wu, Y. (2016). Exploring the limits of language modeling. arXiv preprint arXiv:1602.02410.

Jwo, D. J., Biswal, A., & Mir, I. A. (2023). Artificial neural networks for navigation systems: A review of recent research. *Applied Sciences (Basel, Switzerland), 13*(7), 4475. DOI: 10.3390/app13074475

Kane, R. L., & Parsons, T. D. (2019). Technological developments in assessment. In Goldstein, G., Allen, D. N., & DeLuca, J. (Eds.), *Handbook of Psychological Assessment* (4th ed., pp. 573–592). Elsevier B.V., DOI: 10.1016/B978-0-12-802203-0.00020-1

Kannimuthu, S., & Premalatha, K. (2024). Human Resource Analytics: Performance Analysis of Machine Learning Algorithms for Employee Attrition Analysis. *Proceedings of the 2024 Asia Pacific Conference on Computing Technologies, Communications and Networking*, 105–113. DOI: 10.1145/3685767.3685785

Kaur, J., & Saini, J. R. (2014). Emotion detection and sentiment analysis in text corpus: a differential study with informal and formal writing styles. *International Journal of Computer Application, ISSN*, 0975-8887.

Kawamleh, S. (2022). Against explainability requirements for ethical artificial intelligence in health care. *AI and Ethics, 3*(3), 901–916. DOI: 10.1007/s43681-022-00212-1

Kelly, S., Kaye, S.-A., & Oviedo-Trespalacios, O. (2023). What factors contribute to the acceptance of artificial intelligence? A systematic review. *Telematics and Informatics, 77*, 101925. DOI: 10.1016/j.tele.2022.101925

Kenji, S. (2011). Artificial Neural Networks. *Methodological Advances and Biomedical Applications*, (April), 2011.

Kerber, A., Roth, M., & Herzberg, P. Y. (2021). Personality types revisited–a literature-informed and data-driven approach to an integration of prototypical and dimensional constructs of personality description. *PLoS One, 16*(1), e0244849. DOI: 10.1371/journal.pone.0244849 PMID: 33411758

Kessler, R. C., Berglund, P., Demler, O., Jin, R., Merikangas, K. R., & Walters, E. E. (2005). Lifetime Prevalence and Age of Onset Distributions of DSM-IV Disorders in the National Comorbidity Survey Replication. *Archives of General Psychiatry, 62*(6), 593–602. DOI: 10.1001/archpsyc.62.6.593 PMID: 15939837

Khanam, Z., Alwasel, B., Sirafi, H., & Rashid, M. (2021). Fake news detection using machine learning approaches. *IOP Conference Series. Materials Science and Engineering, 1099*(1), 012040. DOI: 10.1088/1757-899X/1099/1/012040

Khan, F. M., Kumar, A., Puppala, H., Kumar, G., & Gupta, R. (2021). Projecting the Criticality of COVID-19 Transmission in India Using GIS and Machine Learning Methods. *Journal of Safety Science and Resilience = An Quan Ke Xue Yu Ren Xing (Ying Wen), 2*(2), 50–62. DOI: 10.1016/j.jnlssr.2021.05.001

Khan, J. Y., Khondaker, M. T. I., Afroz, S., Uddin, G., & Iqbal, A. (2021). A benchmark study of machine learning models for online fake news detection. *Machine Learning with Applications, 4*, 100032. DOI: 10.1016/j.mlwa.2021.100032

Khanlou, N., & Wray, R. (2014). A whole community approach toward child and youth resilience promotion: A review of resilience literature. *International Journal of Mental Health and Addiction, 12*(1), 64–79. DOI: 10.1007/s11469-013-9470-1 PMID: 24523668

Khan, S. M., Shafi, I., Butt, W. H., Diez, I. de la T., Flores, M. A. L., Galán, J. C., & Ashraf, I. (2023). A Systematic Review of Disaster Management Systems: Approaches, Challenges, and Future Directions. *Land (Basel)*, *12*(8), 1–37. DOI: 10.3390/land12081514

Khan, S., Naseer, M., Hayat, M., Zamir, S. W., Khan, F. S., & Shah, M. (2022). Transformers in vision: A survey. *ACM Computing Surveys*, *54*(10s), 1–41. DOI: 10.1145/3505244

Khaspuria, G., Ranjan, A., Sahil, , Soni, P., & Dadhich, K. (2024). Natural Disaster Mitigation Strategies: A Comprehensive Review. *Journal of Scientific Research and Reports*, *30*(8), 20–34. DOI: 10.9734/jsrr/2024/v30i82221

Khatri, A., Labajo, D., & Rege, M. (2021). Reliability of NLP models to predict human sentiments. *Issues in Information Systems*, *22*(4), 33–48. DOI: 10.48009/4_iis_2021_34-50

Kishwar, A., & Zafar, A. (2023). Fake news detection on Pakistani news using machine learning and deep learning. *Expert Systems with Applications*, *211*, 118558. DOI: 10.1016/j.eswa.2022.118558

Kline, S., Taylor, B., & Huang, Z. (2023). AI and Neurodivergent Learning: The Case for Personalization. *Technology and Education*, *22*(4), 501–520.

Ko, H., Lu, Y., Yang, Z., Ndiaye, N. Y., & Witherell, P. (2023). A framework driven by physics-guided machine learning for process-structure-property causal analytics in additive manufacturing. Journal of Manufacturing Systems, 67, 213-228. ISSN 0278-6125. DOI: 10.1016/j.jmsy.2022.09.010

Koegel, L. K., Ashbaugh, K., & Koegel, R. L. (2020). Improving social communication in ASD through structured interventions. *Journal of Autism and Developmental Disorders*, *50*(4), 1254–1267. https://doi.org/https://doi.org/10.1007/s10803-019-04382-x

Kohrt, B., Asher, L., Bhardwaj, A., Fazel, M., Jordans, M., Mutamba, B., Nadkarni, A., Pedersen, G., Singla, D., & Patel, V. (2018). The role of communities in mental health care in low- and middle-income countries: A meta-review of components and competencies. *International Journal of Environmental Research and Public Health*, *15*(6), 1–31. DOI: 10.3390/ijerph15061279 PMID: 29914185

Kokar, M. M., & Reveliotis, S. A. (1993). Reinforcement learning: Architectures and algorithms. *International Journal of Intelligent Systems*, *8*(8), 875–894. DOI: 10.1002/int.4550080805

Kolaas, K., Berman, A. H., Hedman-Lagerlöf, E., Lindsäter, E., Hybelius, J., & Axelsson, E. (2024). Internet-delivered transdiagnostic psychological treatments for individuals with depression, anxiety or both: A systematic review with meta-analysis of randomised controlled trials. *BMJ Open*, *14*(4), e075796. DOI: 10.1136/bmjopen-2023-075796 PMID: 38569713

Komori-Zevallos, A. R., Montedoro-Garay, F. M., Garcia-Lopez, Y. J., & Quiroz-Flores, J. C. (2023). A Production Model That Combines Lean And Industry 4.0 Principles To Enhance The Productivity Of Small And Medium-Sized Enterprises (SMEs) In Peru's Food Manufacturing Sector [Conference]. 5th Conference On Production Systems and Logistics CPSL 2023-2.Hannover, Germany. DOI: DOI: 10.15488/1525

Kouijzer, M. M. T. E., Kip, H., Bouman, Y. H. A., & Kelders, S. M. (2023). Implementation of virtual reality in healthcare: A scoping review on the implementation process of virtual reality in various healthcare settings. *Implementation Science Communications*, *4*(1), 1–29. DOI: 10.1186/s43058-023-00442-2 PMID: 37328858

Kratzwald, B., Ilic, S., Kraus, M., Feuerriegel, S., & Prendinger, H. (2018). *Deep learning for affective computing: text-based emotion recognition in decision support*. DOI: 10.1016/j.dss.2018.09.002

Kremen, W. S., Elman, J. A., Vuoksimaa, E., Jak, A. J., Mcevoy, L., Holland, D., Lyons, M. J., & Franz, C. E. (2019). P4-363: Improving accuracy of MCI diagnosis and prognosis: A cognitive-genetic approach. *Alzheimer's & Dementia: The Journal of the Alzheimer's Association, 15*(7S_Part_27). DOI: 10.1016/j.jalz.2019.06.4034

Krešňáková, V. M., Sarnovský, M., & Butka, P. (2019). Deep learning methods for Fake News detection. *2019 IEEE 19th International Symposium on Computational Intelligence and Informatics and 7th IEEE International Conference on Recent Achievements in Mechatronics, Automation, Computer Sciences and Robotics (CINTI-MACRo)*, 000143–000148.

Krijn, M., Emmelkamp, P. M. G., Olafsson, R. P., & Biemond, R. (2004). Virtual reality exposure therapy of anxiety disorders: A review. *Clinical Psychology Review*, *24*(3), 259–281. DOI: 10.1016/j.cpr.2004.04.001 PMID: 15245832

Krishna, E. P., Praveena, N., Manju, L., Malathi, N., Giri, R. K., & Preetha, M.E. S. Phalguna Krishna. (2024). IoT-Enabled Wireless Sensor Networks and Geospatial Technology for Urban Infrastructure Management. *Journal of Electrical Systems*, *20*(4s), 2248–2256. DOI: 10.52783/jes.2395

Krogh, A. (2008). What are artificial neural networks? *Nature Biotechnology*, *26*(2), 195–197. DOI: 10.1038/nbt1386 PMID: 18259176

Kuang, W., Yuan, C., Zou, Z., Zhang, J., & Zhang, W. (2024). Autonomous Earthquake Location via Deep Reinforcement Learning. *Seismological Research Letters*, *95*(1), 367–377. DOI: 10.1785/0220230118

Ku, D. Y., & Patel, A. (2020). Kinematic state estimation using multiple DGPS/MEMS-IMU sensors. *IEEE Sensors Letters*, *4*(12), 1–4. Advance online publication. DOI: 10.1109/LSENS.2020.3040661

Kulasegaram, K., & Paradis, E. (2014). Better data ≫ Bigger data. *Medical Teacher*, *36*(11), 1008–1009. DOI: 10.3109/0142159X.2014.917761 PMID: 25072309

Kulkarni, N., & Kaleel, M. (2019). *Customer Behaviour Prediction MSc Research Project MSc Data Analytics*.

Kumar, A., Ekbal, A., Kawahra, D., & Kurohashi, S. (2019). Emotion helps Sentiment: A Multi-task Model for Sentiment and Emotion Analysis. *Proceedings of the International Joint Conference on Neural Networks, 2019-July*. DOI: 10.1109/IJCNN.2019.8852352

Kumar, K. S. (2012). A Survey on Association Rule Mining using Apriori Algorithm. *International Journal of Computer Applications*, *45*(5), 47–50. DOI: 10.5120/5060-7390

Kumar, S., Asthana, R., Upadhyay, S., Upreti, N., & Akbar, M. (2020). Fake news detection using deep learning models: A novel approach. *Transactions on Emerging Telecommunications Technologies*, *31*(2), e3767. DOI: 10.1002/ett.3767

Kwak, A. S., Israelsen, J. O., Morrison, C. T., Bambauer, D. E., & Surdeanu, M. (2022). *Validity assessment of legal will statements as natural language inference*. DOI: 10.18653/v1/2022.findings-emnlp.447

Kyem, P. A. K., & Kyem, P. A. K. (2021). *The Development of Community-Based GIS Applications. Managing Natural Resource Conflicts with Participatory Mapping and PGIS Applications*, 101-133.

Labis, P. M. (2024). Machine Learning for Disaster Risk Reduction - Review and Research Directions. *Sciences, June*, 0–10.

Labrador-Fernández, J. (2023). Implicaciones éticas de la Inteligencia Artificial en las Ciencias de la Educación. Revista Arbitrada Interdisciplinaria Koinonía, 8(16), 1-3. DOI: 10.35381/r.k.v8i16.2545

Lal, T. N., Schröder, M., Hinterberger, T., Weston, J., Bogdan, M., Birbaumer, N., & Schölkopf, B. (2004). Support vector channel selection in BCI. *IEEE Transactions on Biomedical Engineering*, *51*(6), 1003–1010. DOI: 10.1109/TBME.2004.827827 PMID: 15188871

Lalwani, P., Mishra, M. K., Chadha, J. S., & Sethi, P. (2022). Customer churn prediction system: A machine learning approach. *Computing*, *104*(2), 271–294. DOI: 10.1007/s00607-021-00908-y

Lambora, A., Gupta, K., & Chopra, K. (2019). Genetic algorithm- A literature review. *2019 International Conference on Machine Learning, Big Data, Cloud and Parallel Computing (COMITCon)*, 380–384.

Lam, S. U., Xie, Q., & Goldberg, S. B. (2023). Situating meditation apps within the ecosystem of meditation practice: Population-based survey study. *JMIR Mental Health*, *10*(1), 1–7. DOI: 10.2196/43565 PMID: 37115618

Lan, L., Sikov, J., Lejeune, J., Ji, C., Brown, H., Bullock, K., & Spencer, A. E. (2023). A systematic review of using virtual and augmented Reality for the diagnosis and treatment of psychotic disorders. *Current Treatment Options in Psychiatry*, *10*(2), 87–107. DOI: 10.1007/s40501-023-00287-5 PMID: 37360960

Lan, Z., & Turchin, A. (2023). Impact of possible errors in natural language processing-derived data on downstream epidemiologic analysis. *JAMIA Open*, *6*(4), ooad111. Advance online publication. DOI: 10.1093/jamiaopen/ooad111 PMID: 38152447

Lawson, W., Beckett, A. E., & Fernandes, R. (2021). Embracing cognitive diversity: A framework for inclusive pedagogy in higher education. *Educational Research Review*, *34*, 100392. https://doi.org/https://doi.org/10.1016/j.edurev.2021.100392

Le Cunff, A. L., Giampietro, V., & Dommett, E. (2024). Neurodiversity and cognitive load in online learning: A systematic review with narrative synthesis. *Educational Research Review*, *43*(June 2023), 100604. DOI: 10.1016/j.edurev.2024.100604

Lebowitz, B. D. (2000). A public health approach to clinical therapeutics in psychiatry: Directions for new research. *Dialogues in Clinical Neuroscience*, *2*(3), 309–314. DOI: 10.31887/DCNS.2000.2.3/blebowitz PMID: 22034446

Lee, E. E., Torous, J., & Choudhury, M. D. (2021). Artificial Intelligence for Mental Healthcare: Clinical Applications, Barriers, Facilitators, and Artificial Wisdom. *Biological Psychiatry: Cognitive Neuroscience and Neuroimaging*, *6*(9), 856–864. DOI: 10.1016/j.bpsc.2021.02.001 PMID: 33571718

Lee, J. (1995). Machine performance monitoring and proactive maintenance in computer-integrated manufacturing: Review and perspective. *International Journal of Computer Integrated Manufacturing*, *8*(5), 370–380. DOI: 10.1080/09511929508944664

Lee, J. H., & Choi, M. J. (2020). Augmented reality for social skills training in adolescents with autism spectrum disorder: A pilot study. *Journal of Autism and Developmental Disorders*, *50*(6), 2211–2220. DOI: 10.1007/s10803-019-04094-3 PMID: 30864059

Lee, K., Ashton, M. C., & Gallucci, A. (2013). Sex, Power, and Money: Prediction from the Dark Triad and Honesty–Humility. *European Journal of Personality*, *27*(2), 169–184. Advance online publication. DOI: 10.1002/per.1860

Lee, S., & Choi, J. (2004). Landslide susceptibility mapping using GIS and the weight-of-evidence model. *International Journal of Geographical Information Science*, *18*(8), 789–814. DOI: 10.1080/13658810410001702003

Lefranc, G. (2023). Multirobot Allocation In A Flexible Manufacturing System, Using Reinforcement Learning For Decision-Making, Case of Study. *Procedia Computer Science*, *221*, 41–48. DOI: 10.1016/j.procs.2023.07.006

Lei, Z., & Su, W. (2019). Research and application of a rolling gap prediction model in continuous casting. *Metals*, *9*(3), 380. DOI: 10.3390/met9030380

Letzring, T. D., Wells, S. M., & Funder, D. C. (2006). Information quantity and quality affect the realistic accuracy of personality judgment. *Journal of Personality and Social Psychology*, *91*(1), 111–123. DOI: 10.1037/0022-3514.91.1.111 PMID: 16834483

Levis, B., Sun, Y., He, C., Wu, Y., Krishnan, A., Bhandari, P. M., Neupane, D., Imran, M., Brehaut, E., Negeri, Z., Fischer, F. H., Benedetti, A., & Thombs, B. D. (2020). Accuracy of the PHQ-2 alone and in combination with the PHQ-9 for screening to detect major depression: Systematic review and meta-analysis. *Journal of the American Medical Association*, *323*(22), 2290. DOI: 10.1001/jama.2020.6504 PMID: 32515813

Levis, M., Leonard Westgate, C., Gui, J., Watts, B. V., & Shiner, B. (2021). Natural language processing of clinical mental health notes may add predictive value to existing suicide risk models. *Psychological Medicine*, *51*(8), 1382–1391. DOI: 10.1017/S0033291720000173 PMID: 32063248

Levitan, I. B., & Kaczmarek, L. K. (2015). *The neuron: cell and molecular biology*. Oxford University Press. DOI: 10.1093/med/9780199773893.001.0001

Li, C., Zhang, Y., Niu, X., Chen, F., & Zhou, H. (2023). Does artificial intelligence promote or inhibit on-the-job learning? Human reactions to AI at work. *Systems*, *11* (3), 114. DOI: 10.3390/systems11030114

Li, D., Zhou, J., & Liu, Y. (2021). Recurrent-neural-network-based unscented Kalman filter for estimating and compensating the random drift of MEMS gyroscopes in real time. *Mechanical Systems and Signal Processing, 147*(107057), 107057. DOI: 10.1016/j.ymssp.2020.107057

Li, H., Chen, Q., Zhong, Z., Gong, R., & Han, G. (2022). E-word of mouth sentiment analysis for user behavior studies. *Information Processing & Management, 59*(1), 102784. DOI: 10.1016/j.ipm.2021.102784 PMID: 34580562

Li, H., Zhang, R., Lee, Y., Kraut, R. E., & Mohr, D. C. (2023). Systematic review and meta-analysis of AI-based conversational agents for promoting mental health and well-being. *Digital Medicine, 6*(1), 236. Advance online publication. DOI: 10.1038/s41746-023-00979-5 PMID: 38114588

Li, J. P., Haq, A. U., Din, S. U., Khan, J., Khan, A., & Saboor, A. (2020). Heart disease identification method using machine learning classification in e-healthcare. *IEEE Access : Practical Innovations, Open Solutions, 8*, 107562–107582. DOI: 10.1109/ACCESS.2020.3001149

Li, J., Pan, S., Huang, L., & Zhu, X. (2019). A machine learning based method for customer behavior prediction. *Tehnicki Vjesnik (Strojarski Fakultet), 26*(6), 1670–1676. DOI: 10.17559/TV-20190603165825

Limone, P., & Toto, G. A. (2021). Psychological and emotional effects of digital technology on children in COVID-19 Pandemic. *Brain Sciences, 11*(9), 1126. DOI: 10.3390/brainsci11091126 PMID: 34573148

Linaki, E. (2020). The definition of intangible and tangible cultural assets in digital technology. Retrieved from https://www.researchgate.net/publication/344396747

LinkedIn. (n.d.). *Entornos híbridos: Estrategias para integrar infraestructuras*. LinkedIn. https://es.linkedin.com/pulse/entornos-h%C3%ADbridos-estrategias-para-integrar-infraestructuras-kvbmc

Lipton, Z. C. (2016). The Mythos of Model Interpretability. *Proceedings of the 2016 ICML Workshop on Human Interpretability in Machine Learning*, 96-100. https://arxiv.org/abs/1606.03490

Li, S., Wang, G., & Luo, Y. (2022). Tone of language, financial disclosure, and earnings management: A textual analysis of form 20-F. *Financial Innovation, 8*(1), 43. DOI: 10.1186/s40854-022-00346-5

Litvan, I., MacIntyre, A., Goetz, C. G., Wenning, G. K., Jellinger, K., Verny, M., Bartko, J. J., Jankovic, J., McKee, A., Brandel, J. P., Chaudhuri, K. R., Lai, E. C., D'Olhaberriague, L., Pearce, R. K. B., & Agid, Y. (1998). Accuracy of the clinical diagnoses of Lewy body disease, Parkinson disease, and dementia with Lewy bodies: A clinicopathologic study. *Archives of Neurology, 55*(7), 969. DOI: 10.1001/archneur.55.7.969 PMID: 9678315

Liu, B. (2011). Supervised Learning. In *Web Data Mining. Data-Centric Systems and Applications*. Springer., DOI: 10.1007/978-3-642-19460-3_3

Liu, J., Ye, J., Silva Izquierdo, D., Vinel, A., Shamsaei, N., & Shao, S. (2023). A review of machine learning techniques for process and performance optimization in laser beam powder bed fusion additive manufacturing. *Journal of Intelligent Manufacturing, 34*(8), 3249–3275. DOI: 10.1007/s10845-022-02012-0

Liu, Q., Wan, H., & Yu, H. (2023). The Application of Deep Learning in Human Resource Management: A New Perspective on Employee Recruitment and Performance Evaluation. *Academic Journal of Management and Social Sciences, 3* (1), 101–104. DOI: 10.54097/ajmss.v3i1.9573

Liu, S., Zhang, Y., & Huang, J. (2021). Personalized mental health intervention using machine learning: A systematic review. *Health Informatics Journal, 27*(3), 1460458221999187. DOI: 10.1177/1460458221999187

Liu, Y., Lv, B., Wang, Y., & Huang, W. (2020). An end-to-end stereo matching algorithm based on improved convolutional neural network. *Mathematical Biosciences and Engineering, 17*(6), 7787–7803. DOI: 10.3934/mbe.2020396 PMID: 33378920

Li, X., Zhang, J., Du, Y., Zhu, J., Fan, Y., & Chen, X. (2023). A novel deep learning-based sentiment analysis method enhanced with emojis in microblog social networks. *Enterprise Information Systems, 17*(5), 2037160. DOI: 10.1080/17517575.2022.2037160

Li, Z., Yang, Z., Shen, C., Xu, J., Zhang, Y., & Xu, H. (2019). Integrating shortest dependency path and sentence sequence into a deep learning framework for relation extraction in clinical text. *BMC Medical Informatics and Decision Making, 19*(S1), 22. Advance online publication. DOI: 10.1186/s12911-019-0736-9 PMID: 30700301

Llamas, J., Mendoza, O., & Graff, M. (2022). Enfoques regulatorios para la inteligencia artificial (IA). Revista chilena de derecho, 49(3), 31-62. DOI: 10.7764/R.493.2

López, I. (2014). *Justicia ambiental.* EUNOMÍA. Revista en Cultura de la Legalidad, (6), 261-268.

Lopez, O. L., Litvan, I., Catt, K. E., Stowe, R., Klunk, W., Kaufer, D. I., Becker, J. T., & DeKosky, S. T. (1999). Accuracy of four clinical diagnostic criteria for the diagnosis of neurodegenerative dementias. *Neurology, 53*(6), 1292–1292. DOI: 10.1212/WNL.53.6.1292 PMID: 10522887

Loughran, T., & McDonald, B. (2020). Textual analysis in finance. *Annual Review of Financial Economics, 12*(1), 357–375. DOI: 10.1146/annurev-financial-012820-032249

Lu, B., Ott, M., Cardie, C., & Tsou, B. K. (2011). Multi-aspect sentiment analysis with topic models. *Proceedings - IEEE International Conference on Data Mining, ICDM*, 81–88. DOI: 10.1109/ICDMW.2011.125

Lu, C. T., Wang, L. L., Shen, J. H., & Lin, J. A. (2021). Image enhancement using deep-learning fully connected neural network mean filter. *The Journal of Supercomputing, 77*(3), 3144–3164. DOI: 10.1007/s11227-020-03389-6

Luckin, R. (2021). AI in Education: Opportunities and Challenges for Educators. *AI & Society, 36*(1), 1–11. https://doi.org/https://doi.org/10.1007/s00146-020-01050-0

Luo, M., Guo, L., Yu, Y., & Zhang, L. (2020). The psychological and mental impact of coronavirus disease 2019 (COVID-19) on medical staff and general public: A systematic review and meta-analysis. *Psychological Medicine*, •••, 1–10. DOI: 10.1017/S0033291720001200 PMID: 32563745

Luong, M. T., Pham, H., & Manning, C. D. (2015). Effective approaches to attention-based neural machine translation. arXiv preprint arXiv:1508.04025. DOI: 10.18653/v1/D15-1166

Lusardi, A. (2019). Financial literacy and the need for financial education: Evidence and implications. *Swiss Journal of Economics and Statistics, 155* (1), 1–8. DOI: 10.1186/s41937-019-0027-5

Luther, D. (2023). Manufacturing Supply Chain Explained. In Oracle NetSuite. https://www.netsuite.com/portal/resource/articles/erp/manufacturing-supply-chains.shtml

Lyons, A. C., Kass-Hanna, J., Liu, F., Greenlee, A. J., & Zeng, L. (2020). *Building financial resilience through financial and digital literacy in South Asia and Sub-Saharan Africa*. ADBI Working Paper Series.

Lyons, A. C., & Kass-Hanna, J. (2021). A methodological overview to defining and measuring "digital" financial literacy. *Financial Planning Review, 4* (2), e1113. DOI: 10.1002/cfp2.1113

Ma, W., & Lai, Y. (2021). Using Virtual Reality for Neurodiversity Training: A Case Study in Higher Education. *Virtual Reality & Intelligent Hardware, 3*(2), 192–204. https://doi.org/https://doi.org/10.1016/j.vrih.2021.04.003

Madan, B. S., Najma, U., & Rana, D. P. PK, J. K., Sathyakala, S., & Boopathi, S. (2024). Empowering Leadership in Higher Education: Driving Student Performance, Faculty Development, and Institutional Progress. In Best Practices for Behavior Intervention in Special Education (pp. 191-221). IGI Global.

Mahesh, B. (2020). Machine learning algorithms-a review. International Journal of Science and Research (IJSR).[Internet], 9(1), 381-386.

Mahir, E. M., Akhter, S., Huq, M. R., & Associates. (2019). Detecting fake news using machine learning and deep learning algorithms. *2019 7th International Conference on Smart Computing & Communications (ICSCC)*, 1–5.

Mahmood, M. A., Tazeen, M. F., & Abid, A. (2021). The effectiveness of health belief model-based interventions in changing health behavior: A systematic review. *Journal of Public Health, 43*(1), 97–108. DOI: 10.1093/pubmed/fdz043

Maier, H. R., & Dandy, G. C. (1998). The effect of internal parameters and geometry on the performance of back-propagation neural networks: An empirical study. *Environmental Modelling & Software, 13*(2), 193–209. DOI: 10.1016/S1364-8152(98)00020-6

Makhnevich, A., Marefat, A., & Darwish, A. (2021). Evaluating the impact of the AI-driven digital therapy platform Youper on users' emotional health: A pilot study. *BMC Psychiatry, 21*(1), 305. DOI: 10.1186/s12888-021-03437-0 PMID: 34118905

Ma, L., Liu, Y., Zhang, X., Ye, Y., Yin, G., & Johnson, B. A. (2019). Deep learning in remote sensing applications: A meta-analysis and review. [ISPRS]. *ISPRS Journal of Photogrammetry and Remote Sensing, 152*, 166–177. DOI: 10.1016/j.isprsjprs.2019.04.015

Maldonado-Romo, J., & Aldape-Pérez, M. (2021). Sustainable circular micro index for evaluating virtual substitution using machine learning with the path planning problem as a case study. *Sustainability (Basel), 13*(23), 13436. DOI: 10.3390/su132313436

Maliene, V., Grigonis, V., Palevičius, V., & Griffiths, S. (2011). Geographic information system: Old principles with new capabilities. *URBAN DESIGN International, 16*(1), 1–6. DOI: 10.1057/udi.2010.25

Mandave, P. (2013). Data mining using Association rule based on APRIORI algorithm and improved approach with illustration. *International Journal of Latest Trends in Engineering & Technology : IJLTET, 3*(2), 107–113.

Manuel Serra, J., Allen Baumes, L., Moliner, M., Serna, P., & Corma, A. (2007). Zeolite Synthesis Modelling with Support Vector Machines: A Combinatorial Approach. *Combinatorial Chemistry & High Throughput Screening, 10*(1), 13–24. DOI: 10.2174/138620707779802779 PMID: 17266513

Manzini, T., Perali, P., & Murphy, R. R. (2024). Three Challenges in Utilizing Machine Learning to Predict Human Behavior from Observational Data. *Companion of the 2024 ACM/IEEE International Conference on Human-Robot Interaction*, 737–739.

Maples, N. J., & Derlega, V. J. (2021). Virtual reality exposure therapy for post-traumatic stress disorder: A systematic review. *Computers in Human Behavior, 115*, 106610. DOI: 10.1016/j.chb.2020.106610

Maqsood, H., Maqsood, M., Yasmin, S., Mehmood, I., Moon, J., & Rho, S. (2022). Analyzing the stock exchange markets of EU nations: A case study of brexit social media sentiment. *Systems, 10*(2), 24. DOI: 10.3390/systems10020024

Markert, C., Sasangohar, F., Mortazavi, B., & Fields, S. (2021). The Use of Telehealth Technology to Support Health Coaching for Older Adults: Literature Review. *JMIR Human Factors, 8*(1), e23796. DOI: 10.2196/23796 PMID: 33512322

McGurk, K. A., Dagliati, A., Chiasserini, D., Lee, D., Plant, D., Baricevic-Jones, I., Kelsall, J., Eineman, R., Reed, R., Geary, B., Unwin, R. D., Nicolaou, A., Keavney, B. D., Barton, A., Whetton, A. D., & Geifman, N. (2020). The use of missing values in proteomic data-independent acquisition mass spectrometry to enable disease activity discrimination. *Bioinformatics (Oxford, England), 36*(7), 2217–2223. DOI: 10.1093/bioinformatics/btz898 PMID: 31790148

McKay Price, S., Doran, J., Peterson, D., & Bliss, B. (2012). Earnings conference calls and stock returns: The incremental informativeness of textual tone. *Journal of Banking & Finance, 36*(4), 992–1011. DOI: 10.1016/j.jbankfin.2011.10.013

Mechler, R. (2016). Reviewing estimates of the economic efficiency of disaster risk management: Opportunities and limitations of using risk-based cost–benefit analysis. *Natural Hazards, 81*(3), 2121–2147. DOI: 10.1007/s11069-016-2170-y

Mehendale, N. (2020). Facial emotion recognition using convolutional neural networks (FERC). *SN Applied Sciences, 2*(3), 446. DOI: 10.1007/s42452-020-2234-1

Memiş, E., Akarkamçı, H., Yeniad, M., Rahebi, J., & Lopez-Guede, J. M. (2024). Comparative Study for Sentiment Analysis of Financial Tweets with Deep Learning Methods. *Applied Sciences (Basel, Switzerland), 14*(2), 588. DOI: 10.3390/app14020588

Mendoza-Urdiales, R. A., Núñez-Mora, J. A., Santillán-Salgado, R. J., & Valencia-Herrera, H. (2022). Twitter sentiment analysis and influence on stock performance using transfer entropy and EGARCH methods. *Entropy (Basel, Switzerland), 24*(7), 874. DOI: 10.3390/e24070874 PMID: 35885097

Mennis, J., & Mason, M. J. (2011). People, places, and adolescent substance use: Integrating activity space and social network data for analyzing health behavior. *Annals of the Association of American Geographers*, *101*(2), 272–291. DOI: 10.1080/00045608.2010.534712

Meredith, L. S., Humphrey, N., Orlando, M., & Camp, P. (2002). Knowledge of health care benefits among patients with depression. *Medical Care*, *40*(4), 338–346. DOI: 10.1097/00005650-200204000-00009 PMID: 12021689

Metri, J., Ramirez, N., Ramirez, M., & Baigts, D. (2021). Artificial Intelligence for Sensory Acceptability Prediction. At L.R.Hernandez & M.A. Serrano-Meneses (Eds.), Technology, Science and Culture: A Global Vision, Volume IV 2021 (pp. 329-344). UNESCO Hydrometeorological Risks.

Metz, R. K. (2023, junio 29). *The power of machine learning in geospatial analysis: Enhancing decision-making and insights*. Medium. https://ryankmetz.medium.com/the-power-of-machine-learning-in-geospatial-analysis-enhancing-decision-making-and-insights-287cf1bebc51

Meyer, A., Graham, D., & Sharp, M. (2022). The Digital Divide in Education: Addressing the Barriers to Equitable Access. *Computers & Education*, •••, 176.

Meyer, B., Von Hoebel, J., & Dyb, G. (2020). Digital therapy for mental health: A systematic review. *Journal of Medical Internet Research*, *22*(3), e18371. DOI: 10.2196/18371

Miao, T., Li, Q., Hu, X., Wu, X., Wu, W., & Xiao, D. (2022). Virtual rotating MEMS gyrocompassing with honeycomb disk resonator gyroscope. *IEEE Electron Device Letters*, *43*(8), 1331–1334. DOI: 10.1109/LED.2022.3185985

Milani, L., Grumi, S., & Di Blasio, P. (2019). Positive effects of videogame use on visuospatial competencies: The impact of visualization style in preadolescents and adolescents. *Frontiers in Psychology*, *10*(1), 1–9. DOI: 10.3389/fpsyg.2019.01226 PMID: 31231275

Miłkowski, P., Karanowski, K., Wielopolski, P., Kocoń, J., Kazienko, P., & Zięba, M. (2023). Modeling uncertainty in personalized emotion prediction with normalizing flows. *2023 IEEE International Conference on Data Mining Workshops (ICDMW)*. DOI: 10.1109/ICDMW60847.2023.00103

Mira, A., Bretón-López, J., Enrique, Á., Castilla, D., García-Palacios, A., Baños, R., & Botella, C. (2018). Exploring the incorporation of a positive psychology component in a cognitive behavioral internet-based program for depressive symptoms. results throughout the intervention process. *Frontiers in Psychology*, *9*(1), 1–9. DOI: 10.3389/fpsyg.2018.02360 PMID: 30555384

Miralles, I., Granell, C., Díaz-Sanahuja, L., Van Woensel, W., Bretón-López, J., Mira, A., Castilla, D., & Casteleyn, S. (2020). Smartphone Apps for the Treatment of Mental Disorders: Systematic Review. *JMIR mHealth and uHealth*, *8*(4), e14897. DOI: 10.2196/14897 PMID: 32238332

Mirko, D., Gursesli, M. C., & Maria, F. (2023). Cognitive Empathy and the Dark Triad: A Literature Review. *European Journal of Investigation in Health, Psychology and Education*, *3*(11), 2642–2680. PMID: 37998074

Mishra, S., Diesner, J., Byrne, J., & Surbeck, E. (2015, August). Sentiment analysis with incremental human-in-the-loop learning and lexical resource customization. In *proceedings of the 26th ACM conference on Hypertext & social media* (pp. 323-325). DOI: 10.1145/2700171.2791022

Mishra, S., Shukla, P., & Agarwal, R. (2022). Analyzing machine learning enabled fake news detection techniques for diversified datasets. *Wireless Communications and Mobile Computing, 2022*(1), 1575365. DOI: 10.1155/2022/1575365

Mishra, S., Tripathy, H. K., Kumar Thakkar, H., Garg, D., Kotecha, K., & Pandya, S. (2021). An explainable intelligence driven query Prioritization using balanced decision tree approach for multi-level psychological disorders assessment. *Frontiers in Public Health, 9*, 795007. Advance online publication. DOI: 10.3389/fpubh.2021.795007 PMID: 34976936

Mnih, V., Silver, D., & Graves, A. (2015). Human-level control through deep reinforcement learning. *Nature, 518*(7540), 529–533. DOI: 10.1038/nature14236 PMID: 25719670

Mohammad, S. M. (2017). *Challenges in Sentiment Analysis.*, DOI: 10.1007/978-3-319-55394-8_4

Mohanty, A., Venkateswaran, N., Ranjit, P., Tripathi, M. A., & Boopathi, S. (2023). Innovative Strategy for Profitable Automobile Industries: Working Capital Management. In *Handbook of Research on Designing Sustainable Supply Chains to Achieve a Circular Economy* (pp. 412–428). IGI Global.

Monnet, J. M., Chanussot, J., & Berger, F. (2011). Support vector regression for the estimation of forest stand parameters using airborne laser scanning. *IEEE Geoscience and Remote Sensing Letters, 8*(3), 580–584. DOI: 10.1109/LGRS.2010.2094179

Monti, F., Frasca, F., Eynard, D., Mannion, D., & Bronstein, M. M. (2019). Fake news detection on social media using geometric deep learning. *arXiv Preprint arXiv:1902.06673.*

Montoya Herrera, L. F. (2020). LSTM Neural Networks for Remaining Useful Life Estimation of Turbofan Engines.

Morales, J. (2025). Elon Musk's xAI launches Grok 3, expanding AI capabilities with deep search and reasoning. Computerworld. Retrieved from https://www.computerworld.com/article/3826765/musks-xai-launches-grok-3-expanding-ai-capabilities-with-deep-search-and-reasoning.html

Morales, S. P. M. de. (2020). Big data and Sentiment Analysis considering reviews from e-commerce platforms to predict consumer behavior. *MSc Business Research, June.*

Morán-Ortega, S.-A., Ruiz-Tirado, S.-G., Simental-López, L.-M., & Tirado-López, A.-B. (2024). Barreras de la Inteligencia Artificial generativa en estudiantes de educación superior. Percepción docente. *Revista De Investigación En Tecnologías De La Información, 12*(25), 26–37. DOI: 10.36825/RITI.12.25.003

Morency, L. P., Mihalcea, R., & Doshi, P. (2011). Towards multimodal sentiment analysis: Harvesting opinions from the web. *ICMI'11 - Proceedings of the 2011 ACM International Conference on Multimodal Interaction*, 169–176. DOI: 10.1145/2070481.2070509

Mossman, D., Wygant, D. B., & Gervais, R. O. (2012). Estimating the accuracy of neurocognitive effort measures in the absence of a "gold standard.". *Psychological Assessment, 24*(4), 815–822. DOI: 10.1037/a0028195 PMID: 22545695

Mostafa, S., Nasir, K., Tarokh, M. J., Abbasimehr, H., Toosi, K. N., & Tarokh, M. J. (2011). A Neuro-Fuzzy Classifier for Customer Churn Prediction A Neuro-Fuzzy Classifier for Customer Churn Prediction Mostafa Setak. *Article in International Journal of Computer Applications, 19*(8), 975–8887. https://www.researchgate.net/publication/236625411

Mouraz, C. P., Almeida, R. M., & Silva, J. M. (2022). Combining cluster analysis and GIS maps to characterise building stock: Case study in the historical city centre of Viseu, Portugal. *Journal of Building Engineering*, *58*, 104949. DOI: 10.1016/j.jobe.2022.104949

Mpolomoka, D. L. (2021). Using Participatory Approaches To Improve Environmental Sustainabiluity Inselected Parts of Zambia. *Milungushi University Multidisciplinart Journal, 2*(1).

Mühlberger, A., Herrmann, M. J., Wiedemann, G., Ellgring, H., & Pauli, P. (2001). Repeated exposure of flight phobics to flights in virtual reality. *Behaviour Research and Therapy*, *39*(9), 1033–1050. DOI: 10.1016/S0005-7967(00)00076-0 PMID: 11520010

Mukhtar, N., & Khan, M. A. (2018). Urdu sentiment analysis using supervised machine learning approach. *International Journal of Pattern Recognition and Artificial Intelligence*, *32* (2), 1–15. DOI: 10.1142/S0218001418510011

Mukhtar, N., Khan, M. A., & Chiragh, N. (2018). Lexicon-based approach outperforms supervised machine learning approach for Urdu Sentiment Analysis in multiple domains. *Telematics and Informatics*, *35* (8), 2173–2183. Advance online publication. DOI: 10.1016/j.tele.2018.08.003

Mulvenna, M. D., Bond, R., Delaney, J., Dawoodbhoy, F. M., Boger, J., Potts, C., & Turkington, R. (2021). Ethical issues in democratizing digital phenotypes and machine learning in the next generation of digital health technologies. *Philosophy & Technology*, *34*(4), 1945–1960. DOI: 10.1007/s13347-021-00445-8 PMID: 33777664

Munari, B. (2016). *Cómo nacen los objetos?: Apuntes para una metodología proyectual*. Editorial GG.

Munezero, M., Montero, C. S., Sutinen, E., & Pajunen, J. (2014). Are they different? Affect, feeling, emotion, sentiment, and opinion detection in text. *IEEE Transactions on Affective Computing*, *5*(2), 101–111. DOI: 10.1109/TAFFC.2014.2317187

Munikar, M., Shakya, S., & Shrestha, A. (2019). *Fine-grained sentiment classification using BERT. In 2019 Artificial Intelligence for Transforming Business and Society (AITB)* (Vol. 1). IEEE.

Muris, P., & Otgaar, H. (2023). Self-Esteem and Self-Compassion: A Narrative Review and Meta-Analysis on Their Links to Psychological Problems and Well-Being. *Psychology Research and Behavior Management, 16*, 2961–2975. DOI: 10.2147/PRBM.S402455 PMID: 37554304

Nagappan, S. D., & Daud, S. M. (2021). Machine learning predictors for sustainable urban planning. *International Journal of Advanced Computer Science and Applications*, *12*(7). Advance online publication. DOI: 10.14569/IJACSA.2021.0120787

Naik, N., Hameed, B. M. Z., Shetty, D. K., Swain, D., Shah, M., Paul, R., Aggarwal, K., Ibrahim, S., Patil, V., Smriti, K., Shetty, S., Rai, B. P., Chlosta, P., & Somani, B. K. (2022). Legal and ethical consideration in artificial intelligence in healthcare: Who takes responsibility? *Frontiers in Surgery*, *9*(1), 1–13. DOI: 10.3389/fsurg.2022.862322 PMID: 35360424

Najafi-Zangeneh, S., Shams-Gharneh, N., Arjomandi-Nezhad, A., & Hashemkhani Zolfani, S. (2021). An improved machine learning-based employees attrition prediction framework with emphasis on feature selection. *Mathematics*, *9* (11), 1226. DOI: 10.3390/math9111226

Nakisa, B., Rastgoo, M. N., Rakotonirainy, A., Maire, F., & Chandran, V. (2018). Long short term memory hyperparameter optimization for a neural network based emotion recognition framework. *IEEE Access : Practical Innovations, Open Solutions*, *6*, 49325–49338. DOI: 10.1109/ACCESS.2018.2868361

Nandwani, P., & Verma, R. (2021). A review on sentiment analysis and emotion detection from text. *Social Network Analysis and Mining*, *11* (1), 1–19. DOI: 10.1007/s13278-021-00776-6 PMID: 34484462

Nardo, M., Petracco-Giudici, M., & Naltsidis, M. (2016). Walking down wall street with a tablet: A survey of stock market predictions using the web. *Journal of Economic Surveys*, *30*(2), 356–369. DOI: 10.1111/joes.12102

Naylor, C., Bell, A., & Parsonage, M. (2021). The role of technology in mental health services: A systematic review. *The Lancet. Psychiatry*, *8*(2), 98–106. DOI: 10.1016/S2215-0366(20)30466-7

Ndulue, O., & Naslund, J. A. (2024). Digital interventions for alcohol use disorders: A narrative review of opportunities to advance prevention, treatment, and recovery. *Psychiatry Research Communications*, *100183*(3), 100183. Advance online publication. DOI: 10.1016/j.psycom.2024.100183

Neethu, M. S., & Rajasree, R. (2013). Sentiment analysis in twitter using machine learning techniques. *2013 4th International Conference on Computing, Communications and Networking Technologies, ICCCNT 2013*. DOI: 10.1109/ICCCNT.2013.6726818

Nelson, D., Inghels, M., Kenny, A., Skinner, S., McCranor, T., Wyatt, S., Phull, J., Nanyonjo, A., Yusuff, O., & Gussy, M. (2023). Mental health professionals and telehealth in a rural setting: A cross-sectional survey. *BMC Health Services Research*, *23*(1), 200–226. DOI: 10.1186/s12913-023-09083-6 PMID: 36849933

Neumann, M., Niessen, A. S. M., & Meijer, R. R. (2023). Predicting decision-makers' algorithm use. *Computers in Human Behavior*, *145*(107759), 107759. DOI: 10.1016/j.chb.2023.107759

Ngeljaratan, L., Bas, E. E., & Moustafa, M. A. (2024). Unmanned Aerial Vehicle-Based Structural Health Monitoring and Computer Vision-Aided Procedure for Seismic Safety Measures of Linear Infrastructures. *Sensors (Basel)*, *24*(5), 1450. DOI: 10.3390/s24051450 PMID: 38474985

Ngumimi, I. K., Soo-Hyung, K., & Min, J. (2023). A Review of Machine Learning and Deep Learning Approaches on Mental Health Diagnosis. *In Healthcare*, 11(3): 285. DOI: 10.3390/healthcare11030285

Nijhawan, R., Sharma, H., Sahni, H., & Batra, A. (2017, December). *A deep learning hybrid CNN framework approach for vegetation cover mapping using deep features.* In 2017 13th international conference on signal-image technology & internet-based systems (SITIS) (pp. 192-196). IEEE. DOI: 10.1109/SITIS.2017.41

Noor, K., Siddiquee, E. A., Sarma, D., Nandi, A., Akhter, S., Hossain, S., . . . Hossain, M. S. (2017, June). Performance analysis of a surveillance system to detect and track vehicles using Haar cascaded classifiers and optical flow method. In 2017 12th IEEE Conference on Industrial Electronics and Applications (ICIEA) (pp. 258-263). IEEE. DOI: 10.1109/ICIEA.2017.8282853

Norrman, F. (2020). *Predicting employee attrition with machine learning on an individual level, and the effects it could have on an organization.*

Nuhu, A. A., Zeeshan, Q., Safaei, B., & Shahzad, M. A. (2023). Machine learning-based techniques for fault diagnosis in the semiconductor manufacturing process: A comparative study. *The Journal of Supercomputing*, *79*(2), 2031–2081. DOI: 10.1007/s11227-022-04730-x

Nurunnabi, M., Almusharraf, N., & Aldeghaither, D. (2020). Mental health and well-being during the COVID-19 pandemic in higher education: Evidence from G20 countries. *Journal of Public Health Research, 9* (1_suppl), jphr-2020.

NV5 Geospatial. (n.d.). *ENVI deep learning*. NV5 Geospatial. https://www.nv5geospatialsoftware.com/Products/ENVI-Deep-Learning

O'Riordan, M. A., & Bearer, C. (2020). Correspondence on statistical rigor and kappa considerations: Which, when, and clinical context matters. *Pediatric Research*, *88*(1), 6–6. DOI: 10.1038/s41390-020-0880-z PMID: 32225173

Obermeyer, Z., Powers, B., Vogeli, C., & Mullainathan, S. (2019). Dissecting racial bias in an algorithm used to manage the health of populations. *Science*, *366*(6464), 447–453. DOI: 10.1126/science.aax2342 PMID: 31649194

Obrenovic, B., Gu, X., & Godinic, D. (2024). Generative AI and human–robot interaction: Implications and future agenda for business, society and ethics. *AI & Society*, •••, 1–14. DOI: 10.1007/s00146-024-01889-0

Odgers, C. L., & Jensen, M. R. (2020). Adolescent development and growing divides in the digital age. *Dialogues in Clinical Neuroscience*, *22*(2), 143–149. DOI: 10.31887/DCNS.2020.22.2/codgers PMID: 32699514

OECD. (2023). *The potential impact of Artificial Intelligence on equity and inclusion in education.* OECD iLibrary. https://www.oecd-ilibrary.org/education/the-potential-impact-of-artificial-intelligence-on-equity-and-inclusion-in-education_181source

Okuno, F., Yoshimoto, K., Takahashi, H., & Hamagishi, G. (2024, March). Evaluation method of matching points for improving accuracy of position estimation using monocular endoscope. In *Endoscopic Microscopy XIX* (Vol. 12820, pp. 49–55). SPIE., DOI: 10.1117/12.3001311

Olawade, D. B., Wada, O. Z., Odetayo, A., David-Olawade, A. C., Asaolu, F., & Eberhardt, J. (2024). Enhancing mental health with Artificial Intelligence: Current trends and future prospects. *Journal of Medicine, Surgery, and Public Health*, *3*, 100099. DOI: 10.1016/j.glmedi.2024.100099

Olmos, F. G. (2023, Junio 2). *Expansión.* México avanza en la adopción de la IA y así se está usando en las empresas.

Openshaw, S. (1992). Some suggestions concerning the development of artificial intelligence tools for spatial modelling and analysis in GIS. *The Annals of Regional Science*, *26*(1), 35–51. DOI: 10.1007/BF01581479

Orellana Cordero, F. R. (2020). Desarrollo de un modelo de análisis de improductividad en una línea de ensamble de refrigeración para empresas de manufactura en el Ecuador.

Osmanoğlu, U. Ö., Atak, O. N., Çağlar, K., Kayhan, H., & Can, T. (2020). Sentiment Analysis for Distance Education Course Materials: A Machine Learning Approach. *Journal of Educational Technology and Online Learning*, *3* (1), 31–48. DOI: 10.31681/jetol.663733

Østergaard, S. D., Seidler, Z., & Rice, S. (2023). The ICD-11 opens the door for overdue improved identification of depression in men. *World Psychiatry; Official Journal of the World Psychiatric Association (WPA)*, *22*(3), 480–481. DOI: 10.1002/wps.21124 PMID: 37713575

Pachar, S., Dudeja, D., Batra, N., Tomar, V., Bhimavarapu, J. P., & Singh, A. K. (2023). Recent Developments in Machine Learning Predictive Analytics for Disaster Resource Allocation †. *Engineering Proceedings*, *59*(1), 1–10. DOI: 10.3390/engproc2023059019

Pal, S., Mukhopadhyay, S., & Suryadevara, N. (2021). Development and progress in sensors and technologies for human emotion recognition. *Sensors (Basel)*, *21*(16), 5554. DOI: 10.3390/s21165554 PMID: 34451002

Panos, G. A., & Wilson, J. O. (2020). Financial literacy and responsible finance in the FinTech era: Capabilities and challenges. []. Taylor & Francis.]. *European Journal of Finance*, *26* (4–5), 297–301. DOI: 10.1080/1351847X.2020.1717569

Pan, T. (2021). Performance Evaluation Method of Enterprise Human Resource Management Based on Machine Learning. *2021 IEEE International Conference on Industrial Application of Artificial Intelligence (IAAI)*, 100–105. DOI: 10.1109/IAAI54625.2021.9699954

Paredes Rodríguez, A. M., Bravo Bastidas, J. J., Osorio Gómez, J. C., Peña Orozco, D. L., & González Feliu, J. (2023). Fuzzy AHP TOPSIS Methodology for Multicriteria ABC Inventory Classification. *Journal of Engineering*, *2023*(1), 7661628. DOI: 10.1155/2023/7661628

Pargent, F., & Albert-von der Gönna, J. (2018). Predictive modeling with psychological Panel data. *Zeitschrift für Psychologie*, *226*(4), 246–258. DOI: 10.1027/2151-2604/a000343

Park, D., Lee, H., & Young-Jin, K. (2021). Development of machine learning model for diagnostic disease prediction based on laboratory tests. *Scientific Reports*, *11*(1), 7567. Advance online publication. DOI: 10.1038/s41598-021-87171-5 PMID: 33828178

Park, S. Y., Nicksic Sigmon, C., & Boeldt, D. (2022). A Framework for the Implementation of Digital Mental Health Interventions: The Importance of Feasibility and Acceptability Research. *Cureus*, *14*(9), e29329. DOI: 10.7759/cureus.29329 PMID: 36277565

Parra Garcés, V. (2020). Nanocomposites analysis through machine learning. Universidad de los Andes. Disponible en: http://hdl.handle.net/1992/48944

Parsazadeh, M., Sharma, S., & Dahotre, N. (2023). Towards the next generation of machine learning models in additive manufacturing: A review of process dependent material evolution. Progress in Materials Science, 135, 101102. ISSN 0079-6425. DOI: 10.1016/j.pmatsci.2023.101102

Parsons, T. D., & Rizzo, A. A. (2008). Affective outcomes of virtual reality exposure therapy for anxiety and specific phobias: A meta-analysis. *Journal of Behavior Therapy and Experimental Psychiatry*, *39*(3), 250–261. DOI: 10.1016/j.jbtep.2007.07.007 PMID: 17720136

Patwardhan, A. S. (2017). Multimodal mixed emotion detection. *Proceedings of the 2nd International Conference on Communication and Electronics Systems, ICCES 2017, 2018-Janua* (Icces), 139–143. DOI: 10.1109/CESYS.2017.8321250

Paul, T. (2022). Role of Technology in the field of Clinical Psychology. *Indian Journal of Behavioral Research & Therapy*, *3*(1 & 2), 110–120. https://www.researchgate.net/publication/370779032

Peng, S., Cao, L., Zhou, Y., Ouyang, Z., Yang, A., Li, X., Jia, W., & Yu, S. (2022). A survey on deep learning for textual emotion analysis in social networks. *Digital Communications and Networks*, *8*(5), 745–762. DOI: 10.1016/j.dcan.2021.10.003

Pereira, L. A. S. (2024). *Using machine learning to predict employee turnover: A case study of the Willis Towers Watson Lisbon Hub* [PhD Thesis].

Perez, R.A.; Lilkendey, J.T.; and, Koh, S.W. (1994). Machine learning for a dynamic manufacturing environment. ACM SIGICE Bulleting, 19(3). pp. 5.9. DOI: 10.1145/182063.182067

Pérez, J. S., Boulton, M., & Van Melle, E. (2021). Natural language processing in mental health: A systematic review. *Journal of Medical Internet Research*, *23*(6), e25492. DOI: 10.2196/25492 PMID: 33944788

Peters, D., Calvo, R. A., & Ryan, R. M. (2018). Designing for motivation, engagement, and well-being in digital experience. *Frontiers in Psychology*, *9*(1), 1–15. DOI: 10.3389/fpsyg.2018.00797 PMID: 29892246

Peters, R. D. (2000). Predicting the end of term status of community college general psychology students. *Journal of College Student Retention*, *2*(2), 109–114. DOI: 10.2190/1V94-J04T-8U8C-DNY2

Petri, L. D. P. S., Moutinho, E. A., Silva, R. P., Capelini, R. M., Salustiano, R., Ferraz, G. M. F., Neto, E. T. W., Villibor, J. P., & Pinto, S. S. (2020). A portable system for the evaluation of the degree of pollution of transmission line insulators. *Energies*, *13*(24), 6625. DOI: 10.3390/en13246625

Philippas, N. D., & Avdoulas, C. (2021). Financial literacy and financial well-being among generation-Z university students: Evidence from Greece. In *Financial Literacy and Responsible Finance in the FinTech Era* (pp. 64–85). Routledge. DOI: 10.4324/9781003169192-5

Philippe, T. J., Sikder, N., Jackson, A., Koblanski, M. E., Liow, E., Pilarinos, A., & Vasarhelyi, K. (2022). Digital Health Interventions for Delivery of Mental Health Care: Systematic and Comprehensive Meta-Review. *JMIR Mental Health*, *9*(5), e35159. DOI: 10.2196/35159 PMID: 35551058

Pinto, A. M., Rocha, L. F., & Moreira, A. P. (2013). Object recognition using laser range finder and machine learning techniques. *Robotics and Computer-integrated Manufacturing*, *29*(1), 12–22. DOI: 10.1016/j.rcim.2012.06.002

Ponce, P., Anthony, B., Bradley, R., Maldonado-Romo, J., Méndez, J. I., Montesinos, L., & Molina, A. (2024). Developing a virtual reality and AI-based framework for advanced digital manufacturing and nearshoring opportunities in Mexico. *Scientific Reports*, *14*(1), 11214. DOI: 10.1038/s41598-024-61514-4 PMID: 38755242

Pontikas, C., Tsoukalas, E., & Serdari, A. (2020). A map of assistive technology educative instruments in neurodevelopmental disorders. *Disability and Rehabilitation. Assistive Technology*, *17*(7), 738–746. DOI: 10.1080/17483107.2020.1839580 PMID: 33125855

Poria, S., Chaturvedi, I., Cambria, E., & Hussain, A. (2016). Convolutional MKL based multimodal emotion recognition and sentiment analysis. *Proceedings - IEEE International Conference on Data Mining, ICDM*, 439–448. DOI: 10.1109/ICDM.2016.0055

Poria, S., Mazumder, N., Cambria, E., Hazarika, D., Morency, L. P., & Zadeh, A. (2017). Context-dependent sentiment analysis in user-generated videos. *ACL 2017 - 55th Annual Meeting of the Association for Computational Linguistics, Proceedings of the Conference (Long Papers)*, *1*, 873–883. DOI: 10.18653/v1/P17-1081

Poria, S., Cambria, E., Howard, N., Huang, G. B., & Hussain, A. (2016). Fusing audio, visual and textual clues for sentiment analysis from multimodal content. *Neurocomputing*, *174*, 50–59. DOI: 10.1016/j.neucom.2015.01.095

Poria, S., Peng, H., Hussain, A., Howard, N., & Cambria, E. (2017). Ensemble application of convolutional neural networks and multiple kernel learning for multimodal sentiment analysis. *Neurocomputing*, *261*, 217–230. DOI: 10.1016/j.neucom.2016.09.117

Porter, M. E. (1985). *Competitive advantage*. The Free Press.

Pradhan, B., Tehrany, M. S., & Jebur, M. N. (2016). A New Semiautomated Detection Mapping of Flood Extent from TerraSAR-X Satellite Image Using Rule-Based Classification and Taguchi Optimization Techniques. *IEEE Transactions on Geoscience and Remote Sensing*, *54*(7), 4331–4342. DOI: 10.1109/TGRS.2016.2539957

Pranav, E., Kamal, S., Chandran, C. S., & Supriya, M. (2020). Facial emotion recognition using deep convolutional neural network. *2020 6th International Conference on Advanced Computing and Communication Systems (ICACCS)*, 317–320.

Priyanka, N., Sethi, S., Sahai, A., Srivastava, A., Sambathkumar, M., & Boopathi, S. (2024). Reality for Human Experience in AI in the Digital Economy. In *Multidisciplinary Applications of Extended Reality for Human Experience* (pp. 374–400). IGI Global., DOI: 10.4018/979-8-3693-2432-5.ch016

Psichology Today. (2024, 09 21). *The Emotional Lives of Psychopaths What do they really feel?* Retrieved from psychologytoday: https://www.psychologytoday.com/us/blog/shadow-boxing/202104/the-emotional-lives-of-psychopaths

Pusey, M., Wong, K. W., & Rappa, N. A. (2020). Resilience interventions using interactive technology: A scoping review. *Interactive Learning Environments, 30*(10), 1940–1955. DOI: 10.1080/10494820.2020.1772837

Qamar, N. (2022). THE IMPACT OF MACHINE LEARNING ON HUMAN RESOURCE MANAGEMENT PRACTICE. *NeuroQuantology: An Interdisciplinary Journal of Neuroscience and Quantum Physics, 20* (13), 2424.

Qian, L., Luo, Z., Du, Y., & Guo, L. (2009). *Cloud computing: An overview. In Cloud Computing:* First International Conference, CloudCom 2009, Beijing, China, December 1-4, 2009. [Springer Berlin Heidelberg.]. *Proceedings, 1*, 626–631.

Qian, R., Zhao, Y., Feng, R., Yang, W., & Zhang, Z. (2023). An algorithm for single view occlusion area detection in binocular stereo matching. *International Journal of Pattern Recognition and Artificial Intelligence, 37*(2), 2350003. DOI: 10.1142/S0218001423500039

Quigley, F., Moorhead, A., Bond, R., Zheng, H., & McAloon, T. (2019). A virtual reality training tool to improve weight-related communication across healthcare settings. In *Proceedings of the 31st European Conference on Cognitive Ergonomics* (9),19–22. DOI: 10.1145/3335082.3335121

RAE. (2023). *Diccionario de la lengua española.* Retrieved from https://dle.rae.es

Raffel, C., Shazeer, N., Roberts, A., Lee, K., Narang, S., Matena, M., Zhou, Y., Li, W., & Liu, P. J. (2020). Exploring the limits of transfer learning with a unified text-to-text transformer. *Journal of Machine Learning Research, 21*(1), 5485–5551.

Rai, R., Tiwari, M. K., Ivanov, D., & Dolgui, A. (2021). Machine learning in manufacturing and industry 4.0 applications. *International Journal of Production Research, 59*(16), 4773–4778. DOI: 10.1080/00207543.2021.1956675

Rajaraman, V. (2014). JohnMcCarthy — Father of artificial intelligence. *Resonance, 19*(3), 198–207. DOI: 10.1007/s12045-014-0027-9

Ramírez-Hernández, L. R., Rodríguez-Quinoñez, J. C., Castro-Toscano, M. J., Hernández-Balbuena, D., Flores-Fuentes, W., Rascón-Carmona, R., Lindner, L., & Sergiyenko, O. (2020). Improve three-dimensional point localization accuracy in stereo vision systems using a novel camera calibration method. *International Journal of Advanced Robotic Systems, 17*(1), 1729881419896717. DOI: 10.1177/1729881419896717

Ramos, G. (2024, Febrero). *Foro Global sobre la ética de la IA2024.* Retrieved from https://www.unesco.org/es/artificial-intelligence/recommendation-ethics

Ramshaw, G., McKeown, A., Lee, R., Conlon, A., Brown, D., & Kennedy, P. J. (2022). Introduction of technology to support young people's care and mental health: A rapid evidence review. *Child and Youth Care Forum, 52*(3), 509–531. DOI: 10.1007/s10566-022-09700-1 PMID: 35966643

Ranganathan, H., Chakraborty, S., & Panchanathan, S. (2016). Multimodal emotion recognition using deep learning architectures. *2016 IEEE Winter Conference on Applications of Computer Vision, WACV 2016*. DOI: 10.1109/WACV.2016.7477679

Ranjan, N., Kumar, R., Kumar, R., Kaur, R., & Singh, S. (2023). Investigation of Fused Filament Fabrication-Based Manufacturing of ABS-Al Composite Structures: Prediction by Machine Learning and Optimization. *Journal of Materials Engineering and Performance*, *32*(10), 4555–4574. DOI: 10.1007/s11665-022-07431-x

Rao, G., Huang, W., Feng, Z., & Cong, Q. (2018). LSTM with sentence representations for document-level sentiment classification. *Neurocomputing*, *308*, 49–57. DOI: 10.1016/j.neucom.2018.04.045

Rashidiani, S., Doyle, T. E., Samavi, R., Duncan, L., Pires, P., & Sassi, R. (2022). Textionnaire: An NLP-based questionnaire analysis method for complex and ambiguous task decision support. *2022 IEEE 21st International Conference on Cognitive Informatics & Cognitive Computing (ICCI*CC).*

Rathore, R., & Rathore, S. P. S. (2024). Machine Learning Applications in Human Resource Management: Predicting Employee Turnover and Performance. *International Journal for Global Academic & Scientific Research, 3* (2), 48–59. DOI: 10.55938/ijgasr.v3i2.77

Raya, A. P. (2019). *Fundamentos de la IA para el ajedrez.* Retrieved from https://idus.us.es/bitstream/handle/11441/93632/1/Portillo%20Raya%20Alberto%20TFG.pdf?sequence=1

Raygoza-L, M. E., Orduño-Osuna, J. H., & Murrieta-Rico, F. N. (2023). Domestic Policies for Sustainable and Economic Development in Countries With Emerging Economies: A Case Study of Mexico. In *Governance Quality, Fiscal Policy, and the Path to a Low-Carbon Future: Perspectives From Developing Economies*. IGI Global., DOI: 10.4018/978-1-6684-9272-7.ch006

Real-Moreno, O., Castro-Toscano, M. J., Rodrízuez-Ouiñonez, J. C., Hernández-Balbuena, D., Flores-Fuentes, W., & Rivas-Lopez, M. (2018, October). Implementing k-nearest neighbor algorithm on scanning aperture for accuracy improvement. In IECON 2018-44th Annual Conference of the IEEE Industrial Electronics Society (pp. 3182-3186). IEEE. DOI: 10.1109/IECON.2018.8592804

Real-Moreno, O., Rodriguez-Quiñonez, J. C., Sergiyenko, O., Basaca-Preciado, L. C., Hernandez-Balbuena, D., Rivas-Lopez, M., & Flores-Fuentes, W. (2017, June). Accuracy improvement in 3D laser scanner based on dynamic triangulation for autonomous navigation system. In 2017 IEEE 26th International Symposium on Industrial Electronics (ISIE) (pp. 1602-1608). IEEE. DOI: 10.1109/ISIE.2017.8001486

Reger, G. M., Holloway, K. M., Candy, C., Rothbaum, B. O., Difede, J., Rizzo, A. A., & Gahm, G. A. (2011). Effectiveness of virtual reality exposure therapy for active-duty soldiers in a military mental health clinic. *Journal of Traumatic Stress*, *24*(1), 93–96. DOI: 10.1002/jts.20574 PMID: 21294166

Reis, J. C., Correia, A., Murai, F., Veloso, A., & Benevenuto, F. (2019). Explainable machine learning for fake news detection. *Proceedings of the 10th ACM Conference on Web Science*, 17–26. DOI: 10.1145/3292522.3326027

Renault, T. (2017). Intraday online investor sentiment and return patterns in the US stock market. *Journal of Banking & Finance*, *84*, 25–40. DOI: 10.1016/j.jbankfin.2017.07.002

Reséndiz-Flores, E. O., Navarro-Acosta, J. A., & García-Calvillo, I. D. (2022). Smart fault detection and optimal variables identification using Kernel Mahalanobis Distance for industrial manufacturing processes. *International Journal of Computer Integrated Manufacturing*, *35*(9), 942–950. DOI: 10.1080/0951192X.2022.2027019

Revathy, G., Alghamdi, S. A., Alahmari, S. M., Yonbawi, S. R., Kumar, A., & Haq, M. A. (2022). Sentiment analysis using machine learning: Progress in the machine intelligence for data science. *Sustainable Energy Technologies and Assessments*, *53*, 102557. DOI: 10.1016/j.seta.2022.102557

Reyes, A., Escobar, J. W., & Londoño, J. C. (2023). Methodology for the supplier selection and valuation problem by using a combined grouping and multi-criteria approach. *International Journal of Logistics Systems and Management*, *46*(2), 174–205. DOI: 10.1504/IJLSM.2023.134401

Rezaeianzadeh, M., Tabari, H., Arabi Yazdi, A., Isik, S., & Kalin, L. (2014). Flood flow forecasting using ANN, ANFIS and regression models. *Neural Computing & Applications*, *25*(1), 25–37. DOI: 10.1007/s00521-013-1443-6

Ribeiro, M. T., Singh, S., & Guestrin, C. (2016, August). " Why should i trust you?" Explaining the predictions of any classifier. In *Proceedings of the 22nd ACM SIGKDD international conference on knowledge discovery and data mining* (pp. 1135-1144). DOI: 10.18653/v1/N16-3020

Richards, H. J., Bruhn, A. R., Lucente, S. W., & Casey, J. O. (2014). Reliability of the comprehensive scoring system with the early memories procedure. *Sensoria A Journal of Mind Brain and Culture*, *10*(2), 14. DOI: 10.7790/sa.v0i0.404

Riddle, P., Segal, R., & Etzioni, O. (1994). Representation design and brute-force induction in a boeing manufacturing domain. *Applied Artificial Intelligence*, *8*(1), 125–147. DOI: 10.1080/08839519408945435

Riley, R. D., Ahmed, I., Debray, T. P. A., Willis, B. H., Noordzij, J. P., Higgins, J. P. T., & Deeks, J. J. (2015). Summarising and validating test accuracy results across multiple studies for use in clinical practice. *Statistics in Medicine*, *34*(13), 2081–2103. DOI: 10.1002/sim.6471 PMID: 25800943

Risi, N., Aimar, A., Donati, E., Solinas, S., & Indiveri, G. (2020). A spike-based neuromorphic architecture of stereo vision. *Frontiers in Neurorobotics*, *14*, 568283. DOI: 10.3389/fnbot.2020.568283 PMID: 33304262

Riva, G., Baños, R. M., Botella, C., Wiederhold, B. K., & Gaggioli, A. (2012). Positive technology: Using interactive technologies to promote positive functioning. *Cyberpsychology, Behavior, and Social Networking*, *15*(1), 69–77. DOI: 10.1089/cyber.2011.0139 PMID: 22149077

Rizzo, A., Parsons, T. D., Lange, B., Kenny, P., Buckwalter, J. G., Rothbaum, B. O., & Reger, G. (2011). Virtual reality goes to war: A brief review of the future of military behavioral healthcare. *Journal of Clinical Psychology in Medical Settings*, *18*(2), 176–187. DOI: 10.1007/s10880-011-9247-2 PMID: 21553133

Rocha Neto, H. G. D., Cavalcanti, M., & Telles Correia, D. (2022). Kappa accuracy of prototypical diagnosis and ICD-10 criteria for mental disorders: A cross-sectional study in a real-life setting. *European Psychiatry : the Journal of the Association of European Psychiatrists*, *65*(S1), S162–S162. DOI: 10.1192/j.eurpsy.2022.434

Rocha Neto, H. G., Sinem, T. B., Koiller, L. M., Pereira, A. M., de Souza Gomes, B. M., Veloso Filho, C. L., Cavalcanti, M. T., & Telles-Correia, D. (2022). Intra-rater kappa accuracy of prototype and ICD-10 operational criteria-based diagnoses for mental disorders: A brief report of a cross-sectional study in an outpatient setting. *Frontiers in Psychiatry*, *13*, 793743. Advance online publication. DOI: 10.3389/fpsyt.2022.793743 PMID: 35308869

Rodriguez-Gallegos, A. Y., Mendez-Gonzalez, L. C., Hernandez-Holguin, A. I., & Rodriguez-Picon, L. A. (2023). Failure Detection System Controlled by a Mixed Reality Interface. At L. C. Méndez-González et al. (eds.), Innovation and Competitiveness in Industry 4.0 Based on Intelligent Systems, (pp. 21-42), EAI/Springer Innovations in Communication and Computing, DOI: 10.1007/978-3-031-29775-5_2

Rodriguez, W., Angle, P., & Snyder, M. (2021). Can Machine Learning Enhance Human Learning in Times of Disruption? *Ubiquitous Learning. International Journal (Toronto, Ont.)*, *14* (1).

Rohini, V., Thomas, M., & Latha, C. A. (2016, May). Domain based sentiment analysis in regional Language-Kannada using machine learning algorithm. In *2016 IEEE International Conference on Recent Trends in Electronics, Information & Communication Technology (RTEICT)* (pp. 503-507). IEEE.

Rojas, M., Hermosilla, G., Yunge, D., & Farias, G. (2022). An Easy to Use Deep Reinforcement Learning Library for AI Mobile Robots in Isaac Sim. *Applied Sciences (Basel, Switzerland)*, *12*(17), 8429. DOI: 10.3390/app12178429

Rojas-Rodriguez, A., Chiwo, F. S., Arcos-Gutierrez, H., Ovando-Vazquez, C., & Garduño, L. E. (2023). Development and Evaluation of a Machine Learning Model for the Prediction of Failures in an Injection Moulding Process. At L. C. Méndez-González et al. (eds.), Innovation and Competitiveness in Industry 4.0 Based on Intelligent Systems, (pp. 101-121). EAI/Springer Innovations in Communication and Computing, DOI: 10.1007/978-3-031-29775-5_5

Roldán, P. N. (2017, febrero 17). *Economipedia*. Retrieved from https://economipedia.com/definiciones/finanzas.html

Romero Bravo, G. J., Espinoza Maza, J., Macgluf Issasi, A., Suárez Álvarez, Á., & Rodríguez Rodríguez, L. A. (2021). Aplicación de Machine Learning en la Industria 4.0 en tiempos de pandemia. *Interconectando Saberes*, (11). Advance online publication. DOI: 10.25009/is.v0i11.2692

Rosenstock, I. M. (1974). Historical origins of the Health Belief Model. *Health Education Monographs*, 2(4), 328–335. DOI: 10.1177/109019817400200403

Rudin, C. (2019). Stop explaining black box machine learning models for high stakes decisions and use interpretable models instead. *Nature Machine Intelligence*, *1*(5), 206–215.

Ruíz, E. (2024). Las implicaciones del uso de la inteligencia artificial en la academia. Gaceta Universidad Abierta y a Distancia UNAD. Gaceta Extraordinaria. https://gaceta.unadmexico.mx/historico-anual/95-2023/julio-septiembre- 2023/educacion/205-las-implicaciones-del-uso-de-la-inteligencia-artificial-en-la-academia

Ruiz, S. A., Beltrán, E. R., Olivares, J. A. P., Mora, J. L. O., & Valadez, J. O. V. (2023). DIAGNÓSTICO DE FALLAS INTERMITENTES EN SISTEMAS AUTOMÁTICOS DE MANUFACTURA USANDO MACHINE LEARNING (DIAGNOSIS OF INTERMITTENT FAILURES IN AUTOMATIC MANUFACTURING SYSTEMS USING MACHINE LEARNING). Pistas Educativas, 44(144).

Russell, S., & Norvig, P. (2020). *Artificial intelligence: a modern approach.*

Ryu, A., & Lee, J. H. (2021). The role of digital mental health interventions in the COVID-19 pandemic: A review of effectiveness. *Asian Journal of Psychiatry*, *54*, 102458. DOI: 10.1016/j.ajp.2020.102458

Sabour, S., & Ghassemi, F. (2016). The validity and reliability of a signal impact assessment tool: Statistical issue to avoid misinterpretation. *Pharmacoepidemiology and Drug Safety*, *25*(10), 1215–1216. DOI: 10.1002/pds.4061 PMID: 27696610

Sabri, M. F., Anthony, M., Law, S. H., Rahim, H. A., Burhan, N. A. S., & Ithnin, M. (2024). Impact of financial behaviour on financial well-being: Evidence among young adults in Malaysia. *Journal of Financial Services Marketing*, *29* (3), 788–807. DOI: 10.1057/s41264-023-00234-8

Sadeh-Sharvit, S., Camp, T. D., Horton, S. E., Hefner, J. D., Berry, J. M., Grossman, E., & Hollon, S. D. (2023). Effects of an Artificial Intelligence Platform for Behavioral Interventions on Depression and Anxiety Symptoms: Randomized Clinical Trial. *Journal of Medical Internet Research*, *25*, e46781. DOI: 10.2196/46781 PMID: 37428547

Sahoo, S. R., & Gupta, B. B. (2021). Multiple features based approach for automatic fake news detection on social networks using deep learning. *Applied Soft Computing*, *100*, 106983. DOI: 10.1016/j.asoc.2020.106983

Sailunaz, K., & Alhajj, R. (2019). Emotion and sentiment analysis from Twitter text. *Journal of Computational Science*, *36*, 101003. DOI: 10.1016/j.jocs.2019.05.009

Sakib, S., Ahmed, N., Kabir, A. J., & Ahmed, H. (2018). An Overview of Convolutional Neural Network: Its Architecture and Applications. doi:DOI: 10.20944/preprints201811.0546.v4

Saldaña, M., Gálvez, E., Navarra, A., Toro, N., & Cisternas, L. A. (2023). Optimization of the SAG Grinding Process Using Statistical Analysis and Machine Learning: A Case Study of the Chilean Copper Mining Industry. *Materials (Basel)*, *16*(8), 3220. DOI: 10.3390/ma16083220 PMID: 37110055

Sánchez-Rada, J. F., & Iglesias, C. A. (2019). Social context in sentiment analysis: Formal definition, overview of current trends and framework for comparison. *Information Fusion*, *52*, 344–356. DOI: 10.1016/j.inffus.2019.05.003

Sangeetha, J. M., & Alfia, K. J. (2024). Financial stock market forecast using evaluated linear regression based machine learning technique. *Measurement. Sensors*, *31*(100950), 100950. DOI: 10.1016/j.measen.2023.100950

Santamaria-Granados, L., Munoz-Organero, M., Ramirez-Gonzalez, G., Abdulhay, E., & Arunkumar, N. (2018). Using deep convolutional neural network for emotion detection on a physiological signals dataset (AMIGOS). *IEEE Access : Practical Innovations, Open Solutions*, *7*, 57–67. DOI: 10.1109/ACCESS.2018.2883213

Sari, K. S. A. P., & Dantes, N., N. D., & Arum MWP, D. (. (2017). Behavioral counseling with engineering modeling and technique token economy to improve confidence. *Bisma The Journal of Counseling*, *1*(2), 18. DOI: 10.23887/128182017

Sarkar, P., & Etemad, A. (2020). Self-supervised ECG representation learning for emotion recognition. *IEEE Transactions on Affective Computing*, *13*(3), 1541–1554. DOI: 10.1109/TAFFC.2020.3014842

Sarker, I. H. (2021). Machine Learning: Algorithms, Real-World Applications and Research Directions. *SN Computer Science*, *2*(3), 160. DOI: 10.1007/s42979-021-00592-x PMID: 33778771

Sarver, D. E., McCabe, K. A., & Mashburn, J. A. (2021). Classroom engagement strategies for ADHD learners: A longitudinal study. *Journal of Attention Disorders*, *25*(9), 1237–1248. https://doi.org/https://doi.org/10.1177/1087054720974822

Saucedo-Dorantes, J. J., Jaen-Cuellar, A. Y., & Elvira-Ortiz, D. A. (2020). Smart Monitoring Based on Novelty Detection and Artificial Intelligence Applied to the Condition Assessment of Rotating Machinery in the Industry 4.0. At L. Romeral-Martinez; R.A. Osornio-Rios; and M. Delgado-Prieto (Eds.), New Trends in the Use of Artificial Intelligence for the Industry 4.0,(pp. 63-76). IntechOpen. ISBN 978-1-83880-141-0

Saxena, A. (2022). An Introduction to Convolutional Neural Networks. *International Journal for Research in Applied Science and Engineering Technology*, *10*(12), 943–947. DOI: 10.22214/ijraset.2022.47789

Scallan, P. (2003). *Process planning: the design/manufacture interface*. Elsevier. DOI: 10.1016/B978-075065129-5/50003-9

Schemmer, M., Heinz, D., Baier, L., Vössing, M., & Kühl, N. (2021). Conceptualizing digital resilience for AI-based information systems. *European Conference on Information Systems*. https://api.semanticscholar.org/CorpusID:235428356

Schinka, J. A. (1976). Confidence in personality assessment. *Psychological Reports*, *38*(1), 62–62. DOI: 10.2466/pr0.1976.38.1.62 PMID: 1250932

Schoeller, F., Bertrand, P., Gerry, L. J., Jain, A., Horowitz, A. H., & Zenasni, F. (2019). Combining virtual reality and biofeedback to foster empathic abilities in humans. *Frontiers in Psychology*, *9*(1), 1–5. DOI: 10.3389/fpsyg.2018.02741 PMID: 30804868

Schork, N. J. (2019). ARTIFICIAL INTELLIGENCE AND PERSONALIZED MEDICINE. *Precision medicine in Cancer therapy*, 178: 265–283. .DOI: 10.1007/978-3-030-16391-4_11

Schouten, K., & Frasincar, F. (2015). Survey on aspect-level sentiment analysis. *IEEE Transactions on Knowledge and Data Engineering*, *28* (3), 813–830.

Schröter, J. (2019). Artificial Intelligence and the Democratization of Art. In A. Sudmann (Ed.), *The democratization of artificial intelligence. Net politics in the era of learning algorithms* (pp. 297-311). transcript. Retrieved from https://mediarep.org/server/api/core/bitstreams/1c616f3d-1cfa-4d99-87cc-d0282770c540/content

Sculley, D., Holtz, S., Guyon, I., & Eick, S. (2015). Machine Learning: The High-Interest Credit Card of Technical Debt. Proceedings of the 28th International Conference on Machine Learning, 151-158. https://www.aaai.org/ocs/index.php/ICML/ICML15/paper/view/10088

Seabra Lopes, L., & Camarinha-Matos, L. M. (1995). A machine learning approach to error detection and recovery in assembly. Proceedings 1995 IEEE/RSJ International Conference on Intelligent Robots and Systems. Human Robot Interaction and Cooperative Robots, Pittsburgh, PA, USA, 3, 197-203. DOI: 10.1109/IROS.1995.525884

Sechrest, L., Gallimore, R., & Hersch, P. D. (1967). Feedback and accuracy of clinical predictions. *Journal of Consulting Psychology, 31*(1), 1–11. DOI: 10.1037/h0020977

Sekhar, K. Ch., Ingle, R. B., Banu, E. A., Rinawa, M. L., Prasad, M. M., & Boopathi, S. (2024). Integrating VR and AR for Enhanced Production Systems: Immersive Technologies in Smart Manufacturing. In *Advances in Computational Intelligence and Robotics* (pp. 90–112). IGI Global. DOI: 10.4018/979-8-3693-6806-0.ch005

Sen, K., Prybutok, G., & Prybutok, V. (2022). The use of digital technology for social wellbeing reduces social isolation in older adults: A systematic review. *SSM, Population Health*, 17(1), 1-9 101020. DOI: 10.1016/j.ssmph.2021.101020

Şen, T. Ü., & Bakal, G. (2023). A transfer learning application on the reliability of psychological drugs' comments. *2023 International Conference on Smart Applications, Communications and Networking (SmartNets)*. DOI: 10.1109/SmartNets58706.2023.10215681

Senthil, T., Puviyarasan, M., Babu, S. R., Surakasi, R., Sampath, B., & Associates. (2023). Industrial Robot-Integrated Fused Deposition Modelling for the 3D Printing Process. In *Development, Properties, and Industrial Applications of 3D Printed Polymer Composites* (pp. 188–210). IGI Global.

Serrano, F., & Defior, S. (2018). Dyslexia beyond reading: Addressing creativity and verbal reasoning strengths. *Annals of Dyslexia, 68*(1), 59–75. https://doi.org/https://doi.org/10.1007/s11881-018-0158-0

Set, Z. (2020). Social Malicious Personalities: The Dark Triad. In *Psikiyatride Güncel Yaklaşımlar-Current Approaches in Psychiatry* (pp. 12(3):318-329)DOI: 10.18863/pgy.629950

Shabani, S., Samadianfard, S., Sattari, M. T., Mosavi, A., Shamshirband, S., Kmet, T., & Várkonyi-Kóczy, A. R. (2020). Modeling pan evaporation using gaussian process regression k-nearest neighbors random forest and support vector machines; Comparative analysis. *Atmosphere (Basel), 11*(1), 66. DOI: 10.3390/atmos11010066

Shabbir, J. (2018). Artificial intelligence and its role in near future. Retrieved from https://arxiv.org/abs/1804.01396

Shafaee, A., Issa, H., Agne, S., Baumann, S., & Dengel, A. (2014). Aspect-based sentiment analysis of amazon reviews for fitness tracking devices. In Trends and Applications in Knowledge Discovery and Data Mining: PAKDD 2014 International Workshops: DANTH, BDM, MobiSocial, BigEC, CloudSD, MSMV-MBI, SDA, DMDA-Health, ALSIP, SocNet, DMBIH, BigPMA, Tainan, Taiwan, May 13-16, 2014. Revised Selected Papers 18 (pp. 50-61). Springer International Publishing.

Shafi, I., Mazhar, M. F., Fatima, A., Alvarez, R. M., Miró, Y., Espinosa, J. C. M., & Ashraf, I. (2023). Deep learning-based real time defect detection for optimization of aircraft manufacturing and control performance. *Drones (Basel)*, *7*(1), 31. DOI: 10.3390/drones7010031

Shah, S., Akshata Bhat, M., Sumitra Singh, M., Arya Chavan, M., & Aryan Singh, M. (2024). Sentiment Analysis. *International Journal of Progressive Research in Engineering Management and Science (Ijprems)*, *4* (4), 1542–1546. https://www.doi.org/10.58257/IJPREMS33384 PMID: 38834214

Shaikh, J., & Patil, R. (2020). Fake news detection using machine learning. *2020 IEEE International Symposium on Sustainable Energy, Signal Processing and Cyber Security (iSSSC)*, 1–5.

Shao, X., & Shi, Y. (2020). Neural adaptive control for MEMS gyroscope with full-state constraints and quantized input. *IEEE Transactions on Industrial Informatics*, •••, 1–1. DOI: 10.1109/TII.2020.2968345

Shapi, M. K. M., Ramli, N. A., & Awalin, L. J. (2021). Energy consumption prediction by using machine learning for smart building: Case study in Malaysia. *Developments in the Built Environment*, *5*(100037), 100037. DOI: 10.1016/j.dibe.2020.100037

Shariatnia, S., Ziaratban, M., Rajabi, A., Salehi, A., Abdi Zarrini, K., & Vakili, M. (2022). Modeling the diagnosis of coronary artery disease by discriminant analysis and logistic regression: A cross-sectional study. *BMC Medical Informatics and Decision Making*, *22*(1), 85. Advance online publication. DOI: 10.1186/s12911-022-01823-8 PMID: 35351098

Sharma, R., & Dhingra, L. (2024). Advancing Human Resource Strategies with Deep Learning: Predictive Analytics for Improving Employee Retention Rates. *2024 2nd World Conference on Communication & Computing (WCONF)*, 1–4.

Sharma, S., & Chaudhary, P. (2023). *Machine learning and deep learning*. Quantum Comput. Artif. Intell. Train. Mach. Deep Learn. Algorithms Quantum Comput, 71-84.

Sharma, D., & Kumar, S. (2022). A machine learning framework for personalized mental health interventions: A systematic review. *Journal of Biomedical Informatics*, *126*, 103961. DOI: 10.1016/j.jbi.2022.103961

Sharma, M. K., Anand, N., Roopesh, B. N., & Sunil, S. (2021). Digital resilience mediates the healthy use of technology. *The Medico-Legal Journal*, *90*(4), 195–199. DOI: 10.1177/00258172211018337 PMID: 34259055

Sharma, R. (2020). Rethinking Disability Services in Higher Education. *Disability & Society*, *37*(5), 1049–1064.

Sharma, U., Saran, S., & Patil, S. M. (2020). Fake news detection using machine learning algorithms. [IJCRT]. *International Journal of Creative Research Thoughts*, *8*(6), 509–518.

Shataee, S., Kalbi, S., Fallah, A., & Pelz, D. (2012). Forest attribute imputation using machine-learning methods and ASTER data: Comparison of k-NN, SVR and random forest regression algorithms. *International Journal of Remote Sensing*, *33*(19), 6254–6280. DOI: 10.1080/01431161.2012.682661

Shaw, M. J., Park, S., & Raman, N. (1992). Intelligent Scheduling with Machine Learning Capabilities: The Induction of Scheduling Knowledge. *IIE Transactions*, *24*(2), 156–168. DOI: 10.1080/07408179208964213

Shawn, S., Cryan, J., Wenger, E., Zheng, H., Hanocka, R., & Zhao, B. Y. (2023). Glaze: Protecting Artists from Style Mimicry by Text-to-Image Models. Retrieved from https://glaze.cs.uchicago.edu

Sheela, L. J. (2016). A Review of Sentiment Analysis in Twitter Data Using Hadoop. *International Journal of Database Theory and Application*, *9*(1), 77–86. DOI: 10.14257/ijdta.2016.9.1.07

Sheikh, A., Jacob, J., Vostanis, P., Ruby, F., Spuerck, I., Stankovic, M., Morgan, N., Mota, C. P., Ferreira, R., Eruyar, Ş., Yılmaz, E. A., Fatima, S. Z., & Edbrooke-Childs, J. (2024). What Should Personalised Mental Health Support Involve? Views of Young People with Lived Experience and Professionals from Eight Countries. *Administration and Policy in Mental Health*, *51*(5), 753–768. DOI: 10.1007/s10488-024-01382-2 PMID: 38907740

Shekhovtsov, A., & Flach, B. (2018). Feed-forward propagation in probabilistic neural networks with categorical and max layers. In *International conference on learning representations*.

Shimamoto, M., Ishizuka, K., Ohtani, K., Inada, T., Yamamoto, M., Tachibana, M., Kimura, H., Sakai, Y., Kobayashi, K., Ozaki, N., & Ikeda, M. (2024). Machine learning algorithm-based estimation model for the severity of depression assessed using Montgomery-Asberg depression rating scale. *Neuropsychopharmacology Reports*, *44*(1), 115–120. DOI: 10.1002/npr2.12404 PMID: 38115795

Shinde, P. P. (2018, AGOSTO 18). *A review of machine learning and deep learning applications.* https://ieeexplore.ieee.org/abstract/document/8697857

Siabato, W. (2018). *Sobre la evolución de la información geográfica: Las bodas de oro de los SIG.* Cuadernos de Geografía. *Cuadernos de Geografia*, *27*(1), 1–9. DOI: 10.15446/rcdg.v27n1.69500

Sicoly, F. (1992). Estimating the accuracy of decisions based on cutting scores. *Journal of Psychoeducational Assessment*, *10*(1), 26–36. DOI: 10.1177/073428299201000102

Silverman, B. W., & Jones, M. C. (1989). E. Fix and J.L. Hodges (1951): An Important Contribution to Nonparametric Discriminant Analysis and Density Estimation: Commentary on Fix and Hodges (1951). *International Statistical Review / Revue Internationale de Statistique, 57*(3), 233–238. DOI: 10.2307/1403796

Simon, H. A. (2013). *Administrative Behavior* (4th ed.).

Singh, D. K., & Rawat, N. (2023). *Decoding optimization algorithms for convolutional neural networks in time series regression tasks*. Mecs-press.org. https://www.mecs-press.org/ijitcs/ijitcs-v15-n6/v15n6-4.html

Singh, A. K. (2025). Importance of digital technologies in economic sector of selected countries. In Ahmadi, A., Mna, A., & Slimani, A. (Eds.), *Renewable Energy and the Economic Welfare of Society* (pp. 335–370). IGI Global Scientific Publishing., DOI: 10.4018/979-8-3693-7580-8.ch015

Singh, A. K., Srinivasan, S., & Jyoti, B. (2025). Implications of digitalization on public health sector across countries. In Wickramasinghe, N. (Ed.), *Digitalization and the Transformation of the Healthcare Sector* (pp. 359–394). IGI Global Scientific Publishing., DOI: 10.4018/979-8-3693-9641-4.ch013

Singha, S. K., & Singh, S. J. (2015). Analysis and optimization of sand casting defects with the help of artificial neural network. *International Journal of Research in Engineering and Technology*, *4*(5), 24–29. DOI: 10.15623/ijret.2015.0405005

Skovgaard-Olsen, N., Singmann, H., & Klauer, K. C. (2018). Corrigendum to "The relevance effect and conditionals" [Cognition 150 (2016) 26–36]. *Cognition, 170*, 338. DOI: 10.1016/j.cognition.2017.09.002 PMID: 28918984

Smiles, A. L., Chavanakunakorn, K. D., Omarali, B., Oh, C., & Farkhatdinov, I. (2023, September). Implementation of a Stereo Vision System for a Mixed Reality Robot Teleoperation Simulator. In Annual Conference Towards Autonomous Robotic Systems (pp. 494-502). Cham: Springer Nature Switzerland. DOI: 10.1007/978-3-031-43360-3_40

Smith, K. A., Willis, R. J., & Brooks, M. (2000). An analysis of customer retention and insurance claim patterns using data mining: A case study. *The Journal of the Operational Research Society, 51*(5), 532–541. DOI: 10.1057/palgrave.jors.2600941

Socher, R., & Pennington, J. (2011). Semi-Supervised Recursive Autoencoders for Predicting Sentiment Distributions. *InProceedings of the 2011 Conference on Empirical Methods in Natural Language Processing*, 151–161. DOI: 10.3109/03630269.2012.746943

Soleymani, M., Garcia, D., Jou, B., Schuller, B., Chang, S. F., & Pantic, M. (2017). A survey of multimodal sentiment analysis. *Image and Vision Computing, 65*, 3–14. DOI: 10.1016/j.imavis.2017.08.003

Sorgente, A., & Lanz, M. (2019). The multidimensional subjective financial well-being scale for emerging adults: Development and validation studies. *International Journal of Behavioral Development, 43* (5), 466–478. DOI: 10.1177/0165025419851859

Specht, M., Szostak, B., Lewicka, O., Stateczny, A., & Specht, C. (2023). Method for determining of shallow water depths based on data recorded by UAV/USV vehicles and processed using the SVR algorithm. *Measurement, 221*, 113437. DOI: 10.1016/j.measurement.2023.113437

Speer, A. B., Christiansen, N. D., Robie, C., & Jacobs, R. R. (2022). Measurement specificity with modern methods: Using dimensions, facets, and items from personality assessments to predict performance. *The Journal of Applied Psychology, 107*(8), 1428–1439. DOI: 10.1037/apl0000618 PMID: 34582241

Stachl, C., Pargent, F., Hilbert, S., Harari, G. M., Schoedel, R., Vaid, S., Gosling, S. D., & Bühner, M. (2020). Personality Research and Assessment in the Era of Machine Learning. *European Journal of Personality, 34*(5), 613–631. DOI: 10.1002/per.2257

Stead, R., & Fekken, C. G. (2014). Agreeableness at the Core of the Dark Triad of Personality. *EBSCO*, Vol 12, Issue 4-A, p131.

Stein, B., Hoeft, F., & Richter, C. G. (2024). Stress, resilience, and emotional well-being in children and adolescents with specific learning disabilities. *Current Opinion in Behavioral Sciences, 58*, 101410. DOI: 10.1016/j.cobeha.2024.101410 PMID: 39157263

Suarez, F. F., Cusumano, M. A., & Fine, C. H. (1995). An empirical study of flexibility in manufacturing. *MIT Sloan Management Review*.

Suleiman, Z., Shaikholla, S., Dikhanbayeva, D., Shehab, E., & Turkyilmaz, A. (2022). Industry 4.0: Clustering of concepts and characteristics. *Cogent Engineering, 9*(1), 2034264. DOI: 10.1080/23311916.2022.2034264

Sullivan, A. H., Sweeney, M. E., & Jackson, J. (2021). Designing Accessible Digital Health Solutions for Mental Health: A User-Centric Approach. *Journal of Technology in Behavioral Science*, *6*(2), 92–105. DOI: 10.1007/s41347-020-00230-y

Sun, C., Qiu, X., Xu, Y., & Huang, X. (2019). How to fine-tune BERT for text classification? In Chinese Computational Linguistics: 18th China National Conference, CCL 2019, Kunming, China, October 18–20, 2019 [Springer International Publishing.]. *Proceedings*, *18*, 194–206.

Sweeney, M. E., & Sullivan, A. H. (2021). Evaluating the inclusivity of machine learning tools in mental health interventions: A call for diversity in research. *Journal of Health Psychology*, *26*(7), 1069–1076. DOI: 10.1177/1359105320958850

Syahnur, K. N. F., Surianto, D. F., & Dharsana, M. T. (2023). Analysis Of Retail Sector Market Reaction In Indonesia On Social Media And Investor Sentiment. *Journal Manajemen Bisnis*, *10*(2), 371–383. DOI: 10.33096/jmb.v10i2.72

Taunk, K., De, S., Verma, S., & Swetapadma, A. (2019). A brief review of nearest neighbor algorithm for learning and classification. *2019 International Conference on Intelligent Computing and Control Systems (ICCS)*, 1255–1260. DOI: 10.1109/ICCS45141.2019.9065747

Tay, L.-Y., Tai, H.-T., & Tan, G.-S. (2022). Digital financial inclusion: A gateway to sustainable development. *Heliyon*, *8* (6), e09766. DOI: 10.1016/j.heliyon.2022.e09766 PMID: 35785228

Terra, M., Baklola, M., Ali, S., & El-Bastawisy, K. (2023). Opportunities, applications, challenges and ethical implications of Artificial Intelligence in Psychiatry: A narrative review. *The Egyptian Journal of Neurology, Psychiatry and Neurosurgery*, *59*(1), 80–87. DOI: 10.1186/s41983-023-00681-z

Terry, W. S. (1984). A "forgetting journal" for memory courses. *Teaching of Psychology*, *11*(2), 111–112. DOI: 10.1207/s15328023top1102_17

Thakkar, A., Ankita, G., & Avinash, D. S. (2024). Artificial intelligence in positive mental health: a narrative review. *ational Institutes of Health*, .DOI: 10.3389/fdgth.2024.1280235

The Devastator. (2023). *NLP Mental Health Conversations* [Data set].

The impact of artificial intelligence in education and teaching - Universidad ORT Uruguay. (n.d.). https://ie.ort.edu.uy/blog/el-impacto-de-la-inteligencia-artificial-en-la-educacion-y-enla-docencia

Thet, T. T., Na, J. C., & Khoo, C. S. G. (2010). Aspect-based sentiment analysis of movie reviews on discussion boards. *Journal of Information Science*, *36* (6), 823–848. DOI: 10.1177/0165551510388123

Timbers, T., Campbell, T., & Lee, M. (2023, December 23). *Chapter 7 regression I: K-nearest neighbors*. Datasciencebook.Ca. https://datasciencebook.ca/regression1.html

Toms, S., Parker, B., Tucker, C., & Rubalcava, R. (2022). *Python for ArcGIS Pro: Automate cartography and data analysis using ArcPy, ArcGIS API for Python, Notebooks, and pandas*. Packt Publishing Ltd.

Tonia, T., Buitrago-Garcia, D., Peter, N. L., Mesa-Vieira, C., Li, T., Furukawa, T. A., Cipriani, A., Leucht, S., Low, N., & Salanti, G. (2023). Tool to assess risk of bias in studies estimating the prevalence of mental health disorders (RoB-PrevMH). *BMJ Mental Health, 26*(1), e300694. DOI: 10.1136/bmjment-2023-300694 PMID: 37899074

Tonioni, A., Rahnama, O., Joy, T., Stefano, L. D., Ajanthan, T., & Torr, P. H. (2019). Learning to adapt for stereo. In *Proceedings of the IEEE/CVF Conference on Computer Vision and Pattern Recognition* (pp. 9661-9670).

Torous, J., & Keshavan, M. (2018). A digital future for mental health care: A brief overview of the current state of technology in mental health. *Psychiatric Services (Washington, D.C.), 69*(3), 250–255. DOI: 10.1176/appi.ps.201700456 PMID: 29334879

Torralba, J., Hidalgo, J., & Jiménez-Morales, A. (2013). Powder injection moulding: Processing of small parts of complex shape. *International Journal of Microstructure and Materials Properties, 8*(1/2), 87–96. DOI: 10.1504/IJMMP.2013.052648

Tosi, F., Aleotti, F., Poggi, M., & Mattoccia, S. (2019). Learning monocular depth estimation infusing traditional stereo knowledge. In *Proceedings of the IEEE/CVF Conference on Computer Vision and Pattern Recognition* (pp. 9799-9809). DOI: 10.1109/CVPR.2019.01003

Tosti, B., Corrado, S., & Mancone, S. (2024). Using chatbots and conversational agents for the promotion of Well-Being and Mental Health in Adolescents: Limitations and Perspectives. *Journal of Inclusive Methodology and Technology in Learning and Teaching*, 1–10. https://www.inclusiveteaching.it

Townsend, J. C., & Smith, W. J. (1964). *Predicting decision making behavior from personality and cognitive variables*. Defense Technical Information Center. DOI: 10.21236/AD0610521

Tripp Corbin, G. I. S. P. (2015). *Learning ArcGIS Pro*. Packt Publishing Ltd.

Trivedi, P., Ravisankar, A., Kumar, G. P., Krishnamoorthy, V., Shukla, S., & Boopathi, S. (2024). Finance and Technology-Integrated Digital Economy: Circular Age of a Sustainable Digital Revolution. In *Navigating the Circular Age of a Sustainable Digital Revolution* (pp. 91–120). IGI Global. DOI: 10.4018/979-8-3693-2827-9.ch004

Troussas, C., Virvou, M., Espinosa, K. J., Llaguno, K., & Caro, J. (2013, July). Sentiment analysis of Facebook statuses using Naive Bayes classifier for language learning. In *IISA 2013* (pp. 1–6). IEEE. DOI: 10.1109/IISA.2013.6623713

Trujillo-Hernández, G., Rodríguez-Quiñonez, J. C., Flores-Fuentes, W., Sanchez-Castro, J. J., Real-Moreno, O., Castro-Toscano, M. J., & Hernández-Balbuena, D. (2024). Optical 3D Scanning System in Medical Applications. In Scanning Technologies for Autonomous Systems (pp. 159-196). Cham: Springer Nature Switzerland. DOI: 10.1007/978-3-031-59531-8_6

Trujillo-Hernández, G., Rodríguez-Quiñonez, J. C., Ramírez-Hernández, L. R., Castro-Toscano, M. J., Hernández-Balbuena, D., Flores-Fuentes, W., . . . Mercorelli, P. (2019, October). Accuracy improvement by artificial neural networks in technical vision system. In IECON 2019-45th Annual Conference of the IEEE Industrial Electronics Society (Vol. 1, pp. 5572-5577). IEEE. DOI: 10.1109/IECON.2019.8927596

Trujillo-Hernández, G., Flores-Fuentes, W., Rodríguez-Quiñonez, J. C., Hernández-Balbuena, D., Real-Moreno, O., Miranda-Vega, J. E., & Bhateja, V. (2022). 3D reconstruction of human body biometry. In *Optoelectronic Devices in Robotic Systems* (pp. 195–225). Springer International Publishing., DOI: 10.1007/978-3-031-09791-1_8

Truong, Q., Nguyen, M., Dang, H., & Mei, B. (2020). Housing price prediction via improved machine learning techniques. *Procedia Computer Science, 174*, 433–442. DOI: 10.1016/j.procs.2020.06.111

Tumasjan, A., Sprenger, T., Sandner, P., & Welpe, I. (2010, May). Predicting elections with twitter: What 140 characters reveal about political sentiment. In *Proceedings of the international AAAI conference on web and social media* (Vol. 4, No. 1, pp. 178-185). DOI: 10.1609/icwsm.v4i1.14009

Tu, Y., & Peng, C. (2021). An ARMA-Based digital twin for MEMS gyroscope drift dynamics modeling and Real-Time compensation. *IEEE Sensors Journal, 21*(3), 2712–2724. DOI: 10.1109/JSEN.2020.3028140

Twenge, J. M., Joiner, T. E., Rogers, M. L., & Martin, G. N. (2018). Increases in depressive symptoms, suicide-related outcomes, and suicide rates among U.S. Adolescents after 2010 and links to increased new media screen time. *Clinical Psychological Science, 6*(1), 3–17. DOI: 10.1177/2167702617723376

Tzirakis, P., Trigeorgis, G., Nicolaou, M. A., Schuller, B., & Zafeiriou, S. (2016). End-to-End Multimodal Emotion Recognition using Deep. *Neural Networks, 14* (8), 1–9. ArXiv1704.08619v1

U.S. Department of Commerce. (2024). Japan - Protecting Intellectual Property. U.S. Department of Commerce. Retrieved from https://www.trade.gov/country-commercial-guides/japan-protecting-intellectual-property

U.S. Supreme Court. (2021). Supreme Court ruling on copyright eligibility for AI-generated works. Retrieved from https://www.supremecourt.gov

Ullal, M. S., Nayak, P. M., Dais, R. T., Spulbar, C., & Birau, R. (2022). Investigating the nexus between Artificial Intelligence and machine learning technologies in the case of Indian services industry. *Business: Theory and Practice, 23* (2), 323–333. DOI: 10.3846/btp.2022.15366

UNESCO. (2019). Review of the potential of artificial intelligence to accelerate progress towards SDG 4-Education 2030. In *UNESDOC Digital Library* (206 EX/44 + REV.CORR.). United Nations Educational, Scientific and Cultural Organization. Retrieved July 29, 2024, from https://unesdoc.unesco.org/ark:/48223/pf0000367373_spa

UNESCO. (2021). *Informe mundial sobre la inclusión educativa: América Latina y el Caribe.* https://www.unesco.org

UNESCO. (2021). Recommendation on the Ethics of Artificial Intelligence. Retrieved from https://unesdoc.unesco.org/ark:/48223/pf0000380455

UNESCO. (2023). *The Impact of AI on Equity in Education: A Global Perspective.*

United Nations Industrial Development Organization. (2019). Industrial Development Report 2020. Industrializing in the digital age. Vienna.

United Nations Industrial Development Organization. (2024). *Industrial Development Report 2024*. Turning Challenges into Sustainable Solutions. The New Era of Industrial Policy.

Universidad, O. R. T. Uruguay. (n.d.). El impacto de la inteligencia artificial en la educación y en la docencia. https://ie.ort.edu.uy/blog/el-impacto-de-la-inteligencia-artificial-en-la-educacion-y-en-la-docencia

Urbas, U., Vlah, D., & Vukašinović, N. (2021). Machine learning method for predicting the influence of scanning parameters on random measurement error. *Measurement Science & Technology*, *32*(6), 065201. DOI: 10.1088/1361-6501/abd57a

Usama, M., Qadir, J., Raza, A., Arif, H., Alvin Yau, K. L., Elkhatib, Y., Hussain, A., & Al-Fuqaha, A. (2019). Unsupervised Machine Learning for Networking: Techniques, Applications and Research Challenges. *IEEE Access : Practical Innovations, Open Solutions*, *7*, 65579–65615. DOI: 10.1109/ACCESS.2019.2916648

Usman, A. (2017). Integrated disaster risk management in Indian environment: Prediction, prevention and preparedness. *GHTC 2017 - IEEE Global Humanitarian Technology Conference, Proceedings*, *2017-Janua*, 1–6. DOI: 10.1109/GHTC.2017.8239246

Usman, A., Dutta, R., Usman, A., Azmee, F., & Divakar, B. P. (2014). Fire disaster management in trains using a new technique of water pipelines-first aid mechanism. *Procedia Engineering*, *78*(Dm), 112–119. DOI: 10.1016/j.proeng.2014.07.046

Uyeda, A. M., Curtis, J. R., Engelberg, R. A., Brumback, L. C., Guo, Y., Sibley, J., Lober, W. B., Cohen, T., Torrence, J., Heywood, J., Paul, S. R., Kross, E. K., & Lee, R. Y. (2022). Mixed-methods evaluation of three natural language processing modeling approaches for measuring documented goals-of-care discussions in the electronic health record. *Journal of Pain and Symptom Management*, *63*(6), e713–e723. DOI: 10.1016/j.jpainsymman.2022.02.006 PMID: 35182715

Vafeiadis, T., Diamantaras, K. I., Sarigiannidis, G., & Chatzisavvas, K. C. (2015). A comparison of machine learning techniques for customer churn prediction. *Simulation Modelling Practice and Theory*, *55*, 1–9. DOI: 10.1016/j.simpat.2015.03.003

Vaithianathan, V., Shastri, D. S., Nandakumar, V., Misba, M., Kumar, P. S. V. V. S. R., & Boopathi, S. (2024). NEP Policy Implementation Strategies in Education in India. In *Educational Philosophy and Sociological Foundation of Education* (pp. 325–356). IGI Global., DOI: 10.4018/979-8-3693-3587-1.ch013

Vang-Mata, R. (Ed.). (2020). *Multilayer perceptrons: theory and applications*. Nova Science Publishers.

Vasanthi, A., & Satish Kumar, K. (2016). Understanding Conspicuous Gravity Low Over the Koyna–Warna Seismogenic Region (Maharashtra, India) and Earthquake Nucleation: A Paradigm Shift. *Pure and Applied Geophysics*, *173*(6), 1933–1948. DOI: 10.1007/s00024-015-1237-0

Vaswani, A., Shazeer, N., Parmar, N., Uszkoreit, J., Jones, L., Gomez, A. N., Kaiser, L., & Polosukhin, I. (2017). Attention Is All You Need. *Advances in Neural Information Processing Systems*, ●●●, 30.

Velastegui-Montoya, A., Montalván-Burbano, N., Carrión-Mero, P., Rivera-Torres, H., Sadeck, L., & Adami, M. (2023). Google Earth Engine: A global analysis and future trends. *Remote Sensing (Basel)*, *15*(14), 3675. DOI: 10.3390/rs15143675

Vera, F. (2023). Integrating Artificial Intelligence in Higher Education: Challenges and opportunities. *Transform*, *4*(1), 17–34.

Vergara Villegas, O. O., Nandayapa, M., Sossa Azuela, J. H., Cossio Franco, E. G., & Rubin Linares, G. T. (2021). Artificial Intelligence for Industry 4.0 in Iberoamerica. *Computación y Sistemas*, *25*(4), 761–773. DOI: 10.13053/cys-25-4-4056

Verhoeven, T., de Graaf, S., & van Dijk, L. (2023). Teacher Training for AI-Enhanced Education: Addressing the Challenges and Opportunities. *Education and Information Technologies*, *28*(5), 3215–3236.

Verma, M. (2018). Artificial intelligence and its scope in different areas with special reference to the field of education. *International Journal of Advanced Educational Research*, *3*(1), 5–10. https://files.eric.ed.gov/fulltext/ED604401.pdf

Vezilić, B., Gajić, D. B., Dragan, D., Petrović, V., Mihić, S., Anišić, Z., & Puhalac, V. (2018). Binary classification of images for applications in intelligent 3D scanning. *Intelligent Distributed Computing*, *XI*, 199–209. DOI: 10.1007/978-3-319-66379-1_18

Vibe Fersum, K., O'Sullivan, P. B., Kvåle, A., & Skouen, J. S. (2009). Inter-examiner reliability of a classification system for patients with non-specific low back pain. *Manual Therapy*, *14*(5), 555–561. DOI: 10.1016/j.math.2008.08.003 PMID: 18838331

Vijaya Lakshmi, V., Mishra, M., Kushwah, J. S., Shajahan, U. S., Mohanasundari, M., & Boopathi, S. (2024). Circular Economy Digital Practices for Ethical Dimensions and Policies for Digital Waste Management. In *Harnessing High-Performance Computing and AI for Environmental Sustainability* (pp. 166–193). IGI Global., DOI: 10.4018/979-8-3693-1794-5.ch008

Vivar, J. M. F., & Peñalvo, F. J. G. (2023). Reflections on the ethics, potentials and challenges of Artificial Intelligence in the framework of Quality Education (SDG4). Comunicar: Revista científica de comunicación y educación, (74), 37-47.

Voukelatou, V., Gabrielli, L., Miliou, I., Cresci, S., Sharma, R., Tesconi, M., & Pappalardo, L. (2021). Measuring objective and subjective well-being: Dimensions and data sources. *International Journal of Data Science and Analytics*, *11* (4), 279–309. DOI: 10.1007/s41060-020-00224-2

Walkowiak, E. (2023). Digitalization and inclusiveness of HRM practices: The example of neurodiversity initiatives. *Human Resource Management Journal, November 2022*, 578–598. DOI: 10.1111/1748-8583.12499

Walkowiak, E. (2021). Neurodiversity of the workforce and digital transformation: The case of inclusion of autistic workers at the workplace. *Technological Forecasting and Social Change*, *168*(March), 120739. DOI: 10.1016/j.techfore.2021.120739

Wang, D., Liu, Z., Shao, S., Wu, X., Chen, W., & Li, Z. (2023, October). Monocular Depth Estimation: A Survey. In IECON 2023-49th Annual Conference of the IEEE Industrial Electronics Society (pp. 1-7). IEEE. DOI: 10.1109/IECON51785.2023.10311687

Wang, Y., Sun, S., Chen, X., Zeng, X., Kong, Y., Chen, J., Guo, Y., & Wang, T. (2021). Short-term load forecasting of industrial customers based on SVMD and XGBoost. *International Journal of Electrical Power and Energy Systems, 129*(September 2020), 106830. DOI: 10.1016/j.ijepes.2021.106830

Wang, K., Varma, D. S., & Prosperi, M. (2018). A systematic review of the effectiveness of mobile apps for monitoring and management of mental health symptoms or disorders. *Journal of Psychiatric Research, 107*, 73–78. DOI: 10.1016/j.jpsychires.2018.10.006 PMID: 30347316

Wang, L., Ma, C., Feng, X., Zhang, Z., Yang, H., Zhang, J., Chen, Z., Tang, J., Chen, X., Lin, Y., Zhao, W. X., Wei, Z., & Wen, J. (2024). A survey on large language model based autonomous agents. *Frontiers of Computer Science, 18*(6), 186345. DOI: 10.1007/s11704-024-40231-1

Wang, Y., Zhou, P., Yao, C. W., Wang, H. Y., & Lin, B. (2024). Development of a novel multi-line laser triangulation scanning system based on the rotary diffraction grating. *Measurement, 225*, 113988. DOI: 10.1016/j.measurement.2023.113988

Wani, A., Joshi, I., Khandve, S., Wagh, V., & Joshi, R. (2021). Evaluating deep learning approaches for covid19 fake news detection. *Combating Online Hostile Posts in Regional Languages during Emergency Situation: First International Workshop, CONSTRAINT 2021, Collocated with AAAI 2021, Virtual Event, February 8, 2021, Revised Selected Papers 1*, 153–163.

Wankhade, M., Rao, A. C. S., & Kulkarni, C. (2022). A survey on sentiment analysis methods, applications, and challenges. In *Artificial Intelligence Review* (Vol. 55, Issue 7). Springer Netherlands. DOI: 10.1007/s10462-022-10144-1

Warmuth, M. K., Liao, J., Rätsch, G., Mathieson, M., Putta, S., & Lemmen, C. (2003). Active learning with support vector machines in the drug discovery process. *Journal of Chemical Information and Computer Sciences, 43*(2), 667–673. DOI: 10.1021/ci025620t PMID: 12653536

Warrier, U., Warrier, A., & Khandelwal, K. (2023). Ethical considerations in the use of artificial intelligence in mental health. *The Egyptian Journal of Neurology, Psychiatry and Neurosurgery, 59*(1), 36. DOI: 10.1186/s41983-023-00735-2

Waza, K., Antonnette, G., Stephen, Z., Inoue, K., Sasaki, M., Okajima, Y., & Ishibashi, Y. (2005). A pilot study on using the patient health questionnaire in the primary care setting in japan. *General Medicine (Saga), 6*(1), 9–16. DOI: 10.14442/general2000.6.9

Weber, W., Habel, U., & Schneider, F. (2008). Structural brain abnormalities in psychopaths—A review. *Behavioral Sciences & the Law, 26*(1), 7–28. DOI: 10.1002/bsl.802 PMID: 18327824

Wei, C. P., & Chiu, I. T. (2002). Turning telecommunications call details to churn prediction: A data mining approach. *Expert Systems with Applications, 23*(2), 103–112. DOI: 10.1016/S0957-4174(02)00030-1

Wei, J., Chu, X., Sun, X. Y., Xu, K., Deng, H. X., Chen, J., Wei, Z., & Lei, M. (2019). Machine learning in materials science. *InfoMat, 1*(3), 338–358. DOI: 10.1002/inf2.12028

Weiner, B. J., Lewis, C. C., Stanick, C., Powell, B. J., Dorsey, C. N., Clary, A. S., Boynton, M. H., & Halko, H. (2017). Psychometric assessment of three newly developed implementation outcome measures. *Implementation Science : IS, 12*(1), 108. Advance online publication. DOI: 10.1186/s13012-017-0635-3 PMID: 28851459

Weissman, G., Yadav, K., Madden, V., Courtright, K., Hart, J., Asch, D., Schapira, M., & Halpern, S. (2018). Numeracy and understanding of quantitative aspects of predictive models: A pilot study. *Applied Clinical Informatics, 09*(03), 683–692. DOI: 10.1055/s-0038-1669457 PMID: 30157500

Wheeler, L., Rosenthal, M., & Lund, L. (2022). Training Mental Health Professionals in Machine Learning: A Path Forward for Innovative Care. *Frontiers in Psychology, 13*, 765712. DOI: 10.3389/fpsyg.2022.765712

Wiebke, B., & Christopher, H. J. (2018). Using Machine Learning to Advance Personality Assessment and Theory. *Personality and Social Psychology Review, 23*(2), 190–203. Advance online publication. DOI: 10.1177/1088868318772990 PMID: 29792115

Wittkampf, K. A., Naeije, L., Schene, A. H., Huyser, J., & van Weert, H. C. (2007). Diagnostic accuracy of the mood module of the Patient Health Questionnaire: A systematic review. *General Hospital Psychiatry, 29*(5), 388–395. DOI: 10.1016/j.genhosppsych.2007.06.004 PMID: 17888804

Work.r. (2023). *Work.r.* Retrieved from Enterprise IA platform: https://irp- cdn.multiscreensite.com/1e240500/files/uploaded/Workr-Aplicaciones-Cartera- %20Brochure.pdf

World Health Organization. (2022). Mental health: strengthening our response. https://www.who.int/news-room/fact-sheets/detail/mental-health-strengthening-our-response

Wu, Y., Schuster, M., Chen, Z., Le, Q. V., Norouzi, M., Macherey, W., Krikun, M., Cao, Y., Gao, Q., Macherey, K., & Klingner, J. (2016). Google's neural machine translation system: Bridging the gap between human and machine translation. arXiv preprint arXiv:1609.08144.

Xia, Y., Li, J., & Wang, Q. (2020). Unsupervised machine learning for mental health analysis: An overview. *Journal of Biomedical Informatics, 109*, 103529. DOI: 10.1016/j.jbi.2020.103529

Xiong, L., & Yao, Y. (2021). Study on an adaptive thermal comfort model with K-nearest-neighbors (kNN) algorithm. *Building and Environment, 202*, 108026. DOI: 10.1016/j.buildenv.2021.108026

Xu, D., Yang, W., & Ma, L. (2018). Repurchase prediction based on ensemble learning. *Proceedings - 2018 IEEE SmartWorld, Ubiquitous Intelligence and Computing, Advanced and Trusted Computing, Scalable Computing and Communications, Cloud and Big Data Computing, Internet of People and Smart City Innovations, SmartWorld/UIC/ATC/ScalCom/CBDCo*, 1317–1322. DOI: 10.1109/SmartWorld.2018.00229

Xuan, L., He, X., Yi, Y., Shen, A., Yang, X., Dong, J., & Dong, S. (2024). Research on IMU-Based Motion Attitude Acquisition and Motion Recognition. *IEEE Sensors Journal, 24*(13), 20786–20793. DOI: 10.1109/JSEN.2024.3394903

Xu, M., & Brown, S. (2023). AI and the Automation of Administrative Tasks in Education. *Computers in Human Behavior*, •••, 143.

Xu, Z., Gao, Y., & Jin, Y. (2014). Application of an optimized SVR model of machine learning. *International Journal of Multimedia and Ubiquitous Engineering*, *9*(6), 67–80. DOI: 10.14257/ijmue.2014.9.6.08

Yacouby, R., & Axman, D. (2020). Probabilistic extension of precision, recall, and F1 score for more thorough evaluation of classification models. *Proceedings of the First Workshop on Evaluation and Comparison of NLP Systems*. DOI: 10.18653/v1/2020.eval4nlp-1.9

Yadav, N. B. (2023). Harnessing customer feedback for product recommendations: An aspect-level sentiment analysis framework. *Human-Centric Intelligent Systems*, *3*(2), 57–67. DOI: 10.1007/s44230-023-00018-2

Yaden, D. B., Eichstaedt, J. C., & Medaglia, J. D. (2018). The future of technology in positive psychology: Methodological advances in the science of well-being. *Frontiers in Psychology*, *9*(1), 1–11. DOI: 10.3389/fpsyg.2018.00962 PMID: 29967586

Yamazaki, F., Liu, W., & Yamazaki, F. (2016). Remote Sensing Technologies for Post-Earthquake Damage Assessment: a Case Study on the 2016 Kumamoto Earthquake. *ASIA Conference on Earthquake Engineering, 6ACEE*, 22–24. https://www.researchgate.net/publication/307351403

Yang, B., & Cardie, C. (2014). Context-aware learning for sentence-level sentiment analysis with posterior regularization. *52nd Annual Meeting of the Association for Computational Linguistics, ACL 2014 - Proceedings of the Conference, 1* (2008), 325–335. DOI: 10.3115/v1/P14-1031

Yang, K., Xu, H., & Gao, K. (2020). CM-BERT: Cross-modal bert for text-audio sentiment analysis. In *Proceedings of the 28th ACM International Conference on Multimedia* (pp. 521–528). DOI: 10.1145/3394171.3413690

Yeager, D. S., & Dweck, C. S. (2012). Mindsets that promote resilience: When students believe that personal characteristics can be developed. *Educational Psychologist*, *47*(4), 302–314. DOI: 10.1080/00461520.2012.722805

Yeasmin, S., Nowshin, N., & Chowdhury, T. A. (2024). Identifying Human Dark Triad from Text Data Through Machine Learning Models. *nternational Journal of Research and Innovation in Applied Science*, 9(6), 89-104.

Yin, Y., Sunderrajan, A., Huang, X., Varadarajan, J., Wang, G., Sahrawat, D., & Ng, S. K. (2019, November). *Multi-scale graph convolutional network for intersection detection from gps trajectories.* In *Proceedings of the 3rd ACM SIGSPATIAL International Workshop on AI for Geographic Knowledge Discovery* (pp. 36-39). DOI: 10.1145/3356471.3365234

Yiran, Y., & Srivastava, S. (2019, June). Aspect-based Sentiment Analysis on mobile phone reviews with LDA. In *Proceedings of the 2019 4th International Conference on Machine Learning Technologies* (pp. 101-105). DOI: 10.1145/3340997.3341012

Yonaba, H., Anctil, F., & Fortin, V. (2010). Comparing Sigmoid Transfer Functions for Neural Network Multistep Ahead Streamflow Forecasting. *Journal of Hydrologic Engineering*, *15*(4), 275–283. DOI: 10.1061/(ASCE)HE.1943-5584.0000188

Young, M. S., & Schinka, J. A. (2001). Research validity scales for the NEO-PI-R: Additional evidence for reliability and validity. *Journal of Personality Assessment, 76*(3), 412–420. DOI: 10.1207/S15327752JPA7603_04 PMID: 11499455

Yuei-An, L., Kar, S. K., & Chang, L. (2010). Use of high-resolution formosat-2 satellite images for post-earthquake disaster assessment: A study following the 12 may 2008 Wenchuan Earthquake. *International Journal of Remote Sensing, 31*(13), 3355–3368. DOI: 10.1080/01431161003727655

Zadeh, A., Chen, M., & Cambria, E. (2017). Tensor Fusion Network for Multimodal Sentiment Analysis. *ArXiv:1707.07250v1*. DOI: 10.18653/v1/D17-1115

Zadeh, M. R., Amin, S., Khalili, D., & Singh, V. P. (2010). Daily Outflow Prediction by Multi Layer Perceptron with Logistic Sigmoid and Tangent Sigmoid Activation Functions. *Water Resources Management, 24*(11), 2673–2688. DOI: 10.1007/s11269-009-9573-4

Zawacki-Richter, O., Marin, V. I., Bond, M., & Gouverneur, F. (2019). Systematic review of research on artificial intelligence applications in higher education: Emerging trends and future research directions. *International Journal of Educational Technology in Higher Education, 16*(1), 39. https://doi.org/https://doi.org/10.1186/s41239-019-0171-0. DOI: 10.1186/s41239-019-0171-0

Zeng, N., Pope, Z., Lee, J. E., & Gao, Z. (2018). Virtual reality exercise for anxiety and depression: A preliminary review of current research in an emerging field. *Journal of Clinical Medicine, 7*(3), 1–13. DOI: 10.3390/jcm7030042 PMID: 29510528

Zhang, H., Sobreviela, G., Pandit, M., Chen, D., Sun, J., Parajuli, M., Zhao, C., & Seshia, A. A. (2021). A low-noise high-order mode-localized MEMS accelerometer. Journal of Microelectromechanical Systems: A Joint IEEE and ASME Publication on Microstructures. *Journal of Microelectromechanical Systems, 30*(2), 178–180. DOI: 10.1109/JMEMS.2021.3057260

Zhang, S. (2021). Challenges in KNN classification. *IEEE Transactions on Knowledge and Data Engineering, 34*(10), 4663–4675. DOI: 10.1109/TKDE.2021.3049250

Zhang, T., Zheng, J., & Zou, Y. (2023). Weighted voting ensemble method for predicting workpiece imaging dimensional deviation based on monocular vision systems. *Optics & Laser Technology, 159*, 109012. DOI: 10.1016/j.optlastec.2022.109012

Zhang, X., Li, S., Yang, J., Wang, Y., Huang, Z., & Zhang, J. (2022). Tactile Perception Object Recognition Based on an Improved Support Vector Machine. *Micromachines, 13*(9), 1538. DOI: 10.3390/mi13091538 PMID: 36144161

Zhang, Y., & Ling, C. (2018). A strategy to apply machine learning to small datasets in materials science. *npj Computational Materials, 4*(1), 25.

Zhang, Y., Wang, Q., Chen, X., Yan, Y., Yang, R., Liu, Z., & Fu, J. (2022). The prediction of spark-ignition engine performance and emissions based on the SVR algorithm. *Processes (Basel, Switzerland), 10*(2), 312. DOI: 10.3390/pr10020312

Zhang, Y., Zhao, X., Sun, Y., Liu, M., & Shi, D. (2013). Waveform classification based on wavelet transform and k-means clustering. *In2013 International Conference on Machine Learning and Cybernetics, 4*, 14–17.

Zhao, M., & Chen, J. (2015). Improvement and comparison of weighted k nearest neighbors classifiers for model selection. *Journal of Software Engineering*, *10*(1), 109–118. DOI: 10.3923/jse.2016.109.118

Zhao, M., Qi, Y., Zhang, X., Li, B., Jiang, Z., & Wei, X. (2022). Noise Analysis and Performance Improvement of a MEMS Fabry-Pérot Seismic Accelerometer. *IEEE Sensors Journal*, *22*(1), 365–372. DOI: 10.1109/JSEN.2021.3128534

Zhu, X., Jiang, F., Guo, C., Wang, Z., Dong, T., & Li, H. (2023). Prediction of melt pool shape in additive manufacturing based on machine learning methods. *Optics & Laser Technology*, *159*, 108964.

Zucchetti, A., Nibbio, G., Altieri, L., Bertorni, L., & Calzavara-Pinton, I. (2024). Artificial intelligence applications in mental health: The state of the art. *alian. Journal of Psychiatry*, (1). Advance online publication. DOI: 10.36180/2421-4469-2024-5

Zucco, C., Calabrese, B., Agapito, G., Guzzi, P. H., & Cannataro, M. (2020). Sentiment analysis for mining texts and social networks data: Methods and tools. In *Wiley Interdisciplinary Reviews: Data Mining and Knowledge Discovery* (Vol. 10, Issue 1). DOI: 10.1002/widm.1333

About the Contributors

María E. Raygoza-L. has a PhD in Chemistry and Energy from the Engineering Institute of the Universidad Autónoma de Baja California and a bachelor's degree in industrial engineering with a specialization in Manufacturing. The thesis topic developed was "Technological Management for the transfer of Renewable Energies", carrying out an academic stay at the Colegio de la Frontera Norte where she investigated public policies for the energy transition and sustainable development of Mexico, achieving a methodology that aims the integration of the main actors that must contribute to sustainable development. She currently works as a full-time professor and researcher at the Universidad Politécnica de Baja California and as a subject professor at the Universidad Autónoma de Baja California, with a specialty in Energy Engineering and Renewable Energies, contributing to the training of quality human resources. With extensive experience in the food and aerospace industries, his research interests span renewable energy, energy efficiency, environmental studies, energy management systems, and the integration of public policies for sustainable development and transition through social energy innovation and emerging technology applications such as Artificial Intelligence, especially Machine Learning. He is currently pursuing a Second bachelor's degree in psychology, further expanding his academic horizons in relation to social applications of public policies and social innovation that have an impact on sustainable development, addressing issues of energy security, sustainability, renewable energy and water stress, focusing their research to the fulfillment of the Sustainable Development Goals (SDGs).

Jesus Heriberto Orduño-Osuna received his B.Sc. in Mechatronics Engineering from the Universidad Politécnica de Baja California (UPBC) in 2020 and his M.Sc. in Computational Sciences and Applied Mathematics from Universidad Internacional de La Rioja, Mexico (UNIR Mexico) in 2024. His professional experience spans the Post-Harvest Automation and manufacturing sectors, where he has specialized in automation processes, vision systems, and industrial robotics. Currently, he works as a part-time lecturer at Universidad Politécnica de Baja California, where he teaches courses related to Mechatronics, Manufacturing, Energy Systems, and Computer Science. His teaching is complemented by his active research interests in Machine Learning, Automation and Control of Industrial Processes, Microcontrollers, and Digital Signal Processing (DSP), areas in which he seeks to advance the integration of intelligent systems in industrial applications.

Abelardo Mercado-Herrera is a Doctor of Science and Master of Science from the National Institute of Astrophysics, Optics and Electronics, specializing in Astrophysics, Postdoctorate in Astrophysics from the Institute of Astronomy of the National Autonomous University of Mexico, Electronics Engineer from the Autonomous University of Baja California . He is a specialist in the mathematical-statistical description of stochastic and/or deterministic processes, nonlinear systems, complex systems, chaos theory, among others, as well as their application to physical phenomena such as astronomy, economics, finance, telecommunications, science social etc., in order to determine the underlying dynamics in such processes, and if necessary, their connection with real physical variables and possible prediction. He has worked on the development of various electronic circuits, interfaces and programs to carry out electrical tests in the industry, as well as in scientific instrumentation, applied to telemetry, infrared polarimetry, optics and spectroscopy. He has also specialized in image analysis, measurement techniques and noise reduction.

Roxana Jimenez-Sanchez is a Doctoral candidate in International Industrial Safety at Applied Pedagogical University, Monterrey Campus. She holds a Master's degree in Education with a specialization in Organizational Development from the Centro de Enseñanza Técnica y Superior (CETYS) University, Mexicali campus. She is a Chemical Environmental Engineer by profession, graduated from the Instituto Tecnológico de Mexicali. She has worked in the field of implementing international standards in both the private industry and the public sector, as well as implementing analytical techniques in potable and wastewater laboratories contributing to the livestock industry. She is a Federal Labor Inspector and a classroom teacher. Dedicated to teaching for 17 years, she has been at the Universidad Politécnica de Baja California in the Manufacturing Engineering area, with research contributions related to Sustainability, Environment, and Industrial Safety and Hygiene.

Fabian N. Murrieta-Rico received B.Eng. and M.Eng. degrees from Instituto Tecnológico de Mexicali (ITM) in 2004 and 2013 respectively. In 2017, he received his PhD in Materials Physics at Centro de Investigación Científica y Educación Superior de Ensenada (CICESE). He has worked as an automation engineer, systems designer, as a university professor, and as postdoctoral researcher at Facultad de Ingeniería, Arquitectura y Diseño from Universidad Autónoma de Baja California (UABC) and at the Centro de Nanociencias y Nanotecnología from Universidad Nacional Autónoma de México (CNyN-UNAM), currently he works as professor at the Universidad Politécnica de Baja California. His research has been published in different journals and presented at international conferences since 2009. He has served as reviewer for different journals, some of them include IEEE Transactions on Industrial Electronics, IEEE Transactions on Instrumentation, Measurement and Sensor Review. His research interests are focused on the field of time and frequency metrology, the design of wireless sensor networks, automated systems and highly sensitive chemical detectors.

***

Subramani A. K. completed his Ph.D. in Business Administration at Manonmaniam Sundaranar University, Tirunelveli, India. A teaching career spanning 18 years has been amassed, during which 74 articles have been published in international journals, including two papers in ABDC-A category journals, eight papers in the ABDC-B category, and two papers in the ABDC-C category. Additionally, 28 papers were published in Scopus-indexed journals, and eight in Web of Science-indexed journals. Numerous papers have been presented at international and national conferences. Over 31 courses have been completed from the Swayam Online platform, covering HR, Business Analytics, and Intellectual Property domains. NPTEL Star Certificates have been received, including those for NPTEL Evangelist, Motivated Learner, Enthusiast, and Discipline Star. Peer review has been conducted for several journals published by Inderscience and Emerald publishers. Research interests are focused on human resource management and organizational behavior. The ‘Teaching and Research Excellence Award’ from Grabs

Charitable Trust, India, and the 'Distinguished Teacher Award-Management' from MTC Global Trust have been awarded. Resource person roles have been fulfilled in many training workshops on SPSS and AMOS software packages in reputed educational institutions across the country.

Vishnu Achutha Menon is an independent journalist, writer, researcher, and an Indian percussionist. He is a recipient of the Junior Scholarship the Ministry of Culture awarded. His research interests are film studies, verbal & nonverbal communication, south Asian performances, Natyasastra, media studies, media analysis techniques, Laban Movement Analysis, and Ethnomusicology.

Bhavya Agarwal is currently pursuing a B A (H) in Psychology from the Department of Humanities and Social Science, Graphic Era (Deemed to be University) Dehradun. As a graduate student, his research interest began in the fields of psycho-oncology, community psychology, and applied social psychology so that I can understand complex human behavior. Currently, he is a member of the American Psychological Association as a Student Affiliate.

Aliza Ahsan is a student of B.tech (Computer Science) at Jamia Millia Islamia. She has certifications in Unix, Mongo DB, Artificial Intelligence/Machine Learning, Information security etc. She has worked on several projects on Web development, Network intrusion detection, Automation solution using RPA tools etc. Her research interest lies in AI/ML and information security.

Nawab Akram is currently associated with Magadh Professional Institute, Danapur, Patna, as Associate Professor. He has also served International School of Management (ISM), Khagaul, Patna, as Assistant Professor and Head of the Department (Marketing Management) for more than six years. Dr. Akram has a very long association of more than fifteen years with L. N. Mishra Institute of Economic Development and Social Change, Patna, as Assistant Professor (Visiting). Dr. Akram has rich experience of more than fifteen years in the field of teaching and academic research. His area of research and teaching are fundamentals of Marketing Management, Marketing Research, Marketing Analytics, Rural Marketing, Services Marketing, and Strategic Management. After completing his MBA, Dr. Akram worked with many of Multinational Corporations and he has heap of corporate experience of more than eight years in the field of marketing and sales. Dr. Nawab Akram is awarded with Ph.D. and he is UGC NET qualified also. He has published several research papers and authored many books, organized national conferences as Conference Secretary at ISM, Patna. He has attended various conferences, seminars, FDPs, QIPs, Refresher Courses and Orientation Programs organized by reputed universities like University of Hyderabad, Jamia Millia Islamia, Banaras Hindu University, etc. in India.

Sampath Boopathi is an accomplished individual with a strong academic background and extensive research experience. He completed his undergraduate studies in Mechanical Engineering and pursued his postgraduate studies in the field of Computer-Aided Design. Dr. Boopathi obtained his Ph.D. from Anna University, focusing his research on Manufacturing and optimization. Throughout his career, Dr. Boopathi has made significant contributions to the field of engineering. He has authored and published over 225 research articles in internationally peer-reviewed journals, highlighting his expertise and dedication to advancing knowledge in his area of specialization. His research output demonstrates his commitment to conducting rigorous and impactful research. In addition to his research publications, Dr. Boopathi has also been granted one patent and has three published patents to his name. This indicates his innovative thinking and ability to develop practical solutions to real-world engineering challenges. With 17 years of academic and research experience, Dr. Boopathi has enriched the engineering community through his teaching and mentorship roles.

Saurabh Chandra is an accomplished legal scholar and currently serves as an Associate Professor & Assistant Dean at the School of Law, Bennett University, India. With a robust academic foundation, he earned his Bachelor of Arts and Bachelor of Laws (B.A. LL.B.) from the prestigious Aligarh Muslim University (A.M.U.). He further specialized in Business Laws, completing his Master of Laws (LL.M.) at the renowned National Law School of India University (NLSIU), Bangalore. Dr. Chandra's academic journey culminated in a Doctorate in Law (Ph.D.) from the esteemed National Law University. In addition to his impressive legal education from some of India's most prestigious institutions, Dr. Saurabh Chandra also holds a master's degree in Management with specialization in Human Resources in Management and another in Journalism and Mass Communication. With 16 years of teaching experience, he has established himself as an expert in Business and Corporate Laws, with numerous publications to his credit. Dr. Saurabh Chandra is a distinguished expert in Business and Corporate Laws, with a profound understanding of the intricate legal frameworks that govern the business world. His expertise is complemented by a robust portfolio of scholarly publications, which underscores his commitment to advancing knowledge in his field. Throughout his career, Dr. Chandra has consistently produced high-impact research that addresses critical issues in Business and Corporate Law. His publications are widely recognized for their depth of analysis, innovative perspectives, and practical relevance. He has contributed extensively to leading journals, where his work has often set new standards for legal scholarship. His publications not only reflect his scholarly rigor but also his dedication to addressing contemporary challenges faced by businesses and policymakers. By bridging the gap between theoretical insights and real-world applications, Dr. Chandra's work serves as a valuable resource for academics, legal practitioners, and industry professionals alike. Dr. Chandra brings a wealth of experience in Academic Administration, having successfully held various key administrative positions. His diverse qualifications and extensive experience make him a distinguished figure in the academic and legal communities.

Amit Chauhan is Coordinator of Forensic Science in Department of Life Sciences, School of Sciences, CHRIST (Deemed to be University) Bangalore-Karnataka, India He has completed his education in B.Sc. (Physics, Mathematics, Statistics) M.Sc. Forensic Science, P.G. C. F. Sc., M.A. Criminology & Police Administration, P. G. D. Adv. F. Sc. R & T., Ph. D Forensic Science. As a faculty, He has worked in Amity Institute of Forensic Science, Amity University, Delhi since September 2015- August 2021 and Continuing his service in Department of Life Sciences (Forensic Sciences) in CHRIST University, Bengaluru since August 2021 to till date. He is Member of Indian Police Foundation, International society of Forensic Genetics-Germany, British Association for Forensic Anthropology, Balkan Academy of Forensic Sciences- Egypt etc. He is a certified Doping Control Officer from World Anti-Doping Agency (Montreal- Canada), National Anti-Doping Control Agency- Delhi, India and has worked a Doping Control Officer in various international sports. He has published 170 international publications and 14 book chapters in reputed journals i.e., Scopus, web of science, Thomson Reuters, EBSCO. He has published Six books entitled "A textbook on Cyber Forensic" and "A textbook on Engineering Biology", "A textbook for Artificial intelligence for Cyber security", A textbook of Fundamentals of data Science and Python", "A Textbook of Medical Microbiology" and "A textbook on "Medicinal Chemistry- Durg discovery and development" are in publication process. He has an innovation track with 72 patents including 4 design patents from UK (India, U. K.) and 5 patents are under evaluation in various countries.

E. Ivette Cota-Rivera received an M.Eng. from the Autonomous University of Baja California (UABC, Universidad Autonoma de Baja California). In 2016, she received her Ph.D. in Energy Engineering and is currently working as a pro fessor in the Energy Engineering program at the Polytechnic University of Baja California (UPBC, Universidad Politecnica de Baja California). Her research interests include energy efficiency and sustainability, artificial intelligence, energy science and technology, as well as programming.

Lakhwinder Kaur Dhillon is Head & Associate Professor Department of Accounts & Finance, Amity Business School. She is an MBA, M.Phil., Ph.D. and has done Certificate Course in strategic Management (IIT-Delhi). Her research interest includes corporate Finance, Micro Finance and Financial Inclusion, Capital market and financial system, Corporate Social Responsibility and spirituality in Management.

Angela Jean Mary E. is currently serving as the Head of the Department of Humanities and the Student Welfare Office Coordinator at St. Claret College (Autonomous). With a robust educational background and a passion for teaching, she has significantly contributed to the field of higher education over the past 13 years. Ms. Angela holds a Master's degree in Psychology, a discipline that has equipped her with deep insights into human behavior and cognitive processes and she has also successfully cleared the State Eligibility Test for teaching in Higher education.

Wendy Flores-Fuentes received the bachelor's degree in electronic engineering from the Universidad Autónoma de Baja California in 2001, the master's degree in engineering from Technological Institute of Mexicali in 2006, and the Ph.D. degree in science, applied physics, with emphasis on Optoelectronic Scanning Systems for SHM, from Universidad Autónoma de Baja California in June 2014. By now she is the author of 24 journal articles in Elsevier, IEEE Emerald and Springer, 13 book chapters and 6 books in Intech, IGI global, Lambert Academic and Springer, 35 proceedings articles in IEEE ISIE, IEEE IECON, the World Congress on Engineering and Computer Science, and the VII International Conference on Industrial Engineering ARGOS 2014. Recently, she has organized and participated as Chair of Special Session on ''Machine Vision, Control and Navigation'' at IEEE ISIE 2015-2019 and IECON 2018-2019. She holds 1 patent of Mexico and 1 patent of Ukraine. She did receive the award of "Best session presentation" in WSECS2013 in San-Francisco, USA. She did receive as coauthor the award of "Outstanding Paper in the 2017 Emerald Literati Network Awards for Excellence".

Martha González graduated with a Bachelor's degree in Fine Arts from the Autonomous University of Baja California, and holds a Master's degree in Arts from the same university, graduating with honors. Recipient of the Santander Meta program scholarship to attend the Augmented Reality bootcamp at CENTRO, Mexico City in 2023. Full-time professor since 20022, affiliated with the Animation and Visual Effects Engineering department at the Polytechnic University of Baja California. With 18 years of teaching experience at the undergraduate level in academic programs related to art, architecture, design, and animation. Interested in researching new technologies, innovation, sustainability, and teaching related to the artistic field.

Neha Issar is a proactive professional boasting over 8 years of experience in the field of Research Analysis, with more than 4 years dedicated to Academic work. Her expertise extends to training in various areas, including Database & Query Language, Data Science using R, Data Analytics for Business Decision-making, an Overview of Python, Management Information Systems, and Enterprise Resource Planning. Ms. Issar earned her commerce degree and holds a postgraduate diploma in Finance and Marketing from JIMS Kalkaji. She is currently pursuing a PhD in Management from Amity Business School.Her career journey includes serving as a Data Researcher at S&P Global Market Intelligence Pvt. Ltd., working on projects for prominent organisations such as Aditya Birla Group and Times Internet in et.com. She also held the position of Senior Manager Investor Relations at Opera Gratia Pvt Ltd. In addition to her extensive experience, Ms. Issar holds several valuable certifications, including IBM Spark and Scala certification, IBM Data Science certification, IBM Microservice Architecture and Service certification and Cybersecurity certification. She is also a practitioner in IBM Enterprise Design Thinking. Her research interests include Metaverse, Blockchain, Green Finance.

Shubha Jain is currently pursuing a Bachelor's in psychology with honors at Graphic Era (Deemed to be University). As a bachelor's student, she started to do research in pursuit of knowledge and more understanding of the psychological field. Her research started with exploring various subjects and analyzing data.

Aravindhan K. received his B.E. degree in Electrical and Electronics Engineering and Masters in Power Systems Engineering in the year 2011 and 2013 respectively. He has been working as an Assistant professor at St. Joseph's College of Engineering in the department of Electrical and Electronics Engineering since 2013 and he has almost 10 years of experience in the teaching field. He is currently pursuing Ph.D. in Anna University. His subject of interest includes Electric Circuits, Power systems engineering, Electrical Machines, Electrical Drives and Control, Control Systems, Smart grid, Machine Learning and Data Science.

Fehmina Khalique is a PhD and MBA in human resource management and currently works as an Professor with Lloyd Business School, Noida, India. She is associated with academics for the past 17 years in the area of human resource management, business communication and cross-cultural management. She has 4 years of industry experience and enjoys being a corporate trainer, counsellor and consultant. Her research interest lies in cross cultural management, international management, human resource management, human capital management and she has to her credit several high-quality research papers. She is also involved in copywriting, editing and reviewing of research articles. In her free time, she likes to do some social work and deliver free career counseling sessions for students on honing their inter-personal and corporate-communication skills.

Shueb Ali Khan is a distinguished academician serving as a Professor at Kunwar Satyavira College of Engineering and Management in Bijnor. With over 21 years of experience in administration, teaching, and research, he is a seasoned educator and scholar. Dr. Khan completed his schooling at GIC Saharanpur and earned his B.Tech in Computer Engineering from MJP Rohilkhand University, Bareilly. He then pursued an M.Tech from Shobhit University, Meerut, and later obtained a Ph.D. in Computer Engineering from Sri Venkateshwara University, Gajroula.Dr. Khan has made significant contributions to his field, including publishing a book, participating in numerous Faculty Development Programs (FDPs), and authoring more than 10 research papers in reputable national and international journals. His research interests encompass a wide range of topics, including Steganography, Artificial Intelligence, Parallel Computing Systems, Operating Systems, Cloud Computing, Internet and Human Behavior, and the Internet of Things (IoT).

Sateesh Kumar T. K. is presently working as an Asst. Professor at Kristu Jayanti College Autonomous, Bengaluru, India.

S. Logeswaran is an Assistant Professor in the Department of Civil Engineering, KPR Institute of Engineering and Technology.

Kavitha M. is affiliated with Sathyabama Institute of Science and Technology, where she specializes in Electrical Engineering. She earned her Bachelor of Engineering degree in Electrical and Electronics Engineering from Bharathi University, Coimbatore, and her Master of Engineering degree in Power Systems Engineering from Anna University, Chennai. She completed her PhD at Sathyabama Institute of Science and Technology. Currently, Dr. Kavitha serves in the Department of Electrical and Electronics Engineering at Sathyabama Institute of Science and Technology, Chennai. She has published over 15 papers in international journals and presented 30 papers at international conferences. In 2019, she was recognized as an NPTEL Discipline Star. She is currently leading a project funded by Unnat Bharat Abhiyan (UBA) with a grant of Rs. 1 lakh. Additionally, she has published five product patents.

Vijayalakshmi M. is a PhD Research Scholar and an Assistant Professor at Saveetha School of Managment, Saveetha Institute of Medical and Technical Sciences, Saveetha University, Chennai, Tamil Nadu, India. She holds Master of Philosophy (M.Phil) from Sri Venkateshwara University, Tirupati and MBA from Anna University, Chennai. Her exceptional academic performance earned her first-class honors with distinction, along with the prestigious top rank. With a decade of experience in academics, she has made significant contributions to the students' community. Her dedication to scholarly research is evident through her publications in impact factor journals and springer nature. She has leveraged her experiences by receiving Best Paper Award at a prestigious international conference. She has served as a resource person for VIT University, extending her expertise to benefit the wider academic community.

Fabian N. Murrieta-Rico received B.Eng. and M.Eng. degrees from Instituto Tecnológico de Mexicali (ITM) in 2004 and 2013 respectively. In 2017, he received his PhD in Materials Physics at Centro de Investigación Científica y Educación Superior de Ensenada (CICESE). He has worked as an automation engineer, systems designer, as a university professor, and as postdoctoral researcher at Facultad de Ingeniería, Arquitectura y Diseño from Universidad Autónoma de Baja California (UABC) and at the Centro de Nanociencias y Nanotecnología from Universidad Nacional Autónoma de México (CNyN-UNAM), currently he works as professor at the Universidad Politécnica de Baja California. His research has been published in different journals and presented at international conferences since 2009. He has served as reviewer for different journals, some of them include IEEE Transactions on Industrial Electronics, IEEE Transactions on Instrumentation, Measurement and Sensor Review. His research interests are focused on the field of time and frequency metrology, the design of wireless sensor networks, automated systems and highly sensitive chemical detectors.

Chidambararaj N. was born in the year 1981. He completed his Diploma in Electrical and Electronics Engineering in the year 2000. He received his B.E degree in Electrical and Electronics Engineering in the year 2003 and proceeded with pursuing Masters in Power Systems Engineering and graduated in the year 2005 and finally completed his Ph.D in Sathyabama University in deregulated power market in the year 2018. He has been working as an Associate professor at St. Joseph's College of Engineering in the department of Electrical and Electronics Engineering since 2005 and he has almost 20 years of experience in the respective field. His subject of interest includes Electric Circuits, Control Systems, Power Systems Engineering, Engineering Electromagnetism, Digital signal processing and Machine design and his core research is on Deregulated Power System. He has published more than 30 research articles in the field of Congestion Management and also he had published four text books in Engineering subjects.

Ullal Akshatha Nayak is a dedicated and accomplished academic professional with over a decade of teaching experience in the fields of Computer Science and Information Science Engineering. Since August 2018, she has served as an Assistant Professor in the Department of Information Science and Engineering at Nitte Meenakshi Institute of Technology (NMIT), Bengaluru, which is an autonomous institution affiliated with Visvesvaraya Technological University (VTU). She has a track record of excellence in teaching a wide range of courses, guiding students in both technical and non-technical areas, and achieving significant results in the subjects she has handled.Prior to her tenure at NMIT, Mrs. Nayak contributed her expertise as a Senior Software Trainer at IANT (Institute of Advanced Network Technology) and held assistant professor positions at Srinivas School of Engineering and Karavali Institute of Technology, where she consistently demonstrated her ability to inspire and educate students.

Rajagopal R. completed his Ph.D. degree at Anna University, Chennai. He has received his post-graduation and bachelors from the Computer Science and Engineering also from Anna University, Chennai. He served as a reviewer for Inder Science and other prestigious Scopus-indexed journals. He serves on the editorial board of the International Journal of Mathematical Sciences, Engineering, and Sciences Publication. Dr. Rajagopal has published in more than 50 international journals and presented in many international conferences. His research interests include cloud computing, high performance computing, big data analytics, and algorithms.

Luis Roberto Ramirez Hernandez graduated on Universidad Autonoma de Baja California as Mechatronic Enginner, after he realize a master degree on Insituto de Ingenieria of the UABC as master of science of enginnering. After, he realize a ph degree on Universidad Autonoma de Baja California as doctor of science of engineering. Their Principals investigations is the study of stereoscopic vision system and calibration of cameras.

Julio C. Rodríguez-Quiñonez (Senior Member, IEEE) is currently Full-Time Researcher-Professor in the Engineering Faculty of the Autonomous University of Baja California, and a member of the National Research System Level 1. He has been thesis director of several Doctor's and Master's degree students. He holds three patents referred to dynamic triangulation method, has been editor of 4 books, and Author of the Book "Graphical Programming using LabVIEW: Fundamentals and Advanced Techniques", Guest Editor of Measurement, IEEE Sensors Journal, International Journal of Advanced Robotic Systems and Journal of Sensors, written 92 papers indexed in Scopus (h-index 19. He participated as a reviewer and

Session Chair of IEEE ISIE conferences in 2014 (Turkey), 2015 (Brazil), 2016 (USA), 2017 (UK), 2019 (Canada), IECON 2018 (USA), IECON 2019 (Portugal), ISIE 2020 (Netherlands), ISIE 2021 (Japan) and ISIE 2022 (USA). Since 2016 is Senior Member of IEEE. His current research interests include automated metrology, stereo vision systems, control systems, robot navigation and 3D laser scanners.

Srinivasan S. is working as a Professor in Department of Advanced Computing Sciences, AMET Deemed-to-be University, Chennai with more than 26 years of experience in Teaching, Research, Academic and Administration in prestigious institutions. His Qualifications are MCA., ME(CSE)., Ph.D in Computer Science in 2019. He guided Ph.D research scholars with authored around 50 publications which includes books, book chapters, research papers in various journals, conferences along with published more than 10 patents and editor in many peer-reviewed journals, books. He acted as organizing committee member in conferences, editorial board member in peer reviewed journals and also session chair for many conferences along with knowing the process of NAAC, NBA and ISO. He got best research paper awards, best teacher awards, best Faculty Inspiration – Bronze Award from Infosys. He is life member of CSI, IAENG, ISTE. He has delivered several invited talks on IT latest technologies in various institutions across many states of India. He had conducted over 60 FDPs, workshops, seminars and webinars. He played role as Infosys Campus Connect – SpoC in different institutions. He has finished AWS Technical Professional course and 75 MOOC certification courses.

Ajay K. Singh is working as an Associate Professor (Research) in the Department of Humanities and Social Science, Graphic Era (Deemed to be University) Dehradun. He has worked as Assistant Professor (Economics) in the School of Liberal Arts & Management, DIT University Dehradun for 6 years. He did Post-Doctorate Research with EDI of India, Ahmedabad, Gujarat (India). He received MPhil (Economics) from DAVV Indore (India), and PhD (Economics) from IIT Indore (India. He has published several research papers in the diversified area such as climate change, agricultural productivity, assessment of food security; estimation of GFSI, development of environmental sustainability index (ESI) and its association with socio-economic indicators; measurement and determinants of entrepreneurship ecosystem, and dimension of sustainable development and its interlinkages with economic development.

Gagan Singh is an accomplished Assistant Professor at Meerut Institute of Technology, Meerut, with a distinguished 7-year tenure in academia. He holds both a B.Tech degree in Civil Engineering and an MBA from Dr. APJ Abdul Kalam University. Alongside his robust academic background, he brings 5 years of invaluable industry experience. His scholarly achievements are notable, having presented six research papers at esteemed international conferences, published four papers in UGC CARE-listed journals, and secured a patent. Furthermore, Mr. Singh is a respected member of the organizing committee for numerous academic conferences, reflecting his deep dedication to the advancement of his field.

Ravindra Singh has completed his Ph.D. from Banaras Hindu University (Institute of Eminence). Currently, he is working as an Assistant Professor of Psychology at Magadh University, Bodh Gaya, Bihar. Earlier he was Indian Youth Ambassador at BRICS. Beijing China 2017 during Indian delegation. Dr. Singh has also worked as a Junior Project Fellow, on a funded project, in New Delhi in 2018. He worked as a Former Teaching Assistant at IIT, BHU. Recently he was the Youth Member at G20 Youth for India's G-20 Presidency. He served as an Independent Reviewer in Frontiers in Psychology Journal and a Member of APA, an International Affiliate, Washington DC, and also a member of the prestigious National Academy of Psychology, (NAOP) India. He has 7 Years of teaching experience and supervising 3 PhD candidates along with many posts' graduate thesis supervision.

K. Sujatha obtained her Ph.D. degree in April 2018 from Krishna University, located in Machilipatnam, Andhra Pradesh. Her research journey focused on the realms of Data Mining, Machine Learning, and Deep Learning. Her research interests are reflected in her extensive academic output, which includes more than 40 publications in high-impact international journals indexed by reputable databases such as Scopus and ESCI. Additionally, she has presented her research findings at more than 20 international conferences sponsored by distinguished organizations such as IEEE, Springer, and Taylor and Francis. Dr. Sujatha has been actively involved in continuous professional development, having attended several AICTE-sponsored workshops and Faculty Development Programs (FDPs). She is also a member of the prestigious Indian Association of Engineers (IAE), a lifetime member of the International Association of Engineering and Technology for Skill Development (IAETSD), member in IEEE and member in CSI. With over 16 years of teaching experience, Dr. Sujatha currently serves as an Associate Professor at Vignan's Foundation for Science, Technology and Research, located in Vadlamudi, Guntur District, Andhra Pradesh. Her commitment to education and research is evident in her impressive accomplishments and contributions to the field of engineering and technology.

Deepak Sundrani is BE(Civil), ME(Construction Management), MBA(Marketing), LLB, PhD (Management). His date of birth is 07 November 1966. He is working as an Associate Professor in Nicmar University since June 200.9

Bhoopathy V. is a Professor at Sree Rama Engineering College. He holds a Ph.D. in Information and Communication Engineering from Anna University and has over 16 years of academic experience. Specializing in wireless sensor networks and IoT, he has published numerous research papers and holds several international patents. Dr. Bhoopathy is a recipient of prestigious awards for his contributions to education and research. His commitment to academia continues to drive innovation and excellence in the field of computer science.

Vijayagopal V. received his M.Sc., and M. Phil from Madurai Kamaraj University. Madurai, Tamil Nadu, India. He received his PhD, degree in Computer Science from Bharathidasan University, Tiruchirapalli, Tamil Nadu, India. He is now working as an Assistant Professor, Faculty of Computer Applications, Dr.M.G.R. Educational and Research Institute, Maduravoyal, Chennai, Tamil Nadu, India. His research interested in Adhoc Networks, MANET. He has published more than 5 technical papers at various National / International Conferences and Journals.

Karthik Kumar Vaigandla obtained B.Tech in Electronics and Communication Engineering from Jayamukhi Institute of Technological Sciences, Warangal, Telangana, India in 2008 and M.Tech in Embedded systems from Ramappa Engineering College, Warangal, Telangana, India in 2011. Currently working as a Assistant Professor in Electronics and Communication Engineering from Balaji Institute of Technology and Science, Telangana, India. Total Teaching experience of 16 years and 5 years research experience. He has the credit of having published 17 research papers in Web of Science/ Scopus Indexed journals, 50 papers in UGC CARE listed/refereed/reviewed journals and 5 papers presented in international conferences. Current research interest includes on wireless communications,

Communication networks, and signal processing. He is the Life Member of Indian Society for Technical Education (ISTE), Teaching and Education Research Association (TERA), International Association of Engineers (IAENG) and Middle East Association of Computer Science and Engineering (MEACSE).

Index

A

B

C

D

E

F

G

H

I

K

L

M

N

P

R

S

T

V

W

www.ingramcontent.com/pod-product-compliance
Lightning Source LLC
LaVergne TN
LVHW062104300926
849161LV00010B/2370